Climate Change Management

Series editor

Walter Leal Filho, Faculty of Life Sciences, Research and Transfer Centre, Hamburg University of Applied Sciences, Hamburg, Germany

More information about this series at http://www.springer.com/series/8740

Walter Leal Filho · Bettina Lackner
Henry McGhie
Editors

Addressing the Challenges in Communicating Climate Change Across Various Audiences

 Springer

Editors
Walter Leal Filho
Faculty of Life Sciences
Hamburg University of Applied
  Sciences
Hamburg, Germany

Henry McGhie
Manchester Museum
University of Manchester
Manchester, UK

Bettina Lackner
Doctoral Program Climate Change
University of Graz
Graz, Austria

ISSN 1610-2002   ISSN 1610-2010   (electronic)
Climate Change Management
ISBN 978-3-319-98293-9   ISBN 978-3-319-98294-6   (eBook)
https://doi.org/10.1007/978-3-319-98294-6

Library of Congress Control Number: 2018950934

This Springer imprint is published by the registered company Springer Nature Switzerland AG
The registered company address is: Gewerbestrasse 11, 6330 Cham, Switzerland

Preface

It is widely known that a better understanding of what climate change is, and what it means to both people and nations, is necessary. It is also a fact that the identification of approaches, processes, methods and tools which may help to better communicate it is needed. There is also a perceived need to showcase successful examples of how communication on matters related to climate change across society and stakeholders can take place, so as to catalyse the sort of cross-sectoral action needed to address the phenomena and its many consequences.

Aware of these facts, the International Climate Change Information Programme (ICCIP), led by the Hamburg University of Applied Sciences (Germany), has engaged on a wide range of information and communication initiatives, around the world. Apart from many events on matters related to various aspects of climate change, it has been trying to reach a variety of stakeholders, and to bring climate change close to them.

The book *Addressing the Challenges in Communicating Climate Change Across Various Audiences* is a further step on this direction. It has been prepared with a view to provide a comprehensive assessment of the potentials, means and methods to communicate climate change across various groups. Apart from chapters handing the theory and applications of climate change communication, the book contains also papers handling the ways museums communicate and handle climate change, providing an overview of the unique contributions museums can provide to an understanding of climate change. Altogether, this book meets a perceived need for publications which specifically aim at documenting and disseminating the wealth of experiences of climate change communication available today.

We hope this book will be useful to specialists on climate change communication, but also to environmental managers, policy makers, planners researchers and students, as we continue to work together to address the challenges related to climate change communication.

Hamburg, Germany   Walter Leal Filho
Graz, Austria   Bettina Lackner
Manchester, UK   Henry McGhie
Winter 2018/2019

Contents

An Overview of the Challenges in Climate Change Communication Across Various Audiences

Walter Leal Filho

Abstract There is a consensus about the need to foster information and communication on matters related to climate change. Yet, realising the goal of informing people and passing across climate change messages are difficult tasks, exacerbated by the fact that different audiences have different requirements. Based on the need to shed light into this problem, this paper outlines some of the challenges seen when communicating climate change across audiences. It also suggests some measures which may be helpful in reaching different groups, so as to meet their specific information needs.

1 Introduction

Climate change is a matter of great concern, and due to its scope and far ranging implications at the global (Lenton et al. 2008; IPCC 2014) at the local levels (e.g. Leal Filho et al. 2017), it can be regarded as one of the major challenges of modern times. One which is accompanied by much political discussion (e.g. Edwards 2019), as well as is sometimes charged with emotions (Chapman et al. 2017). The theme has for sometime now taken a key place in the media reporting (Mazur 1998).

In the context of climate change, the literature refers to previous works which look at aspects of perceptions (Leal Filho 2009), understanding (Leiserowitz 2007, 2011) and social learning (Harvey et al. 2012).

Whereas a great deal of attention and efforts have been devoted towards an understanding of the dynamics of climate variability and global change, as well as the various physio-chemical-geological phenomena that influence it (Caldeira and Wickett 2003), comparatively little efforts have been made in respect of climate change information and communication.

W. Leal Filho (✉)
Faculty of Life Sciences, Research and Transfer Centre
"Sustainable Development and Climate Change Management",
Hamburg University of Applied Sciences, Ulmenliet 20,
21033 Hamburg, Germany
e-mail: walter.leal2@haw-hamburg.de

© Springer Nature Switzerland AG 2019
W. Leal Filho et al. (eds.), *Addressing the Challenges in Communicating Climate Change Across Various Audiences*, Climate Change Management,
https://doi.org/10.1007/978-3-319-98294-6_1

This imbalance means that the substantial amount of information and knowledge on aspects of climate change mitigation and adaptation, is not matched by a similar body of expertise on climate change communication, despite the fact that the latter is central to any attempts to modify the human behaviours and man-led actions which have been largely responsible for the climate-related problems seen today. As a result, there is a certain degree of confusion among the public about what climate change really means and how it affects people.

An example of this reality is seen if one considers the fact that even though CO_2 emissions may be reduced by means of a greater use of renewable energy, only a minority of the public in geographical areas outside Europe, is actively engaged in using renewable energy as a means to meet their energy needs. The same applies to transport, which is responsible for a substantial amount of the greenhouse gases emitted every year: not many people see it as necessary to make more use of public transport, or to engage in more environmental friendly means of transportation such as the use of electric cars, despite the obvious benefits such changes of behaviour would bring about to their own ecological footprint.

One element which is also largely overlooked, is the fact that as a result of different cultural models, cultural heritage, traditions or social pressures, different audiences require a different approach so as to be reached by means of climate change related messages. For instance, whereas an educated public in rich countries may be interested at, and concerned with the long term implications of CO_2 emissions and the need to reduce them by means of mitigation efforts over years and decades, for most people in developing nations -especially the poorest countries in the official

ODA List[1]—climate change impacts and the means to handle them in the short term are the priority. This is so for 2 main reasons:

1. the limited resources, both financial and technological seen in most developing countries, restrict the ability of poor countries to deeply engage in climate change mitigation. Indeed, in many of them, the amounts of CO_2 emissions are rather modest when compared to their industrialised counterparts, as a result of low levels of industrialisation. This makes adaptation efforts which render results in the short term more attractive—and better to communicate—, than those which focus on results expected over longer time-spans;
2. climate change leads to impacts observed in central areas such agriculture, land use, crop productive or availability of water supplies among others, all of which pose an additional burden to the economies of developing counties which are already under pressure, and hence need to be handled immediately. This, in turn, means that communication efforts aimed at informing, educating and communicating elements related to these issues are more likely to be successful, since people can immediately related to them. These are not issues which are

---

[1]The List of Official Devellopment Asssistance (ODA) recipients, contains the names of all low and middle income countries, as well as the Least Developed Countries (LDCs) as defined by the United Nations (UN) as eligible to receive official development assistance.

theoretical or abstract. These are real matters, to which people can immediately relate to, since they impact their livelihoods.

The above state of affairs suggests that, under a developing country perspective, climate change messages such as those related to mitigation efforts may have a lesser impact due to other competing (and more pressing) priorities, which have a direct (and immediate) relevance to the livelihood of millions of people.

However, the above illustrated trends do not necessarily need to be perceived as completely negative. To the extent that the messages addressing matters and measures which need to be undertaken in the long-term are not really able to fully mobilise communities in developing countries, those which address priority areas and have a direct related to people's livelihoods are more likely to reach the desired audiences, and yield the expected benefits in respect of both better information and motivation of people to act.

But communicating about climate change is not without challenges. Apart from knowledge on the theoretical aspects of climate change communication, finding the right balance in conveying the right messages is important (Ereaut and Segnit 2006). Therefore, upfront knowledge of the difficulties inherent in talking about climate change to different types of public using a various types of communication tools and strategies (Nerlich et al. 2010), is necessary.

The next section describes some of challenges seen in communicating climate change, whereas the final section of this paper illustrates some ways via which a grater sense of awareness and motivation to engage on climate action may be achieved.

2 Challenges in Communicating Climate Change

One the main features of the subject matter of climate change, is its **complexity**. This is characterised by the fact that:

i. climate change relates to complex scientific information, data and models, whose comprehension is far from simple, and requires a good understanding of how facts and figures are translated into observed trends;

ii. it entails information from the domain of climate science and natural sciences on the one hand, but also from social sciences (especially economics and politics, but also elements such as ethics) on the other;

iii. it has concrete implications to both economic growth and development, but also to livelihoods

iv. its impacts are wide ranging in both rural and urban contexts;

v. apart from links with areas such as floods, erosions and extreme events, climate change is also associated with impacts to human health, from the spread of mosquito-born diseases, to phenemona such as the urban heat islands (Leal Filho et al. 2018).

Table 1 Some of the challenges seen in communicating climate change

Item	Impacts
Complexity of climate messages	Due to their wide scope, climate-related messages may not be fully understood or may confuse the audiences
Geographical distribution and focus	It is often reported on climate chage impacts in the developing world, thus often not being regarded as of direct importance to some people in industrialised nations
Variety of themes	Many themes are under the umbrella of "climate change", among others CO_2 emissions, droughts, extreme events, transport, energy production and use, etc. making the understanding challenging at times to some audiences
Responsibility for the theme	Many people perceive climate change as a matter which governments and companies need to address, and do not immediately see the role of the individual in the process
Uncertainty	Since it is not absolutely certain that some impacts are specifically due to climate change, a "wait and see" approach is often adopted
Lack of specialised reporting	Whereas newsmedia have reporters specialising on social affairs, economics or sports, very few have staff training on matters related to climate change
Competing themes	Climate change is not as rated as highly as political themes, wars, conflicts or even some major sport events

Source Author

Therefore, the challanges seen in communicating climate change, and the barriers associated with it (Lorenzoni et al. 2007) are manifold. Some of them are outlined in Table 1.

Despite the fact that there is a wide spread belief that the impacts of climate change communication are limited towards educating or informing audiences, much more can be achieved by means of well focused climate change messages. A duly prepared climate change communication programme should be able to achieve results such as:

(a) inform people about specific risks climate change may pose to their own surroundings, hence making a connnection between a phenomena which is global in nature, but whose impacts are also local, so they can take action needed. This is required in sober way, since the emphasis on doomsday messages is known to be far less efficient, then the presentation of positive ones (Carvalho and Burgess 2005; O'Neill, Nicholson-Cole 2009);

(b) persuade people to engage on a reflection on (i) how the policies their countries follow may be associated with climate change and (ii) how their own behaviours may influence the global climate and mobilise people to become more involved in implementing climate change mitigation measures on their own in the way they use transport or energy, and in respect of climate change adaptation by using resources more wisely (e.g. water use in households).

The rather limited impact that many climate change communication efforts performed to date by some government agencies and international organisations across the world have had, may be explained by the fact that they have ignored one or more of the factors outlined in Table 1. Attempts to address the deep roots of climate change by focusing on running initiatives intended to foster greener attitudes amongst the public are helpful, but do not suffice on their own. An encouragement of changes in attitudes cannot be achieved overnight, or by focusing on one element only. Indeed, a full engagement from the public on matters related to climate change cannot be expected, unless a set of measures are combined with the aim to:

(a) inform
(b) educate
(c) encourage and
(d) monitor

action. Unfortunately, due to a variety of reasons, which vary from limited resources to the limited amount of time available, most of the climate change communication processes performed to date have focused on items "a" and "b", seldom engaging with "c" and rarely with "d". As a result, only a few of them have led to the expected results. Therefore, a change of thinking is needed, and it is important to add the "monitor" component to climate change communication plans, so as to really ensure they have achieved what is expected from them.

It is admittedly very difficult to achieve an across-the-board and full public engagement with matters related to climate change. Society is too diversified to allow this from happening, and among countries (or even within nations) the cultural gaps are even deeper.

However, it is perfectly possible to reach specific audiences, provided care is exercised when approaching them. The reasons for this are two-fold. Firstly, each audience has its own mental framework and has long-established priorities of their own, which differ them from others. Even though some of the priorities may be the same among various audiences (e.g. well-being of people), there are many which are quite specific and unique to them. Secondly, the best means to reach the various audiences also differ. Whereas schoolchildren are easily reached in the classroom, politicians get their information from the media on the one and, and from specialist circles on the other. Providing information via events and vehicles which specific audiences normally use and trust, is an efficient way to reach them.

As an attempt to foster information and communication on climate change, the International Climate Change Programme (ICCIP) was established in 2008 by the Hamburg University of Applied Sciences in Germany. ICCIP aims to:

(a) disseminate the latest findings from scientific research on climate change, including elements related to its environmental, social, economic and policy aspects in a way that allow them to be understood by the non-specialist audience. This takes place by means of books, book chapters, journal articles and information via the media;

(b) undertake education, communication and awareness-raising projects on matters related to climate change in both industrialised and developing countries in cooperation with UN agencies, universities, scientific institutions, government bodies, NGOs and other stakeholders;

(c) network people and organisations to find ways to discuss the problems, barriers, challenges and chances and potentials related to communication on climate change.

Based on the fact that current internet technologies can provide a substantial contribution in terms of disseminating information and the latest scientific findings on climate change research in a fast and efficient way, ICCIP also organises a variety of Symposia round the world, which have been dealing with a variety of issues. For instance:

- Symposia on Climate Change in Africa
- Symposia on Climate Change in Latin America
- Symposia on Climate Change and Coastal Zone Management
- Symposia on Climate Change and Health
- Symposia on Climate Change Resilience
- Symposia on Climate Change Communication

and many others. Last but not least, ICCIP encourages more networking and information exchange, and catalyse new cooperation initiatives and new projects, in a true attempt to reach a wide audience, which includes:

1. members of NGOs working with climate change;
2. researchers at universities and research centres;
3. teaching staff at universities;
4. representatives from companies;
5. representatives from UN and national development and aid agencies working with climate change adaptation and funding/executing projects on the ground;
6. members of social movements;
7. project officers and consultants, as well as other people interested on the topic.

ICCIP uses this cross-sectoral approach since it believes that this wide range of participants helps to outline the need for, and the usefulness of integrated approaches towards climate change adaptation, hence contributing to the further consolidation of this thematic area. In order o foster the cause of climate change communication across audiences, ICCIP is undertaking a five-years project (Box 1), involving a wide range of stakeholders.

Box 1 The project: "Climate Change Communication Across Audiences" (2018–2022)
Communication, information and training on matters related to climate change are perceived as important in order to mobilise people and to catalyse action. Over the years, the International Climate Change Information

Programme https://www.haw-hamburg.de/en/ftz-nk/programmes/iccip/ has been organising a range of events across all continents, reaching a wide range of stakeholders: https://www.haw-hamburg.de/en/ftz-nk/programmes/iccip/ events.html. It has also been spearheading the world's leading peer reviewed book series on climate change https://www.springer.com/series/ 8740 as part of a large outreach exercise. Since 2008, over 50,000 climate scientists, members of NGOs, government officials and practitioners from over 100 countries have taken part on, or benefitted from ICCIP's activities.

ICCIP is presently running a major, long-term study, which analyses how climate change communication takes place across various audiences. The focus of the research is on the developing countries which are part of the DAC list: http://www.oecd.org/dac/financing-sustainable-development/ development-finance-standards/daclist.htm.

The study is being performed over a 5 years period (from 1st June 2018 to 30th May 2022), and will involve a wide range of stakeholders across the developing world, where there is a chronic paucity of research and data on aspects of climate change information and communication.

The study is structured around various stages, and will entail the organisation of specialist events and collection of data across a variety of audiences. These include:

- media professionals
- policy and decision-makers
- scientists
- members of NGOs
- farmers
- workers at international aid agencies
- industry representatives

In order to systematise the data collection, simple and straightforward instruments will be deployed, containing a set of very specific questions which were developed by a panel of experts. These will be sent to each group, on a step-by-step approach, for completion. The results of each study will be published in peer-reviewed journals via open access, so that they are readily available for a wide audience.

Further details on how to take part on the study will be released by means of spefic calls for contributions.

Table 2 describes some of the most important audiences in the climate change communication framework, and the mind-settings which need to be considered, in attempts to reach them. The list is by no means exhaustive. Many other groups

Table 2 Some of the audiences to be reached via climate change communication initiatives

Audience	Mind-settings
Politicians	Focused on laws, policies and their socio-economic implications
Scientists	Focused on the reliability and robustness of scientific information, and the technical support needed to tackle climate related issues and problems
Goverment officials	Focused on the implementation of the policies set by the government of the day
Farmers	Focused on land use, crop productivity and yields
Media representatives	Focused on reporting and public information
Industry officials	Focused on the impacts of their specific sectors (e.g. transport, production, logistics, etc.)
Teachers	Focused on the curriculum-related information of pupils
University teaching staff	Focused on the information and education/training of university students
The "general public"	Too wide a focus and range of interests, very difficult to cluster into single frames

Source Author

could be added. Rather, it is meant to illustrate the fact that each audience has a particular focus. In addition, it illustrates the various angles under which they perceive, treat and handle matters related to climate change. Knowledge about their mind-settings is important, since it allows a better understanding of the potentials, as well as the limitations of their actions and help to explain the scope of their engagement and commitments to matters related to climate change.

The complexity of each audience—and the vagueness of the groups which may be classified under the term "general public"—illustrate why it is so difficult to build a consolidated societal support to climate change. The range of mental frameworks and personal interests of each audience is so wide, that organising a coherent information and communication programme to reach all of them in one goal is a daunting task.

Instead of that, audience-specific information and communication programmes, which "speak the language" of each audience and formulate messages which are consistent with their interests, are likely to yield greater benefits. In order to support this debate, ICCIP recently produced the "Handbook of Climate Change Communication" (Leal Filho et al. 2018), spread over three volumes, which provides a unique overview of the theory, methodologies and best practices in climate change communication from around the world.

3   Moving Forward

It has to be admitted that even though is part of the reality of millions of people in the developing world, climate change is regarded by for many people in rich countries as a problem of lower ranking, especially when put against matters such as economic development, international conflicts, politics or health issues, which tend to take most of the media's attention. The general perception of many people in the rich world is that since their countries have the resources and access to the technology needed to cope with climate change and its impacts, climate change is indeed a matter of concern, but not one which does not occupy the top place in their list of priorities.

Indeed, it is fair to say that, the fact that climate issues may be important to rich countries and that it is of relevance to future generations, it tends to be overshadowed by debates on the price of oil, unemployment rates, levels of taxation or refugees. These "day-to-day" issues tend to get more attention, and are often more comfortable to handle, than climate matters which affect people and countries far away from them.

There is also a danger that the debate and discussions on climate change are centralised around disasters and catastrophes—as opposed to what may be achieved through the improvement of the built and natural environment—, which tend to get the attention from the media and a great deal of the public, but which also may lead to alienation (and in the worst case, to fear). Therefore, finding a balance between what to communicate, to whom and how, is vital to any attempt to successfully communicate about climate change. So, what is the way forward? How can climate change be better communicated across various audiences?

The answer to this question is not simple, but four measures may be put to use, to help to achieve this goal:

1. avoid the focus on negative messages: it is certainly correct to refer to the damages caused by climate change to crop yields, properties or infra-structure, but it is equally important to illustrate examples of succesful actions or projects which have led to improvements in livng conditions and/or infra-structured;
2. use a constructive approach: apart from stating the due facts and figures on and about climate change, is it important to specific to the different audiences how they may contribute towards addressing the problem. For instance, politicians should support policies which may help to reduce CO_2 emissions, or to increase the resilience of cities, ports and facilities to a changing climate; or farmers should try new crops or adjust the seedings times in order to avoid droughts. The range of actions the various audiences may take is quite comprehensive, but they are often not that obvious tot hem, hence requiring specific information;
3. selected the best tools to reach specific groups: as Table 2 has shown, the mind-settings of the various audiences differ, and so do the range of tools to reach them. For instance, one of the best ways to reach schoolchildren is by relating climate change to components taught in the curriculum, since teachers would have little time otherwise. To the same measure, many farmers in

develoing countrie cannot be reached by specific media such as newspapers, but may be informed by means of radio and the extension workers who regularly visit them;

4. find ways to monitor progress: most climate change information programmes are short-lived (e.g. a few weeks, months or years) and seldom make provisions to ascertain the progress reached by them. It is thus important that, in trying to reach various audiences, mechanisms are put in place in order to monitor progress and identifiy the extent to which the expected results were reached.

The ultimate evidence of success of any climate change information initiative is the mobilisation of people to take action in handling the climate problems which concern them, i.e. moving away of theorising, and into active handling. If a climate communication initiative has led to such a mobilisation, then chances are that it has worked successfully.

References

Caldeira K, Wickett ME (2003) Anthropogenic carbon and ocean pH: the coming centuries may see more ocean acidification than the past 300 million years. Nature 425:365

Carvalho A, Burgess J (2005) Cultural circuits of climate change: An analysis of representations of "dangerous" climate change in the UK broadsheet press 1985–2003. Risk Anal 25:1457–1469

Chapman DA, Lickel B, Markowitz EM (2017) Reassessing emotion in climate change communication. Nat Clim Change 7:850–852

Edwards PN (2019) A vast machine: computer models, climate data, and the politics of global warming. MIT Press, Cambridge

Ereaut G, Segnit N (2006) Warm words: how are we telling the climate story and can we tell it better?. Institute for Public Policy Research, London

Harvey B, Ensor J, Carlile L, Garside B, Patterson Z, Naess LO (2012) Climate change communication and social learning—review and strategy development for CCAFS. CCAFS working paper 22. CAFS, Wageningen

IPCC (2014) Climate change 2014: synthesis report. Contribution of working groups I, II and III to the fifth assessment report of the Intergovernmental Panel on Climate Change. In: Core Writing Team, Pachauri RK, Meyer LA (eds) IPCC, Geneva, Switzerland, p 151

Leal Filho W (2009) Communicating climate change: challenges ahead and action needed. Int J Clim Change Strategies Manag 1(1):6–18. https://doi.org/10.1108/17568690910934363

Leal Filho W, Echevarria Icaza L, Emanche VO, Quasem Al-Amin A (2017) An evidence-based review of impacts, strategies and tools to mitigate urban heat Islands. Int J Environ Res Public Health 14(12). https://doi.org/10.3390/ijerph14121600

Leal Filho W et al (2018) Handbook of climate chance communication, vol 1. Theory Clim Change Commun, Springer, Berlin

Leiserowitz A (2007) Public perception, opinion and understanding of climate change—current patterns, trends and limitations. United Nations Development Programme, New York

Leiserowitz A et al. (2011) Climate change in the American mind: Americans' global warming beliefs and attitudes. le Project on Climate Change Communication, New Haven. Available at http://environment.yale.edu/climate/files/ClimateBeliefsMay2011.pdf

Lenton TM, Held H, Hall J, Lucht W, Rahmstorf S, Schellnhuber H (2008) Tipping elements in the earth's climate system. Proc Natl Acad Sci USA 105:1786–1793

Lorenzoni I, Nicholson-Cole SA, Whitmarsh L (2007) Barriers perceived to engaging with climate change among the UK public and their policy implications. Glob Environ Change 17:445–459

Mazur A (1998) Global environmental change in the news. 1987–90 vs. 1992–96. Int Soc 13 (4):457–472

Nerlich B, Koteyko N, Brown B (2010) Theory and language of climate change communication. Wiley Interdiscip Rev Clim Change 1(1):97–110

O'Neill S, Nicholson-Cole S (2009) "Fear won't do it"- promoting positive engagement with climate change through visual and iconic representations. Sci Commun 30(3):355–379

Climate Change Engagement: A Different Narrative

Henry McGhie

If we change nothing, nothing will change

Abstract Climate change engagement presents a number of challenges to museums, which tend to be most comfortable in dealing with the topics in which they are expert, and focus on presenting information. This chapter will explore some of the challenges and 'letting go' that could help museums reposition themselves to engage people more constructively with climate change and related issues, to embrace a more future-focused frame, and to focus more effectively on the connection between thinking–feeling–doing, and on inspiration, in order to encourage, inspire and realize positive futures. More generally, it will explore how museums could work to develop a more positive and inclusive vision of the future, as an alternative, rather than an antidote, to that presented in mass media, and to work with people at local and global levels to create and enact that narrative. The chapter proposes a set of 15 'shoulds' for museums, to help global museums of any scale or subject to support climate action constructively.

1 Introduction

Climate change is no longer confined as a scientific issue: it is an issue of mass concern, with a high level of awareness in the political and public consciousness. In the *Global Risks Report 2018*, produced by the World Economic Forum (WEF), environmental risks had risen in importance so that the top three risks (maximum likelihood against maximum impact) all belonged to this category, namely *extreme weather* (1 in terms of ranking of all risks); *natural disasters* (2); *failure of climate-change mitigation and adaptation* (3). The other two environmental risks, *man-made environmental disasters* (7) and *biodiversity loss and ecosystem collapse* (8) were not far behind. All of these are linked with climate change and its impacts, as are the top three societal risks, *water crises* (5); *large-scale involuntary migration* (6) and *food crises* (9).

H. McGhie (✉)
Manchester Museum, University of Manchester, Manchester, UK
e-mail: henry.mcghie@manchester.ac.uk

© Springer Nature Switzerland AG 2019
W. Leal Filho et al. (eds.), *Addressing the Challenges in Communicating Climate Change Across Various Audiences*, Climate Change Management,
https://doi.org/10.1007/978-3-319-98294-6_2

13

Yet, environmental issues fell far behind other issues of concern to the business and political communities on a nation-by-nation basis, contained in the report *Global Risks of Highest Concern for Doing Business*, based on the same risks in the *Global Risks Report 2018*.[1] *Failure of climate-change mitigation and adaptation* ranked 13th in the UK, 25th in the US and 22nd in China, while *Extreme weather events* ranked 18th, 10th and 15th in the in the UK, US and China respectively. Globally, *Failure of climate-change mitigation and adaptation* ranked 22nd out of 28 risks.

In the preface to the *Global Risks Report 2018*, Klaus Schwalbe and Børge Brende of the WEF note that "Multistakeholder dialogue remains the keystone of the strategies that will enable us to build a better world." Yet, multi-stakeholder initiatives are not easily co-ordinated or led, and form a nested and scaled challenge at all levels, whether at an institutional, regional, national or international level. Indeed, international organisations were established to deal with the challenges that their members could not deal with alone (Blokker 2017: 36–7). This chapter will explore how museums can position themselves within global issues, and facilitate partnerships that can hopefully support and enable climate action at all levels.

1.1 Building the Future

In response to a number of global challenges, 193 Member States attending the UN Sustainable Development Summit in New York in December 2015 adopted a new global development framework, 'Transforming Our World: the 2030 Agenda for Sustainable Development'. The Agenda is delivered through 17 Sustainable Development Goals (SDGs) towards an extremely ambitious 15-year programme for 'the future we want'. The '2030 Agenda', as it is sometimes called, builds on the previous Decade of Sustainability that focussed on the Millennium Development Goals, and which focussed on developing countries. The SDGs are seventeen interlinked goals with 169 targets that address environmental, social and economic sustainability, with the principle to 'leave no-one behind'; they have a truly global horizon and emphasise the interconnectedness of the world, socially, economically and environmentally. Climate change forms the basis of Sustainable Development Goal 13, 'take urgent action to combat climate change and its impacts', but is heavily entangled with most, if not all, of the other goals, and a principle of the SDGs is that they cannot be disentangled. The Paris Agreement of December 2015 was a major step in working towards SDG 13, but that really was only a first step. The 2030 Agenda calls on all nations, civil society, business and the public to contribute towards this global programme. Repeatedly, the phrase '[we] cannot do it alone' appears, whether from President Barack Obama's address to the UN

[1]http://reports.weforum.org/global-risks-2018/global-risks-of-highest-concern-for-doing-business-2018/ (accessed 26 March 2018).

Climate Summit in 2014),[2] in the UK Clean Growth Strategy (2017) ("we cannot achieve this through Government action alone. We must harness the ingenuity and determination of all our people and businesses across the country"),[3] or, perhaps most notably, from Ban Ki-moon at the 'Tackling climate, development and growth' session in Davos in 2015: "It's not only government. Government cannot do it alone. The UN cannot do it alone. There should be full partnership... then we should have civil society coming together. Even one normal citizen—they have a role to play."[4] All of these words can be interpreted as a political version of 'we need your help'.

In the face of growing nationalism (acknowledged in the *Global Risks Report*: the 2018 WEF was entitled 'Creating a Shared Future in a Fractured World'), a number of concerning events have taken place that threaten to undermine progress towards a sustainable future. President Trump's approach to environmental issues is the most notable, while the withdrawal of the UK from the EU could result in a weakening of environmental protections, and increased deregulation at the expense of the environment.[5] More generally, the UK's response to Agenda 2030 has been slow, and the SDGs have been given little visibility by government. They are co-ordinated by the Department for International Development, and have not been promoted by the Department for Digital, Culture, Media and Sport, which oversees the cultural sector in the UK. The Government's approach to the SDGs has been criticized by the International Development Committee, who report they are "deeply concerned at the lack of a strategic and comprehensive approach to the implementation of the Goals... It also reflects a worrying absence of commitment to ensure proper implementation of the SDGs across Government" (International Development Committee 2016: 4).

What does all this mean? Put simply, it means that economically driven politics at the national level may be unlikely to address global environmental challenges that will have a profound effect on all aspects of society and the environment. Challenging times call for things to be done differently, as has already been seen in relation to climate change action in the US, as reported in the *Global Risks Report 2018*:

> The risk that political factors might disrupt efforts to mitigate climate change was high-lighted last year when President Trump announced plans to withdraw the United States from the Paris Agreement... In addition, many US businesses, cities and states have

---

[2]https://obamawhitehouse.archives.gov/blog/2014/09/23/president-obama-no-nation-immune-clim ate-change (accessed 26 March 2018).

[3]https://www.gov.uk/government/uploads/system/uploads/attachment_data/file/651916/BEIS_The_ Clean_Growth_online_12.10.17.pdf (accessed 26 March 2018).

[4]https://www.weforum.org/agenda/2015/01/24-quotes-on-climate-change-from-davos-2015/ (accessed 26 March 2018).

[5]https://socenv.site-ym.com/page/EUWithdrawalBill (accessed 26 March 2018).

pledged to help deliver on the country's emissions reduction targets. This kind of network of subnational and public–private collaboration may become an increasingly important means of countering climate change and other environmental risks, particularly at a time when nation-state unilateralism appears to be ascendant" (WEF 2018: 13).

2 Calls for Public Engagement and Broad Partnerships Around Climate Change and Sustainability

The importance of mass participation in climate action is recognised in a number of high-level policies and strategies, which promote greater public awareness, education and action around climate change and environmental issues (Table 1). Post-Paris, some of these calls have begun to be been operationalized through the UNFCCC Action for Climate Empowerment, and UNESCO/UNFCCC initiatives, however, at a national level, at least in the UK, they have been given little direct profile.

Table 1 Calls for greater public engagement with environmental issues from key strategy documents

UN Framework Convention on Climate Change (1992)[a]	Article 6: Parties shall… promote… (i) the development and implementation of educational and public awareness programmes on climate change and its effects (ii) public access to information on climate change and its effects (iii) public participation in addressing climate change and its effects and developing adequate responses…
Paris Climate Agreement (2015)[b]	Article 12: Parties shall cooperate in taking measures, as appropriate, to enhance climate change education, training, public awareness, public participation and public access to information, recognizing the importance of these steps with respect to enhancing actions under this Agreement
UN Sustainable Development Goals (2015)[c]	e.g. 4.7 By 2030, ensure that all learners acquire the knowledge and skills needed to promote sustainable development, including, among others, through education for sustainable development and sustainable lifestyles, human rights, gender equality, promotion of a culture of peace and non-violence, global citizenship and appreciation of cultural diversity and of culture's contribution to sustainable development 11.4 Strengthen efforts to protect and safeguard the world's cultural and natural heritage. 12.8 By 2030, ensure that people everywhere have the relevant information and awareness for sustainable development and lifestyles in harmony with nature.

(continued)

Table 1 (continued)

	13.3 Improve education, awareness-raising and human and institutional capacity on climate change mitigation, adaptation, impact reduction and early warning.
UK National Ecosystem Assessment (2011)[d]	A move to sustainable development will require an appropriate mixture of regulations, technology, financial investment and education, as well as changes in individual and societal behavior and adoption of a more integrated, rather than conventional sectoral, approach to ecosystem management. This will need the involvement of a range of different actors—government, the private sector, voluntary organisations and civil society at large…
A Green Future: Our 25 Year Plan for Nature (HM Government 2018)[e]	Internationally, we will lead the fight against climate change… …Our goal is to see more people from all backgrounds involved in projects to improve the natural world. We will make 2019 a year of action for the environment, putting children and young people at its heart. This year of green action will provide a focal-point for organisations that run environmental projects, and will encourage wider participation. A series of public engagement activities for 2019 will link to initiatives on waste reduction, cleaner air or other aspects of pro-environmental behaviour. We will look to get the business community and voluntary sectors involved in these activities, and urge them, with the education sector, to develop their own initiatives throughout the year to engage communities and raise awareness. We expect 2019 to be the foundation of a five-year programme that will help turn the commitments in this 25 Year Environment Plan into action

[a]http://unfccc.int/resource/docs/convkp/conveng.pdf. Accessed 6 April 2018
[b]http://unfccc.int/resource/docs/2015/cop21/eng/l09r01.pdf. Accessed 6 April 2018
[c]https://sustainabledevelopment.un.org/sdgs. Accessed 6 April 2018
[d]http://uknea.unep-wcmc.org. Accessed 6 April 2018
[e]https://assets.publishing.service.gov.uk/government/uploads/system/uploads/attachment_data/file/693158/25-year-environment-plan.pdf. Accessed 6 April 2018

The Tokyo Protocol was developed and approved by the world's science museum networks in 2017 'on the role of science centres and science museums in support of the United Nations Sustainable Development Goals', with a focus on science centres and museums as educators, to engage with communities, to serve as platforms for discourse and exchange, to facilitate partnerships and collaborations, to embrace technological innovations, serve as trusted links and valued communicators, support other museums and centres to achieve progress with the SDGs, accept the responsibility to serve as catalysts better understanding and coordinated actions within communities throughout the world by stimulating tolerance and critical thinking, and to support collective worldwide public STEM activities. Rather surprisingly, no specific mention is made of climate change, although this is the subject of a particular SDG.[6] The Tokyo Protocol also takes a narrow focus, emphasizing the role of museums in science education and engagement, rather than connecting the full range of their resources and opportunities—collections, staff expertise, partnership possibilities and in support of climate-related research—to connect with climate change action.

As already mentioned, multi-stakeholder agreements of all kinds present particular challenges to traditional governance structures, whether locally, regionally or internationally. The Talanoa approach, adopted in 2018 to help accelerate action and move 'further forward faster' together is a welcome contribution. 'Talanoa' is a Fijian word describing a process of inclusive and transparent dialogue during which participants share stories, build trust and empathy, and strive to "make wise decisions for the collective good." Talanoa is based on mutual respect and critical observations and blaming others are considered inconsistent with its principles.[7] The Talanoa approach is one that museums could usefully connect with, in 2018 and beyond, although museums are as yet slow to connect with the SDGs or Paris Agreement. Nevertheless, we have to start somewhere, be prepared to enter new territory, try things out, be prepared to fail and learn from mistakes to do things better.

2.1 What the Public Think—and Do—About Climate Change

The European Social Survey (2016–17) found that the UK has a mid-range level of concern relating to climate change among 18 countries surveyed, with the highest levels of concern found in Germany (86% of respondents worried), and low levels of concern in Israel (53%) and the Czech Republic (54%).[8] Seventy percent of UK respondents were concerned about climate change, and 24% were 'very' or 'extremely' concerned. Only 2% of respondents in the UK stated that they do not think climate change is happening already. However, 52% in the UK were sceptical that enough governments would act to limit climate change. Levels of concern were higher among those with an interest in politics, and those with left-wing political concerns across Europe. In the UK, concern increased with education level, and was higher among younger generations (although such patterns were not found in many parts of Europe). One of the reports authors, Roger Harding (Head of Public Attitudes at the National Centre for Social Research) stated:

---

[6]https://scws2017.org/tokyo_protocol/ (accessed 26 March 2018).
[7]http://sdg.iisd.org/news/unfccc-launches-talanoa-dialogue-platform-to-boost-climate-ambition/ (accessed 6 April 2018).
[8]http://natcen.ac.uk/our-research/research/european-attitudes-to-climate-change-(1)/ (accessed 26 March 2018).

> Tackling climate change requires a major international effort so it's concerning that so many people doubt that this will happen. To make personal sacrifices to stop climate change people will need more reassurance that others will join them.

While another co-author, Leo Brassi, stated:

> People who want action on climate change need to get better at talking to those who aren't left-wing or interested in politics. Solutions to climate change need support from across society—they won't succeed if the subject is seen as a partisan concern.[9]

2.2   And What People Think Other People Think About Social Problems

Social psychologists have identified a set of intrinsic and self-transcendent values that have been repeatedly found to underpin both concern about social and environmental problems, and action in line with this concern (see Kasser and Ryan 1996; Grouzet et al. 2005; Aasen and Vatn 2018; Crompton and Lennon 2018). Intrinsic and self-transcendent values are in opposition to extrinsic and self-enhancement values, which are concerned with external rewards such as wealth, social status or other forms of power. It is worth noting, for example, that the basis of a great deal of advertising—of all forms (and including advertising by nature conservation charities and museums)—is framed in terms of extrinsic/self-enhancement values.

In the UK, pro-environmental behaviour and sustainable consumption have been found to be positively related to life satisfaction. However, the increase in life satisfaction is linked with self-image rather than direct pro-environmental behaviours. In short, value action gaps exist between people's self-perception of the contribution they make and their actual impact upon the environment (Binder and Blankenberg 2017). It is also worth considering the value-action gaps that museums hold at an institutional level, where well-meaning words of mission statements are not linked with programming or actual environmental outcomes. It is also worth reflecting on the reason why people don't act on climate change at an individual level, and the part that institutions play in maintaining a status quo (see Janes 2016).

As climate change, and other large-scale societal problems, cannot be solved alone, they rely on people's perceptions and attitudes to others. In a survey of 1184 people across Greater Manchester (a region combining Manchester and surrounding conurbations), 85% of people attached greater importance to compassionate values such as social justice, environmental protection, forgiveness and honesty, than they do to selfish values such as wealth and social status. However, 75% of respondents thought that fellow citizens attached less importance to compassionate values than they did themselves; 65% of respondents thought a typical citizen held selfish

---

[9]http://www.natcen.ac.uk/news-media/press-releases/2017/december/survey-reveals-uk-pessimistic-about-international-action-on-climate-change/ (accessed 6 April 2018).

values to be more important that they did themselves. This means there is a perception gap where the majority of people think others are more selfish and less compassionate than they are themselves.[10] Similar results have been found across the UK (Crompton 2016).

Common Cause Foundation, who initiated the values survey work, posit there to be three inter-related challenges facing UK society: to mount proportionate responses to oppressing social and environmental problems (climate change, inequality, child poverty, biodiversity loss); deepen public commitment to civic participation; and to rebuild social cohesion and reshape social institutions to inspire public trust. Values are inter-related in terms of people's own values, perceptions of the values of others, and the values seen to be encouraged by social institutions as a 'values nexus' (Crompton 2016: 8). Interestingly, although museums make frequent claims that they are trusted institutions, and they scored highest in terms of encouraging compassionate values when compared to education, media, government and business, they were still perceived as encouraging compassionate values less and selfish values more than people considered themselves to hold; people considered others' values to lie somewhere between education and media categories (Crompton 2016: 28). Crompton advocated that museums could provide citizens to explore their own values and those of other citizens to address the perception gap and promote deeper and richer civic participation on topics of shared interest (Crompton 2016: 2).

3 How Museums Can Create Public Value Around Climate Change

Museums traditionally develop exhibitions that provide information, often with little regard to what people are intended to do with that information. However, information is not an end in itself, nor is it neutral or impartial (see Kellert 2012: 67–80). Simply providing information on climate change alone has been shown to have limited impact in terms of predicting actual pro-environmental action (Whitmarsh 2011).

Museums and other cultural institutions can create public value by connecting their work with external agendas to promote positive social and environmental outcomes relating to those agendas, rather than an internally focussed agenda (e.g. Janes 2016; Azmat et al. 2018).

Museum codes of ethics exist, but these sometimes exist in isolation from broader perspectives. For example, the UK Museums Association has a Code of Ethics for Museums which includes the following:

---

[10]Shared Values for the City Region. Available at https://valuesandframes.org/downloads/ (accessed 6 April 2018)

Museums and those who work in and with them should... act in the public interest in all areas of work... ...use collections for public benefit –for learning, inspiration and enjoyment.

A clear link can be drawn between acting in 'the public interest', the Global Risks Assessment, the Sustainable Development Goals and other calls for civic engagement with societal challenges.

Museums can help people explore topics and collaborate with audiences to create positive social norms. Social norms are the widespread convergence of the "unplanned, unexpected result of individuals' interaction... that specify what is acceptable and what is not in a society or group" (Bicchieri and Muldoon 2014), or "the unwritten codes and informal understanding that define what we expect of others and what others expect of us" (Young 2015, see also Farrow et al. 2017; Aasen and Vatn 2018). Social norms are important, and cultural institutions have a responsibility and a role to play in supporting collaborative, constructive social norms.

University museums have a particular mandate to challenge the status quo, being, as Holland Cotter said, "institutions that are, at their best, equal parts classroom, laboratory, entertainment center and spiritual gym where good ideas are worked out and bad ideas are worked off", drawing on the latest research, and with a commitment to intellectual rigour (Cotter 2009).

However, museums will need to radically shift the ways in which they connect with visitors, and with the wider world, to accelerate and enable positive climate action. In order to do so, Cameron et al. (2013) proposed nine principles for museums and science centres to adopt to promote understanding and action on climate change (summarized and adapted below):

1. Climate change is too important to deny and too complex to reduce to a single analysis or problem. They proposed that museums should adopt a forward-facing timeframe, and that the challenge of climate change presents a creative opportunity to draw on people's imagination to innovate and propose new directions for themselves and others.
2. The museum sector needs to draw on its heterogeneity to respond to the challenges of climate change. Allowing people to draw on a range of perspectives, ideas, scales and disciplines and to form their own conclusions is important, but it is also important for museums to express their own position in contested positions.
3. Climate change is multi-scalar in space and time, and needs a multi-scalar response. Museum-goers need to be able to connect with the scale that is appreciable to them.
4. Climate change responses should be polycentric, using networks. Museums need to be part of an ecology of partners from different sectors. Museums can be media within larger networks of organisations and can facilitate dialogue across sectors.
5. Climate change responses need porous boundaries, "liquid" organizations and "clumsy" solutions. Museums need to be able to work beyond traditional boundaries (physical and conceptual), to act more meaningfully in a dynamic,

turbulent world. Museums can be flexible spaces for deliberative experience and participation, and thus contribute to true civic action.

6. Engaging citizens' needs "thick" communication, interaction, dialogue, tria-logue–not monologues from the powerful. Museums need to take a radically different standpoint in terms of their relationship with their audiences, going beyond a comfortable expert, broadcasting, position, to one that facilitates users' expression, creativity and participation. This is not to say that expertise does not have a place, but that it is deployed in ways that meet with users' needs.

7. A dirty war has been declared, but it should be resisted, not fought. Climate change action is highly politicised and vested interests, and political inaction, produce inertia. However, closing viewpoints out of discussion is risky in museums. Promoting a high level of critical thinking and decision-making is more effective than censoring out particular viewpoints.

8. Give art a go, as creative explorations can help unlock deadlocks and overcome inertia and a sense of apathy.

9. Build new relations to new publics, going beyond existing audiences to reach constituencies whose interests are outwith those of museums' existing audiences and stakeholders, as new combinations create new opportunities and catalyse new collaborations and shared agendas.

3.1 The Challenge of Climate Change Engagement

Climate change presents particular challenges to public engagement: it is enormous, seemingly beyond the power of any one person or group of people; it is uncertain, and its impacts are impossible to predict fully; it will be experienced in different ways and by different groups of people and wildlife, so defies succinct description; its causes are complex; we all contribute towards it, in different ways. While climate change impacts represent a huge and wide-ranging challenge to society, climate change is not itself a root cause: it is a product of a systemically unsustainable use of resources.

Climate change is perceived to be distant (psychologically distant) in terms of its temporality, geography, uncertainty and socially (Spence et al. 2012). Psychological distance has been linked with the way that people represent concepts mentally and with personal action; perhaps counter-intuitively, increased psycho-logical distance has been found to facilitate choices that are more abstract and linked with people's core values, although psychological closeness was linked to increased concern (Spence et al. 2012). McDonald et al. (2015) investigated per-sonal experience, psychological distance and inclination to take climate action. They found a complex relationship, where direct experience (closeness) did not necessarily lead to action, and that 'the optimal framing of psychological distance depends on (1) the values, beliefs and norms of the audience, and (2) the need to avoid provoking fear and resulting avoidant emotional reactions.

3.2   The Purpose of Climate Change Programming

The underlying concern, task and purpose of climate change engagement and related programming are worth clarifying. Climate change engagement often, and quite clumsily, simply raises awareness of the concern—of the challenge of climate change. However, this is to conflate the topic with what we are trying to achieve and, given some of what has been explored above, poor communication, the enormity of the challenge and the complexity of psychological distance means that raising people's awareness of things they can do nothing about may inhibit rather than promote climate action; it is worse than doing nothing. Howell (2013) investigated the motivations of UK citizens to explore the values, motivation and routes to engagement of those who have adopted lower-carbon lifestyles. Perhaps surprisingly, many interviewees were not interested in climate change or environmental concerns as a primary driver of their own values and behaviour, but about the plight of poorer people who will suffer as a result of climate change. That is to say, their concerns were more around altruism than environmentalism. Howell concluded that there was 'a need for climate change mitigation campaigns to promote a holistic view of a lower-carbon future, rather than a 'to do' list to 'combat climate change''. Instead, we can ask, 'because climate change is an enormous issue that people are presented with but disempowered by?' (the concern), 'what do we need to do?' (the task) 'so that people are empowered to engage with climate action in personally meaningful ways?' (the purpose).

3.3   Redefining Engagement

In museums, 'engagement' is often thought of as a time-bound activity that takes place in museums and is separated from the impact of the engagement, so that it can be unimportant, or even irrelevant, if the engagement leads to any wider impact. In an alternative model, climate change engagement may be defined as 'an ongoing personal state of connection'—as opposed to participation in a time-bound process of engagement—with the issue of climate change, whether in terms of conscious awareness of the issue, or unconscious action that contributes positively towards climate action and, importantly, in people's day to day lives (Lorenzoni et al. 2007: 446, Whitmarsh et al. 2011). This approach is worth reflecting on, as the effectiveness of museum engagement activities—their impact—is defined in terms of their impact beyond the museum in people's daily lives: the impact is the effect of museum activities in supporting people's ongoing engagement with climate change. A state of engagement incorporates a broad range of aspects that constitute what we think, feel and do about climate change—cognitive, affective and behavioral aspects. Simply knowing more about climate change does not necessarily promote action and, as information has an impact on affective and behavioural aspects, may inadvertently disempower action. People need to know what they can do about climate change and care about it to be motivated to take action. Cognitive, affective and behavioural aspects connect with one another in non-linear, non-sequential ways, but are iterative and dialogical. Engaging constructively with all three aspects presents a plausible route towards constructive engagement with the topic in people's daily lives, connecting people's thoughts and concerns, with choices and actions (Lorenzoni et al. 2007; McGhie et al. 2018). Particularly important in the context of behaviour and behaviour change is the role of inspiration, which can be defined as the 'feeling that moves us to action': of turning concerns into deeds. However, people can be inspired by a huge range of things, not necessarily positive. People can be inspired by hatred and by fear. Museums can focus on promoting inspiration towards positive social and environmental outcomes, connecting personal satisfaction and fulfilment with a wider public good. By raising awareness of people's individual and collective viewpoints, museums can support collective awareness and action, reducing the gap between people's perceptions and real-world situations.

3.4 Elements of a New Story—A Set of 'Shoulds' for Museums and the Society They Serve

McGhie et al. (2018) proposed a set of 15 key points for museums to consider in terms of climate change engagement. These asked museums to critically assess what difference they were trying to make to the world through their activities; to disrupt narratives of hopelessness and inevitability; to draw on symbols, images and ideas, and creative experiences; and give people plenty of chances to respond to exhibitions and events. As a supplement to these key points, we propose the following set of 'shoulds' for museums to help them position themselves in relation to real-world issues, such as climate change:

1. Museums should not allow themselves to be irrelevant from society.
2. Museums should acknowledge that what they do normalises people's views on what is acceptable in society.
3. It is not enough to aim to connect people with the museum. The museum should aim to connect people with the world.
4. Museums have a wealth of resources that can contribute positively, often uniquely, towards climate action. They should mobilise these resources to address this, and other societal challenges.
5. Many people are concerned about climate change, but do not see the connection between their own lives and the wider world. Museums should help them make that connection.
6. The museum should help people explore the past, the present and possible futures. Museums should help people understand the options they have beyond the museum in relation to museum topics, and help them understand the implications of their different choices. For example, museums should promote

climate action rather than climate damage, whether or not climate change is foregrounded or even presented.

7. They should develop a vision of a better future, working with people individually and collectively to shape and realise that future. This is a different vision from what is in mass media.

8. They should help people explore pictures bigger than themselves: framing discussions, promoting discussion of future[s] and disrupting a sense of hopelessness, and connecting possible choices with their impacts and consequences.

9. Museums need to keep abreast of research on topics relating to their collections, in the context of a rapidly changing world. They cannot do this alone, but need to work in partnership with researchers and organisations involved in climate action in diverse ways.

10. Museums should aim for deep and rich cultural experiences, rather than shallow, populist or solely commercially driven ones.

11. Museums should acknowledge that they are a medium that is not dependent on mass media or on neoliberal marketing.

12. Museums should be up-front about what they hold to be important, and be able to articulate why they hold particular topics to be important in reference to the real world.

13. Museums should base their decisions and concerns on robust, critically reflective and evaluated information.

14. In aiming to be honest and trustworthy, museums should neither use heavy-hitting information uncritically, nor disguise serious issues. They should deploy serious information and issues sensitively with an aim of promoting constructive engagement.

15. Museums should provide both challenge and support, considered alongside one another. In providing more challenge and more support they will provide people with even deeper, richer and more impactful opportunities to connect with real-world issues and challenges.

4   Considering Climate Change in Interpretation and Engagement—Some Examples

The following case studies are presented as examples of how different narratives might work.

Case study 1

Britain is home to internationally important populations of birds, notably waterfowl (ducks and geese), seabirds and shorebirds. Many of these birds migrate to Britain from the Arctic, while, for some species, important populations are found breeding in Britain. To take one example, the Golden Plover is a medium-sized shorebird that breeds in internationally important numbers in the UK, including on moors near Manchester; the species is protected under the EU Birds Directive. The birds' breeding is affected by the climate, so that the birds breed earlier than they did 50 years ago; weather conditions affect the availability of their food (craneflies that emerge when the birds are breeding), as dry summers reduce the amount of food available the following year (Pearce-Higgins and Yalden 2004; Pearce-Higgins et al. 2010). Windfarms, one of the commonest forms of green energy production, have been erected in increasing numbers on moorland hills around Manchester; windfarms have been found to have a negative effect on the birds, so as people move towards green energy it will be important that this is generated in ways that do not negatively impact on wildlife such as the Golden Plover. A number of projects are underway in the Pennines and the Peak District to help restore the Sphagnum Moss and Heather, some of which were lost as a result of pollution from industrial cities, and overgrazing. Sphagnum Moss will help combat erosion of the peat, and will help absorb water, so will help combat extreme rainfall and reduce the risk of flooding downstream in towns and cities, and provide a valuable ecosystem service; it will also help keep the moors wet, and increase the Golden Plovers' food supply for breeding (Pearce-Higgins 2011; Carroll et al. 2011). This short example aims to demonstrate the connection between people's local natural heritage; the impacts of past, present and future human activities on these birds and habitats; and the benefits of habitat restoration for both people and nature (Fig. 1).

Case study 2

Many Western museums contain highly distinctive coconut-fibre armour from Kiribati, formerly known as the Gilbert Islands, in the Pacific. The nation of Kiribati is likely to be seriously impacted by climate change, through extreme weather events and rising sea levels; the islands' fresh water and crop-growing areas are also at great risk. The islanders may be forced to move elsewhere, risking marginalisation and loss of their cultural identity. The islanders do what they can, and show great enterprise in adapting their food growing methods, but tackling climate

Fig. 1 Taxidermied Golden Plover, a species whose past, present and future are intimately tied to climate change (Manchester Museum, University of Manchester)

change will need everyone to play their part. Museums in developed countries can help tell the story of the people of Kiribati, and protect the cultural heritage from the islands, and to join the dots between personal and political decisions—and indecision—on people's lives. This case study aims to bring 'dead' museum objects to life, and to help give voice to people who are directly impacted by climate change but whose voice is rarely heard (see, especially, Newell et al. 2017).

Case study 3

Museum collections have proved to be essential in answering a wide range of environmental questions. Notable examples have included linking the use of DDT with the laying of thin-shelled eggs by birds of prey in the mid-20th century. Climate change is already causing serious redistribution of plants and animals, and this exacerbates more immediate environmental concerns from, for example, agricultural intensification, over fishing, pollution, and habitat destruction. Climate change increases the pressure and unpredictability of changes on wildlife, threatening wildlife and people alike, whether as a result of novel infectious diseases, or novel crop pests. Museums should be collecting wildlife (in ethical, and scientifically useful ways) as it spreads, and participating in wildlife recording schemes to better understand the spread of wildlife in response to climate change and globalisation. Without specimens to study, collections-based research will come to an end, with profound implications for our understanding of living things, against a backdrop of unprecedented, and unpredictable, environmental change. They should also present these new collections and associated research findings to the public. This case study aims to articulate that museums are engaged in climate change in more ways than simply interpreting existing collections in exhibitions and events: they can help scientists understand the dynamics of a changing world, but they need to continue collecting in order to do so (see, e.g. Krishtalka and Humphrey 2000; Dorfman 2018).

5   Conclusion

Bob Janes has said that everything museums need for climate change engagement is already here: collections, audiences, and the need for museums to connect with real-world issues. All that is needed is the will to direct them effectively and constructively towards climate action. Climate change has shifted from a specialist, scientific, or niche issue, to an issue of pre-eminent importance. However, as explored above, for various reasons, climate change is difficult for sector-by-sector governments and agencies to address. Museums are almost uniquely placed as sites where the past, present and future can be addressed. Climate change, and a raft of sustainability issues, present tremendous challenges to society, yet they also present a real-world opportunity for museums to realise their social potential. Combining their memory function with a broad humanism can contribute practically to securing a future that aims to be better for all, and for the natural environment. In so doing, museums will earn the trust that they value so highly, and will connect with a wider spectrum of society. Climate change is intimately connected with a wide range of social, economic and environmental issues. Tackling these, both locally and globally, and making the connection between local and global development, is a very real opportunity for museums to act as brokers within and beyond national boundaries. In an increasingly compromised global setting, this role will become ever more important. However, global challenges, including climate action and wider sustainability, will not always be led from the top from a national level. This should not deflect museums from doing the right thing, and can help deliver the change that governments find so difficult, and that is required for a more sustainable, equitable future. Museums do not need anyone's permission to start with climate action and in supporting the SDGs, and their participation could help move this agenda forward, within their own sector and with others. Museums can be accelerators, not brakes, in the transformation towards a better future, and that is surely a role worth embracing.

Acknowledgements I am grateful to Sarah Mander (Tyndall Manchester), Asher Minns (Tyndall Centre, Norwich), Ralph Underhill (formerly Public Interest Research Centre) and Tom Crompton (Common Cause Foundation) for useful discussion relating to this chapter. This chapter is dedicated to the memory and inspiration of the late Stephen Kellert.

References

Aasen M, Vatn A (2018) Public attitudes toward climate policies: the effect of institutional contexts and political values. Ecol Econ 146:106–114
Azmat F, Ferdous A, Rentschler R, Winston E (2018) Arts-based initiatives in museums: creating value for sustainable development. J Bus Res 85:386–395
Bicchieri C, Muldoon R (2014) Social norms. In: Zalta EN (ed), The Stanford encyclopedia of philosophy. https://plato.stanford.edu/entries/social-norms/. Accessed 11 May 2018
Binder M, Blankenberg A-K (2017) Green lifestyles and subjective well-being: more about self-image than actual behavior? J Econ Behav Organ 137:304–323
Blokker (2017) Member State Responsibility for Wrongdoings of International Organisations, Beacon of Hope or Delusion. In: Barros AS, Ryngaert C, Wouters J (eds) International Organizations and Member State Responsibility: critical perspectives, 34–47. Volume 28 in series 'Nova et vetera iuris gentium'. Brill, Leiden. Originally published as vol. 12(2) (2015) of Brill Nijhoff's Journal International Organizations Law Review
Cameron FR, Hodge B, Salazar F (2013) Representing climate change in museum space and places. WIREs Clim Change 4(1):9–21
Carroll MJ, Dennis P, Pearce-Higgins JW, Thomas CD (2011) Maintaining northern peatland ecosystems in a changing climate: effects of soil moisture, drainage and drain blocking on craneflies. Glob Change Biol 17:2991–3001
Collins A (2018) The global risks report 2018, 13th edn. World Economic Forum, Geneva
Cotter H (2009) Why university museums matter. Art and Design Review. New York Times 19 Feb 2009. http://www.nytimes.com/2009/02/20/arts/design/20yale.html. Accessed 11 May 2018
Crompton T (2016) Perceptions matter: the common cause UK survey. Common Cause Foundation, UK. https://valuesandframes.org/survey/. Accessed 26 March 2018

Crompton T, Lennon S (2018) Values as a route to widening public concern about climate change. In: Leal Filho W et al (eds) Handbook of climate change communication, vol 1. Springer, Berlin, pp 385–397

Dorfman E (ed) (2018) The future of natural history museums. ICOM Advances in Museum Research, Routledge, Abingdon (UK)

Farrow K, Grolleau G, Ibanez L (2017) Social norms and pro-environmental behaviour: a review of the evidence. Ecol Econ 140:1–13

Grouzet FME, Kasser T, Ahuvia A, Fernandez-Dols JM, Kim Y, Lau S, Ryan RM, Saunders S, Schmuck P, Sheldon KM (2005) The structure of goal contents across fifteen cultures. J Pers Soc Psychol 89:800–816

Howell R (2013) It's *not* (just) "the environment, stupid!" Values, motivations, and routes to engagement of people adopting lower-carbon lifestyles. Glob Environ Change 23(1):281–290

International Development Committee (2016) UK implementation of the sustainable development goals. House of Commons, 2016–17 Session, HC 103. https://publications.parliament.uk/pa/cm201617/cmselect/cmintdev/103/10302.htm. Accessed 26 Mar 2018

Janes RR (2016) Museums without borders, selected writings of Robert R. Janes. Routledge, London

Kasser T, Ryan RM (1996) Further examining the American dream: differential correlates of intrinsic and extrinsic goals. Pers Soc Psychol Bull 22:280–287

Kellert SR (2012) Birthright: people and nature in the modern world. Yale University Press, New Haven

Krishtalka L, Humphrey PS (2000) Can natural history museums capture the future? BioScience 50(7):611–617

Lorenzoni I, Nicholson-Cole S, Whitmarsh L (2007) Barriers perceived to engaging with climate change among the UK public and their policy implications. Glob Environ Change 17:445–459

McDonald RI, Chai HY, Newell BR (2015) Personal experience and the 'psychological distance' of climate change: an integrative review. J Environ Psychol 44:109–118

McGhie HA, Mander SJ, Underhill R (2018) Engaging people with climate change through museums. In: Leal Fihlo W et al. (eds.), Handbook of climate change communication: vol. 3, case studies in climate change communication. Springer, Switzerland, pp. 329–348

Newell J, Robbin L, Wehner K (2017) Curating the future: museums, communities and climate change. Routledge Environmental Humanities Series. Routledge, London

Pearce-Higgins JW (2011) Modelling conservation management options for a southern range-margin population of Golden Plover *Pluvialis apricaria* vulnerable to climate change. Ibis 153:345–356

Pearce-Higgins JW, Dennis P, Whittingham MJ, Yalden DW (2010) Impacts of climate on prey abundance account for fluctuations in a population of a northern wader at the southern edge of its range. Glob Change Biol 16:12–23

Pearce-Higgins JW, Yalden DW (2004) Habitat selection, diet, arthropod availability and growth of a moorland wader: the ecology of European Golden Plover *Pluvialis apricaria* chicks. Ibis 146:335–346

Spence A, Poortinga W, Pidgeon N (2012) The psychological distance of climate change. Risk Anal 32(6):957–972

Whitmarsh L (2011) Scepticism and uncertainty about climate change: dimensions, determinants and change over time. Glob Environ Change 21:690–700

Whitmarsh L, Lorenzoni, O'Neill S (2011) Engaging the public with climate change: behaviour change and communication. Earthscan, London

Young HP (2015) The evolution of social norms. Ann Rev Econ 7:359–387

When Facts Lie: The Impact of Misleading Numbers in Climate Change News

Marlis Stubenvoll and Franziska Marquart

Abstract This study examines how numerical misinformation in the news can lead to a bias in readers' own judgment on climate change issues after a retraction. Building on theories of the continued influence effect and anchoring, the experimental research investigates the link between inaccurate facts, biased estimations, and the evaluation of climate change policies and risks. The results indicate that presenting participants with a low number on the carbon footprint of commuting traffic induces a bias into their own estimated values. This effect appears regardless of the participants' level of issue involvement. However, the study finds no subsequent effect of this bias on participants' policy support or perceived threat of climate change. The results are discussed in light of anchoring and misinformation theories. The paper proposes media literacy as a fruitful avenue to a more accurate understanding of climate change in view of a factually flawed representation of climate change in the news.

1 Introduction

The journalistic usage of the word "post-truth" increased by 2000% in 2016, mirroring the concern that instead of objective facts, personal beliefs dominate political debates (Flood 2016). This trend has also surfaced in environmental news. More and more frequently, anthropogenic climate change has been questioned by prominent political figures (Kenny 2016), and climate scientists have to withstand unsubstantiated attacks on their work (Lundberg 2017).

M. Stubenvoll (✉)
Graduate School of Communication, University of Amsterdam,
Nieuwe Achtergracht 166, 1018 WV Amsterdam, The Netherlands
e-mail: marlis.stubenvoll@gmx.at

F. Marquart
Amsterdam School of Communication Research, University of Amsterdam,
Nieuwe Achtergracht 166, 1018 WV Amsterdam, The Netherlands

© Springer Nature Switzerland AG 2019
W. Leal Filho et al. (eds.), *Addressing the Challenges in Communicating Climate Change Across Various Audiences*, Climate Change Management,
https://doi.org/10.1007/978-3-319-98294-6_3

How does this spread of incorrect facts about climate change through the mass media affect the public? Brüggemann (2017) argues that citizens are unable to grasp the phenomenon of climate change by common sense alone. The public is therefore strongly dependent on mediated information. However, if the electorate is frequently exposed to inaccurate facts, this poses a threat to good governance on climate change issues and may influence citizens to support policies that may be harmful (Hochschild and Einstein 2015).

This study empirically addresses this concern. The aim of this paper is to test the effects of inaccurate numbers in the news on beliefs on and attitudes towards climate change mitigation issues. A refined understanding on how misinformation might influence public opinion is vital for evaluating and counteracting campaigns against climate science. Numerical information is in the spotlight of this study, as it plays a central role in the mass-mediated discourse on CO_2 emissions, temperatures or sea level rise in climate change news.

Two areas of research inform this study: While anchoring theory provides a framework for understanding how irrelevant numbers can distort people's judgments (for a review, see Furnham and Boo 2011), misinformation studies have shown that inaccurate information in the news continues to promote divisive biases among readers even if it is retracted (for a review, see Lewandowsky et al. 2012). By integrating those areas as a first step, we develop an operational framework to investigate how retracted numbers can bias perceptions of climate change mitigation issues. By employing an experimental design, this study tests a causal link between the exposure to a retracted high or low number and the participants' own estimations of the carbon footprint of commuting traffic. This estimation, in turn, is expected to mediate the respondents' evaluation of a mitigation policy and climate change threats.

The results contribute to a deeper understanding of the effects of inaccurate numbers on the news audience and help address the critical area of climate change policy support. By taking a closer look at the underlying processes of anchoring, this study increases our knowledge on how to respond to misinformation and thus promote an accurate public understanding of the science of climate change.

2 Theoretical Framework

2.1 Misinformation and Its Effects

More than 30 years of research on misinformation has examined the powerful and lasting effects of inaccurate information on people's beliefs about political matters (Garrett et al. 2013; Nyhan and Reifler 2010). Researchers investigating this phenomenon acknowledge that the line between "right" and "wrong" in political debates is often blurred (Kuklinski et al. 1998). A struggle over how to interpret events and problems is inherent in political communication—therefore facts, while giving the impression of neutrality, are often ideologically loaded (Carvalho 2007). To separate correct from incorrect claims, academic studies have consequently defined misinformation as "information that is initially assumed to be valid but is later corrected or retracted" (Ecker et al. 2014, p. 292).

Seifert (2014) coined the term "continued influence effect" to describe how people fail to discount the initially presented, inaccurate information when they make further inferences. The phenomenon persists over time and occurs both when the retraction is presented immediately after the misinformation as well as with a delay (Feinholdt 2016) and is equally present for people who can report the correction (e.g., Seifert 2014).

Recent studies have examined cognitive heuristics as the underlying cause of the continued influence effect, most prominently the accessibility and applicability of the presented misinformation. Cognitive science shows that more accessible information has greater influence on people's judgments, as it is more easily retrieved from memory (Tversky and Kahneman 1973). When misinformation is readily available, because it was mentioned earlier as part of the storyline (accessibility), and when it plausibly fills a gap in a causal chain of events (applicability), it will then be influential despite corrections (Seifert 2014).

However, the effectiveness of a correction also seems to depend on whether or not it is consistent with the receiver's worldview or represents a threat to it (Nyhan and Reifler 2010; Ecker et al. 2014). These findings are also in-line with persuasion research. Rather than accommodating new information, people tend to resist persuasion in order to defend their existing worldviews and protect their free will (Sagarin et al. 2002; Zuwerink Jacks and Cameron 2003).

2.2 Misinformation in Climate Change Communication

Climate change communication poses an important subject for study in terms of the continued influence of misinformation. False claims about the scientific consensus of climate change can cast doubt on its existence among news consumers (van der Linden et al. 2017). Even highly involved subjects might become less certain about the causes of climate change after being exposed to biased information (Cho et al. 2011). However, more research is required to determine the impact of different types of inaccurate information on climate change beliefs.

Little attention has been paid to the role of numbers in messages about climate change. So far, research indicates a stronger impact of statistical information on climate change beliefs as compared to narratives (Hart 2013). Therefore, this study aims at deepening the understanding of how numbers can shape people's beliefs on climate change, drawing on findings from anchoring theory.

2.3 Anchoring Effects

Anchoring as a form of misinformation. Anchoring, as first systematically explored by Tversky and Kahneman (1974), describes how wrong or irrelevant numbers distort future judgments. In their classic study, Tversky and Kahneman let participants turn a wheel of fortune that arrived at either a high or low number. When later asked about the number of African countries that are a member of the United Nations, the participants' responses leaned towards this irrelevant number—the anchor.

Anchoring research has produced insights into diverse fields, such as jurisdiction (Englich et al. 2006) and purchasing decisions (Mussweiler et al. 2000). Studies examine the strength of the anchoring effect by comparing the average estimations of groups that are presented with a numerical anchor to the estimations of a control group (Jacowitz and Kahneman 1995).

So far, it is less understood how anchoring effects play out in the context of news consumption. This is surprising since bridging studies about mass-mediated misinformation and anchoring studies could provide a fruitful avenue for research. Both areas share similarities in their experimental set-up in which initial information is declared wrong and unfit for later judgments. Taking into account anchoring's robust effects (Furnham and Boo 2011) and the repeatedly shown impact of other forms of misinformation on a news audience (Lewandowsky et al. 2012), this study expects a similar effect of numerical anchors in climate change news on the audience's estimations:

H1: Numerical anchors about climate change presented in a news item skew the respondents' own estimations on the presented issue in direction of the anchor value.

Limits of anchoring in climate change communication. Different theoretical models have been tested to explain the occurrence of anchoring, some of which show similarities to the processes underlying the continued influence effect. The "insufficient adjustment" model suggests that people start from the numerical anchor and insufficiently adjust their estimation into the direction in which they suspect the right value, therefore arriving in the vicinity of the anchor (Tversky and Kahneman 1974). In contrast, selective accessibility in anchoring suggests that people generate arguments that are in line with the anchor, which then skew their own estimation (Mussweiler et al. 2000). Knowledge that is coherent with an inaccurate piece of information is activated, which in return shapes later judgments. This line of thinking parallels assumptions of the continued influence effect as described above.

Attitude change theory emerged as a third approach to anchoring, which shares ideas on selective accessibility. However, it also takes into account that people might counterargue the presented value instead of solely generating confirming arguments (Wegener et al. 2001). This theoretical strand links well to findings on the continued influence effect, as both suggest that pre-existing attitudes play a strong role in the processing of inaccurate information. In the context of this study, people who already have strong ideas on climate change might be more motivated and better equipped to counterargue anchors. To test this assumption, I investigate whether issue involvement acts as a moderator on anchoring effects. Issue involvement—also referred to as issue importance—is an important predictor of people's ability to resist persuasive attempts, as it is a central indicator for the strength of pre-existing attitudes (Zuwerink Jacks and Cameron 2003).

However, studies on anchoring show that numerical misinformation might still have the power to influence even highly involved people (Englich et al. 2006; Mussweiler et al. 2000). I, therefore, hypothesize that the influence of inaccurate information will be less pronounced for people who show strong involvement in the topic of climate change, but that high issue involvement does not negate anchoring effects:

H2: Respondents' issue involvement in climate change moderates the strength of the anchoring effect on the respondents' estimation.

Impact of anchoring on policy preferences. Public support is crucial to ensure that costly climate change mitigation measures are politically viable. Therefore, I test whether anchoring can have an impact on the support for a specific climate change mitigation measure, in this case, the compaction of cities to decrease CO_2 emissions from traffic. In addition, this study also tests if anchored high or low estimations affect people's threat perception of climate change.

If a certain problem is expressed as more urgent through e.g. the attribution of a high share of CO_2 emissions to it, this might alarm people about the threat of climate change and heighten policy support. This assumption is consistent with the selective accessibility model of anchoring. However, coming from an attitudinal perspective, it is also possible that people counterargue the anchor and access information that will shift their preferences to the other direction. Lastly, since climate change attitudes are mostly stable (e.g. McCright et al. 2016; Nisbet et al. 2013), policy preferences could be resistant to the influence of inaccurate facts.

As no clear prediction can be made, the possible effect is investigated in a supplementary question:

SQ: To which extent does the anchoring effect alter climate change policy support and the risk perception of climate change?

When combined, these hypotheses and the additional question represent a model of moderated mediation, as can be seen in Fig. 1.

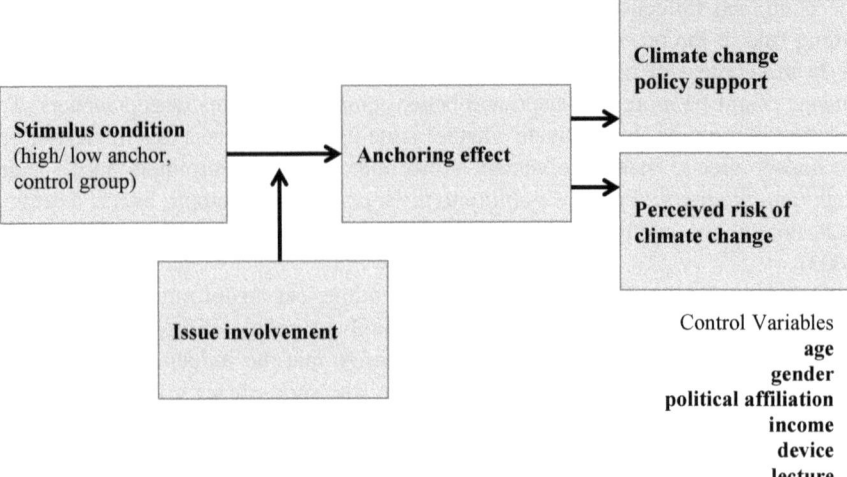

Fig. 1 Conceptual model

3 Methodology

3.1 Experimental Design

To test the presented model, an experimental between-subjects design was implemented (e.g. Feinholdt 2016; McCright et al. 2016). One group was presented with a low anchor (LA) manipulation, one group with a high anchor (HA), while the control group (CG) saw the article without any numerical anchors. The stimuli have been pre-tested among a snowball sample of environmentally conscious Austrians.

Data collection took place in April 2017. 55 participants enrolled at the University of Life Sciences were recruited through e-mail and 83 participants answered the survey in a lecture on environmental communication (ages 19–52, $M = 24.08$, $SD = 4.90$, 60.3% females). All questions and materials were presented in German. Considering this background, this sample is likely to have above-average awareness of climate change issues. Therefore, this model test provides rather conservative estimates of the investigated effects.

3.2 Research Procedure

Students were informed that they would take part in a study on environmental news. They were randomly assigned to a stimulus or control group ($n = 46$ per group). After reading the article, the groups in the numerical misinformation conditions were exposed to a retraction, followed by an estimation question about the CO_2

emissions caused by commuters, which all three groups had to answer. Then, participants indicated their policy support and assessed the general threat of climate change, followed by questions about issue involvement and the quality of the article. Lastly, they answered questions on their demographics and political background before being debriefed.

3.3 Stimulus Material

The newspaper article about problems caused by the commute belt draws on an original published article (Putschögl 2015). The text was shortened, formatted in the style of an online article, and an author line, date, and publishing section were added (see also Fig. 2). An important criterion for the choice of this issue was its relevance, as the growing CO_2-emissions from traffic pose a challenge to climate change mitigation in Vienna (Magistrat der Stadt Wien 2009).

In the LA and HA conditions, respondents read a statement on CO_2 emissions by commuting traffic in Vienna ("17% of CO_2 emissions of Vienna are due to commuting traffic"; "50% of CO_2 emissions of Vienna are due to commuting traffic"). The anchors were set at the 15th and 85th percentiles of estimations of the pre-test group, as proposed by Jacowitz and Kahneman (1995).

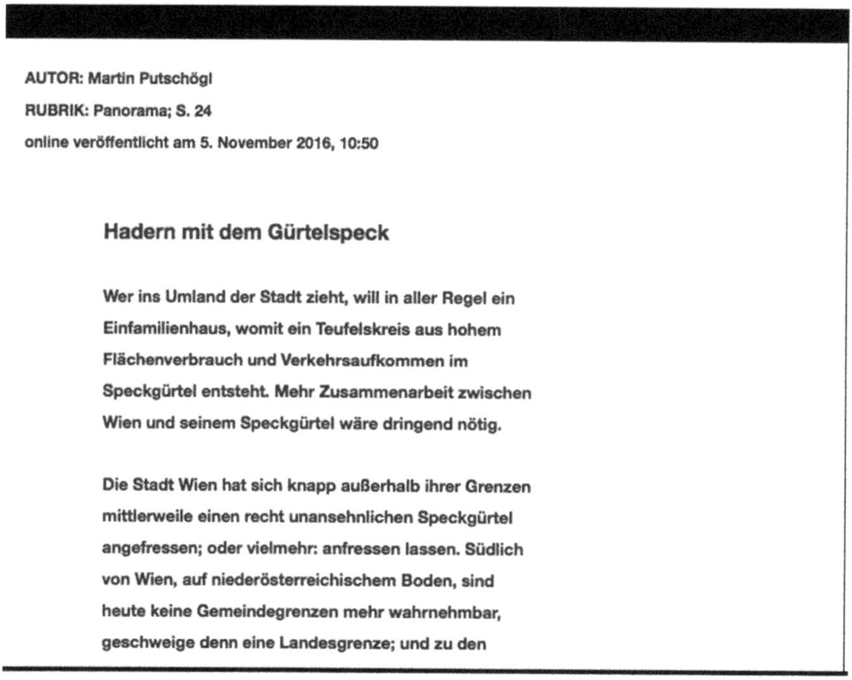

AUTOR: Martin Putschögl

RUBRIK: Panorama; S. 24

online veröffentlicht am 5. November 2016, 10:50

Hadern mit dem Gürtelspeck

Wer ins Umland der Stadt zieht, will in aller Regel ein Einfamilienhaus, womit ein Teufelskreis aus hohem Flächenverbrauch und Verkehrsaufkommen im Speckgürtel entsteht. Mehr Zusammenarbeit zwischen Wien und seinem Speckgürtel wäre dringend nötig.

Die Stadt Wien hat sich knapp außerhalb ihrer Grenzen mittlerweile einen recht unansehnlichen Speckgürtel angefressen; oder vielmehr: anfressen lassen. Südlich von Wien, auf niederösterreichischem Boden, sind heute keine Gemeindegrenzen mehr wahrnehmbar, geschweige denn eine Landesgrenze; und zu den

Fig. 2 Screenshot of the stimulus as seen by participants in the online survey

3.4 Measurement

Mediator. Following the procedure of previous anchoring studies (for a review, see Furnham and Boo 2011), participants were asked to give their own estimation on the retracted number in an open question ("Which percentage of the total CO_2 emissions of the city of Vienna is caused by commuting traffic?", $M = 31.62$, $SD = 19.64$, range: 3–80). In addition, they had to indicate how certain ($M = 3.20$, $SD = 1.44$) they felt about their estimation on a seven-point Likert scale.

Dependent variables. All dependent variables and the moderator were measured on a seven-point Likert scale. Two items measured the extent to which the respondents favor a specific policy ($M = 2.39$, $SD = 1.39$) and whether they think it is effective ($M = 2.86$, $SD = 1.20$). Specifically, students had to indicate whether they would support the construction of affordable living space in order to attract commuters to permanently move to Vienna, even if this means eradicating green spaces. This subject was chosen because the losses (less green space $M = 6.43$, $SD = 0.85$) and the gains (less CO_2 from traffic, $M = 6.14$, $SD = 1.14$) involved were both highly important to the pre-test group, mirroring a balanced interest in both issues. This policy is also of political relevance: The city of Vienna names the densification of the city as one of the most important climate change mitigation strategies. (Magistrat der Stadt Wien 2009). The two items measuring participants' threat perception of climate change (Cronbach $\alpha = 0.78$, $M = 6.34$, $SD = 0.96$) were adapted from Leiserowitz (2006), e.g., "I am concerned about climate change".

Moderator. Issue involvement (Cronbach $\alpha = 0.74$, $M = 4.07$, $SD = 1.56$) was measured with two survey items based on a study by Zuwerink Jacks and Cameron (Zuwerink Jacks and Cameron 2003; e.g., "My attitude toward climate change helps define who I am as a person.")

Control variables. Participants indicated their age and gender. In addition, they answered questions on their political leaning on a 11-point scale ($M = 3.13$, $SD = 1.42$), on their voting intention (coded as indicator variable, 35.5% would vote for the Green Party), and their income on a 5-point scale ($M = 3.57$, $SD = 0.97$), as well as the device they have used (61.6% used a smartphone) and whether they completed the survey in the lecture. The quality of the text (Cronbach $\alpha = 0.81$, $M = 4.39$, $SD = 1.15$) was assessed using semantic differentials on a 4-item scale (e.g. "Please indicate how you would describe the previous article: (1) untrustworthy–(7) trustworthy").

4   Results

After a successful randomization test, an analysis of covariance (ANCOVA) tested whether the three experimental groups differed in their estimation of the carbon footprint created by commuting traffic in Vienna (H1).

Participants arrived at different estimations based on their condition, F (14, 137) = 4.08, $p < 0.001$, $\eta_2 = 0.32$. Estimations in the low anchor condition (LA) were significantly lower compared to both the control group (CG), $M_{difference} = -18.03$, $p < 0.001$, 95% CI [-25.21, -10.86] and the high anchor condition (HA), $M_{difference} = -23.16$, $p < 0.001$, 95% CI [-30.34, -15.97], as an LSD post hoc test indicates. Participants in the HA group also estimated the share of commuters' CO_2 emissions to be higher as compared to the CG ($M_{difference} = 5.12$). However, this difference was not significant, $p = 0.160$, 95% CI [-2.05, 12.29]. To rule out alternative explanations, the factors gender, age, income, issue involvement, the device used to fill out the survey, and recruitment (i.e. via e-mail or in the lecture) were controlled for. No differences were found when including the respective controls. Therefore, Hypothesis 1 is partly accepted: Numerical misinformation affects people's own estimation of the respective number, but only for participants in the LA condition (see also Fig. 3).

In a next step, a moderated mediation model was tested in PROCESS (Hayes 2013), using indicator variables for the LA and the HA group, and taking the CG as a reference category. The aim was to see if high involvement in the issue of climate change could limit the anchoring effect (H2), which in turn could affect policy support. Against expectations, issue involvement did not moderate the anchoring effect for participants in the LA condition, because the interaction's direct effect ($a_3 = -1.97$, $p = 0.232$) included zero (bias-corrected bootstrapping, 5000 samples, 95%, CI [-5.21, 1.27]).

A similar pattern occurred when testing the interaction between the HA condition and issue involvement, outlined by a 95% bootstrap confidence interval for the interaction ($a_3 = -1.40$, $p = 0.477$) from -5.30 to 2.49. In other words, the anchoring effect appeared for participants in the LA condition irrespective of their level of issue involvement.

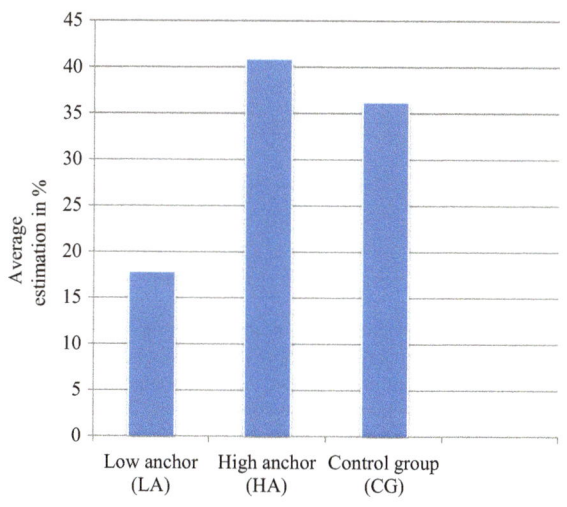

Fig. 3 Average estimation of the percentage of CO_2 emissions in Vienna caused by commuting traffic by experimental condition

In addition, the model of moderated mediation for the LA condition and the HA condition was not supported by the moderated mediation index, with bias-corrected 95% bootstrap confidence intervals of −0.07 to 0.11 and −0.02 to 0.10 respectively (see Fig. 4). Neither was a moderation found for the models of perceived policy effectiveness or perceived threat of climate change. Thus, Hypothesis 2 is rejected.

Further analysis examined whether the participants' own estimations mediate the effect of numerical misinformation on policy support, the perceived effectiveness of this policy, and the threat perception of climate change (H3). Since involvement does not act as a moderator, the model was retested using a simple mediation analysis in PROCESS, and issue involvement was included as an additional control variable. All tests were performed using a 95% bias-corrected bootstrapping procedure based on 5000 samples.

Again, the models show that the LA condition led participants to give lower estimations ($a_1 = -18.387, p < 0.001$). However, there was no evidence that a lower or higher estimation as such affected participants' support of the presented policy ($b = -0.01, p = 0.077$), whether they found it effective ($b = -0.003, p = 0.558$), or whether they felt threatened by climate change ($b = -0.001, p = 0.890$).

Next, the indirect effects of the numerical misinformation in the stimulus on policy support, perceived efficiency, and perceived threat were examined for each

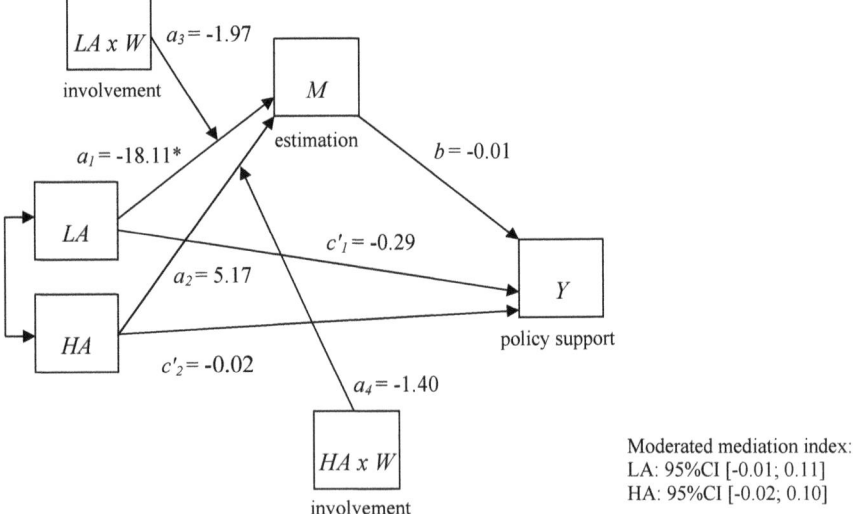

Fig. 4 Moderated mediation model showing the indirect effects of low numerical anchors (LA) and high numerical anchors (HA) on policy support through participants' estimations and moderated by issue involvement. *Note* The analysis was conducted using indicator variables for LA and HA as compared to the control group. The figure shows unstandardized beta coefficients. Bias-corrected bootstrap confidence intervals using 5000 bootstrap samples are shown for the moderated mediation model of the HA and LA condition separately at the bottom of the panel. $*p < 0.05, **p < 0.01, ***p < 0.1. N = 138$

condition individually. Being exposed to the low anchor stimulus led participants to support the presented policy more through their biased estimation, with an unstandardized indirect coefficient of $ab = 0.22$ and a 95% bootstrap confidence interval of 0.005 to 0.52. However, this result has to be interpreted with caution, as the overall model can only explain a marginal level of the examined variance in policy support, $R^2 = 0.06$. Overall, the additional research question (RQ2) arrives at the conclusion that the anchoring effect had no subsequent effect on how strongly people supported the policy, how effective they thought it to be or as how serious people perceived the threat of climate change.

5 Additional Analysis

An additional analysis was conducted to see whether previous findings on anchoring effects could be replicated. Wegener et al. (2001) named source credibility as a possible moderator of anchoring effects. A regression model was tested using the deviation from the anchor as a dependent variable and the experimental condition, article quality, political affiliation, gender, age, income, device, and testing environment as predictors, $F(8,91) = 8.04$, $p < 0.001$. The model accounts for 44% of the variation examined in the participants' deviations from the anchor value ($R_2 = 0.44$). In this model, participants' quality assessment acted as a moderator: The lower they rated the quality of the article, the further their estimations deviated from the anchor, $b = -2.05$, $t = -2.92$, $p = 0.005$, CI $[-6.75, -0.50]$.

6 Discussion

This study extends current knowledge on climate change communication by showing that wrong but retracted numbers in climate change news introduce a bias into people's own estimations. Presenting participants with a low figure on commuters' CO_2 emissions made them arrive at significantly lower estimations despite a prior retraction. This finding is especially relevant in the light of a post-truth situation in political debates on climate change, as disseminating wrong facts on this issue appears to be an effective political strategy.

Research on anchoring effects and the continued influence of misinformation have produced important insights into how people's judgment can be affected by inaccurate information even after a retraction (Furnham and Boo 2011; Lewandowsky et al. 2012). This study indicates that news consumption parallels a number of real-life situations in which these phenomena have been examined (for a review, see Furnham and Boo 2011). Considering the above-average knowledge on environmental issues of the student sample, the finding also supports the assumption that anchoring effects from inaccurate numbers apply to knowledgeable individuals (Englich et al. 2006; Mussweiler et al. 2000). However, no effect could be detected when people were presented with a high anchor. Hence, further research has to clarify if the high anchor might have been set too low.

This study also set out to investigate the moderating role of issue involvement for anchoring effects. In contrast to previous findings from misinformation studies (Cho et al. 2011), the strength of the anchoring effect was not affected by participants' involvement in the issue of climate change. These conflicting findings could stem from different types of misinformation, as it may be more difficult for people to resist persuasion by numbers than by narratives on climate change (Hart 2013). An alternative explanation emerges from persuasion research: Receiving a higher or lower anchor about CO_2 emissions from commuting traffic might not have been perceived as an attack on an individuals' stance on climate change, thus there was a lack of motivation to counterargue the presented misinformation (e.g. Ecker et al. 2014; Sagarin et al. 2002). A second motivator to discount misinformation is the feeling of distrust (Schul and Mayo 2014). Additional analysis revealed that participants who rated the quality of the article as lower were more likely to give estimations that lie further away from the anchor. Therefore, the results support the notion that motivation is a key component in anchoring effects (Wegener et al. 2001).

When numerical misinformation is powerful enough to influence informed and involved people, could these misconceptions pose a threat to public support for climate change mitigation policies? While this study alone cannot give a definite answer, its results indicate that anchoring effects do not change people's policy support or their general threat perception by default. All experimental groups supported measures against commuting to an equal extent, regardless of their biased estimations. From a selective accessibility perspective, this result comes as a surprise: The arguments evoked by the high or low number should also be more present in people's considerations when they later evaluate a policy (Tversky and Kahneman 1973).

There is a number of reasons why biased estimations seem to have no further implications for people's beliefs and attitudes. Firstly, numbers are subject to interpretation as Kuklinski et al. (1998) state. It is important to note that this study's frame of the news article presented commuting traffic in a negative light. As a result, claiming that 17 or 50% of the carbon footprint of Vienna comes from commuting traffic could have had a similar effect—both numbers could be perceived as high in the eyes of students aware of climate change. Moreover, a high level of concern about car emissions among participants ($M = 6.04$, $SD = 1.28$) points to a probable ceiling effect, as it may be difficult to induce further support for traffic reduction. Lastly, existing research on climate change found stable preferences in framing experiments (McCright et al. 2016; Nisbet et al. 2013). Therefore, it is not surprising that a *single* inaccurate fact failed to tip over participants' attitudes.

Nevertheless, there are valid reasons to extend research on anchoring and its effects on public opinion on climate change. People encounter misinformation on climate change multiple times in a real news environment. This repeated exposure might be influential in the long term. In addition, future research should test whether a different sample or policy might change the picture or whether the impact of misinformation only surfaces after some time has elapsed, referred to as the "sleeper effect" (Feinholdt 2016). Future research should also focus on a deeper integration of anchoring into the methodological set-up of misinformation studies. Specifically, testing different forms of retractions, different frames, counter-attitudinal corrections, or new moderators could produce new insights. Climate change communication poses a vital field for these explorations since the audience frequently encounters numbers in the discourse on CO_2 emissions, changing temperatures, or risks of natural disasters.

Lastly, this study raises important ethical concerns about news reports on climate change. On one hand, the findings suggest that post hoc fact checking might be insufficient to eliminate misconceptions (see also Garrett et al. 2013; van der Linden et al. 2017). This further points to a heightened responsibility for media practitioners to carefully research facts before they are distributed among a mass audience. On the other hand, this study shows that more critical individuals might be able to shield themselves from an anchoring bias. Recent research also indicates that being more cautious about news messages on climate change could be induced by prior messages that warn readers about misleading content (van der Linden et al. 2017). This is an especially promising strategy, as targeting newsrooms alone might be insufficient to ensure that the public is accurately informed about climate science and climate mitigation, as news is more and more frequently distributed over social media and comes from various sources. Therefore, the advancement of media literacy could pose a fruitful avenue to counter misinformation, for instance by cultivating the audience's skepticism towards unreliable sources and numbers without substantiation.

7 Limitations

A few limitations need to be considered. While experiments offer the advantage of internal validity, the study did not resemble a real situation of news consumption. Furthermore, it will be necessary to investigate other media, issues, and employ a different sample. Despite the common criticism of student samples, this population was useful here, as it enabled the researcher to reach a great number of environmentally conscious participants. Another difficulty was the reliability of the issue involvement and perceived threat of climate change scales. While measures have proved to be highly reliable in previous studies (Leiserowitz 2006; Zuwerink Jacks and Cameron 2003), they do not appear to translate smoothly into a European context, which should be considered in future studies.

8   Conclusion

This study makes a significant contribution to the field of climate change communication by exploring the effects of inaccurate numbers in the news on people's beliefs on a mitigation issue. Our results suggest that wrong numerical facts in climate change reporting are able to skew the audience's own estimations despite of a retraction. In an experiment, a retracted low number on CO_2 emissions from commuting traffic made participants arrive at a lower estimation of this figure. This so-called anchoring effect was noticeable regardless of participants' involvement in the issue of climate change. However, their policy support and their threat perception of climate change didn't significantly differ from the control group despite of their skewed beliefs. No significant anchoring effect was found when presenting participants with an inaccurate high number.

This study set the stage for further research on the topic of numerical misinformation in climate change news by linking misinformation studies and anchoring studies. Future research can deepen our understanding of the effects of misinformation on the climate change attitudes and beliefs by examining different time spans, corrections and moderators of anchoring effects.

Acknowledgements   Thank you to the professors Martin Schönhart, Monika Kobzina, and Ika Darnhofer from the University of Life Sciences, Vienna, who helped distribute the survey and, of course, to the students who took the time to participate.

References

Brüggemann M (2017) Wissenschafts-Kommunikation im Trump-o-zän: Wie wir alle das post-faktische Zeitalter verhindern können [Science Communication in the "Trumpocene"—How we can prevent the post-factual age]. *klimafakten.de*. Accessed 18 May 2017. https://www.klimafakten.de

Carvalho A (2007) Ideological cultures and media discourses on scientific knowledge: re-reading news on climate change. Public Underst Sci 16(2):223–243. https://doi.org/10.1177/0963662506066775

Cho CH, Martens ML, Kim H, Rodrigue M (2011) Astroturfing global warming: it isn't always greener on the other side of the fence. J Bus Ethics 104(4):571–587. https://doi.org/10.1007/s10551-011-0950-6

Ecker UKH, Lewandowsky S, Fenton O, Martin K (2014) Do people keep believing because they want to? Preexisting attitudes and the continued influence of misinformation. Mem Cogn 42:292–304. https://doi.org/10.3758/s13421-013-0358-x

Englich B, Mussweiler T, Strack F (2006) Playing dice with criminal sentences: the influence of irrelevant anchors on experts' judicial decision making. Pers Soc Psychol Bull 32:188–200. https://doi.org/10.1177/0146167205282152

Feinholdt A (2016) What is done cannot be undone? The role of misinformation in news framing effects. In: Fight or flight: affective news framing effects. Dissertation, Amsterdam School of Communication Research, pp 115–146

Flood E (2016) 'Post-truth' named word of the year by Oxford Dictionaries. The Guardian, 15 November. Accessed 18 Nov 2017. https://www.theguardian.com

Furnham A, Boo HC (2011) A literature review of the anchoring effect. J Socio-Econ 40:35–42. https://doi.org/10.1016/j.socec.2010.10.008

Garrett RK, Nisbet EC, Lynch EK (2013) Undermining the corrective effects of media-based political fact checking? The role of contextual cues and naïve theory. J Commun 63:617–637. https://doi.org/10.1111/jcom.12038

Hart PS (2013) The role of numeracy in moderating the influence of statistics in climate change messages. Public Underst Sci 22(7):785–798. https://doi.org/10.1177/0963662513482268

Hayes AF (2013) Introduction to mediation, moderation, and conditional process analysis. A regression-based approach. The Guilford Press, New York

Hochschild J, Einstein KL (2015) 'It isn't what we don't know that gives us trouble, it's what we know that ain't so': misinformation and democratic politics. Br J Polit Sci 45(3):467–475. https://doi.org/10.1017/s000712341400043x

Jacowitz K, Kahneman D (1995) Measures of anchoring in estimation tasks. Pers Soc Psychol Bull 21(11):1161–1166. https://doi.org/10.1177/01461672952111004

Kenny C (2016) Trump: 'nobody really knows' if climate change is real. CNN politics. Accessed 18 May 2017. http://edition.cnn.com/2016/12/11/politics/

Kuklinski JH, Quirk PJ, Schwieder DW, Rich RF (1998) "Just the facts, ma'am": political facts and public opinion. Ann Am Acad Polit Soc Sci 560:143–154. https://doi.org/10.1177/0002716298560001011

Leiserowitz A (2006) Climate change risk perception and policy preferences: the role of affect, imagery, and values. Clim Change 77(1):45–72. https://doi.org/10.1007/s10584-006-9059-9

Lewandowsky S, Ecker UKH, Seifert CM, Schwarz N, Cook J (2012) Misinformation and its correction: continued influence and successful debiasing. Psychol Sci Public Interest 13 (3):106–131. https://doi.org/10.1177/1529100612451018

Lundberg E (2017) How the blogosphere spread and amplified the daily mail's unsupported allegations of climate data manipulation. Climate feedback. Accessed 18 May 2017. https://medium.com/climate-feedback/

Magistrat der Stadt Wien (2009) Klimaschutzprogramm der Stadt Wien. Fortschreibung 2010–2020 [Climate protection plan of the city of Vienna]. Accessed 19 May 2017. https://www.wien.gv.at/umwelt/klimaschutz/pdf/klip2-lang.pdf

McCright AM, Charters M, Dentzman K, Dietz T (2016) Examining the effectiveness of climate change frames in the face of a climate change denial counter-frame. Top Cogn Sci 8:76–97. https://doi.org/10.1111/tops.12171

Mussweiler T, Strack F, Pfeiffer T (2000) Overcoming the inevitable anchoring effect: considering the opposite compensates for selective accessibility. Pers Soc Psychol Bull 2(9):1142–1150. https://doi.org/10.1177/01461672002611010

Nisbet EC, Hart PS, Myers T, Ellithorpe M (2013) Attitude change in competitive framing environments? Open-/closed-mindedness, framing effects, and climate change. J Commun 63 (4):766–785. https://doi.org/10.1111/jcom.12040

Nyhan B, Reifler J (2010) When corrections fail: the persistence of political misperceptions. Polit Behav 32(2):303–330. https://doi.org/10.1007/s11109-010-9112-2

Putschögl M (2015) Hadern mit dem Gürtelspeck [Quarrels with the commuting belt]. Der Standard. 30 May, p 9

Sagarin BJ, Cialdini RB, Rice WE, Serna SB (2002) Dispelling the illusion of invulnerability: the motivations and mechanisms of resistance to persuasion. J Pers Soc Psychol 83(3):526–541. https://doi.org/10.1037/0022-3514.83.3.526

Schul Y, Mayo R (2014) Discounting information: when false information is preserved and when it is not. In: Rapp DN, Braasch JLG (eds) Processing inaccurate information: theoretical and applied perspectives from cognitive science and the educational sciences. The MIT Press, London, pp 203–221

Seifert CM (2014) The continued influence effect: the persistence of misinformation in memory and reasoning following correction. In: Rapp DN, Braasch JLG (eds) Processing inaccurate information: theoretical and applied perspectives from cognitive science and the educational sciences. The MIT Press, London, pp 39–71

Tversky A, Kahneman D (1973) Availability: a heuristic for judging frequency and probability. Cogn Psychol 5(2):207–232. https://doi.org/10.1016/0010-0285(73)90033-9

Tversky A, Kahneman D (1974) Judgment under uncertainty: heuristics and biases. Science 185 (4157):1124–1131. https://doi.org/10.1126/science.185.4157.1124

van der Linden S, Leiserowitz A, Rosenthal S, Maibach E (2017) Inoculating the public against misinformation about climate change. Glob Chall 1(2). https://doi.org/10.1002/gch2. 201600008

Wegener DT, Petty RE, Detweiler-Bedell B, Jarvis WBG (2001) Implications of attitude change theories for numerical anchoring: Anchor plausibility and the limits of anchor effectiveness. J Exp Soc Psychol 37(1):62–69. https://doi.org/10.1006/jesp.2000.1431

Zuwerink Jacks J, Cameron KA (2003) Strategies for resisting persuasion. Basic Appl Soc Psychol 25(2):145–161. https://doi.org/10.1207/s15324834basp2502_5

Marlis Stubenvoll holds a Master's Degree in the Erasmus Mundus program Journalism, Media and Globalisation from Aarhus University and the University of Amsterdam.

Franziska Marquart is a post-doctoral researcher at the Amsterdam School of Communication Research where she works on an ERC-funded project studying the antecendents and consequences of public opinion towards the European Union. She holds a Ph.D. in Communication Science from the University of Vienna.

From Awareness to Action: Taking into Consideration the Role of Emotions and Cognition for a Stage Toward a Better Communication of Climate Change

Mélodie Trolliet, Thibaut Barbier and Julie Jacquet

Abstract The general public expects relevant, comprehensible and acceptable communication on climate change. Many efforts have been, and are still, being made to make the message clear and comprehensible. This paper focuses on the acceptability of the message by the receiver, to move from awareness of climate change to concrete action, which is seldom discussed in the literature. In order to make a climate change communication assessment, we choose to take as reference the Prochaska's behavioural stage of change model. Our analysis suggests that taking into account emotions and cognition mechanisms is needed in order to accompany people to better process the information and integrate it to move toward action. This paper highlights different commonly used communication practices and the underlying brain mechanisms involved in each one. A better understanding of those mechanisms should help to improve the message's receivability in communication about climate change. In turn, it will help to move from individuals knowledge into concrete action. By this way, we hope to provide inspiration to communicators in order to better accompany people in their process from awareness to action.

M. Trolliet (✉)
Center for Observation, Impacts, Energy, O.I.E., MINES ParisTech,
PSL Research University, 1 Rue Claude Daunesse, CS 10207,
F-06904 Sophia Antipolis Cedex, France
e-mail: melodie.trolliet@mines-paristech.fr

T. Barbier
Center for Processes, Renewables Energies and Energy System, PERSEE,
MINES ParisTech, PSL Research University, 1 Rue Claude Daunesse,
CS 10207, F-06904 Sophia Antipolis Cedex, France
e-mail: thibaut.barbier@mines-paristech.fr

J. Jacquet
Laboratoire Interuniversitaire de Psychologie, LIP-PC2S,
Université Grenoble Alpes, BP 47—38040 Grenoble Cedex 9, France
e-mail: julie.jacquet@univ-grenoble-alpes.fr

© Springer Nature Switzerland AG 2019
W. Leal Filho et al. (eds.), *Addressing the Challenges in Communicating Climate Change Across Various Audiences*, Climate Change Management,
https://doi.org/10.1007/978-3-319-98294-6_4

1   Introduction

1.1   General Context

In a 2017 survey, 61% of worldwide interviewees think climate change is a major threat to their country (Pew research center 2017). This indicates a degree of efficiency in communicating on the climate change. However, citizens seem to have difficulties to engage in tangible responses to climate change. In 2017, a survey of US citizens showed that 11% of them take action towards fighting global warming (Leiserowitz et al. 2017). This number reveals a relative lack of concrete action and raises questions about the ability of the general public to implement changes in response of the current communication.

1.2   Previous Work

Since the late 1990s, a number of analyses of communication on climate change and its impacts (e.g. Moser 2010; Weingart et al. 2000; Wynne 1993) proposed various explanations regarding the global lack of individual response to climate change. Several studies have shown that providing more or better information is not sufficient to shape an effective communication in terms of raising awareness and promoting active engagement (Weingart et al. 2000; Boykoff 2008; Carvalho 2007; Alexander 2008). Extensive work has also been done in the area of health and social psychology to understand why receiving information is often not sufficient to create a change in the behaviour or attitude of a person (Diclemente and Prochaska 1983; Ajzen 1985). Because of the sheer volume of studies on the communication of climate change, this paper does not provide a literature review of the topic. This works aims to provide a better understanding of individual response to climate change thanks to the psychological approach of behavioural changes.

1.3   Scope and Contribution

Let us consider a communication as efficient when it is in the same time (1) comprehensible, (2) inspiring and (3) admissible. In this paper, it is assumed that climate change communication is comprehensible (1): information on climate change is scientifically correct, is correctly emphasized and is available to the general public. It is also assumed that communication is already inspiring (2): people are already convinced by climate change issues. Communication of climate change to skeptics and deniers is dealt by literature (e.g. Dunlap 2013; Bain et al. 2012; Whitmarsh 2011), and is not in the scope of this paper. This paper focus on the admissibility (3) of the climate change message. This paper supposes that a better admissibility of the climate change message will lead to a better understanding of this message, which in turn will help people taking action related to climate change issues. It proposes an investigation of the complex interplay between giving information about climate change, raising awareness about climate change, and eliciting behavioural responses to messages.

1.4 Description of the Paper

In Sect. 2, an assessment of communication on climate change shows that the communication strategy has gone through various stages, from proving that climate change is happening to starting action. We propose to make a link between the stages in climate change communication strategy and the stages of individual behavioural change described in Prochaska's model (Diclemente and Prochaska 1983; Prochaska 2013). This model, which we chose as a reference, will be presented in Sect. 2.1. It consists in six stages, from recognizing the existence of a problem to adopting sustainable new behaviours. Considering the communication of climate change aims at inducing behavioural changes to the audience and following the Prochaska model, the current stages need to take into account emotion and cognition in order to move to the next stage: action. This will be highlighted in Sect. 2.2. Then, analyses of few classical communication advices in order to take action regarding emotion and cognitive mechanisms are provided in Sect. 3. These advices are organized in three main categories: the choose of the information support leading to an integrable message (Sect. 3.1), an efficient knowledge of the receiver leading to a suitable message (Sect. 3.2), and the promotion of a both positive and realistic vision, leading to meaningful message (Sect. 3.3). This aims to provide inspiration to communicators by ways to help people in their process from awareness to action.

2 Assessments of Climate Change Communication

The communication on climate change is considered challenging because of a double complexity: that of understanding climate change itself and that of communicating climate change (Nerlich et al. 2010). The traditional communication is defined as: "the imparting or exchanging of information by speaking, writing, or using some other medium." (Oxford dictionary 2017). Communication in climate change pushes the boundaries of this definition because it aims both at providing information and leading people to adopt concrete actions to take part in the mitigation and adaptation processes (e.g. Nerlich et al. 2010; Moser 2010). All in all, the communication of climate change aims at inducing behavioural changes among the audience.

2.1   Theory of Change and the Different Stages of Change

Change is a dynamic, non-linear process made of several stages (Cloninger 2004; Diclemente and Prochaska 1983). Changing sustainably our habits for behaviors or causes we value, such as the care one would like to give to the environment, imply changes in our awareness, in our attitudes and thoughts, in our attention, in addition to changes in our behaviors.

In 1982, Diclemente and Prochaska developed a "Transtheorical model of change" where six stages of change are described: (1) precontemplation, (2) contemplation, (3) preparation, (4) action, (5) maintenance and (6) relapse or termination/integration. These steps are illustrated on the middle of Fig. 1 coming along with classical individual position relative to change. Going through those stages, people are increasingly ready to change, and the change becomes more sustainable. On the right side of Fig. 1, the underlying process allowing to go up to the next stage is written. Effectively communicating climate change should contribute to help people go through the stages pertaining to the climate-related aspects of their life. The historical main communication strategies about climate change leading to help people go through the stages are presented on the left side of Fig. 1. Each step is described below.

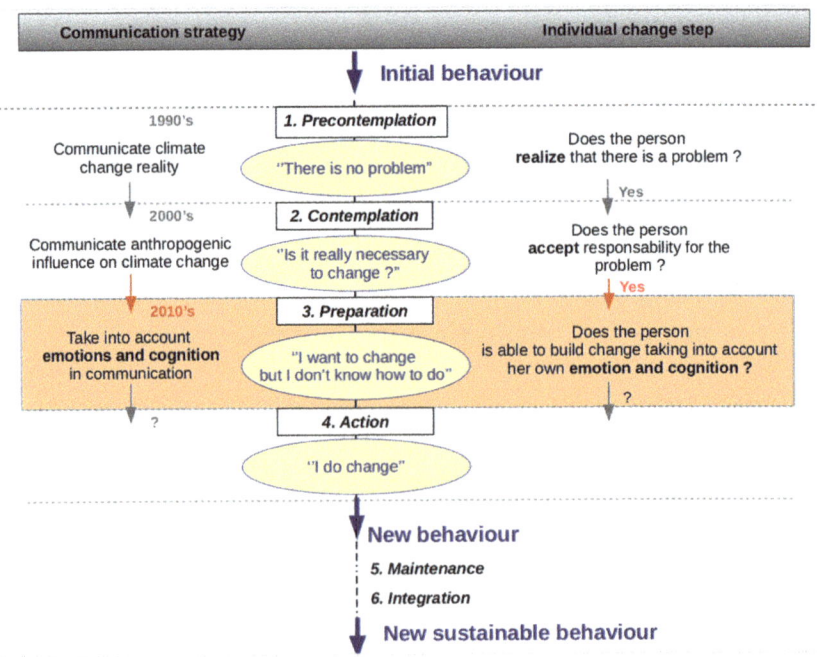

Fig. 1 Historical stages in climate change communication and the corresponding analogy in individual behaviour change

During precontemplation (1), people are not considering changing their behaviour and they are often unaware of any problem in their behaviour. Information and awareness raising are here needed to go up to the second stage. The precontemplation stage was addressed in the domain of climate change since the early 1990s, when communication mostly focused on proving and convincing people that climate change was something real, occurring here and now. The emblematic starting point for the need of this kind of communication is the creation of the IPCC in 1988 (Agrawala 1998).

The contemplation stage (2) is characterized by ambivalence: the person is aware of the existence of the problem and considers changing, but does not feel personally responsible of the problem. Concerning climate change, new communication efforts have been undertaken over the last two decades. They started with raising awareness about the happening of climate change and moved then onto establishing and communicating the anthropogenic contribution to climate change, in other words communication of the anthropogenic responsibility (Nerlich and Koteyko 2009).

When the person finally firmly sets to change, he or she enters the preparation stage (3). He or she intends to change and plans the necessary course of action. This turning point expressed by most of communicators in the climate change area is summarized in the New Scientist (2009): 'it's time to get practical over climate change'. This is currently where we are at regarding climate change. It has been highlighted with the orange coloured square in Fig. 1. In 2013, 78% of US interviewees thought they had enough information on global warming. 83% of them, however, feel that they do not know enough about what they can do to reduce global warming (Leiserowitz et al. 2014). Moving from the contemplation stage to the next stage can be complex because it requires considering not only the information level, but also the way people think and feel about the problem. To address this issue, an analysis of communication strategy taking into account cognitive and emotional barriers for getting into action is proposed in the following section.

In the action stage (4), people have made significant changes in their behaviour. Processes of change such as self-reevaluation, self-liberation, helping relationships, and reinforcement management are particularly active during this stage (Diclemente and Prochaska 1983).

At the maintenance stage (5), behavioural change stabilizes; people are confident in the fact that they are improving and continue to enable change.

The termination/integration stage (6) is described as "0 temptation and 100% self-efficacy" (Prochaska 2013). In this step, people can also relapse and come back to a previous stage. However, relapse does not mean going back to the starting point, it may be in itself a learning process and an evolution within the non-linear and dynamic process of change.

To sum up, providing information may be of particular interest during the stages of precontemplation and contemplation, where people have little awareness of the situation and corresponding problems in their behaviours, or are ambivalent regarding the possibility or will to change. For the next stages, providing information can be insufficient for most of the people. Communicators can improve their relevance by considering the role of emotion and cognition in their communication practices (Roeser 2012).

2.2 Taking into Account Emotion and Cognition for Taking Sustainable Action

2.2.1 Information Is not Enough to Make a Behavioural Change

Psychology can be of great interest to help us to understand why the information can be insufficient. According to Ajzen (1985), attitudes, subjective norms and perceived behavioural control should predict the intention to act and the subsequent behaviour. Thus, commitment into acts aiming at preventing global warming would be determined by (1) the attitude of people toward those acts, whether they consider it as good or not or whether they are in favor or not of such behaviours, by (2) the subjective norm associated with such behaviours, whether significant pairs adopt such behaviours and the motivation to comply with them, and by (3) perceived behavioural control, which is the perceived capacity to adopt those behaviours. These three components should be predictive of an intention to act.

Informing people about climate change may mainly address the first component (1) of Ajzen's model but may address poorly others.

The second component (2) deals with societal norms. It is noteworthy in the example of the lack of reaction of people facing a fire alarm (Seitter 2017). In that case, people have the knowledge of how to act when the fire alarm is ringing and the training of evacuation is regularly done, but the reaction is often not immediate, people are usually waiting for someone else moving. This phenomena has been described in terms of "bystander effect" in social psychology (Darley and Latané 1968). The societal norm will encourage the action depending, among others, to the interest of this topic in people's daily life, their current acts, and their social interactions. In the same way, studies have shown the lack of presence of climate change in small talks: surprisingly, this thematics are few tacked by people. 67% of interviewees in the USA never or rarely talk about global warming with family and friends, as illustrated in Fig. 2 (Leiserowitz et al. 2017). We see on Fig. 2 that the trend in the presence of global warming in discussions seems to be constant from 2011 to 2017. Marshall (Marshall 2015) affirms that the topic is actively avoided in discussion.

The lack of climate change in daily talk could be explained by three factors: it is considered to divide people, to lead to a conversation break and to bring anxiety in a conversation. A reason that could explain this is the bipolarization of the climate debate. People are either convinced by the existence of climate change or skeptical. Eradicating the doubt on the existence of climate change seems to have eradicated in the same time the possibility to doubt on our comprehension of the climate system. This phenomenon can also be amplified because scientific explanation of

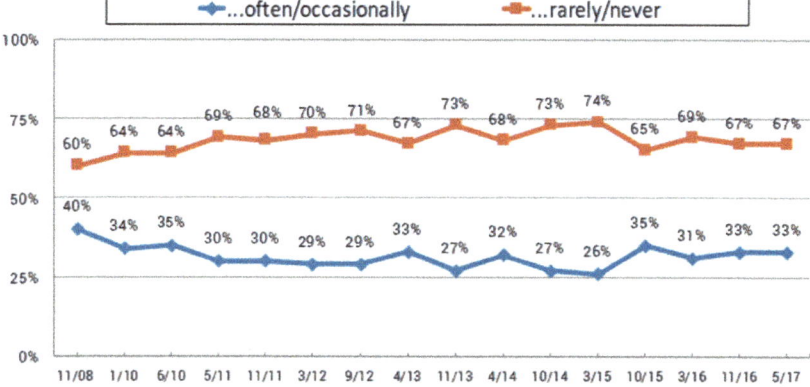

Fig. 2 Results to the question "How do you often discuss global warming with your family and friends?" (Leiserowitz et al. 2017)

climate change is complex (Roeser 2012). But many doubts are legitimates as the climate system is a complex system. As a consequence, most people do not know what to say about climate change because they are not expert and are afraid of saying something wrong. This bipolarisation of the debate, which leaves no place to any doubt or wonders, results into no place for exchanges and discussions. Finally, bringing anxiety prevents people affected by climate change from talking about it as it makes them uncomfortable (Marshall 2015). Communication need to provide information that is easy to understand, that opens a space for debate, and that address people's emotions.

2.2.2 Emotional States Faced to Climate Change Informations

Analyses of general public's conceptions of climate change in the USA highlight that emotions are widely involved when people face climate change communication (Leiserowitz et al. 2017). For example, in a study made on 1226 people, 46% of people answered "very or moderated" in the category helpless, 42% in the category afraid, 38% in the category angry, to the question "How strongly do you feel each of the following emotions when you think about the issue of global warming?".

According to Ockwell et al. (2009), existing communication approaches often fail to meaningfully engage people as they do not consider the implicit values and emotions of individuals. The role of emotion and cognition mechanism has also already been discussed for moving to the next step of behavioural change: action. Thus, next section proposes some classical communication advices in order to take action, taking into account emotion and cognitive mechanisms.

3 Analyses of Some Classical Communication Advices in Order to Take Action Regarding Emotion and Cognitive Mechanisms

The communication advices are tackled considering three aspects of an efficient admissible message. The message needs to be integrable, suitable and meaningful. First, the support of communication and the message need to be easily processed by the receiver (Sect. 3.1). Then, the message needs to be adapted to the receiver's stage of awareness and readiness for change (Sect. 3.2), and to help people by promoting both positive and realistic vision (Sect. 3.3).

We propose some classical advices for communication taking into account underlying mechanisms in emotional and cognitive functioning. This aims to provide inspiration to communicators by ways to help people in their process from awareness to action.

3.1 Making an Integrable Message: Choosing the Information Support

3.1.1 Be Consistent

Say what do you mean, mean what do you say, and do what you say you will do

Climate change is widely accepted, and largely diffused. Most people are aware of best practice concerning mitigation. However, we are also constantly subject to contradictory behaviours or information. In a study made in Switzerland, Stoll-Kleemann et al. (2001) showed that most people associate high energy scenario for future with threatening images. There is also a consensus for the goodness of low energy future. At the same time, almost nobody in the survey was prepared to take personal actions which are necessary to achieve such a future, based on mitigated climate change.

Why taking into account emotion and cognition?

When a person desires something and finds it unachievable or when people are facing contradicting choices and cannot choose between them, they will experience the pressure of an aversive emotional state (Festinger 1957, Bem 1967). In order to come back to an acceptable emotional state, people experiencing this inconsistency may block out or distance themselves from a certain information. This can explain why the climate information will not play a major role in the decision making process as much as it should be. More precisely, people will develop an attitude that matches with their actions, such as the unwillingness to let go of one's habits and comfort in the name of climate change, or the adoption of the belief that one's action if insignificant to change the order of things. Other examples include the faith in some form of managerial fix, or the rely on people's distrust on the government and believe that even if they change, such individual change would not be matched by a reliable governmental response (Stoll-Kleemann et al. 2001).

Dissonance will decrease if people adopt behaviors that are coherent with what they are aware of. Communicators can contribute to this process by proposing actions that people can handle. Moreover, communication and information given by communicators need to be coherent in itself in order to be consistently integrated by the people.

Involved mechanism: Cognitive dissonance

People subject to this cognitive dissonance seek to resolve it. In case of people cannot resolve it, the two powerful strategies people used in order to decrease dissonance have been identified by Stoll-Kleemann et al. (2001): denial and displacement strategies. These strategy will help to maintain coherent meaning systems and "act powerful to maintain the gap between attitude and behaviour with regard to climate change norms". Moreover, it will minimize the guilt caused by the knowledge that their actions adversely affect the climate, and to justify inaction in response to the uncomfortable implications of climate change mitigation action (Hillman 2004). Based on the analysis of focused-group, it has been suggested nine "barriers of denial" that may occur in the person's discourse and they illustrated it with examples, summarized in Table 1 (Stoll-Kleemann et al. 2001).

Understanding such strategy will allow the communicator to build a message that will be more easily processed by people even if inter-individual differences exist.

3.1.2 Pick up Carefully Your Terminology

Sow the wind and reap the whirlwind

The terminology is often a source of misunderstanding leading to unappropriate reaction. It is well known that the attribution of a meaning to a word depends both on the fields and on the audience. In climate change field it can be illustrated by

Table 1 Strategies in order to deal with cognitive dissonance

Barriers of denial	Typical reaction
Metaphor of displaced commitment	I protect the environment in other ways
To condemn the accuser	You have no right to challenge me
Denial of responsibility	I am not the main cause of this problem
Rejection of blame	I have done nothing so wrong as to be destructive
Ignorance	I simply don't know the consequences of my actions
powerlessness	I am only an infinitesimal being in the order of things
Fabricated constraints	There are too many impediments
"After the flood"	What is the future doing for me?
Comfort	It is too difficult for me to change my behaviour

Hulme (2007) or Bell (1994) studies. Typically, climate change is often communicated with a language of alarmism inducing fear feeling in various audiences (Bonnici 2007; Ereaut and Segnit 2006). As an illustrative example, nine of the ten major U.K. national newspapers covering the parution of the IPCC Working Group I report published articles including the adjectives "catastrophic", "shocking", "terrifying", or "devastating" (Hulme 2007). None of these words were present in the original IPCC document.

Why taking into account emotion and cognition?

Thoughts, words and emotions are closely linked together. In the process of human thoughts described by Cloninger (2004), the immediate perception of a situation leads to the emergence of emotions and words, which leads to actions and new emotions. This process takes place within a few seconds. As a consequence, the emotional response to the information will depend on how we firstly perceive it, and to the words associate to the perception. The resulting emotions and actions of the perceived climate change communication are induced by the words used. For example, fear-related messages can lead to counter-productive behaviors, particularly when the person does not perceive she is able to change (Witte and Allen 2000 for a meta-analysis). Such messages may change attitudes and verbal expressions of concern but not necessarily increase active and appropriate engagement or a creative behaviour change (Ruiter et al. 2001). Moreover, fear is an emotion that induces stress-related responses within the organism, leading to automatic reactions such as fight or flight or freeze (Selye 1956). Those reactions may be adaptive in some situations such as fighting with a dangerous animal, but won't be effective in new situations requiring calm and positive emotions for creative responses (Fredrikson 2001; Cloninger 2004).

Involved mechanism: emergency reaction

In decision making, two systems are mainly involved: the limbic system and the prefrontal cortex. The limbic system corresponds to the center of emotions. The prefrontal cortex inhibits automatic actions which are not coherent with people's goals. It enables to select and organize actions in time and space to reach a goal. In a normal functioning, the limbic system and the prefrontal cortex work together in order to use emotional information to think and act in coherence with the rational though. However, when the emotion perceived is considered as too important by the limbic system, it shortcuts the prefrontal cortex. This shortcut is a vital adaptative process which allows to resolve automatically crisis situation without using the reasoning, which is a slower process. As a conclusion, communication needs to be realistic in order for people to be plan and adopt adequate behaviors in their life and in cooperation with others, but should be positive and adapted to each one perceived capacity and stage of awareness in order to allow for the expression of creativity.

3.1.3 Talk to Both Rational and Emotional Brain

A picture is worth a thousand words

Since centuries, the use of accurate metaphors and images is used in order to teach values or behaviours (e.g. greek mythology, celtic legends…). However, in the Western cultural context, we are used to think that the cognitive response is in a binary way: rational or irrational. This kind of representation has a long-standing tradition coming from Descartes's philosophy (Damasio 1994). Emotions tends to be placed in the irrational response, contrary to the rational thought (Roeser 2012). Climate change information is scientifically based and scientists who often give the most accurate and up to date information make it in a rather rational way (Marshall 2015).

Why taking into account emotion and cognition?

Good reasoning and decision making involve both emotions and reason (Damasio 1994). This can be explained by the structure of the brain in two hemispheres also called in common language the emotional brain and the rational brain. The emotional brain has a mostly associative logic. In other words, it considers that element symbolizing reality is reality. As a consequence, metaphors and images talk directly to the emotional part of brain. Using these tools on the communication can be an efficient way to communicate a rational information involving at the same time the emotional brain (Hassol 2008).

Involved mechanism: emotional and rational brain functions

In a schematic view, the right hemisphere is often considered as dealing with creative and emotional, spatial and body awareness. It will function by associations, analogies and will deal with memory of events and the feelings associated with. The left hemisphere deals with logic, speech and language (Damasio and Damasio 1989). It changes the perception into logical, semantic and phonetic representation. Theoretically, the two hemispheres can work independently. If all the stimuli contained in the information perceived is in the specific language of one of the two hemispheres, this hemisphere will process it alone. However, it has been suggested and supported that analytic reasoning cannot be effective unless it is guided by emotion and affect (Slovic et al. 2004). Nevertheless, emotional reactions without the rational facilities will lead to inappropriate responses (Davidson et al. 2000). As a consequence, both emotional and rational capabilities are necessary for efficient response.

3.1.4 Propose Concrete and Feasible Actions

Practice makes perfect

Many messages highlight the necessary link between a small action and huge environmental issues. Environmental message related to climate change will be more efficient if it encourages actions that could be taken to deal with it. People need to perceive they are able to do such actions (see Sect. 2). In order to do so, providing information about the actions that could be taken to reduce the issue can be efficient (Xue et al. 2016). By promoting feasible actions without promising unrealistic results, it is possible to take more and more action in a quiet emotional state because it will reduce cognitive dissonance (see Sect. 3.1.1). However trying to have a big success with a first action can be highly discouraging and frustrating.

Increasing progressively the exigence of the actions that are promoted will be more efficient than proposing an action strongly frustrating in one time.

Why taking into account emotion and cognition?

The action linked to climate change mitigation and adaptation can be highly frustrating. Acting in a sustainable way suppose to choose to renounce to some immediate pleasures to guarantee an efficient answer to the frustrating assertion. An example of small frustration can results from waiting for a moment for a carpool instead of taking one's own car. If we are not trained to manage frustration, we are psychologically incompetent for making frustrating choices. Additionally, when life circumstances of people make the recommendation difficult to take into action, the frustration will increase, and negative feelings may emerge (Guttman 2000).

Involved mechanism: frustration management

Classically, it is advocated that all the psychic activity answers to a principle of pleasure-displeasure. Psychism has to avoid displeasure and promote pleasure (Freud 2003). Displeasure corresponds to an increase of the psychism tension, pleasure to decrease of this tension. The frustration deals with the repression mechanism that contributes to manage the pleasure and displeasure mechanism. A part of the tension generated by urges will turn back in the unconscious system, generating frustration. In order to manage in the most efficient way the repressed energy associated with the frustration, taking action with small steps that are perceived as under the capacity of the person can be an efficient possibility.

As a consequence, the energy added in the system stays tolerable. Such shift in one's action does not come without any frustration because it still implies facing our current behaviours and its drawbacks, and temporarily giving up some advantages and pleasures associated with our previous behaviour. Thus it is essential to make both strong communication on the advantage of frustrating change induced by climate change recommendation and promotion of feasible actions, that allows to the psychism to stay with a tolerable quantity of energy.

3.2 Making a Suitable Message: Knowing the Receiver

In the 1980s, the public was considered to need education from experts in science. This remains present in our conception of communication (Royal Society 1985; House of Lords Science and Society 2000). However, the notion of expertise itself has been denounced and the distinction between expert and non-expert has been problematized (Evans and Collins 2007). The exploration of exchanges from individual climate perceptions to experts in science (bottom-up) rather than experts disseminating their knowledge (top-down) are now studied. Going from dissemination of information to dialogues can provide many information to the communicator in climate related issue communication:

– To assess in which stage of behavioural change the person is. Thus, the communicator will efficiently adapt the information to guarantee its good integrability. He will pick up correctly his terminology depending on the stage and on the emotional and personal status of the receiver. He will avoid messages leading to blockages and to a decrease of resources within the person and he can propose achievable actions generating a correct amount of frustration.
– To question people's values. Studies show that one important factor of motivation for getting into action is the respect of personal values. In the XVI century, Spinoza already explained that in order to change an undesirable desire, the more efficient way is to replace the original desire by a bigger one rather than trying to stop the undesirable one. It is counterproductive to fight against the original desire. Basing the communication on people values could help to create a higher desire that could replace the seek for immediate pleasure. In order to question values and meaning, the communicator has to be himself someone who has found meaning and purpose in the cause he's defending (Frankl 1959).
– To build a collective approach for facing climate change. Climate change is a global and widespread issue. It cannot be created by few individuals but can eventually emerge from a myriad of vision, ideas and alternatives (Moser 2007). Communicators can serve the emergence of such a vision, by promoting dialogue with and between people, invite people in envisioning a future worth fighting for, drawing up pathways and supporting each other in working toward this goal (Moser 2007).

3.3   Providing a Meaningful Message: Promoting Both Positive and Realistic Vision

Think positive, act constructive

We are naturally highly optimistic most of the time (Lovallo and Kahneman 2003). According to most estimations, 80% of the population display an optimism bias: "we overestimate the likelihood of positive events, and underestimate the likelihood of negative events" (Sharot 2011). This bias has a lot of positive impacts. An optimist person generates much more enthusiasm, is globally happier and more friendly. It enables people to be resilient when facing difficult situations or challenging goals (Lovallo and Kahneman 2003). Their chances to get over depression are greater. A lack in optimistic bias is key symptom to the depression (Frances 1994).

Why taking into account emotion and cognition?

As any illusion, optimistic bias can be harmful. Studies shows that auto-illusion is omnipresent in various domains. For example, businessmen estimate the probability that their small company survives after 5 years to 81%, but the reality is only a 35% (Lovallo and Kahneman 2003). People are aware that divorce rates are nearly 50% in the Western world, but couples who are about to get married estimate their own likelihood of divorce as negligible (Sharot et al. 2007). 25% of students in the USA considered themselves to be in the top 1% (Lovallo and Kahneman 2003). Regarding climate change, in the USA, 25% of people affirm that climate change will harm the people in their country in a great deal, but only 16% consider it will harm within their own community or family. (Leiserowitz et al. 2017). Optimistic bias puts the threat at distance. This can reduce our action capacity against the threat. Moreover, we also have a tendency to overestimate the potential of technical solutions. As an illustration, 38% of interviewees in the USA believe that new technologies can solve global warming without having making noticeable changes in their lives (Leiserowitz et al. 2014). Kahneman (2011) affirmed that optimism bias is both a blessing and an important risk. Therefore, one has to be aware and vigilant when considering the future. There is a need to have a balance between optimism and realism, between goals and forecasts, as well as between what is inevitable and the own responsibility. Fostering true hope for a long-term future enhances the natural optimism of the human being, without erasing fear, doubts, apprehension and uncertainties. It allows to face more widely reality. Sharing this view can help to fill the gap between illusion and reality.

Involved mechanism: Optimistic bias

The more or less optimistic behaviour is linked to different belief updating between high and low optimist. The belief updating of optimist people is asymmetric: the process to a selected update of failure and diminished neural coding of undesirable information regarding the future. The region of the frontal cortex sensitive to negative estimation errors is less used by high optimists and more by low ones. (Sharot et al. 2011). Additionally, it will push the trend to take credit for positive outcomes and attributing negative outcomes to external factors, with few consideration of the real cause.

4 Conclusion and Perspective

4.1 Take Home Message

This paper highlights the various stages crossed by climate change communication, from communicate climate change reality to the intent of taking action. In order to understand the current situation, this study chose Prochaska and Diclimente model as a comprehensive model, consisting in six stages from denial to sustainable action. Our analysis suggests that we are currently at a stage where communication about climate change needs to take into account emotion and cognition in order to encourage action. Different commonly used communication strategies have been discussed along with corresponding underlying mechanisms. Mechanisms such as cognitive dissonance, emergency reaction, emotional and rational brain functions, frustration management and optimism bias have been discussed. A better understanding of those mechanisms could help to improve the message's admissibility, which is a message that is integrable, suitable and meaningful. In turn, it will help to move from individuals knowledge into concrete action. More precisely, if people have a minimum amount of information, a realistic assessment of the threat, a clear goal, an understanding of the strategies to reach that goal, and frequent feedbacks that allows them to see that they are moving in the right direction, they will be more ready to take action in a sustainable way. This can be supported by collective oriented approach that must be dealt with dialogues to build collective and individually adapted solutions. It does not exist unique solution or best practice in using these solutions. However, knowledge of all these mechanisms and idea of how to go through will help to stay flexible, adaptive and to react in a smart way depending on the receiver. In any case, information will gain in relevance when complemented by other kinds of action, at both collective and individual level, that would address other determinants of change such as self-efficacy.

4.2 Future Prospects

This work has proposed some advices in coherence with what is known in the communication field. Further works may include a complete literature review of advices in order to build a clustering of efficient communication practices. This work can be completed with a collection of detailed interviews of people involved in the climate change communication field thanks to methodologies coming from social sciences. The efficiency of each advice cluster could be studied in the light of emotion and cognition mechanisms involved in it. This clustering would be an efficient tool to unify and improve the different communication techniques.

Additionally, an integrative approach of the person should be taken into account in order to study the sustainability of the individual change.

References

Agrawala S (1998) Structural and process history of the Intergovernmental panel on climate change. Clim Change 39(4):621–642

Ajzen I (1985) From intentions to actions: a theory of planned behavior. In: Action control. Springer, Berlin, (pp 11–39)

Alexander R (2008) Framing discourse on the environment. Routledge, London

Bell A (1994) Media (mis)communication on the science of climate change. Public Unders Sci

Bem D (1967) Self-perception: an alternative interpretation of cognitive dissonance phenomena. Psychol Rev 74:183–200

Bonnici T (2007) Climate of fear: stark warning. The Sun, pp 26

Boykoff MT (2008) Media and scientific communication: a case of climate change. Geol Soc 305:11–18

Bain PG, Hornsey MJ, Bongiorno R, Jeffries C (2012) Promoting pro-environmental action in climate change deniers. Nat Clim Change 2(8):600

Carvalho A (2007) Ideological cultures and media discourses on scientific knowledge: re-reading news on climate change. Public Unders Sci 16(2):223–243

Cloninger CR (2004) Feeling good. Oxford University Press, New York

Darley JM, Latané B (1968) Bystander intervention in emergencies: diffusion of responsibility. J Pers Soc Psychol 8(4):377–383

Damasio H, Damasio AR (1989) Lesion analysis in neuropsychology. Oxford University Press, USA

Damasio AR (1994) Descartes' error: emotion, rationality and the human brain

Davidson RJ, Jackson DC, Kalin NH (2000) Emotion, plasticity, context, and regulation: perspectives from affective neuroscience. Psychol Bull 126(6):890

DiClemente CC, Prochaska JO (1983) Stages and processes of self-change of smoking: toward an integrative model of change. J Consult Clin Psychol 51(3):390–395

Dunlap RE (2013) Climate change skepticism and denial: an introduction. Am Behav Sci 57 (6):691–698

Ereaut G, Segnit N (2006) Warm words: how we are telling the climate story and can we tell it better

Evans R, Collins HM (2007) Rethinking expertise. University of Chicago Press

Festinger L (1957) A theory of cognitive dissonance. Stanford University Press

Frances A (1994) Diagnostic and statistical manual of mental disorders: DSM-IV. American Psychiatric Association

Frankl VE (1959) The spiritual dimension in existential analysis and logotherapy. J Individ Psychol 15(2):157

Fredrikson BL (2001) The role of positive emotions in positive psychology: the broaden and built theory of positive emotions. Am Psychol 56(3):218–226

Freud S (2003) Beyond the pleasure principle. Penguin UK

Guttman N (2000) Public health communication interventions. Sage

Hassol SJ (2008) Improving how scientists communicate about climate change. EOS Trans Am Geophys Union 89(11):106–107

Hillman M (2004) How we can save the planet. Penguin UK

House of Lords Science and Society (2000) Select Committee on Science and Technology, HL Paper 38, Session 1999–2000, Third Report, London

Hulme M (2007) Newspaper scare headlines can be counter-productive. Nature 445(7130):818

Kahneman D (2011) Thinking, fast and slow. p. 255

Leiserowitz A, Maibach E, Roser-Renouf C, Feinberg G, Rosenthal S, Marlon J (2014) Climate change in the American mind: Americans' global warming beliefs and attitudes in November, 2013. Yale University and George Mason University. Yale Project on Climate Change, New Haven, CT

Leiserowitz A, Maibach E, Roser-Renouf C, Rosenthal S, Cutler M (2017) Climate change in the American mind: November 2016. Yale University and George Mason University. Yale Program on Climate Change Communication, New Haven, CT

Lovallo D, Kahneman D (2003) Delusions of success. Harvard Bus Rev 81(7):56–63

Marshall G (2015) Don't even think about it: why our brains are wired to ignore climate change. Bloomsbury Publishing, USA

Moser S (2007) More bad news: the risk of neglecting emotional responses to climate change information. In Moser S, Dilling L (eds), Creating a climate for change: communicating climate change and facilitating social change. Cambridge University Press, Cambridge, pp. 64–80. https://doi.org/10.1017/cbo9780511535871.006

Moser SC (2010) Communicating climate change: history, challenges, process and future directions. Wiley Interdiscip Rev Clim Change 1(1):31–53

Nerlich B, Koteyko N (2009) Compounds, creativity and complexity in climate change communication: the case of 'carbon indulgences'. Glob Environ Change 19(3):345–353

Nerlich B, Koteyko N, Brown B (2010) Theory and language of climate change communication. Wiley Interdiscip Rev Clim Change 1(1):97–110

New Scientist Prepare for a climate-changed world, say engineers (2009). New Scientist

Ockwell D, Whitmarsh L, O'Neill S (2009) Reorienting climate change communication for effective mitigation: forcing people to be green or fostering grass-roots engagement? Sci Commun 30(3):305–327

Oxford Dictionary (2017). Available on https://en.oxforddictionaries.com

Pew Research Center (2017) Global attitudes survey. Q17a-h

Prochaska JO (2013) Transtheoretical model of behavior change. In: Encyclopedia of behavioral medicine. Springer, New York, pp. 1997–2000

Royal Society Public Understanding of Science (1985). The Royal Society, London

Roeser S (2012) Risk communication, public engagement, and climate change: a role for emotions. Risk Anal 32(6):1033–1040

Ruiter RA, Abraham C, Kok G (2001) Scary warnings and rational precautions: a review of the psychology of fear appeals. Psychol Health 16(6):613–630

Seitter KL (2017) Being as disciplined in our engagement with society as we are in our scientific research. EMS Annu Meet Abs 14(EMS2017-862):2017

Selye H (1956) The stress of life. McGraw-Hill, New York

Slovic P, Finucane ML, Peters E, MacGregor DG (2004) Risk as analysis and risk as feelings: some thoughts about affect, reason, risk, and rationality. Risk Anal 24(2):311–322

Sharot T (décembre 2011) The Optimism Bias. Current Biology 21(23):R941–945.https://doi.org/10.1016/j.cub.2011.10.030

Sharot T, Riccardi AM, Raio CM, Phelps EA (2007) Neural mechanisms mediating optimism bias. Nature 450(7166):102

Sharot T, Korn CW, Dolan RJ (2011) How unrealistic optimism is maintained in the face of reality. Nat Neurosci 14(11):1475–1479

Stoll-Kleemann S, O'Riordan S, Jaeger CC (2001) The psychology of denial concerning climate mitigation measures: evidence from Swiss focus groups. Glob Environ Change 11(2):107–117. ISSN 0959-3780. https://doi.org/10.1016/S0959-3780(00)00061-3

Whitmarsh L (2011) Scepticism and uncertainty about climate change: dimensions, determinants and change over time. Glob Environ Change 21(2):690–700

Witte K, Allen M (2000) A meta-analysis of fear appeals: implications for effective public health campaigns. Health Educ Behav 7(5)

Weingart P, Engels A, Pansegrau P (2000) Risks of communication: discourses on climate change in science, politics, and the mass media. Public Unders Sci 9(3):261–283

Wynne B (1993) Public uptake of science: a case for institutional reflexivity. Public Unders Sci 2 (4):321–337

Xue W, Hine DW, Marks AD, Phillips WJ, Nunn P, Zhao S (2016) Combining threat and efficacy messaging to increase public engagement with climate change in Beijing, China. Clim Change 137(1–2):43–55

Strengthening Personal Concern and the Willingness to Act Through Climate Change Communication

Kuthe Alina, Körfgen Annemarie, Stötter Johann, Keller Lars, Riede Maximilian and Oberrauch Anna

Abstract The climate change awareness of teenagers, as the leading generation for future development affects how societies will be able to cope with climate change. Especially teenagers' concerns about climate change and their willingness to act in a climate-friendly manner are factors climate change communication has to address. This study investigates why teenagers do not feel concerned about climate change or are not willing to act in a climate-friendly manner in order to identify strategies regarding how to strengthen these aspects in target group-oriented climate change communication formats. Using a mixed-method approach, a quantitative analysis of questionnaires answered by 760 13–16-year old teenagers was validated by a qualitative analysis of interviews with selected respondents. The findings suggest that those teenagers who are not concerned about climate change believe that climate change will happen only in the future. Furthermore, they do not recognize the interconnections or feedbacks regarding climate change between the components of the global system. The group of teenagers who are not willing to act question their own impact and ability to influence the effects of climate change. The findings are discussed, in order to identify implications for climate change communication tailored to the needs of the target group.

1 Introduction

Teenagers are one of the main audiences climate change communication should focus on. Young people will be affected by climate change more than any generation before, so it is their future that is at stake (Corner et al. 2015; Ojala 2012). Thus, numerous studies have ascertained various facets of teenagers' climate change awareness and how it can be addressed in climate change communication tailored to this specific target group (Corner et al. 2015).

K. Alina (✉) · K. Annemarie · S. Johann · K. Lars · R. Maximilian · O. Anna
University of Innsbruck, Innrain 52f, 6020 Innsbruck, Austria
e-mail: alina.kuthe@uibk.ac.at

© Springer Nature Switzerland AG 2019
W. Leal Filho et al. (eds.), *Addressing the Challenges in Communicating Climate Change Across Various Audiences*, Climate Change Management,
https://doi.org/10.1007/978-3-319-98294-6_5

65

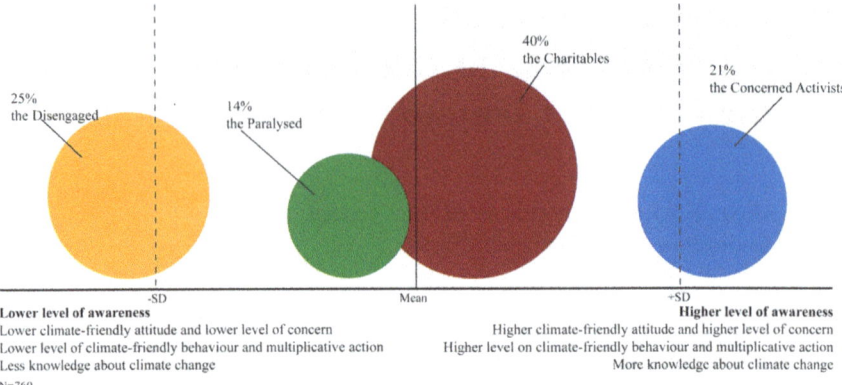

Fig. 1 Basis of the study: the four groups of teenagers regarding their climate change awareness
(Kuthe et al., subm.)

However, a study by Kuthe et al. (submitted for publication) showed that the young generation cannot be considered to be one *uniform* generation, on the basis of a sample of 760 teenagers in Germany and Austria. The study identified four different clusters of teenagers, differing in various aspects of climate change awareness, such as their attitudes towards climate change, their knowledge of the causes and effects of climate change, and their climate-friendly behaviour. Figure 1 illustrates the four clusters, found in the analysis and the distribution of the teenagers among the groups.

The first cluster of teenagers is called the *Disengaged*, forming the most unaware group as its members are not interested in the issue or in engaging in a societal solution dealing with climate change. The teenagers of the second group, called the *Paralysed*, feel concerned by climate change, but do not act in a climate-friendly manner. The *Charitables* form the third and largest group; as the group being the antipode of the *Paralysed*, they do not feel concerned about climate change while showing a high level of climate-friendly behavior. The fourth group is called the *Concerned Activists* and the members thereof are the most aware of climate change. These findings show the variety of teenagers' awareness and thus the diversity of the target groups that climate change communication has to address (Kuthe et al. submitted).

Drawing upon these findings, the present study aims to investigate the strengths and weaknesses of the different groups of teenagers and how to address them in climate change communication. Assuming that if society wants to be able to tackle climate change, it is crucial to generate a democratic majority (O'Neill and Nicholson-Cole 2009; WBGU 2011), this study will focus on two groups: the *Charitables* and the *Paralysed,* and examine the following questions:

– Why do some teenagers—called the *Paralysed*—not act in a climate-friendly manner? What do they need in climate change communication to change their behaviour into a more climate-friendly behaviour?
– Why do some teenagers—called the *Charitables*—not feel concerned? What do they need in climate change communication to feel more concerned?

The gap between attitude, concern and action has been identified in previous research (e.g. Wibeck 2013; Ockwell et al. 2009; Boyes and Stanisssstreet 2012). This study highlights the need to gain a better understanding of the relevance of these factors in the context of climate communication, with a special focus on identifying and understanding the different strengths and weaknesses of teenagers. The results will help to enhance the strategies of climate change communication tailored to the needs of these target groups.

2 Concern About Climate Change and Its Impacts

Whether a young person is aware of the impacts of climate change and is able to relate them to his or her own life influences the concern he or she feels about climate change (Smith and Leiserowitz 2014). This is even more important, as people who feel affected by climate change in their present or future life (e.g. a perceived threat from natural disasters) are more likely to behave in a climate-friendly manner than those who do not (Semenza et al. 2008). Teenagers' concept of how their own life is or will be affected is mainly shaped by their concept of climate change in total (Schuler 2011). That is why numerous studies have analysed teenagers' knowledge and conceptions of climate change and its consequences. These studies show that teenagers often have misconceptions regarding the causes (Bofferding and Kloser 2014; Karpudewan et al. 2014) and effects of climate change (Schuler 2011). Possible reasons lie in the complex, multidimensional, intangible and invisible character of climate change (Wibeck 2013).

This explains why teenagers often find it difficult to identify their own individual concerns caused by climate change. Numerous teenagers, for example, perceive climate change as being most likely to affect people mainly in distant places and in the far future (Roeser 2012). As a consequence, they feel psychologically distant from climate change (Spence et al. 2012) and climate change is often relegated in teenagers' mind behind other issues that are more pressing in everyday life (Pidgeon 2012).

3 Implementing Climate-Friendly Behaviour

Studies show that whether an individual is able to change his/her behaviour depends, among other factors, on his/her level of knowledge of the issue (Frick et al. 2004). Based on this, he/she decides what kind of action might influence the situation. Whether a person implements these actions actually depends on even more factors which many psychological models try to explain, e.g. Hines et al. (1986/1987), Kollmuss and Agyeman (2002) and Bamberg and Möser (2007). This study will focus on certain factors: One of them is termed *locus of control*, describing the feeling that climate change is still controllable (external locus of control) and that one's own behaviour has a positive impact (internal locus of control) (Hines et al. 1986/1987). Another predictor of climate-friendly behaviour is so-called social knowledge or social feedback (Bamberg and Möser 2007; Kaiser and Fuhrer 2003; Kollmuss and Agyeman 2002). This greatly depends on social-isation, which comprise two different forms: the first form encompasses the motives and intentions of others, deriving from the observation of other people; the second form comprises the social norms everyone has grown up with (Kaiser and Fuhrer 2003). Furthermore, a sense of personal responsibility is an influencing factor, as people who feel responsible for mitigating climate change are more willing to act in a climate-friendly manner than those who pass on the responsibility to adults, politicians, or economists (Fielding and Head 2012; Hines et al. 1986/1987).

4 Method

4.1 Previous Study

This study is based on the results of a previous study, which mainly consisted of a two-step hierarchical cluster analysis of 760 teenagers in Austria and Germany (Kuthe et al. submitted). The data were collected by means of a questionnaire that surveyed the climate change awareness of teenagers. The sample, which is not representative, consisted of teenagers 13–16 year of age, from 14 different schools in Germany and Austria; 53% of the respondents were female and 47% were male.

4.2 Mixed-Method Approach

In order to analyse the two groups, i.e. of the *Charitables* and the *Paralysed* in more detail, a mixed-method approach is used in this study (Spicer 2011), so the research design comprises two phases:

- **Phase 1—quantitative analysis**: this phase mainly focuses on the question of how the teenagers of the respective group differ from the teenagers of the other groups.
- **Phase 2—qualitative analysis**: this phase attempts to understand the reasons behind the patterns, found in phase 1.

Phase 1

First, this study used the dataset that was also used in the cluster-analysis (Kuthe et al. submitted). As mentioned above, this dataset contains the responses of 760 teenagers in total, 103 which belong to the group named *Paralysed* and 305 to the group named *Charitables*. In order to get to know the teenagers of the two groups better, this study examined items used in the cluster analysis as well as items not yet used in the cluster analysis. In order to investigate the characteristic of the group members, the items captured different attitudes and dimensions of climate change awareness:

- To capture the **personal concern** of young people caused by climate change, two items were used, covering concern in one's own life now and in the future.
- Teenagers' feelings regarding their **locus of control** were examined through three items asking about the internal and external loci of control.
- **Items on attitude** consisted of aspects such as interest in the topic of climate change and willingness to act.
- To capture who should be held responsible for mitigating climate change in the eyes of the young people, they were asked about different groups of people and their **responsibility** regarding this issue.
- In order to ascertain which forms and factors of **social knowledge** or social feedback are important for the teenagers, five items were used, which asked about the social conditions that might increase their willingness to behave in a more climate-friendly manner.

In phase 1 a two-sided t-test for independent samples was used to compare the participants of both cases with the participants of the other corresponding groups. In order to assess the equality of variance, the Levene-test was used and for inhomogeneous variations a correction in accordance with of the t-value according to Welch was conducted (Good 2013). Phase 1 was carried out using RStudio 1.0.153.

Phase 2

Second, semi-structured interviews with a selection of the participants from phase 1 enabled a better understanding of the teenagers' perceptions. For this purpose, 20-min semi-structured interviews were conducted individually with five teenagers from each of the *Charitables* (four girls and one boy) and the *Paralysed* (three girls and two boys). These teenagers were selected randomly based on two criteria: first, they had to be a member of the relevant group, and second, they had to be willing to participate. The interview protocol probed teenagers' attitudes according to their group membership. The interviewees belonging to the *Charitables* group were asked about different impacts of climate change and how they relate these to their own life (Mochizuki and Bryan 2015). The interviewees from the *Paralysed* group were asked about different climate-friendly behaviours and how they implement them in their own life (Anderson 2012). The interviews were recorded electronically using a digital recorder, and fully transcribed (Bryman 2012; Rapley 2010). In phase 2, each transcript was then analysed using structuring content analysis (Krippendorff 2013; Mayring 2015) for the attitude and patterns of argumentation, using the analysis software MaxQDA12. In order to be able to quote some statements of the students, they were translated into English by the authors of the study.

5 Results—The Paralysed

5.1 Quantitative Analysis

In comparing the teenagers belonging to the Paralysed group (N = 103) with the teenagers belonging to the other three groups (N = 657), it is apparent, that they differ significantly (*p < .05; **p < .01) regarding some of the aspects. First, on average they feel less responsibility to act than the other groups (diff$_M$ = −.43**). Moreover, they attribute less responsibility to act to the adults (diff$_M$ = −.48**), economist (diff$_M$ = −.38**) and politicians (diff$_M$ = −.26*) as well.

Furthermore the *Paralysed* feel more concerned about climate change than the others, today (diff$_M$ = 1.37**) and their future lives (diff$_M$ = .38**). They also have a significantly higher level of disinterest (diff$_M$ = −.74**) and a lower mean score in willingness to act (diff$_M$ = −.62**). Regarding the locus of control as to adaptation and mitigation, they think, more than the others, that society is neither able to adapt to (diff$_M$ = −.35**) nor mitigate its impacts (diff$_M$ = .31*). When it comes to willingness to act in a climate-friendly manner and the role of societal knowledge, on the one hand, the motives and intentions of others, just as the laws (diff$_M$ = .46**) or habits of society (diff$_M$ = .41**) play a higher role in their decision to behave in a climate friendly manner. On the other hand, ensuring justice towards animals and other people (diff$_M$ = −.37**) and the fairness towards other regions on earth (diff$_M$ = −.36**) play a significantly smaller role.

5.2 Qualitative Analysis

Knowledge about climate-friendly behaviour

When the interviewees were asked about the climate-friendly behaviour they know about, all but one first mentioned "travelling by car less" (P1). Some of them added alternatives like "travelling by bus, instead of car" (P4), "we carpool a lot" (P3) or

"I don't travel by car that often, but by bicycle" (P2). Additionally, two of the interviewees named buying regional products (P4: "buying products on the regional market"), recycling/avoiding waste (P3: "at home, we separate waste, such as plastic, paper etc.") and using a photovoltaic system for power generation. In addition, one interviewee cited "eating less meat" (P1).

Implementation of climate-friendly behaviour

In the second part of the interview, the members of the *Paralysed* group were asked about different climate-friendly behaviour and how they integrate them into their own everyday life. It is apparent that especially so-called low-cost actions, like turning off heat or water, are implemented by the teenagers: all but one affirm that they take this measure in their everyday life (P1: "I turn off the heating just because I don't want to consume that much gas or power"). Concerning the high-cost actions (in this study we asked about being vegetarian and riding a bike instead of driving the car) they vary a lot. Two of the interviewees say that they do not eat meat, but not out of climate-friendly concerns, but due to ethical concern for animals (P1: "but less because I want to be climate-friendly, and more because I love animals"). Three of the interviewees in this group are non-vegetarian and cannot imagine stopping to eat meat, because "it is too tasty" (P5). When these young people were asked whether they talk with their family and friends about the issue of climate change, the majority of them said that they don't discuss such issues with them (P3: "we don't talk about issues like that, normally we talk about issues which are more interesting."). Only two of the interviewees say that they sometimes discuss such issues with their parents, but mainly after having talked about the issue at school or "when we drive through the city and see a poster" (P2). Furthermore they do not inform themselves actively, so they neither visit presentations/exhibitions about climate change, nor watch related broadcasts on television.

Factors affecting the implementation of climate-friendly behaviour

Social Environment: The main reasons to implement climate-friendly behaviour in their everyday life, mentioned by the interviewees are their parents and education, as nearly all of the interviewed young people say that their parents want them to implement such (P2: "First of all, because my mum would scold me") or remind them to do so (P3: "my mum sometimes has to remind me a little bit, to turn off the heating because sometimes I forget to"). Furthermore, it is often the responsibility of the parents to provide the means to enable the teenagers to behave in a climate-friendly manner, such as a facility for separating waste (P3: "in our house, we have a little pantry, and there are some boxes where we can separate our plastic"). Another factor named by the teenagers, is their friends' behaviour. Some of the teenagers concede, for example, that if their friends would behave in such manner, maybe they would change their behaviour too (P3: "I don't want to [...] maybe if one of my friends went, then it would be nicer"). Another factor mentioned by the teenagers is uncertainty about the reaction of their friends (P5: "actually I don't know, if they are interested"; P1: "maybe just because it is uncool").

Routine/Laziness: A dominant factor the interviewees cited as preventing them from behaving in a climate-friendly manner is their routine or laziness. For example, when they were asked why they do not go to school by bike, they say "because it is more uncomfortable" (P1) or "I'm just not used to it" (P3).

External locus of control: When the interviewees were asked how much they think climate change is still controllable, they are all very sceptical (P5: "it is practically not influenceable... so I do not really think that we are able to influence it that much anymore"). Although they all believe that society can still influence climate change a bit, all but one emphasised and also questioned, whether society would want to influence the degree of climate change, if it has to react right now and everyone has to engage (P4: "regarding the situation now, I don't think we can influence it that strongly because then everybody would just have to start right know and I don't think everyone is willing to do that").

Internal locus of control: This view is also shared concerning the internal locus of control. All of the teenagers emphasised that their own contribution is only very small (P3: "not really, if so, only little") and they stressed the importance of the cooperation and engagement of everyone to achieve that a change would become apparent (P4: "well, I alone can't affect anything; there would have to be many more people, before you can see something is changing"; P2: "but, it just doesn't have such an impact; if one person changes his behaviour, this means everyone ought to be involved").

The role of action knowledge: When prompted, the teenagers said that they knew a wide range of climate-friendly behaviour. When they were asked where this knowledge originates from, the majority of them could not really explain and answered "just because I heard it somehow" (P1). Furthermore, one interviewee added the consumption of resources as one criterion of climate-unfriendly behaviour (P2: "whenever it comes to the consumption of something, for example oil").

Responsibility: When the interviewees were asked, who they think is responsible for reducing the degree of climate change, they agree that it is everyone's responsibility. Two of the teenagers underlined that climate change affects everybody and so everyone has to engage (P3: "we are all living on this earth, [...] and so it affects everyone, thus everybody is responsible"). One of the interviewees excludes people living in developing countries from this responsibility, "because they have to build schools first to be able to learn something about the issue" (P5).

6 Results—The Charitables

6.1 Quantitative Analysis

When comparing the teenagers belonging to the *Charitables* group (N = 305) with the teenagers of the other three groups (N = 455), it is apparent that they differ significantly ($*p < .05$; $**p < .01$) in some aspects. First, the analysis shows that the members of this group feel significantly less concerned about climate change in their own life (diff$_M$ = $-.69^{**}$). Interestingly, this only relates to their present life; regarding their concern about climate change in their future life, they do not differ from the other teenagers significantly (diff$_M$ = .11). Furthermore, they are more interested than the others in the topic of climate change (diff$_M$ = $-.36^{**}$), and are more willing to act (diff$_M$ = .36**). The *Charitables* appear to have a significantly higher level of locus of control, believing that society is able to adapt to the impacts of climate change (diff$_M$ = $-.37^{**}$). Regarding the attribution of responsibility, they do not differ from the other groups in average. When it comes to willingness to act in a climate-friendly manner, the *Charitables* are mainly willing to act, if social norms are stressed, such as responsibility (diff$_M$ = .23**) and fairness (diff$_M$ = .31**) towards other people and play a significantly more important role. On the contrary, they need significantly fewer laws and rules telling them to act in a climate-friendly manner than the other teenagers (diff$_M$ = $-.24^*$).

6.2   Qualitative Analysis

Knowledge of the impacts of climate change

When the teenagers where asked about the effects of climate change, all but one first cited changing temperature (C1: "temperatures are rising"; C3: "temperatures are changing"). Further they cited some more concrete effects, such as "melting glaciers" (C3), environmental disasters (C1), "less snow in winter" (C5) and "a rise in sea-level" (C3). In total, it is apparent, that teenagers know numerous possible climate change consequences, but upon closer inspection the effects seem to be mainly on a quite abstract level. Furthermore it is apparent that media and the news play an important role in forming a concept of the impacts and affection of climate change; as one said: "you can watch it on the news" (C2).

One interviewee stands out for citing "deforestation" (C2) and "plastic in the ocean" (C2). Even after he/she was asked again if he/she really thought that plastic in the ocean is an effect of climate change, he/she said: "yes, because fish are going to die for this reason". This shows that he/she is not able to differentiate between the causes and effects of climate change or even between climate change and other environmental issues in general.

Concern about the impacts of climate change in their own lives

In the second part of the interview, we presented different effects of climate change and asked the teenagers, whether they think that these impacts affect them personally or not. It is apparent that the more abstract the impact is, the harder it is for them to identify their own concern. When talking about very abstract impacts such as extreme *weather events, rising sea-levels* or the *melting of the glaciers and the polar ice caps* they all agree that it will not affect them, mainly because they do not live close to the sea or a glacier (C5: "and I'm actually not affected, because I don't live at the North Pole"; C2: "because there is no glacier close to where we live"). When talking about forest fires and floods, the teenagers also mainly brought up arguments regarding where they live, such as "the place where we live is very hilly". Two of them think that forest fires could affect them, but they do not really know how this could really relate to their lives, "maybe we need wood from the forest" (C2). It is also noticeable that they obviously find it hard to assess the risk of such events happening (C3: "yes maybe, because there is a forest, and it could burn, but I am not sure whether that is likely to happen"). When the impacts are formulated more concretely, the teenagers can relate the impacts, such as environmental refugees, to their own life even better. Also the fact that they already know these phenomena and therefore have some understanding of what they could mean for their own life helps them (C4: "environmental refugees could affect me, [...] there are already some refugees living in my village").

Factors affecting concern about climate change

Systemic relations: From the outset, it is obvious that teenagers have difficulties building systemic connections within the global climate-human system in general and the ecological system in particular. This becomes particularly apparent when talking about the extinction of species as one result of climate change. Nearly all of them are not able to recognise the ecological consequences, so they do not consider it to be a serious problem. Like one of the interviewees said: "it would not be so bad if some of the animals died" (C4), while one said while talking about what would happen, if bees declined: "because then the flowers won't be pollinated anymore, and then everything would be grey, and maybe that wouldn't be a good situation for the people as well, because it's not that beautiful anymore" (C2).

Spatial scale: All the teenagers think that climate change will especially affect people living in other regions or other countries than their home countries. Teenagers think, that regions will be affected are, for example, islands (C3: "islands like the Maldives are going to be flooded"), other continents (C3: "in warmer countries like Africa"), or countries next to the sea (C5: "in countries where the sea is not that far, like in the Netherlands").

Time scale: All teenagers also think that climate change is somehow happening in the future (C4: "it doesn't affect me now, but maybe someday later"). Only one of them thinks that it could affect him/her, when he/she is "an adult, maybe having children, in twenty or thirty years" (C1). All the others think that climate change is going to happen not before their children are born (C5: "I really think that it will take at least sixty or seventy years for climate change to happen", C4: "I don't think it will affect me, if ever, my children will be affected").

7 Implications for Climate Change Communication

In this study we attempted to examine the factors that prevent teenagers from feeling concerned about climate change or from acting in a climate-friendly manner in order to identify strategies to strengthen these aspects in target group-oriented climate change communication formats.

7.1 Implications for Climate Change Communication with the Paralysed Group

The results provide valuable information on how to address the *Paralysed* group and strengthen group members' willingness to act in a climate-friendly manner in climate change communication. One of their main barriers is the low level of their external locus of control. Therefore, it could be a good way to point out and stress the various scenarios (IPCC 2014), in order to make them aware of societal scope of influence and to underline the importance of societal action today. Another barrier often mentioned by the teenagers was their low level of internal control, so the teenagers do not believe that their behaviour contributes to mitigating climate change. This factor has also been identified as an important determinant of climate-friendly behaviour in other studies, like e.g. Fielding and Head (2012) and Ojala (2015). Based on these results, it could be promising to strengthen the feeling of agency and the effectiveness of single contributions (Anderson 2012). In order to achieve this, possible and effective courses of action should be communicated, that go beyond the typical suggestions, such as turning off lights or taking the bus.

In analysing the interviews, it also became apparent that the interviewed teenagers often use phrases like "you should do this" or "I should do that", which shows that they strongly connect the issue to a moralised and value-based component. They also often use the word "less", as in, for example "driving cars less", "producing less garbage", or "eating less meat" which shows, that they often relate climate-friendly behaviour with a abstaining from or reducing something. This highlights the urgent need for a more positive framing of sustainable and climate-friendly behaviour, one that stresses the potential positive aspects of these measures on the personal level as well (Ojala 2012; O'Neill and Nicholson-Cole 2009).

Another important aspect is the upbringing by and the influence of the parents, which underscores the importance of pursuing a holistic approach that involves the parents of teenagers in climate change communication. Furthermore, social acceptance and solidarity play an important role. Thus, a promising approach could be to initiate actions in schools, clubs, or other institutions, that aim at implementing and realising climate-friendly measures collectively (Mochizuki and Bryan 2015; Ockwell et al. 2009; Jugert et al. 2016). As one of the interviewees (P4) proposes: "Maybe as a separate subject in school, in which we go outside with the whole class or school and collect some garbage or grow some vegetables in the garden." One other interviewee (P5) added: "Maybe we just need a trend […] so more people would join these actions, and teenagers would heed them more."

7.2 Implications for Climate Change Communication with the Charitables Group

These results can help to align climate change communication to the strengths and weaknesses of the *Charitables* group. First of all, it is apparent that they have promising starting position because they are interested in the issue and are quite willing to act. Their only weakness is their low level of concern. This is probably due to the fact that they do not really know what impacts climate change could have on their own lives. Hence, it could be promising to discuss and visualise the impacts of climate change on their own region and on their own life (Schweizer et al. 2013; Wibeck 2013), ideally connected to situations or impacts they already experienced, so that they can relate to them (Spence et al. 2012). Thereby, it is absolutely necessary not to frighten or scare them, which, as studies have shown, can be counterproductive to engagement (O'Neill and Nicholson-Cole 2009; Smith and Leiserowitz 2014).

Further, it is crucial to strengthen the understanding of ecological interrelations and dependencies, because the majority of teenagers are not able to connect the consequences with regard to the different levels and regions, and their concepts related to climate change are very fragmented. Climate change is only one of the Grand Challenges, which are all complex in nature, multidimensional and require systemic thinking in order to be able to solve or even deal with them (Buckler and Creech 2014; Reid et al. 2010). Therefore, teenagers need this understanding and skill to deal with not only climate change, but also with many other societal problems now and in the future (Kagawa and Selby 2013; Sinatra et al. 2012).

Additionally, it can be promising for the *Charitables* to not frame climate change only as an environmental issue but also a societal one, because societal responsibility and fairness plays important roles for the teenagers. When stressing the societal aspects of climate change, its causes and effects may enhance the engagement of those who do not feel responsible for protecting nature, but for defending people's lives and their environment (Howell and Allen 2016).

8   Discussion

The current research provides valuable information about climate change communication tailored to the needs of the target groups, nevertheless a number of limitations must be acknowledged. The use of self-reported measures regarding attitude and behaviour is not ideal because the issue of climate change is susceptible to bias and desirability, as it is strongly normative. Although the teenagers were encouraged to say the truth, it cannot be ruled out that some of the teenagers were influenced in their responses especially during the interviews (Rapley 2010). Furthermore, some of the scales were measured with only two or three items. So these results should be confirmed in future studies, especially in respect to ensuring a larger and more representative sample. Especially because we have to realise that there may have been a positive selection within the sample based on the selection of the teenagers participating in the study, a group restricted to persons involved in the project k.i.d.Z.21-Austria.

Furthermore, this study raises a few new questions worth for investigate for future studies. One particularly interesting question is how young people in the different groups develop in projects such as k.i.d.Z.21-Austria and other formats of climate change communication. Furthermore it would be interesting to see whether if it is possible to transfer these results to other groups such as adults, especially with regard to the fact that the *Paralysed* showed similarities to the *Cautious* group of adults found in a study by Metag et al. (2015). Another question that arose, particularly in the interviews with the *Charitables*, is what needs to happen, such that a person has the feeling of being affected by climate change and its consequences. Especially regarding the fact that it is nearly impossible, to clearly attribute any incident to climate change only (IPCC 2014). Furthermore it could be interesting to develop some communication formats tailored to the different groups in collaboration with the respective groups in order to account for their needs more closely.

9 Conclusion

This study shows the different strengths and weaknesses of teenagers belonging to two groups and the urgent need to address them in climate change communication. The findings suggest that those teenagers who are not concerned about climate change believe that climate change will happen only in the future. Furthermore, they do not recognize the interconnections or feedbacks regarding climate change between the components of the global system. The group of teenagers who are not willing to act question their own impact and ability to influence the effects of climate change. Furthermore we can say that many teenagers find it hard to relate climate change to their own life and to derive their own involvement and opportunities to participate. This shows that one main task of climate change communication is to assess and enable new positive and personal approaches to the issue of climate change and the importance of trivial messages going beyond *it's getting warmer* and *you shouldn't travel by car anymore*.

Acknowledgements We would like to thank the Austrian Climate and Energy Fund (KR14AC7K11819), which finances k.i.d.Z.21-Austria, project number: B464783. Special thanks go to everyone who has contributed to the success of the project, especially to the teenagers involved in this study.

References

Anderson A (2012) Climate change education for mitigation and adaptation. J Educ Sustain Dev 6(2):191–206

Bamberg S, Möser G (2007) Twenty years after Hines, Hungerford, and Tomera: a new meta-analysis of psycho-social determinants of pro-environmental behaviour. J Environ Psychol 27(1):14–25

Bofferding L, Kloser M (2014) Middle and high school students' conceptions of climate change mitigation and adaptation strategies. Environ Educ Res 21(2):275–294

Boyes E, Stanisstreet M (2012) Environmental education for behaviour change: which actions should be targeted? Int J Sci Educ 34(10):1591–1614

Bryman A (2012) Social research methods, 4th ed. Oxford University Press, Oxford

Buckler C, Creech H (2014) Shaping the future we want: UN decade of education for sustainable development. Final report; 2014: UN Decade of education for sustainable development (2005–2014) Final report

Corner A, Roberts O, Chiari S, Völler S, Mayrhuber E, Mandl S, Monson K (2015) How do young people engage with climate change? The role of knowledge, values, message framing, and trusted communicators. WIREs Clim Change 6:523–534

Fielding KS, Head BW (2012) Determinants of young Australians' environmental actions: the role of responsibility attributions, locus of control, knowledge and attitudes. Environ Educ Res 18(2):171–186

Frick J, Kaiser FG, Wilson M (2004) Environmental knowledge and conservation behavior: exploring prevalence and structure in a representative sample. Personal Individ Differ 37(8): 1597–1613

Good PI (2013) Introduction to statistics through resampling methods and R, 2nd ed. Wiley, Hoboken

Hines J, Hungerford H, Tomera A (1986/1987) Analysis and synthesis of research on responsible environmental behavior: a meta analysis. J Environ Educ 18(2): 1–8

Howell RA, Allen S (2016) Significant life experiences, motivations and values of climate change educators. Environ Educ Res 48:1–19

IPCC (2014) Climate change 2014: summary for policymakers

Jugert P, Greenaway KH, Barth M, Büchner R, Eisentraut S, Fritsche I (2016) Collective efficacy increases pro-environmental intentions through increasing self-efficacy. J Environ Psychol 48:12–23

Kagawa F, Selby D (2013) Ready for the storm: education for disaster risk reduction and climate change adaptation and mitigation. J Educ Sustain Dev 6(2):207–217

Kaiser FG, Fuhrer U (2003) Ecological behavior's dependency on different forms of knowledge. Appl Psychol Int Rev 52(4):598–613

Karpudewan M, Roth WM, Mohd Nor Syahrir Bin A (2014) Enhancing primary school students' knowledge about global warming and environmental attitude using climate change activities. Int J Sci Educ 37(1):31–54

Kollmuss A, Agyeman J (2002) Mind the gap: why do people act environmentally and what are the barriers to pro-environmental behavior? Environ Educ Res 8(3):239–260

Krippendorff K (2013) Content analysis: an introduction to its methodology, 3rd ed. Sage, London

Kuthe A, Keller L, Körfgen A, Stötter J, Oberrauch A, Höferl KM (submitted) How many young generations are there?—The variety of teenagers' climate change awareness. J Environ Educ

Mayring P (2015) Qualitative Inhaltsanalyse: Grundlagen und Techniken, 12th ed. Beltz Pädagogik, Beltz, Weinheim

Metag J, Füchslin T, Schafer, MS (2015) Global warming's five Germanys: a typology of Germans' views on climate change and patterns of media use and information. Public Understanding of Science

Mochizuki Y, Bryan A (2015) Climate change education in the context of education for sustainable development: rationale and principles. J Educ Sustain Dev 9(1):4–26

Ockwell D, Whitmarsh L, O'Neill S (2009) Reorienting climate change communication for effective mitigation: forcing people to be green or fostering grass-roots engagement? Sci Commun 30(3):305–327

Ojala M (2012) Hope and climate change: the importance of hope for environmental engagement among young people. Environ Educ Res 18(5):625–642

Ojala M (2015) Climate change scepticism among adolescents. J Youth Stud 18(9):1135–1153

O'Neill S, Nicholson-Cole S (2009) "Fear won't do it": promoting positive engagement with climate change through visual and iconic representations. Sci Commun 30(3):355–379

Pidgeon N (2012) Climate change risk perception and communication: addressing a critical moment? Risk Anal 32(6):951–956

Rapley T (2010) Interviews. In: Seale C, Gobo G, Gubrium JF (eds) Qualitative research practice. Sage, London, pp 15–33

Reid WV, Chen D, Goldfarb L, Hackmann H, Lee YT, Mokhele K, Ostrom E, Raivio K, Rockstrom J, Schellnhuber HJ, Whyte A (2010) Earth system science for global sustainability: grand challenges. Science 330(6006):916–917

Roeser S (2012) Risk communication, public engagement, and climate change: a role for emotions. Risk Anal 32(6):1033–1040

Schuler S (2011) Alltagstheorien zu den Ursachen und Folgen des globalen Klimawandels: Erhebung und Analyse von Schülervorstellungen aus geographiedidaktischer Perspektive. Bochumer Geographische Arbeiten, vol 78. Europäischer, Universitätsverlag, Bochum

Schweizer S, Davis S, Thompson JL (2013) Changing the conversation about climate change: a theoretical framework for place-based climate change engagement. Environ Commun J Nat Cult 7(1):42–62

Semenza JC, Hall DE, Wilson DJ, Bontempo BD, Sailor DJ, Lab George (2008) Public perception of climate change. Am J Prev Med 35(5):479–487

Sinatra GM, Kardash CM, Taasoobshirazi G, Lombardi D (2012) Promoting attitude change and expressed willingness to take action toward climate change in college students. Instr Sci 40(1): 1–17

Smith N, Leiserowitz A (2014) The role of emotion in global warming policy support and opposition. Risk Anal 34(5):937–948

Spence A, Poortinga W, Pidgeon N (2012) The psychological distance of climate change. Risk Anal 32(6):957–972

Spicer N (2011) Combining qualitative and quantitative methods. In: Seale C (ed), Researching society and culture 3rd ed. Sage, Los Angeles, pp 479–493

WBGU (2011) Welt im Wandel: Gesellschaftsvertrag für eine große Transformation. WBGU, Berlin

Wibeck V (2013) Enhancing learning, communication and public engagement about climate change-some lessons from recent literature. Environ Educ Res 20(3):387–411

Philippine Private Sector Engagement *Beyond* Climate Change Awareness

Maria Fe Villamejor-Mendoza

Abstract Climate change is a global phenomenon that affects everyone. It is a development issue (World Bank in World development report and climate change. The International Bank for Reconstruction and Development, 2010) that has to be understood in order to be addressed more collectively and comprehensively. There have been numerous studies on the role of governments in climate change mitigation and adaptation. This is because as duty bearers, governments are expected to orchestrate initiatives to address environmental, economic, social and other vulnerabilities of countries. Studies concerning the role of the private sector in climate change are however few. Thus, under the framework of collaborative governance (Ansell and Gash in J Public Adm Res Theor 18:543–571, 2008) where each sector of the society has a role to play, this paper focuses on the roles played by the private sector in addressing the risks and vulnerabilities brought about by climate change. In so doing, it hopes to contribute to the understanding of private sector engagement in climate change beyond awareness raising. Being mainly a desk review of secondary materials and existing studies on the topic at hand, it argues that the private sector in the Philippines has embraced the issue of climate change and has been doing all it can to help address this development concern. It also looks at some of the messages, e.g., the risks of climate change, support needed for climate change solutions, and some climate change issues and concerns, selected business companies have been communicating to the public to understand the content as well as the underpinning philosophy of why private sector initiates activities for climate change.

Paper submitted for presentation at the World Symposium on Climate Change Communication, held in Graz, Austria on 7th–9th February 2018.

M. F. Villamejor-Mendoza (✉)
National College of Public Administration and Governance,
University of the Philippines Diliman, Raul P. de Guzman St. Diliman,
Quezon City 1100, Philippines
e-mail: mvmendoza@up.edu.ph; fevmendo@yahoo.com

© Springer Nature Switzerland AG 2019
W. Leal Filho et al. (eds.), *Addressing the Challenges in Communicating Climate Change Across Various Audiences*, Climate Change Management,
https://doi.org/10.1007/978-3-319-98294-6_6

1 Overview: Climate Change as a Development Issue

Climate change is a global phenomenon that affects the economies and socio-cultural, ecological and other vulnerabilities of countries. Being a development issue (World Bank 2010) and not merely an environmental concern, climate change has become a defining and most challenging sustainable development issue of the twenty first century (Letchumanan 2013).

It is defining in the sense that it is now dictating the pace and nature of economic growth, development and social progress, while potentially becoming the greatest threat to humankind and survival if left unchecked. It is challenging because of its multifaceted nature, affecting almost all sectors and the basic means and lifestyle of human existence (Ibid. p. 2). In seconds, climate change may wipe out communities and the gains of development, economic or otherwise. It does not discriminate against rich or poor; developed or developing country. It affects everyone that comes along its path.

Climate change refers to "the state of the climate that can be identified (e.g., by using statistical tests) by changes in the mean and/or the variability of its properties and that persists for an extended period, typically decades or longer. Climate change may be due to natural internal processes or external forcing, or to persistent anthropogenic changes in the composition of the atmosphere or in land use" (IPCC 2012). Its more obvious manifestations include rising temperature, variability of precipitation, frequency and intensity of typhoons, rise in sea levels, and the risks of more droughts, floods, heat waves, and forest and grassland fires and other disasters (CCC 2011).

In the Asian region, a 2009 study by the Asian Development Bank (ADB), suggests that on average Southeast Asia "is likely to suffer more from climate change than the rest of the world, if no action is taken." Letchumanan (2013) continues that such action should be "a holistic and integrated response, which cannot be fixed by technology or finite human or capital resources alone." In addition, Sano (as cited in Oxfam 2014) suggests that the only global multilateral process that addresses climate change, e.g., the United Nations Framework Convention on Climate Change (UNFCC) and its Kyoto protocol, has to move because "we cannot solve a wholesale problem like climate change with retail solutions at the individual or national level only."

2 Addressing Climate Change in the Philippines

Addressing the risks and hazards of climate change has been considered a post-2015 development agenda (sustainable development goals or SDGs), beyond the concerns of achieving the Millennium Development Goals of generally halving poverty by 2015 (GPDRR 2013 as cited by Hitoshi 2014).

Advocacies and actions towards disaster risk reduction and the building of resilience[1] have been put forward, including specifically, (1) the call on countries to develop nationally agreed standards for hazard risk assessment especially of critical infrastructure like schools, health centers, electricity and water systems, nodal ITC data centers, road and transport systems; (2) the call on the private sector to incorporate disaster risk considerations in risk management practices; and (3) stimulate collaboration among the public and private sectors at the national and local levels in risk management (GPDRR 2013 as cited by Hitoshi 2014).

In the Philippines, in response to the urgency for action on what has essentially become a global crisis, Republic Act 9729, also known as the Climate Change Act of 2009, was passed. It is anchored on the 1987 constitutional provision, which states "it is the policy of the State to afford full protection and the advancement of the right of the people to a balanced and healthful ecology... to fulfill human needs while maintaining the quality of the natural environment for current and future generations." (CCC 2011).

RA 9729 provides the policy framework with which to systematically address the growing threats on community life and its impact on the environment. Among others, it instituted of a Climate Change Commission (CCC), an independent and autonomous body that has the same status as that of a national government agency. The CCC is under the Office of the President and is the "sole policy-making body of the government which shall be tasked to coordinate, monitor and evaluate the programs and action plans of the government relating to climate change" (Sect. 4).

The national climate change framework strategy has recently been translated into a National Climate Change Action Plan (NCCAP), which prioritizes food security, water sufficiency, ecosystem and environmental stability, human security, climate-smart industries and services, sustainable energy, and capacity development as the strategic direction for 2011–2028.

The Philippines through the CCC later crafted the National Framework Strategy for Climate Change 2010–2022 in 2010 and the National Climate Change Action Plan: 2011–2028 in 2011 (CCC 2011).

The Framework envisions a climate risk-resilient Philippines with healthy, safe, prosperous and self-reliant communities, and thriving and productive ecosystems. The goal is to build the adaptive capacity of communities and increase the resilience of natural ecosystems to climate change, and optimize mitigation opportunities towards sustainable development.

---

[1]Disaster resilience is defined as the ability of countries, communities, businesses, and individual households to resist, absorb, recover from, and reorganize in response to natural hazard events, without jeopardizing their sustained socioeconomic advancement and development (ADB 2012) It recognizes the highly dynamic, continually shifting nature of the state of resilience as populations grow and move; capital investments expand; and the frequency and intensity of meteorological, hydrological, and climatological events change as a consequence of climate change. Disaster resilience at all levels of society is a critical component of efforts to achieve sustainable socioeconomic development and poverty reduction.

In addition, it recognizes the value of forming multi-stakeholder participation and partnerships in climate change initiatives, including partnerships with civil society, the private sector and local governments, and especially with indigenous peoples and other marginalized groups most vulnerable to climate change impacts. Also, policy and incentive mechanisms to facilitate private sector participation in addressing adaptation and mitigation[2] objectives shall be promoted and supported (CCC 2011). The plan also provides for a policy environment that will encourage the participation of the private sector to optimize mitigation opportunities towards sustainable development.

3 Research Focus: Private Sector Participation in Climate Change Adaptation and Mitigation

This paper is thus a contribution to the discussions on climate change solutions, and focuses on the participation of the private sector in addressing climate change impacts and risks. It is mainly a desk review of secondary materials and existing studies on the topic at hand. It builds on an earlier research by the author (Villamejor-Mendoza 2014, 2015), which mapped out some of the private sector engagement in climate change mitigation and adaptation in the Philippines.

For this paper, it argues that the private sector in the country, having realized the risks and vulnerabilities of the phenomenon, has embraced the issue of climate change and has been doing all it can to help address this development concern, Thus, its role has been beyond awareness raising, and has substantially acted to address climate change, for its own (business') sake and also for the public's.

Moreover, it looks at some of the messages selected business companies have been communicating to the public, e.g., the risks of climate change, support needed for climate change solutions, and some climate change issues and concerns, in order to understand the content as well as the underpinning philosophy of why private sector initiates activities for climate change.

This cursory look is limited to the number of business companies whose web-pages were accessed by the author from July to September 2017. As such, this covers the voices of a small 'sample' of companies and not the totality of the private sector working out or implementing climate change solutions in the Philippines.

---

[2]Climate change mitigation is a human intervention to reduce the sources or enhance the sinks of greenhouse gases (IPCC 2013). Adaptation is the process of adjustment to actual or expected climate and its effects. In human systems, adaptation seeks to moderate or avoid harm or exploit beneficial opportunities. In some natural systems, human intervention may facilitate adjustment to expected climate and its effects (IPCC 2014).

4 Private Sector Engagement (PSE) in Addressing Climate Change Concerns

As duty bearer, planning to address and acting on climate change has been the domain of government. However, to effectively address the adverse effects of extreme weather events and other manifestations of this global problem, and ensure successful implementation of these initiatives, there is an urgent need to engage the private sector, civil society and ordinary citizens because the government cannot do these enormous tasks alone. According to IFC (2010), the private sector has particular competencies, which can make a unique contribution to adaptation, through innovative technology, design of resilient infrastructure, development and implementation of improved information systems and the management of major projects.

It is also essential to mobilize financial resources and technical capacity, leverage the efforts of governments, engage civil society and community and develop innovative climate services and adaptation technologies (Biagini and Miller 2013).

Unfortunately, governments are not clear on how to engage the private sector. Likewise, private sector efforts at climate change mitigation and adaptation are not widely understood nor seen as good business practice. Also, in general, their participation is impeded by several obstacles (Biagini and Miller 2013).

In addition, relatively few (private) companies have yet considered both the impact of climate change on their existing activities, and perhaps as importantly, the new commercial opportunities that will emerge both domestically and globally. Specifically, corporate climate change (initiative) is perceived to be either irrelevant or at best an extension of their Corporate Social Responsibility (CSR) (IFC 2010).

Financial institutions in both public and private sectors also lack the capacity to evaluate 'innovative projects, goods and services' for climate change mitigation and adaptation. This lack of understanding of specific types of climate change investments and their risk profiles means that banks often find it difficult to develop and structure appropriate financial products. Add to this is a seeming consensus that the bulk of climate change funding would be administered by the government with a lot of the implementation done by Non- Government Organizations (NGOs). Hence there was little incentive or motivation for companies to commit scarce and valuable senior management time to consider opportunities in tackling climate change (IFC 2010).

Nevertheless, the information gaps and awareness inadequacies are recently gradually being bridged and built. There seems to be a greater appreciation of the risks and costs of climate change at present as many countries have identified their own climate change priorities through their National Climate Change Action Plans (NCCAPs). They have also publicized these needs in the form that may encourage business engagement.

Private initiatives are not a substitute for governmental efforts, and indeed, the former are very dependent on the latter for information, supportive policies and regulation, and other support. Some elements of climate change adaptation (and mitigation) are primarily or even exclusively government functions and are likely to remain so, particularly the provision of basic weather and climate information, design and implementation of risk management policies (e.g., building codes, land use restrictions, and insurance regulations), and disaster planning and preparedness. Non-state actors and the citizenry, as a whole, may complement these efforts.

5   Private Sector Engagement (PSE) in the Philippines

The private sector in the Philippines is generally more engaged now to help government address climate change, especially after a series of natural disasters struck the country, the most damaging of which were the Bohol great earthquake in 2013 and the Yolanda (Hainan) super typhoon in 2014. The bottom line is the realization that extreme weather disturbances and weather-related disasters pose and have, in fact, caused financial losses to several businesses. Disrupted production and delivery due to floods, typhoon, and earthquake, for example, resulted in dramatic losses, thus prompting the private sector to think of solutions to sort of minimize, if not totally avoid, such vulnerabilities.

One of the very first climate change adaptation-related steps taken by private sector organizations is to prepare for and manage their exposure to climate change. "Climate proofing" of private sector investments became a natural and logical response. Design and construction of factories, workplaces and economic zones, among others, took into consideration safety and environmental standards meant to protect business from unforeseen climate-related catastrophes (Asian Development Bank 2005).

It did not take long for the sector to realize that its interventions and action must go beyond its own backyard. Private sector organizations were soon engaged also in rehabilitating and making resilient communities mostly under the ambit of their CSR and Corporate Social Investment. Furthermore, there also came "emerging business opportunities in helping others to reduce their climate risks. These include generating new finance, to help fill the massive deficit in available funds for adaptation; designing, manufacturing and distributing goods and services that can help reduce the vulnerability of individuals and communities to climate change; and, providing risk management tools, including insurance" (SEI 2010).

A new model is emerging, e.g., Area Business Continuity Paradigm (ABCP), which embraces and forges various means of collaboration, unified (area scale) management in coordination with public and private sectors, and sharing of critical resources such as water, energy, transportation and communication, to ensure continuity of business operations or promptly restart business after disasters (Hitoshi 2014). New paradigms such as "building back better for next time" or "reconstructing better after (tsunamis and other major natural) disasters" (European Union and UNISDR n.d.; UNWCDRR 2015) have slowly crept into the consciousness of the public, especially those from the disaster-prone and vulnerable countries.

A survey of existing climate change-related programs across the world reveals that the private sector has covered areas such as capacity building, education and training; finance and insurance; food, agriculture, forestry and fisheries; technology and information & communications technology (ICT); water resources; science, assessment, monitoring and early warning; business; and human health (GEF 2012).

In the Philippines, space has been provided for the private sector to participate in the priority activities in the National Climate Change Action Plan (NCCAP) (see Fig. 1; Table 1). Thus they are into food security to ensure the availability of food amidst climate change; water sufficiency to manage the supply, demand and quality of water; the protection and rehabilitation of major ecosystems; as well as reduce the risks of men and women during disasters.

This space has been filled up by the private sector in the country through initiatives such as technology transfer, information, education and communication (IEC) campaigns, capacity building, research and development.

Specifically, based on the initial mapping in 2014, the following are some of the PSE in climate change adaptation and mitigation in the Philippines (Table 2):

Thus, it can be observed that the private sector in the Philippines has embraced the issue of climate change for their business' sake as well as for others. They did it

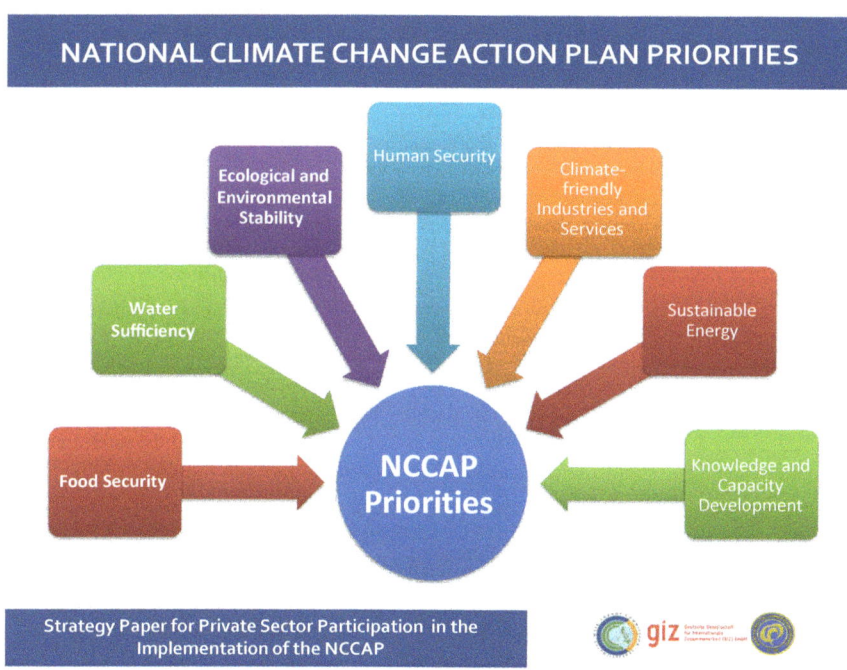

Fig. 1 The national climate change action plan priorities. *Source* CCC (2012) as cited in Villamejor-Mendoza (2015)

Table 1 Possible private sector engagement (PSE) in the NCCAP

Priorities	Outcomes	Possible private sector engagement
1. Food security	The objective of the national strategic priority on food security is to ensure availability, stability, accessibility, and affordability of safe and healthy food amidst climate change	Firms can support the crop-specific vulnerability assessments and integration of climate change in farming and fishing practices and trainings
2. Water sufficiency	In light of climate change, however, a comprehensive review and subsequent restructuring of the entire water sector governance is required. It is important as well to assess the resilience of major water resources and infrastructures, manage supply. demand, and quality of water, and promote conservation	Private sector participation in the rehabilitation and improvement of water infrastructure and management systems will enhance initiatives in sustaining equitable access to safe and affordable water
3. Ecological and environmental stability	Ecosystem resilience and environmental stability during the plan period is focused on achieving one immediate outcome: the protection and rehabilitation of critical ecosystems, and the restoration of ecological services	Promotion of environmentally-sound business practices will contribute much to promoting environmental and ecological stability
4. Human security	The objective of the human security agenda is to reduce the risks of women and men to climate change and disasters	Firms that deliver health and social protection packages, and those that help establish human settlements are key to improving responsive health and social delivery systems
5. Climate friendly industries and services	NCCAP prioritizes the creation of green and eco-jobs and sustainable consumption and production. It also focuses on the development of sustainable cities and municipalities	Climate-Smart Industries and Services promote green industries (those that put heavy emphasis on low-carbon strategies and mitigation efforts), the creation of green jobs, and the development of sustainable cities and municipalities
6. Sustainable energy	NCCAP prioritizes the promotion and expansion of energy efficiency and conservation; the development of sustainable and renewable energy; environmentally sustainable transport; and climate proofing and rehabilitation of energy systems infrastructures	For promotion and use of sustainable energy, the private sector is particularly important as they have great potential for developing, promoting and marketing alternative environmentally sustainable transport and renewable energies

(continued)

Table 1 (continued)

Priorities	Outcomes	Possible private sector engagement
7. Knowledge and capacity development	The priorities of the NCCAP on knowledge and capacity development are: • Enhanced knowledge on the science of climate change; • Enhanced capacity for climate change adaptation, mitigation and disaster risk reduction at the local and community level; and • Established gendered climate change knowledge management accessible to all sectors at the national and local levels	They also have roles to play in cross-cutting strategies such as gendered knowledge management, capacity building, research and development and technology transfer

Source CCC (2010) as cited in Villamejor-Mendoza (2015)

Table 2 Initial mapping of private sector engagement in climate change adaptation and mitigation in the Philippines

Food security

• Setting up of a state-of-the-art egg production facility in Tarlac that would use modern ventilation and climate control mechanisms to ensure high production
• 2014 AgriNegosyo Summit for agri-entrepreneurs and provision of industry-leading technology such as business connectivity solutions and cloud computing and storage services to help them develop and maintain innovative services
• Sustainable large-scale organic moringa agribusiness for the MNFA farmers and partner organizations, which includes technical assistance to the farmers on organic and natural farming techniques for successful implementation of organic methods

Water sufficiency

• The "Adopt-a-Watershed" program in Ipo, begun in 2006, engaged volunteer organizations in reforestation. The 2 concessionaires reforested 110 ha in 2010 and 450 ha in 2011. The reforestation target for 2012 onwards is 900 ha per year
• Wastewater recycling and resource recovery strategies (Anaerobic digestion and recovery of methane gas from wastewater)

Ecological and environmental stability

• Hiring of foresters to provide technical assistance in various reforestation methods; alternative livelihood for villagers (former Kaingeros and charcoal-producers)
• Develop an alternative ecotourism destination near Baguio City where tourists, guests, and visitors can commune with nature, can be entertained with indigenous culture, and can buy fresh products
• Designing, developing and assembling electric vehicles as substitutes for smoke-belching gasoline or diesel-fueled vehicles

(continued)

Table 2 (continued)

- Setting up of the first integrated green charcoal, vermicast, biotech fertilizer and pest control production lines
- PEMC Cares coastal cleanup campaign in Calatagan, Batangas. PEMC partnered with Hands-on Manila to implement the activity, in cooperation with the Conserve and Protect Oceans Foundation (CAP-Oceans)
- SKAL Tourism Personality Awards include a category on environment and agri-tourism

Human security

- Plant rehabilitation and refurbishments to ensure energy security, develop a sustainable energy system, and promote human health and security
- Introduction of a unified number (7622) that may be used for free during disaster—powered by Chikka
- Libreng Tawag Centers for affected residents during typhoon and other natural disasters in some areas in NCR. #StaySafePH shares valuable tips and info on how to be safe during times of calamities and disasters. Selected Globe stores have also offered free charging for customers' mobile phones and gadgets
- Barangay emergency response programme (BERP)/Neighborhood emergency services team (NEST)
- Master planned communities—Veritown Fort in Bonifacio Global City, Filinvest City, Makati CBD, Circuit Makati, Vista City, ARCA South
- Distribution of 200 fiberglass-fishing boats to fishermen in Ormoc City affected by typhoon Yolanda through the 6200 Mission Possible Project in Leyte IV
- Megaworld's percentage distribution of its total donations to charitable causes—6% for calamity response and 1% for environment

Climate smart industries and services

- Investing in a global flash storage solutions to boost disaster recovery, backup and replication processes. The system is used to store and manage designs, keep an eye on budgets, manage suppliers, communicate with customers and control projects; In the event of disaster, getting systems fully up and running again would now require only 20 min as compared to at least half a day to restore just one database to one server in the past. Backups, which used to take 8 h, are now completed within 5 min
  (*Source* Mercurio, R.S., "Ortigas & Co. investing in modern technology with Nimble Storage", Phil Star 21 Sept 2014:B-5)
- Wastewater recycling and resource recovery strategies (Anaerobic digestion and recovery of methane gas from wastewater)
- Development and commercialization of the solar-powered Automated Weather Monitoring System for urban and rural use
- Preparation of an industry roadmap to create an environment where the use of electric vehicles is highly promoted, encouraged and supported by both the Philippine government and society that would lead to eco-friendly and efficient transportation
- The Philippine Green Building Council (PHILGBC) was organized to serve as a single voice in the promotion of holistic and market-based green building practices by developing a green building rating system called Building for Ecologically Responsive Design Excellence (BERDE) Rating System, the Philippines' own green building rating system (*Source* http://philgbc.org/)
- Air conditioning system with comfort-cooling technology, energy-saving and durability features through their comfort cool and efficiently designed heat transfer system. The air conditioning unit has high energy efficiency ratio (EER) compared with other brands with the exact same horsepower
- Use of fiber cement (green building material) for the quick construction of affordable yet long-lasting homes to grapple with destructive

(continued)

Table 2 (continued)

• Clean Fleet Management—Improve the fuel efficiency and attain sustainable greenhouse gas (GHG) reduction on the business operations, specifically its 350 units of fleet

Sustainable energy

• Net metering service and provision of solar energy systems, commercial and residential clients that are within the grid: Households can sell excess electricity generated through renewable energy (i.e., solar power) through credit system
• RAMCAR/Motolite's development of deep cycle batteries for solar panels to generate solar power. A solar facility was set up for testing
• Meralco Power Academy's SOLAR PV Boot Camp—training and information dissemination on new technology in the energy sector

Knowledge and capacity development

• Disaster related computer applications, such as NABABAHA.COM, a system that enables real-time flood simulation
• SHELL ECO-MARATHON (SEM) program on smarter mobility, the engineering student teams whose unique vehicle design and construction, travel the farthest distance using the least amount of energy
• Bayer Young Environmental Envoy (BYEE) Program—College students participate in a five-day Eco-Camp. Resource persons from government, business, academe, the private sector, and non-governmental organizations talk about what their respective sectors are doing for the environment
• In situ education for ecovillage design and community resilience by establishing a doable and replicable model of sustainable living, which shall serve as teaching demonstration area

Climate financing

• Endowments and grants for CCA initiatives
• Financing instruments, e.g. credits and guarantees for projects and public private partnerships
• Technical assistance to projects, e.g., capacity building, consultation
• BPI's funding of research study "Business Risk Assessment and the Management of Climate Change Impacts"
• Assess the vulnerability levels of cities prone to the impacts of climate change. Cities—Cebu, Davao, Baguio and Iloilo (2010); Cagayan de Oro, Zamboanga, Laoag and Dagupan (2012), Angeles, Tacloban, Naga and Bacolod (2013)
• IFC's advisory services
• (BDO) giving the local private sector appropriate financing for sustainable energy investments
• (Manila Water) public-private partnership advisory support from IFC
• $60 million in long-term debt financing in 2003 and another $15 million in equity in 2005 to Manila Water
• BDO Capital's collaboration with four local commercial banks to finance a syndicated term loan for the 67.5 MW wind power project in Pililla, Rizal
• BPI loan financing of Central Mall Biñan solar rooftop project
First solar project that was granted loan financing by BPI
Due to Solar Philippines' BOI accreditation, the project enjoys tax exemptions and credits
Purchase power agreement between Solar Philippines and Premiumlink: energy generated by the solar power plant is sold to the mall for its own consumption at a lower rate
Reduces the mall's electricity bills by Php 100,000

Source Villamejor-Mendoza (2014) and DENR (2013) as cited in Villamejor-Mendoza (2015)

by working on various initiatives and doing all they can, along the key priority areas of the National Climate Change Action Plan, to help address this development concern.

The businesses in our country have been doing research to address the consequences of higher temperatures, drought, flooding, earthquake and the like to ensure steady production and reduce losses. They have been engaged in knowledge sharing and capacity building to help in awareness raising on the dangers, prevention and interventions to mitigate the adverse effects of this phenomenon. In addition, or beyond these awareness-raising activities, they themselves have reduced water consumption, invested in water use technologies and adopted recycling to push for sufficiency and sustainability of water resources. They have master-planned communities and adopted green architecture and green building codes and regulations in order to develop climate smart and resilient communities. They have also harnessed solar and other renewable energy and ventured into climate financing to help those adversely affected by climate change.

6   The Messages in Doing Climate Change Solutions of Some Private Companies

So why do business companies in the country do these climate change solutions and initiatives? What underlying message/s or philosophies do these activities underpin?

A cursory look at snippets of information on some of the climate change solutions and initiatives selected business companies have done in the past reveals the following (Table 3):

As inferred from these snippets, it can be surmised that the private sector in the country, having realized the risks and negative impacts of climate change not only in their own businesses but also in the host communities they serve, has done its part to be resilient, energy efficient in its operations, mindful of better green technology that works for them. They have also integrated sustainability in their operations because "Green" is a strategic element for strengthening the competitiveness of Philippine industries (Gutterer 2015). They have also undertaken in addition to internal efficiency and productivity, good corporate governance and responsibility, projects for environmental protection and conservation in their host communities. They have as well helped in the disaster risk reduction of their wider communities.

ABS CBN, the media conglomerate has its Bantay Kalikasan (Nature Watch) program, estero-cleaning, tree planting and monitoring of extractive industries like mining, programs. It also does disaster rehabilitation programs. Generally, it does all these "to implant in the minds and hearts of Filipinos that taking care of the environment is the path to quality of life, that livelihood does not have to come from environmentally destructive practices, … and to rebuild lives and communities

Table 3 Some messages in the private sector initiatives on climate change

Company/study	Message or take on climate change
WWF-Philippines and a Consortium of Philippine Business Companies-Earth Hour 2012 Video (https://youtu.be/pUNyPq7hgvc)	"We are talking about one future, one planet" "We are in it together" "We can make the world better"
"Greening the Philippine Manufacturing Industry Roadmap. Strengthening Systematic Competitiveness and Fostering Inclusive Growth" (Gutterer 2015)	"Green" as a strategic element for strengthening the competitiveness of Philippine industries "Good business practice = integral part of global business relations" "Green" makes industries resilient to climate change impacts "Green" reduces ecological footprint
Bank of Philippine Islands (http://www.bpifoundation.org/environment/how-ready-are-we-for-climate-change) Founded in 1851, Bank of the Philippine Islands is the first bank in the Philippines and in the Southeast Asian region. BPI is a universal bank and together with its subsidiaries and affiliates, it offers a wide range of financial products and solutions that serve both retail and corporate clients BPI's services include consumer banking and lending, asset management, insurance, securities brokerage and distribution, foreign exchange, leasing, and corporate and investment banking The bank has a network of over 800 branches in the Philippines, Hong Kong and Europe, and close to 3000 ATMs and CDMs (cash deposit machines) The establishment of BPI, originally known as El Banco Español Filipino de Isabel II, ushered in the start of the Philippine banking and finance industry. The bank performed many functions, from providing credit to the National Treasury to printing and issuing currency, making it in effect the country's first Central Bank. BPI proudly carries on this tradition, financing many private and public sector initiatives and enterprises in support of economic growth and nation building BPI is acknowledged as a leading provider of financial services in the Philippines. One of its visions is service to country. It believes that it is its responsibility to be inclusive and responsible in nation building. Through BPI Foundation, BPI is committed to the welfare and sustainability of the communities it	"Progress can't be achieved without transforming our behavior towards the ENVIRONMENT Climate change impacts many aspects of development like food security, livelihood, and governance To better prepare communities, we work with experts in developing solutions for sustainable cities—from helping businesses adapt to climate risk to working with local governments for disaster-readiness We also work with smallholder farmers in increasing their lands' yield through sustainable farming practices" "We can climate-proof our cities if we act decisively. By gearing up for climate change, we can protect businesses and save lives," said Fidelina Corcuera, Executive Director of BPI Foundation

(continued)

Table 3 (continued)

Company/study	Message or take on climate change
serves (https://www.bpiexpressonline.com/p/ 1/776/about-bpi) Its climate change solutions and initiatives are carried out through its corporate social responsibility (CSR) arm, The BPI Foundation	
ABS-CBN Lingkod Kapamilya (https:// corporate.abs-cbn.com/lingkodkapamilya/ operation-sagip/about) ABS-CBN Corporation is the Philippines' leading media and entertainment organization. Primarily involved in television and radio, the Company has since expanded and now owns the leading cinema and music production/distribution outfits in the country and operates the largest cable TV service provider as well ABS-CBN also has business interests in merchandising, licensing, mobile and online multimedia services, publishing, video and audio post production, overseas telecommunication services, money remittance, cargo forwarding, TV shopping services, food and restaurant services, theme park development and management, and property management. These investments are all aimed at making positive changes in the media landscape and strengthening the fiscal position valued by our stakeholders (http:// corporate.abs-cbn.com/about-us) Its climate change solutions and initiatives are carried out through its corporate social responsibility (CSR) arm, its ABS-CBN Foundation	For the Lingkod Kapamilya program of the ABS-CBN Foundation: "Our mission is to conduct disaster relief and humanitarian assistance in calamity-stricken areas with the aim of enabling affected families to rebuild their lives and communities to start sustainable rehabilitation programs" For its Bantay Kalikasan (Nature Watch): "The goal is to implant in the minds and hearts of Filipinos that taking care of the environment is the path to quality of life, that livelihood does not have to come from environmentally destructive practices"
SM and SM Foundation SM is a leading Philippine company that is invested in market leading businesses in retail, banking and property. It also invests in ventures that can capture the high growth opportunities in the emerging Philippine economy. It looks for market leaders or those with potential to become leaders in their chosen sectors that offer synergies and attractive returns and cash flows SM was listed on the Philippine Stock Exchange in 2005, and owns blue chip listed firms SM Prime Holdings, BDO Unibank and China Banking Corporation	"Working together for a sustainable future" "SM recognizes that global sustainability practices will evolve alongside the needs of the various stakeholders of the company which include its customers, employees, its host communities and its investors. In line with our vision to be a catalyst for positive long-term change, we remain committed to do our part in creating an even more sustainable future." Harley T. Sy, President "SM's continued sustainability as a business relies heavily on the partnerships that we've established with the communities that we operate in, as well as our other various stakeholders. It is also dependent on how we

(continued)

Table 3 (continued)

Company/study	Message or take on climate change
SM's retail operations are the country's largest and most diversified with its food (SM Markets, WalterMart, Alfamart), non-food (THE SM STORE) and specialty retail stores which are leading players that provide consumers with an aspirational lifestyle, reliable service, quality products and consistent convenience SM Prime has become one of the largest real estate conglomerates in the country and in Southeast Asia with interests in malls, residences, office buildings, resorts, hotels and convention centers. SM Prime is the Philippines' largest mall developer, both in terms of gross floor area (GFA) and geographical reach where it operates over 50 malls. In China, the company's six malls are thriving in second- and third-tier cities, a strategy that the company will maintain as it grows further in numbers A dominant player in the Philippine residential business, SM Prime has offerings mainly in key cities in Metro Manila— Quezon City, Mandaluyong, Pasay, Pasig, Makati, Paranaque and Taguig. True to its origins, SM Prime caters to the dreams and aspirations of its customers by offering affordable luxury in convenient locations SM has the largest footprint in the Philippines through BDO Unibank, Inc. and China Banking Corporation. BDO is the Philippines' largest bank in terms of total resources, loans and deposits and is also the market leader in most key business segments such as investment banking, asset management, wealth management, remittances, credit cards, insurance and leasing. China Bank is uniquely positioned to service the needs of emerging companies and small- and medium-scale enterprises building on its long-term relationships that date back to the post-war era. Its acquisitions of two banks in the last three years signal China Bank's desire to grow further in scale, market reach and product base (http://www.sminvestments.com/our-company) Its climate change solutions and initiatives are carried out through its corporate social responsibility (CSR) arm, The SM Foundation (http://www.sm-foundation.org)	further enhance the environment for greater progress and sustainability" "SM continues to find ways to provide Filipinos with various opportunities for growth. Equally important is our impact on the environment and the company's thrust to becoming a progressive leader in sustainable business practices in the country. With programs tailored to conserve our precious resources, SM continues to safeguard the environment for future generations"

(continued)

Table 3 (continued)

Company/study	Message or take on climate change
Manila Declaration 2015 is the Philippine private sector's statement of solidarity to lower emissions, to help increase the resiliency of communities against the consequences of a changing climate, and to address collective sustainable development challenges while embedding sustainability into business processes. Signatories to the Declaration are: Philippine Business for Social Progress (PBSP), Philippine Business for the Environment (PBE), Philippine Chamber of Commerce and Industry (PCCI), Management Association of the Philippines (MAP), Federation of Philippines Industries (FPI), and the Financial Executives of the Philippines (FEP) There is also the Water Alliance, which is the private sector's response to addressing the adverse impact of the El Nino phenomenon. Water scarcity is being experienced in many parts of the country, as water withdrawal per person per year has more than doubled in recent years. To better address this problem, the Alliance will develop a roadmap with target and metrics, monitor progress, and evaluate its effectiveness. The other members of the Alliance are: Shell, Coca-Cola FEMSA, LafargeHolcim, Dow Chemical Pacific Ltd., HSBC, Meralco, Lopez Group, Maynilad Water, Manila Water, Unilever, Pepsi Cola Products Philippines, Roxas Holdings, Nestlé, Smart Communications and Splash (http://maniladeclaration.thepbe.org/downloads/Manila%20Declaration%20Brochure.pdf; http://aboitizeyes.aboitiz.com/aboitiz-signs-2015-manila-declaration-joins-water-alliance/)	"Climate change is a real, present and future threat to everyone...has to be addressed with urgency through global and local mitigation and adaptation solutions" "While all sectors of the society must act.... The private sector in partnership with government and other stakeholders, can play a more significant role in bringing effective, long-term solutions ..." "We commit...to develop and implement mitigation strategy as part of our core business, in our operations and in our value chain" "Develop solutions to increase the resilience of our communities resilient..." "Continue to embed sustainability in our business and continue to innovate products and services to address our collective sustainable development challenges"

As inferred from the program profiles and reports found in a handful of open educational resources and websites of companies with climate solutions or initiatives that were accessed by the author from July to September 2017. This cursory analysis is by no means comprehensive or representative of all business companies in the Philippines

through sustainable rehabilitation programs" (https://corporate.abs-cbn.com/lingkodkapamilya/operation-sagip/about).

BPI, the banking pioneer and the largest in the country, believes that "Progress can't be achieved without transforming our behavior towards the ENVIRONMENT." Thus, it "works with experts in developing solutions for sustainable cities—from helping businesses adapt to climate risk to working with local governments for disaster-readiness. It also works with smallholder farmers in increasing their lands' yield through sustainable farming practices." In general, it believes that "We can climate-proof our cities if we act decisively. By gearing up for climate change, we can protect businesses and save lives," (Fidelina Corcuera, Executive Director of BPI Foundation; http://www.bpifoundation.org/environment/how-ready-are-we-for-climate-change).

SM, another conglomerate, which is into retail, property and banking, has integrated environmental concerns in its operations and believes that "SM's continued sustainability as a business relies heavily on the partnerships that it has established with the communities where it operates in, as well as with its other various stakeholders. It is also dependent on how it further enhances the environment for greater progress and sustainability."

"SM continues to find ways to provide Filipinos with various opportunities for growth. Equally important is its impact on the environment and the company's thrust to becoming a progressive leader in sustainable business practices in the country. With programs tailored to conserve our precious resources, SM continues to safeguard the environment for future generations." Generally, its credo is "together, we can work for a sustainable future" (http://www.sminvestments.com/our-company).

SM also takes a leadership role in spearheading the UN Private Sector Alliance for Disaster Resilient Societies (ARISE) initiative in the Philippines. It invited at least 100 companies to make disaster resilience part of its business models. The companies signed a commitment in 2015 to help raise awareness and to implement projects in disaster risk reduction (SM 2015 Creating a Sustainable Future).

Called the 2015 Manila Declaration, it is the Philippine private sector's statement of solidarity to lower carbon emission, to help increase the resiliency of communities against the consequences of a changing climate, and to address collective sustainable development challenges while embedding sustainability into business processes. They also forged a partnership named Water Alliance to address the adverse impact of the El Nino phenomenon. Water scarcity is being experienced in many parts of the country, as water withdrawal per person per year has more than doubled in recent years (http://maniladeclaration.thepbe.org/downloads/Manila%20Declaration%20Brochure.pdf).

The private sector's earlier collaboration in 2012 with the World Wide Fund for Nature- Philippines has produced many information materials that impart messages that advocate, "We are talking about one future, one planet"; "We are in it together"; and "We can make the world better" (https://youtu.be/pUNyPq7hgvc). These reinforce the need to understand climate change as a problem we have to solve together in order to make the future better.

7 Concluding Statement

This paper mapped out some of the climate change solutions and initiatives of the private sector in the Philippines and showed that their engagement in climate change adaptation and mitigation, particularly in such areas as food and human security, water sufficiency, environmental stability, climate-friendly industries, sustainable energy and knowledge sharing and capacity building, has been substantial and far more than lip service. They have embraced the issue of climate change as affecting not only their business but also all of us. They have thus been doing all they can to help address this development concern, because of the belief that "we are in it together" and that the impacts of climate change affect everyone of us.

The business sector has climate-proofed their businesses and cascaded this to the communities they serve, the latter as part of their corporate citizenship and social responsibility. The snippets of information collected above show that the business sector has realized early on that by doing their share amidst the climate vulnerabilities in the country and by tempering corporate greed, a better and sustainable future may be possible. They work out climate solutions because they generally believe "we only have one planet", "people and collective actions are needed" and "partnerships help." Let us hope more stakeholders realize the same, and as global citizens mindful of the dangers and risks of climate change, work together to create a better world.

Despite this rosy picture, more researches are needed to validate the reasons and philosophical underpinnings to private sector engagement in climate change adaptation and mitigation beyond awareness raising. Understanding the latter will more strategically guide this important sector's more meaningful engagement in creatively solving climate change.

References

Ansell C, Gash A (2008) Collaborative governance in theory and practice. J Public Adm Res Theor 18(4):543–571
ADB (2012). Investing in resilience: ensuring a disaster-resistant future. Manila
Asian Development Bank (ADB) (2005) Pacific studies series climate proofing: a risk-based approach to adaptation. Manila
Biagini B, Miller A (2013) Engaging the private sector in adaptation to climate change in developing countries: importance, status, and challenges. Clim Deve 5(3):242–252
CCC (Climate Change Commission) (2010) National framework strategy on climate change
CCC (2011) National climate change action plan. Philippines
CCC (2012) Climate change commission annual report. Philippines
DENR (2013) Climate change adaptation best practices in the Philippines
European Union and United Nations Office for Disaster Risk Reduction (UNISDR) (n.d.) Fires in Europe fuelled by Urbanisation and Climate Change. At http://www.unisdr.org/archive/54477
GEF (Global Environmental Facility) (2012) Private sector engagement in climate change adaptation. Prepared by the GEF secretariat in collaboration with the International Finance

Corporation, GEF/LDCF.SCCF.12/Inf.06, 9 May 2012. https://www.thegef.org/sites/default/files/council-meeting-documents/Note_on_Private_Sector_4.pdf

Hitoshi K (2014) CCS-geo-engineering: the only one reasonable climate geo-engineering technology at present conference paper May 2014 Japan geoscience union meeting 2014 (JpGU2014)

IFC (International Finance Corporation) (2010) Climate change. Filling the financing gap. http://www.ifc.org/wps/wcm/connect/73baa8004aa80b2fa434f69e0dc67fc6/TOS_ClimateChange.pdf?MOD=AJPERES

IPCC (Intergovernmental Panel on Climate Change) (2012) Managing the risks of extreme events and disasters to advance climate change adaptation. Special report. Cambridge University Press. https://www.ipcc.ch/report/srex/

IPCC (2013) Climate change. The physical science basis. https://www.ipcc.ch/report/ar5/wg1/

IPCC (2014) Climate change 2014: impacts, adaptation and vulnerability at https://www.ipcc.ch/report/ar5/wg2/

Letchumanan R (2013) In: Koh K-L, Kelman I, Kibugiand R, Eisma R-L (eds) Adaptation to climate change. ASEAN and comparative experiences. www.worldscientific.com/worldscibooks/10.1142/9642

Sano Y (as cited in Oxfam 2014) We are at war with climate change and hunger: Yeb Sano at https://www.oxfam.org/en/multimedia/video/2014-we-are-war-climate-change-and-hunger-yeb-sano

SEI (Stockholm Environment Institute) (2010) Private sector finance and climate change adaptation. Policy brief at https://www.sei-international.org/mediamanager/documents/Publications/Climate-mitigation-adaptation/Atteridge-Private-sector-finance-PBupdate-101118-

UNWCDRR (United Nations World Conference on Disaster Risk Reduction) (2015) Highlights. Earth Negot Bull 26(12), 16 Mar 2015. http://enb.iisd.org/vol26/enb2612e.html

Villamejor -Mendoza MF (2014) Private sector engagement in climate change mitigation and adaptation: implications in regional governance. Paper presented at the plenary session of the 2014 EROPA international conference on public administration and governance in the context of regional and global integration held 19–24 Oct 2014 at Hanoi, Vietnam

Villamejor-Mendoza MF (2015) Private sector engagement in climate change mitigation and adaptation: implications in regional governance. J Polit Gov Coll Polit Gov 5(2):285–305 (Mahasarakham University, Thailand)

World Bank (2010) World development report and climate change. The International Bank for Reconstruction and Development/The World Bank 1818 H Street NW

PS Initiatives/Cases:

• DENR. 2012. Climate change adaptation best practices in the Philippines
• Philippine star
  • PLDT, Globe networks normal despite heavy rains, 20 Sept 2014:17
  • Concepcion J, Pilipinas now is our time: rise to the rice bucket challenge, 18 Sept 2014, p B-10
  • With power and responsibility to change lives, 22 Aug 2014:I-2
  • Chill at home with Panasonic air conditioners, 20 Sept 2014:D-3
  • Fiber cement offers perfect fit for affordable housing, 15 Aug 2014:B-10
  • Valencia C, Aboitiz putting up egg production facility, 15 Aug 2014:B-8
  • Largest solar rooftop panel powers up mall in Biñan, 21 Sept 2014:B-4
  • PLDT, Globe networks normal despite heavy rains, 20 Sept 2014:17
  • Beltran C, Solar sober, 22 Aug 2014:I-6
  • Torres L, A benchmark of generosity, 10 Aug 2014:G-1
  • Teehankee P, The pepper mill column, excellence in phil tourism, 19 Sept 2014:C-9
• http://www.bdo.com.ph/news-site

- http://philgbc.org/)
- Mercurio RS, Ortigas & Co. investing in modern technology with Nimble storage
- http://www.ifc.org/wps
- http://www.rappler.com/move-ph/39489-project-agos-climate-change
  - http://www.rappler.com/business/industries/173-power-and-energy/64165-solar-power-ph-house holds-net-metering
  - https://www.mybpimag.com/BPI_foundation/BPI_Foundation_Climate_Change_Project.html
  - www.boi.gov.ph
  - http://a-fab.org/climate-change/
  - https://youtu.be/pUNyPq7hgvc

Climate Change Messages

Gutterer B (2015) Greening the Philippine manufacturing industry roadmap. Strengthening systematic competitiveness and fostering inclusive growth. http://industry.gov.ph/wp-content/uploads/2015/03/greening-the-phil-roadmap.pdf
http://aboitizeyes.aboitiz.com/aboitiz-signs-2015-manila-declaration-joins-water-alliance/
http://corporate.abs-cbn.com/about-us
http://maniladeclaration.thepbe.org/downloads/Manila%20Declaration%20Brochure.pdf
http://www.bpifoundation.org/environment/how-ready-are-we-for-climate-change
http://www.sm-foundation.org
http://www.sminvestments.com/our-company
https://www.bpiexpressonline.com/p/1/776/about-bpi
https://corporate.abs-cbn.com/lingkodkapamilya/operation-sagip/about
https://youtu.be/pUNyPq7hgvc

Lessons Learned About the Hindering Factors for Regional Cooperation Towards the Mitigation of Climate Change

Pınar Gökçin Özuyar

Abstract As the importance of climate change mitigation and adaptation increases, tools to assist these ranging from training materials, awareness raising event models to company level cooperation tools are being introduced to various stakeholders. These tools can only be effective by extensive utilisation throughout the globe which requires the communication and awareness raising on climate change. The actual implementation and impact assessment of these tools need to be further investigated. Opportunities and barriers for the use of such tools and whether climate change communication is an enhancing or hindering effect is very important in this investigation. As an example for such a tool, an industrial symbiosis model where an unorthodox regional approach is taken rather than close proximity cooperating companies, has been implemented in the Western Black Sea Region countries. The results of the study include three major barriers; namely, lack of regional policy and relevant legislation, trust among companies and a common working language in the region. The effects of other barriers and possible opportunities that would hinder these barriers are discussed in this study including the lack of regional policies on climate change based on one-to-one interviews with selected company representatives in the region. The lessons learned are significant for similar regional exemplary tools of sustainable development and climate change mitigation practices.

1 Climate Change, Denial or Acknowledgment

As Mc Cright et al. (2016) states over the last three decades, climate change has become publicly defined as an important social problem deserving action and a substantial body of social science research examines the patterns of climate change views in the general publics of countries around the world. This differentiation of

P. G. Özuyar (✉)
School of Business, Ozyegin University, Orman Sok, Alemdag,
Cekmekoy 34794, Turkey
e-mail: pinar.ozuyar@ozyegin.edu.tr

© Springer Nature Switzerland AG 2019
W. Leal Filho et al. (eds.), *Addressing the Challenges in Communicating Climate Change Across Various Audiences*, Climate Change Management,
https://doi.org/10.1007/978-3-319-98294-6_7

the opinion of public among or within countries have led to climate scepticism in turn slowing the climate change adaptation and mitigation efforts. Furthering to this diversion of opinions, the US status on climate change is described as an obstructionist role in international climate negotiations and its effects on global public opinions (Mc Cright et al. 2016).

In climate change communication, the interaction between a population's attitudes and government's policies may differ in various communities. For the individual's or public's opinion's to form studies have been carried out to understand the factors shaping the attitudes. As opinions and attitudes are usually based on knowledge, Thompson (2017) collected the opinions of 746 respondents by a 20 question survey and showed that results reflect certain misconceptions of climate change, and is useful for investigators to begin forming opinions of the public's knowledge regarding the potentially inflammatory topics of climate change, greenhouse gases, and geo-engineering.

Capstick and Pidgeon (2014) argue that two main types of climate change scepticism or denial should be distinguished: epistemic scepticism, relating to doubts about the status of climate change as a scientific and physical phenomenon; and response scepticism, relating to doubts about the efficacy of action taken to address climate change. Whilst each type is independently associated by people themselves with climate change scepticism, the latter is more strongly associated with a lack of concern about climate change. As such, additional effort should be directed towards addressing and engaging with people's doubts concerning attempts to address climate change.

One study carried out by Tvinnereim et al. (2017) support the understanding of the dynamics of the public's opinion on climate change and basing their argument that most public opinion research uses either closed questions about agreement with various pre-determined statements or use open-ended questions eliciting generic associations with climate change, they have used an open-ended survey question in a probability-based Internet survey panel, analysing 4634 textual responses to the question of "what should be done" about climate change by inducing seven topics: Transportation, energy transition, attribution of climate change, emission reduction, the international dimension, lifestyle/consumption and government measures. Results indicated a willingness to accept stronger mitigation action, but also that central and local governments need to facilitate low-carbon choices, bridging policy and individual action to mitigate climate change.

A similar study is carried out by Kaltenborn et al. (2017) by a survey to examine how cultural resources and trust in environmental governance institutions are related to attitudes toward climate change. The belief in climate science turned out to be irrelevant to an individual's political, professional and intellectual orientations, as well as life histories. This result suggest the importance of climate change communication since improved knowledge about the social basis for climate change is an imperative part of futures-oriented expertise, leaving the ground for a higher impact opportunity.

Climate Change communication may take on many forms and has tools including the use of media representations. A comparative analysis for the years

1996–2010 is presented by Schmidt et al. (2013) in 27 countries including among others, countries that have committed themselves to greenhouse gas emission reductions under the Kyoto Protocol such as Germany as well as countries that are strongly affected by the consequences of climate change like India. The analyses show that climate change coverage has increased in all countries. Still, overall media attention levels, as well as the extent of growth over time, differ strongly between countries. Media attention is especially high in carbon dependent countries with commitments under the Kyoto Protocol. Another tool for climate change communication is the use future scenarios. Dulic et al. (2016) studied the challenge to embed community perspectives in a communication process of climate change solutions by 3D interactive simulations using design inquiry as a development process. Similarly, Glass et al. (2015) propose a web-based tool aimed at increasing the adaptive capacity among Nordic homeowners. Based on the results from continuous user-testing and focus group interviews we outline lessons learned and key aspects to consider in the design of tools for communicating complex issues such as climate change effects and adaptive response measures. Films on climate change is yet another tool suggested for climate change communication (Manzo 2017).

2 Industrial Symbiosis as a Tool of Industrial Ecology to Assist Climate Change Mitigation

Industrial ecology is defined as an approach to the industrial design of products and processes and the implementation of sustainable manufacturing strategies. It is a concept in which an industrial system is viewed not in isolation from its surrounding systems seeking to optimize the total materials cycle from virgin material to finished material, to component, to product, to waste product, and to ultimate disposal (Jelinski et al. 1992). Albeit, there are many tools of industrial ecology, all serve to protect industries' financial sustainability–companies' economic performance–and at the same time prevent any damage to the environment including climate change.

Industrial Symbiosis, a sub-discipline and one of the tools of industrial ecology, is defined explicitly in the literature as: "Industrial Symbiosis engages traditionally separate industries in a collective approach to competitive advantage involving physical exchanges of materials, energy, water and/or by-products. The keys to Industrial Symbiosis are collaboration and the synergistic possibilities offered by geographic proximity (Chertow 2000)."

- There are other attributes of Industrial Symbiosis that are generally accepted such as;
- A systems view of the interaction between industrial and ecological systems (Boons and Baas 1997);
- The study of material and energy flows and transformations (Hart et al. 2005; Wolf 2007; Zhao et al. 2008);

- A change from linear (open) processed to cyclical/closed processed, so that the by-product from one industry is used as an input for another (Van Berkel et al. 2009; Frosch and Gallopoulos 1989);
- An effort to reduce industrial systems' environmental impact on ecological systems;
- The idea of making industrial systems emulate more efficient and sustainable natural systems.

Industrial Symbiosis is used to define the physical exchange and shared management of input and output materials by geographically neighbouring firms (Chertow et al. 2008). Industrial Symbiosis is best conceptualized as a process that is established by a group of companies located in proximity to one another (e.g. Eco-industrial parks), where companies can cultivate material, energy and social networks. Companies engaged in Industrial Symbiosis are said to belong to an industrial ecosystem. It also describes partnerships among firms in a region to pursue broader strategies for sustainable industrial development (Baas and Boons 2004).

Industrial Symbiosis is developed as a cooperative, multi-industrial approach to improve economic, social and environmental performance that leads to sustainability. Industrial Symbiosis is created with the economic motivation, such as cost minimization; as well as by environmental ones, such as accessing limited water and energy supplies. Companies engage in Industrial Symbiosis collectively for competitive advantage and simultaneously realize economic and environmental benefits. It is generally assumed that Industrial Symbiosis is a "win-win" situation due to the fact that it both generates economic benefits for the companies involved and reduces environmental impact (Chertow 2007).

Economic benefits include improving resource productivity by; reducing input, production, and waste management costs, and by generating additional income due to value added to by-product streams, reducing resource use, improving relationships with external parties, and by enabling development of new products and their markets and generating new employment, and helping to create a safer and cleaner natural and working environment. Environmental benefits include improving eco-efficiency of production and consumption process mainly by transforming the by-product of one firm into the valuable input of another, which changes the material flows from a linear one into a closed-loop.

Chertow (2000) defined the three types of symbiotic transactions as;

- Type 1: Utilization of waste and by-products as a raw material input: In the Industrial Symbiosis context, a group of companies exchange materials, such that the waste or by-product of one becomes an input for another. It is the exchange of firm-specific materials between two or more parties for use as substitutes for commercial products or raw materials.
- Type 2: Sharing the utilities or access to services: Companies that belong to an industrial ecosystem share the utilities such as energy, water, or wastewater treatment to reduce costs and improve resource productivity and environmental

performance. It is the pooled use of resources such as energy, water, and wastewater.

- Type 3: Cooperating on issues of common interest and sharing secondary services: Companies can cooperate on issues of common interest such as emergency planning, training or sustainability planning. Companies can also share ancillary services, such as transportation, landscaping, and by-product collection.

One of the earlier examples of industrial symbiosis is the Kalundborg example dated from 1980s. In Kalundborg, Denmark an Industrial symbiosis network exists where companies in a region collaborate to use each other's by-products and otherwise share resources. At the center is a coal-fired power plant, which has material and energy links with the community and several other companies. The Kalundborg example is a Type 1 example where all interactions take place in a close proximity (Valentine 2016). The eco-industrial parks are similar practices.

3 Challenges and Issues of Climate Change Mitigation as a Common Goal in the Black Sea Region

The Black Sea Region includes the greater part of 17 countries, six of which border directly on Black Sea. The population of the whole region is some 320 million. Total current Gross Domestic Product (GDP) of the region is US 3.416 billion.

The Black Sea region is the subject of intense debate during the last decades. The changing dynamics of the region, its complex realities, the interests of outsiders and the region's relations with the rest of the world make the region a challenged neighbourhood. Its strategic position, linking north to south and east to west, as well as its oil, gas, transport and trade routes are all important reasons for its increasing significance. The wider Black Sea area is of increasing political and economic importance for the European Union, presenting unique challenges and opportunities. The development of bilateral relations with all the countries, the launching of the European Neighbourhood Policy and the recent EU enlargement (Bulgaria, Romania) has considerably strengthened the European Union's involvement in the area.

However, the region's real priorities and needs are still being largely ignored despite intensified interest in the area. The surrounding countries have fast economic growth and considerable production compared to developed countries, which positively leads to regional development, but also has significant negative environmental impacts. Thus, the region, with all of its challenges and priorities, should be re-evaluated. This will provide all actors involved with a better understanding of what can be done, as well as allowing them to develop innovative approaches to problems, thus enhancing the region's stability and welfare.

Hereby, regional cooperation, sustainable development and environmental protection of the region are evaluated as the most crucial agendas of the Black Sea
region. Our objective is to propose an innovative tool—Industrial Symbiosis–to
promote the sustainable development of the countries in the region through
responsible management of natural and other resources and proper care for the
integrity of the environment. The Black Sea region has enormous wealth in terms of
resources and biodiversity, but it also faces significant environmental problems that
can only be tackled through regional cooperation and with substantial support from
the international community. Moreover, environmental challenges in the region will
increasingly become a matter of immediate concern for the EU following the next
rounds of enlargement (Gavras 2010).

Due to the fact that the countries of the Black Sea region are diverse in terms of
size, economic structure, and level of development, there may be challenges which
are particular to a country or a small sub-set of countries. This diversity and the
consequent difference in outlooks of each country make it difficult to identify issues
that may be considered as challenges for the region as a whole. Instead, it neces-
sitates identifying select macro or broad-based challenges, which then likely take
specific forms for each country, depending upon its particular characteristics.

In environmental and health terms, the Black Sea region suffers from very acute
problems. The Black Sea is subject to pressure from irrigation, industry, fishing,
tourism, power generation, navigation and not the least as the final destination of
urban wastewater. These intensive uses have created severe problems of water
quality and drastically reduced biodiversity in the basin. The pollution ends up in
the Black Sea and affects a very large area. Major problematic areas affecting the
environmental state of the Black Sea have been identified as pollution, loss of
biodiversity and coastal degradation. Furthermore, various effects of climate change
on human health to fishermen have been studied for some time (Ebi et al. 2017;
Delgado and Li 2016) although studies on the Black Sea basin as a whole is scarce.

4 Regional Cooperation in the Black Sea Basin

Promoting regional cooperation in a region divided by the Cold War heritage and
mentality is a difficult task. Especially, bringing the regional countries together to
embrace the benefits of globalization is a real challenge. Common interests such as
trade, transportation, tourism and energy should be considered to overcome the
division within the region (Aydin and Triantaphyllou 2010).

Regional cooperation provides general framework within which innovative
solutions to these problems could be more easily found. As the regional cooperation
in the Black Sea has been essentially an extension of the EU's philosophy that
deeper cooperation with neighbouring countries can provide national as well as
regional stability and growth, serving mutual interests of all countries concerned,
the regional approach in the Black Sea might even be more successful than the

other regions that has already been tied to the EU. There are existing economic networks and nongovernmental organizations on green business and environment in the region promoting regional cooperation in the Black Sea Region. Black Sea Economic Cooperation (BSEC), Black Sea Trust for Regional Cooperation, Black Sea Caspian Business (formerly known as Union of Black Sea and Caspian Confederation Enterprises), The Transport Corridor Europe-Caucasus-Asia (TRACECA) and Black Sea Regional Energy Center are the major existing networks.

The regional cooperation is an issue that may certainly grow in magnitude, as the evolution of relations among neighbours is always of relevance and there is much scope for mutually beneficial cooperation around the Black Sea. Challenges to regional cooperation efforts include; persistence of unresolved conflicts, need to generate trust and political commitment among leaders, lack of financial and institutional resources, need to engage the private sector and civil society, and fragmented nature of regional organizations. Another area is the enhancement and systematization of policy dialogue in key sectors. Its key challenge is to upgrade the level of participation, and the timeliness and quality of the information exchanged within the countries so that the relevance and usefulness of the dialogue is acknowledged by the participants. Sectors such as transport, energy, telecommunications, trade facilitation and the environment lend themselves to discussion and formal exchanges, and possibly coordination.

5 Methodology of the Study

The Black Sea Basin is a region that is in continuous development which in turn yields problems the management of natural resources and industrial activities, waste generation and disposal, energy efficiency, demand on raw material production have become urgent issues. At the same time, it is listed as one of those regions that will be heavily affected by the climate change (Davy et al. 2017) Adapting to climate change, minimizing environmental degradation but still having a prosperous and resilient economy will encourage the sustainable development in the whole of the Black Sea Basin. Nationwide strategies and measures can only have a limited effect on this development. It is evident that only by regional cooperation and coordination in the Basin on innovative technologies and systems; a solution can be achieved for the mutual benefiting of all basin countries.

However, since there are several countries, regions, administrations, organizations and people managing and benefiting the resources, "the tragedy of commons" (a dilemma in which multiple individuals acting independently and solely and rationally consulting their own self-interest will ultimately destroy a shared limited resource even when it is clear that it is not in anyone's long term interest for this to happen (Hardin 1968) is clearly seen in the case of management of natural resource of the Black Sea region and unfortunately it cannot be overcome as a result of many different interests of many stakeholders.

This study entails one of the tools for the mitigation of climate change in the Black Sea Region. Industrial Symbiosis Network for Environmental Protection and Sustainable Development in Black Sea Basin Project (SymNet) was a project supported by the European Union under Joint Operational Programme "Black Sea Basin 2007–2013" and aimed contribute to minimize the environmental degradation while maximizing economic and social development in Black Sea Basin by establishing Industrial Symbiosis system. The overall objective was to create a new system from which both producers and consumers will benefit while decreasing the environmental footprint on Black Sea Basin. With this understanding, stakeholders from Turkey, Bulgaria, Romania and Moldova developed the basic principles of an Industrial Symbiosis program platform, where by enhancing cooperation between industries and decision makers of the Black Sea Basin, a mutual benefit for trade as well as environment can be gained. It was particularly important that local and regional economic development need to be promoted while limiting the negative impacts of exploitation, production, consumption and waste disposal on natural resources.

As an output of the SymNet Project Black Sea Industrial Symbiosis Platform (BSISP) was created aiming to generate a medium to enhance communication and cooperation between the Black Sea basin countries. It is supported by a database of member companies as well as NGOs from the above mentioned countries. BSISP was developed to create a platform for opportunities for the region's companies to network and cooperate; online tools are developed like carbon foot-printing and trade networks optimization to support the regional cooperation and the increasing the use of the platform. BSISP was designed not only as a passive source of information and by using its tools a private determination of alternatives but also a tool for easy communication. The need for creating further tools for the BSISP members resulted in the two tools that were developed: trade optimization tool and the carbon footprint tool.

The first tool of BSISP is the Trade Optimization Tool (TOT) developed as a free tool to indicate the economic costs associated with logistics and transportation of goods and people in the region. The tool aims to present indicative results for comparison of options for a business owner in a very simple and efficient way. It provides the following selection criteria to get better results; type of transport, vehicle type, fuel type, load type, routes including stops and the vehicle occupancy ratio. For the moment, all calculations are based on land transport and a quick comparison is provided for sea and air transportation. The tool has been developed considering real life constraints and limits the options that are not possible in an actual case. During a series of runs with different choices, each result is listed making it easier to compare. The TOT is linked to an online source to use momentary fuel prices in its model.

The second BSISP tool is the Carbon Footprint Tool (CFT) developed as another free tool to indicate the carbon emissions associated with logistics and transportation of goods and people in the region. CFT aims to present indicative results for comparison of options for a business owner in a very simple and efficient way. It provides the following selection criteria to get better results; type of transport, vehicle type, fuel type, load type, routes including stops and the vehicle occupancy ratio. The tool has been developed considering real life constraints and limits the options that are not possible in an actual case. During a series of runs with different choices, each result is listed making it easier to compare.

This study presents the results about the opinion and attitudes of the member companies on cooperation and awareness of industrial symbiosis as a tool for climate change. Their overall perception on climate change and the much known environmental pollution are investigated.

The means to end as to whether climate change communication and understanding assists in wider use of the industrial symbiosis concept or on the contrary by using industrial symbiosis and achieving financial gains resulting in a direct effect on climate change mitigation is analysed.

Grounded theory was selected as the qualitative analysis method as a suitable approach to use, especially when large quantities of unstructured or semi-structured qualitative data need to be analysed (Glaser and Strauss 1967). The detailed write-up of the cases and all the data generated by interviews, and documentation were examined and coded by focusing on the factors that influence adoption. Core-categories, Subcategories and Codes are determined. Rather than an affirmative/negative approach, a more indicating percentage is used.

The below questions were asked to the interviewees without the assistance of keywords that may

- What are the opportunities for industrial symbiosis in the Black Sea Region?
- What are the barriers for industrial symbiosis in the Black Sea Region?

6   Results with Lessons to Take Home

The 112 companies who are among more than 300 BSISP members have been interviewed in person or over video conferencing. All the interviewed companies are SME's operating in either of the following sectors, logistics, energy, manufacturing, tourism. These sectors are the initial sectors selected for the SymNet project and they are the initial members of the BSISP, receiving ample information on industrial symbiosis as a tool for climate change prevention by project bi-monthly newsletters, the project and platform websites, stakeholder meetings and public conferences. All the interviewees were middle to high management employees.

As mentioned before, the detailed write-up of the cases and all the data generated by interviews, and documentation were examined and coded by focusing on the factors that influence adoption. Core-categories and Codes are determined. Rather than an affirmative/negative approach, a more indicating number of companies referring to the code based expression is used.

Table 1 What are the opportunities for industrial symbiosis in the Black Sea Region?

Categories	Codes	Number of companies
Social	Increase in opportunities for intra-regional cooperation in economic, human and social	112
Operational	Increase in the exchange of knowledge to find common solutions to the mutual operational problems	112
Environmental	Protecting the environment	112
Environmental	Improving the environmental performance of the firms engaged in symbiotic actions	112
Financial	Reducing operational costs (e.g. raw material, logistics, and marketing)	112
Financial	Increase gains from improved trade	112
Financial	Increasing competitive advantage	111
Environmental	Assisting in climate change prevention and adaptation	101
Financial	Ensuring financial stability	109
Financial	Increased contribution to economic growth in the region	99
Financial	Opening up new business opportunities	97
Environmental	Reducing use of scarce resources	88
Regulatory/ legislative	Awareness of environmental legislation related to international trade	77
Financial	Earning new revenue	76
	Improving harmonization of quality standards across the region	64
Operational	Improving efficiency of production and consumption processes	51
Social	Generating new employment	46
Operational	Increased disaster preparedness and cooperation during emergencies and extreme weather conditions'	43
Social	Preserving peace and security in the region	23

Analysis of Tables 1 and 2 simultaneously, indicates that companies are aware of the opportunities but hesitate due to barriers. All 112 companies have stated that industrial symbiosis is likely to have a positive impact on financial gains as well as protecting the environment.

A majority of them has established the link between environmental protection and climate change, however the lack of mentioning climate change prevention as an outcome of industrial symbiosis is less relevant than the imminent environmental protection. Further analysis of establishing the links between climate change and the operational aspect of 'disaster preparedness during emergencies and extreme weather conditions' indicates that companies, by already being a member of BSISP

Table 2 What are the barriers for industrial symbiosis in the Black Sea Region?

Categories	Codes	Number of companies
Regulatory/legislative	Different national legislations	112
Regulatory/legislative	Lack of common regional legislation/strategy for industrial symbiosis	112
Regulatory/legislative	Lack of common regional legislation for climate change and environmental pollution	112
Regulatory/legislative	Lack of common regional strategy for climate change and environmental pollution	112
Economic	Initial financial support in creating the industrial symbiosis	112
Economic	Lack of proven long-term examples for industrial symbiosis in the region	112
Economic	Trust in possible partner company	112
Social	Language barriers	110
Regulatory/legislative	Sector based legislative problems	92
Regulatory/legislative	Visa regulations	92
Economic	Lack of common insurance framework	87
Operational	Transparency on information sharing	83
Regulatory/legislative	Different legislations within the country	80
Economic	High production and operational costs	44
Economic	Major economic differences in the region	35
Economic	Corruption	22
Social	Different cultures	23

are aware of the industrial symbiosis related financial gains and environmental protection/climate change prevention but they have not established the operational understanding and have not embedded the industrial symbiosis thinking. For the lingage between climate change and industrial symbiosis, the need for a regulatory regional policy comes forward as a barrier. For the company's interviewed, climate change communication is expected to be done by governments and the responsibility of the private sector is not really understood.

The results on opportunities as well as barriers as a summary suggest that industrial symbiosis as a tool for climate change is only an auxiliary matter dealt as an external project and not embedded to company strategies indicating that climate change communication and awareness among companies of the region are still need to be enhanced. Additionally, for the Black Sea Basin companies that were interviewed an initial financial support mechanism is a trigger. Overall, this type of enhancement of a regional strategy associated can be accomplished by cross border legislation and goals on both industrial symbiosis and climate change.

7 Conclusion

In the Black Sea Region, severe usage of natural resources for economic development and social welfare made negative impact on the region's environment and jeopardizing future generations. Black Sea regional cooperation in the field of industrial symbiosis not only provides financial, operative and social returns but also on a larger scale assists climate change mitigation and adaptation efforts. Nationwide measures and strategies can only have a limited effect on this effort. It is evident that only by regional coordination and cooperation in the Basin on innovative technologies and systems; a solution can be achieved for the mutual-benefiting of all Basin countries.

For cross border industrial symbiosis challenges or even any other type of industrial cooperation, potential challenges exist about regulatory, economic and even social issues. Nevertheless, there is general awareness that Industrial Symbiosis offers generous opportunities to the stakeholders in the partner countries and the region.

In this study there were two folds in the discussion of climate change communication. The first one is the impact of climate change communication and awareness on the way companies perceive and adapt the climate change mitigation tools. The main regulatory barriers indicated in the results need to be resolved by the cooperation of region's governments so that a common policy as well as harmonised legislation is established not only for climate change but also environmental pollution. Overall, companies expected a top-down approach with complete policy and associated legislation to enhance the cooperation with the responsibility of climate change communication left to governments.

On the second fold, the company-level perception of responsibility in climate change and its effect on strategic decisions were reviewed. Companies did not present the need for industrial symbiosis for climate change mitigation but rather they were interested in the possible financial gains. For cooperation on industrial symbiosis in the region between companies and without the guidance of the governments, trust issue is significant. The language barriers and the related 'trust in possible partner company' can be diminished by creating possible regional or sectoral networks where companies of different countries will have the opportunity to interact on a personal and even face-to-face basis. This is where the guidance of governments does not cause a direct impact on the cooperation. A suggestion in non-governmental acknowledgment of the use of climate change tools for businesses is to keep the 'financial' aspect as relevant and significant as the aim for an overall good of the region in respect to climate change communication.

In a summary, the results indicate that there are many opportunities for enhancing regional cooperation by using industrial symbiosis while there are many others that need to be resolved standing in the way. The governments of the region as well as the business networks should collectively work towards enhancing of the first and minimizing of the second. When analysed, most of the points in the lists for opportunities and challenges are those issues that are not due to the culture or language barriers but those than can be relatively easily resolved with clever change in legislation and financial assistance. The main discussions regarding Industrial Symbiosis presented in this study could serve as a primary basis and decision support tool by both public and private decision makers, local and regional policy makers in achieving climate change awareness in the Black Sea region.

References

Aydin M, Triantaphyllou D (2010) A 2020 vision for the Black Sea region: the Commission on the Black Sea proposes. Southeast European and Black Sea Studies 10(3):373–380

Baas LW, Boons FA (2004) An industrial ecology project in practice: exploring the boundaries of decision-making levels in regional industrial systems. J Clean Prod 12:1073–1085

Boons FA, Baas LW (1997) Types of industrial ecology: the problem of coordination. J Clean Prod 5(1–2):79–86

Capstick SB, Pidgeon NF (2014) What is climate change scepticism? Examination of the concept using a mixed methods study of the UK public. Glob Environ Change 24:389–401

Chertow M (2000) Industrial symbiosis: literature and taxonomy. Ann Rev Energ Environ 25: 313–337

Chertow M (2007) Uncovering industrial symbiosis. J Ind Ecol 11(1):11–30

Chertow MR, Ashton WS, Espinosa JC (2008) Industrial symbiosis in Puerto Rico: environmentally related agglomeration economies. Reg Stud 42(10):1299–1312

Davy R, Gnatiuk N, Pettersson L, Bobylev L (2017) Climate change impacts on wind energy potential in the European domain with a focus on the Black Sea. Renew Sustain Energ Rev. (In Press, Corrected Proof) https://doi.org/10.1016/j.rser.2017.05.253

Delgado JA, Li R (2016) The Nanchang communication about the potential for implementation of conservation practices for climate change mitigation and adaptation to achieve food security in the 21st century. Int Soil Water Conserv Res 4:148–150

Dulic A, Angel J, Sheppard S (2016) Designing futures: inquiry in climate change communication. Futures 81:54–67

Ebi KL, Frumkin H, Hess JJ (2017) Protecting and promoting population health in the context of climate and other global environmental changes. Anthropocene 19:1–12

Frosch RA, Gallopoulos NE (1989) Strategies for manufacturing. Sci Am 266:144–152

Gavras P (2010) The current state of economic development in the Black Sea Region. Southeast Eur Black Sea Stud 10(3):263–285

Glaser EG, Strauss AL (1967) The discovery of grounded theory: strategies for qualitative research. Weidenfeld and Nicplson, London, England

Glass E, Ballantyne AG, Neset T-S, Linner B-O, Navarra C, Johansson TO, Rod JK, Goodsite ME (2015) Facilitating climate change adaptation through communication: insights from the development of a visualization tool. Energ Res Soc Scie 10:57–61

Hardin G (1968) The tragedy of the commons. Science 162:1243–1248

Hart A, Clift R, Riddlestone S, Buntin J (2005) Use of life cycle assessment to develop industrial ecologies—a case study: graphics paper. Process Saf Environ Prot 83(4):359–363

Jelinski LW, Graedel TE, Laudise RA, McCall DW, Patel CKN (1992) Industrial ecology: concepts and approaches. Proc Nat Acad Sci USA 89:793–797

Kaltenborn BP, Olve Krange O, Tangeland T (2017) Cultural resources and public trust shape attitudes toward climate change and preferred futures—a case study among the Norwegian public. Futures 89:1–13

Manzo K (2017) The usefulness of climate change films. Geoforum 84:88–94

McCright AM, Marquart-Pyatt ST, Shwom RL, Brechin SR, Allen S (2016) Ideology, capitalism, and climate: explaining public views about climate change in the United States. Energ Res Soc Sci 21:180–189

Schmidt A, Ivanova A, Schafer MS (2013) Media attention for climate change around the world: a comparative analysis of newspaper coverage in 27 countries. Glob Environ Change 23:1233–1248

Thompson JE (2017) Survey data reflecting popular opinions of the causes and mitigation of climate change. Data in Brief 14:412–439

Tvinnereim E, Fløttum K, Gjerstad Ø, Johannesson MP, Nordø AD (2017) Citizens' preferences for tackling climate change. Quantitative and qualitative analyses of their freely formulated solutions. Glob Environ Change 46:34–41

Valentine SV (2016) Kalundborg symbiosis: fostering progressive innovation in environmental networks. J Clean Prod 118:65–77

Van Berkel R, Fujita T, Hashimoto S, Fuji M (2009) Quantitative assessment of urban and industrial symbiosis in Kawasaki (Japan). Environ Sci Technol 43(5):1271–1281

Wolf A (2007) Industrial symbiosis in the Swedish forest industry. Doctoral dissertation thesis, Department of Management and Engineering, Linkoping Institute of Technology, Linkoping, Sweden

Zhao Y, Shang J, Chong C, Wu H (2008) Simulation and evaluation on the eco-industrial system of Changchun economic and technological development zone China. Environ Monit Assess 139:339–349

Avoiding Dispatches from Hell: Communicating Extreme Events in a Persuasive, Proactive Context

Sean Munger

Abstract Extreme weather events, like Hurricane Maria or floods in South Asia, are projecting the impacts of climate change ever more frequently into public consciousness. However, such events with their dramatic imagery risk triggering harmful reactions and behaviors that ultimately make it more difficult to motivate meaningful responses to the climate crisis. Instead of framing extreme events as validations of previous warnings ("We told you so") or harbingers of the future ("It'll be worse next time"), climate change professionals should communicate extreme events as one facet of a balanced choice, with a positive upside, that stakeholders can coalesce behind. Extreme events should be part of a broader context that emphasizes positive results rather than simply avoidance of harm or mitigation of its effects. An increasing trend in literature on climate change communication indicates that positive, forward-thinking messaging is the path to motivate effective action. Communication around extreme events, however, is uniquely susceptible to reinforcing unhelpful narratives of climate change catastrophe. This paper will help communicators get their messages across while avoiding these traps. This paper hopes to inform the field of climate change by highlighting the importance and the great potential contribution of a significant strategic shift in communication strategy.

1 Introduction

The summer of 2017 was a banner season for extreme weather events linked to climate change. The summer began with a series of searing heat waves that set records across various regions of North America, including 122 °F in Palm Springs, 101 °F in Portland, Oregon and 123.8 °F in Mexicali, Mexico (Masters 2017).

S. Munger (✉)
Centric Law, 5885 Meadows Road, Suite 330, Lake Oswego, OR 97305, USA
e-mail: sean@centriclaw.com

© Springer Nature Switzerland AG 2019
W. Leal Filho et al. (eds.), *Addressing the Challenges in Communicating Climate Change Across Various Audiences*, Climate Change Management,
https://doi.org/10.1007/978-3-319-98294-6_8

115

Torrential rains across South Asia in August brought catastrophic flood conditions, affecting 31 million people in India, 8 million in Bangladesh and 1.7 million in Nepal. The dead could not be buried because there was too little dry land on which to dig graves. The floods triggered a boom in the mosquito population, greatly increasing the reach of diseases like malaria and chikungunya (Barker 2017). Most spectacularly so far as Western-oriented press was concerned, the United States and Caribbean was pummeled by three monstrous hurricanes in succession: Harvey, Irma and Maria. The latter storm, a Category 5, wrought significant damage on the U.S.-held island of Puerto Rico, whose electric power infrastructure was regarded as "destroyed" ("Hurricane Maria Updates" 2017). The degree to which these extreme events were caused or exacerbated by climate change was also a subject of public discussion, warranting frequent explanations by climate scientists such as Michael E. Mann (Mann 2017).

Summer 2017 was also an important season for climate change communication. On June 2, the day after U.S. President Donald Trump announced his intention to withdraw the U.S. from the Paris climate accords, widespread condemnation of the decision put climate change on the front pages of newspapers and websites worldwide, and triggered public statements by private companies like Apple, Google, Amazon, Microsoft and IBM (Hunt 2017). In early July *New York Magazine* writer David Wallace-Wells penned an editorial warning that climate "doomsday" would make the Earth essentially uninhabitable in a fairly short period of time (Wallace-Wells 2017). Anecdotally, the Wallace-Wells piece seems to have caught the attention of many people formerly uninvolved in climate change discourse, who spent the summer asking climate change professionals, "Is it really *that* bad?" (Munger 2017).

Media reports on the summer's events tended toward the dramatic, emphasizing widespread devastation and the shock of eyewitnesses at the scale of the disasters. An account of chaos at the San Juan airport included a quote from a stranded passenger: "It's like the end of the world." (Jervis 2017). The broadening of public narratives from on-the-ground disaster to climate change issues still stressed apocalyptic language. Days after the nation of Dominica was ravaged by Maria, Dominican Prime Minister Roosevelt Skerrit told the United Nations, "I come to you straight from the front line of the war on climate change…to deny climate change is to procrastinate while the earth sinks; it is to deny a truth we have just lived!" (Skerrit 2017). Media covered 2017's climate catastrophes like dispatches from embedded correspondents in a war zone. Such coverage of extreme events gives the viewer an impression of the world as a climate-ravaged dystopian Hell, forever teetering on the brink of another dramatic disaster.

Climate change professionals have a duty to communicate candidly and accurately about extreme weather events and their relationship to climate change. However, narratives about extreme events that emphasize apocalyptic themes, whether intentionally or unwittingly, may easily trigger shutdown and disengagement from climate change problems, feelings of powerlessness and futility, and "psychological denial" of the reality of climate change. Professionals communicating about extreme events should be mindful of these unintended consequences.

It is the thesis of this paper that communication regarding extreme weather events is better framed in terms of presenting the events as one facet of a balanced choice, with a positive upside, that stakeholders can coalesce behind. Extreme events should be part of a broader context that emphasizes positive results rather than being deployed simply as cautionary tales, validations of previous warnings or harbingers of the future. Increasing trends in literature on climate change communication indicate that positive, forward-thinking messaging is the path to motivate effective action and responsible policy on climate change adaptation. Communication about extreme events, however, is uniquely susceptible to reinforcing unhelpful narratives of climate change catastrophe. To avoid these traps, communicators should strive to illuminate this broader context.

This paper hopes to inform the field of climate change by highlighting the importance and the great potential contribution of a significant strategic shift in communication strategy. Given the propensity of audiences to react negatively and counterproductively to messaging that reinforces depictions of disaster or does not engage how people typically think about risk, loss and the future, the importance of a new messaging strategy is almost impossible to understate. While the benefits of positive, constructive-focused messaging are well established in climate change literature, most public discourse involving climate change and particularly extreme events still favors depictions of disaster. Climate change communicators do not seem generally to be careful enough in crafting the narratives they choose, as well as choosing their target audiences. The chief rationale for this paper is the need to make climate change communicators more aware of the potential effects of their narratives, as well as how they might seek to change them to motivate meaningful action.

While this paper seeks to outline general principles for better and more productive messaging by climate change communicators, it is not an empirical study and does not present original quantitative research—in other words, the message strategies suggested here have not been "market-tested" under specific conditions. Studies of that nature may be the next step in developing improved messaging; but with these limitations, this paper hopes to generate consideration by climate change communicators of the conceptual implications of framing extreme events in a more proactive and persuasive context.

2 Pictures of Hell: A Counterproductive Communication Strategy

The field of climate change is filled with apocalyptic visions and imagery. A simple Google image search for the phrase "climate change" amply demonstrates this. Among images of temperature maps glowing angry red and orange and polar bears clinging to melting icebergs, the first image that appears in the search, credited to the stock photo site Shutterstock.com, is a three-fold composite showing a firefighter silhouetted against a blazing forest fire, mud cracks in a drought-ravaged landscape, and the eye of a hurricane approaching Florida. This image is the header on NASA's general information website about climate change, thus accounting for the prominence of the image in a search (NASA 2017). Extreme weather events are even more dramatic and memorable, and they connect to personal and cultural associations. One of the iconic images from Hurricane Sandy of 2012, replayed millions of times on YouTube, is a video clip of water flooding violently through the turnstile gates of a New York City subway station. This image closely resembles one of the most-recalled images from the disaster film *Titanic*—one of the most popular motion pictures of all time—in which violent blasts of water burst open stateroom doorways in a rapidly-flooding hallway aboard the doomed ocean liner.

In 2009 the UK sustainability consulting firm Futerra stated the "Hell" problem succinctly:

> The most common message on climate change is that we're all going to hell. That's what climate change looks like when you get right down to it: rising seas, scorched earth, failing food supplies, billions of starving refugees tormented by wild weather. (Futerra 2009)

From a standpoint of effective messaging, however, Futerra points out—equally succinctly—that "Hell doesn't sell…[t]hreats of climate hell haven't seemed to hold us back from running headlong towards it" (Futerra 2009). Environmental communication strategists Ted Nordhaus and Michael Shellenberger observed a decade ago that Americans' attitudes toward climate change have remained remarkably stable over the preceding two decades. Apocalyptic messaging was productive neither in raising the public profile of climate change, nor in motivating widespread public support for addressing it. While majorities of Americans—as well as people around the world—generally support government efforts to address climate change, few rank it as a particular priority, and apocalyptic imaging may well be counterproductive:

> [E]fforts to increase salience [of the climate change issue] through offering increasingly dire prognosis about the fate of the planet (and humanity) have also probably undermined public confidence in climate science. Rather than galvanizing public demand for difficult and far-reaching action, apocalyptic visions of global warming disaster have led many Americans to question the science. (Nordhaus and Shellenberger 2009)

Why? Studies involving the psychology of climate change communications have shown that climate change is a problem that is uniquely inaccessible to how the human mind typically responds to threats and decides upon action. Climate change lacks salience, meaning that it does not rise to the level of threat necessary to trigger humans' fight-or-flight responses; dealing with it requires immediate trade-offs of people's standards of living to ameliorate potential future harms that are perceived as distant and speculative; and climate change seems, to many people, to be contested and uncertain. Some psychologists looking at the problem are profoundly pessimistic, stating bluntly, "I really see no path to success on climate change" (Marshall 2014).

Whether one accepts this pessimistic view—*"Game over!"* as it is sometimes expressed in popular colloquialism[1]—it is clear that climate change meshes awkwardly with human psychological response. The human brain is characterized by two different systems of processing stimuli: experiental processing, which is linked to emotions and survival instincts; and analytical processing, which controls rational evaluation of information. "Despite evidence from the social sciences that the experiental processing system is the stronger motivator for action," argued the Center for Research on Environmental Decisions, "most climate change communication remains geared toward the analytical processing system" (CRED 2009). Essentially, analytical and scientific data on climate change—which is the departure point for many climate change messaging efforts—is processed in the same part of the brain that would be engaged by solving a mathematical equation.

To add another layer of difficulty to the climate change communication problem, even messages designed at engaging the experiental processing system often backfire. Many climate change communicators assume that an audience either doesn't have sufficient awareness of climate change threats (low information), or is not sufficiently outraged about them (high apathy). Perhaps more often, however, audiences *are* aware of the problem, and they are *too* moved by it—to the point where it is psychologically unbearable to think about it. In these cases, risk communication expert Peter Sandman argues that distressed audiences retreat into "psychological denial" to avoid having to think about the terrible consequences of climate change:

> [I]f people are in or near denial about climate change, their failure to act is more deeply psychological. It's not that they are giving priority to other issues; it's more that they can't bear to think about *this* issue. The most crucial risk communication task, then, is to make it more bearable to focus on global warming—to diagnose the reasons why people are so powerfully motivated to avoid the issue, and to change our messaging to reduce their avoidance and make it easier for them to face the issue and take action. (Sandman 2009)

Climate change messaging involving extreme weather events presents special dangers in this area. Another, similar psychological effect called "emotional numbing" occurs when people are exposed repeatedly to emotionally draining situations; the effect has been noted both in people who live near war zones, and, relevant to climate change, people in areas exposed to repeated hurricane threats in a short period of time. The danger of emotional numbing is magnified by a media environment with the capacity to deliver the emotional shocks of extreme weather events over and over again, whether or not in the form of news being reported, climate change warnings, or even movies—like the aforementioned example of the bursting cabin doors in *Titanic*. Numbing and worry are easily-reached thresholds of fatigue when talking about climate change (CRED 2009).

---

[1]The phrase "game over," which comes from video games, has frequently appeared in conjunction with climate change discourse, especially upon the election of Donald Trump as U.S. President (e.g., Carpenter 2016). A popular Medium article pushed back against defeatism in the realm of climate change, arguing "'Game over' is neither realistic nor responsible" (Steffen 2016).

In addition to the problems of psychological denial and emotional numbing, some particular audiences may actually be attracted, rather than repelled or horrified, by images of apocalyptic disaster. The prevalence of disaster in popular culture seems to reveal a deep societal attraction to large-scale destruction. In his seminal environmental history of Los Angeles, historian Mike Davis devoted 80 pages to a survey of the history of "the literary destruction of Los Angeles," which is "often depicted as, or at least secretly experienced as, a victory for civilization." Audiences packing theaters for the 1996 film *Independence Day* literally cheered as the laser beams of fictional aliens annihilated California's largest city; this is but one example of a lengthy tradition in novels, movies and television in which Los Angeles is incinerated by nuclear explosions, leveled by earthquakes or fires, suffers direct hits from comets and asteroids, or is invaded by Japanese, Communists, the Viet Cong, Nazis, terrorists and extraterrestrials (Davis 2006). Given that Los Angeles is but one city that features in a worldwide culture of disaster fiction, it seems clear that we find something thrilling and beautiful in images of apocalypse. While no sane person would actively wish for the kind of world that climate change may give us, neither can we assume that everyone will recoil instinctively from envisioning its features.

For all these reasons, visions of Hell, in Futerra's terminology, "doesn't sell." It is not the "sizzle" in the climate change message, to employ a term from advertising ("Don't sell the sausage, sell the sizzle!"), and, so long as climate change messaging appeals to the analytical rather than experiential mode of processing information, such messaging is probably doomed to failure (Futerra 2009). In the context of communicating about extreme weather events, avoiding Hell is especially difficult, because such messages inherently evoke the same images of disaster—flooded streets, refugees huddled on rooftops, hospital emergency rooms swamped with heat stroke victims—that have the potential to spark psychological denial, emotional numbing or unconscious attraction.

When woven into climate change narratives, extreme weather events often follow three patterns. The first is what might be termed the "I told you so" narrative: extreme weather events as validations of previous predictions or warnings given in the past regarding climate change (e.g., Klein 2017; Snider 2017). A second common narrative might be termed "This is the future": a narrative that extreme events, such as this or that recent specific example, are destined to become more common as the effects of climate change continue to unfold. In popular media, these narratives often employ the phraseology "the new normal" (e.g., Karnad 2017, referencing South Asia floods; Berwyn 2017, referencing simultaneous heat waves in various parts of the globe).

Each of these narratives are constructed for differing purposes and sometimes for specific audiences. Drawing lessons, whether broad or narrow, from extreme weather events in the context of climate change is a necessary action. But in doing so, climate change communicators must be mindful of how—and to what ends—they deploy narratives of extreme weather. All too often they can devolve into "dispatches from Hell."

3 What's Better Than Hell: A More Positive Context

The problem with traditional messages surrounding climate change is fundamentally a simple one: they're almost always negative. The dominance of negative messaging reflects the reasons why the messengers—who are usually much more deeply committed to climate change issues than members of the general public—care about climate change in the first place: they wish to prevent the damages and disasters they anticipate as a result of a warming world. While laudable, this approach ignores a central truth of human behavior: working for positive outcomes generally feels better, and is more sustainable over the long term, than working to prevent negative ones. Some analysts of climate change communication have drawn a parallel with the civil rights movement in the United States from the 1940s to the 1960s. The iconic phrase to emerge from this movement is "I have a dream," the title of Dr. Martin Luther King's August 1963 address that depicted a hopeful and morally compelling future of racial equality. Traditional messaging on climate change, however, is more like, "I have a nightmare," and messengers deploy their nightmares in the hope stakeholders will be motivated to avoid them (Rich 2016).

A fairly simple psychological experiment involving climate change imaging demonstrates that positive and forward-thinking imagery is generally more motivating and appealing than images of disaster. An international study by the UK-based Climate Outreach foundation tested 49 pictures pertaining to climate change and asked participants to rank their reactions based on how the image made them feel, how motivated they would be to change their personal behavior, and other factors. Of the 49 images tested, most were negative, depicting scenes of pollution, disasters and traffic jams. Only five pictures made participants collectively feel more positive than negative. Yet one of these five—a photograph of two children, apparently in the developing world, playing on a rooftop with solar panels—was the only picture that scored higher in the aggregate than the control photograph, classified as a "cliché," of desperate-looking polar bears on a melting iceberg, which had a much lower positive/negative score. Furthermore, the polar bear cliché scored just barely higher in the aggregate than another of the positive photos, showing a group of smiling children holding up solar panels and a flag identifying them as participants in the "Solar Schools" program in the UK.[2] Photographs of extreme weather events in this survey tended to focus on floods, and more often than not depicted street signs being consumed by rushing waters, with no humans visible in the frame. One photo showing a man apparently drowning in a flood polled especially poorly, placing significantly behind the positive children

---

[2]The statistical dominance of the polar bear cliché photo seems to stem mainly from its high score in terms of recognition: a higher number of survey participants recognized it as being emblematic of climate change than they did any other photo. This may not be because of something inherent in the image, but rather the exposure it's gotten in the media. A survey participant noted, "The visual vocabulary of climate change has not changed significantly in years—even organisations like Getty Images are scratching their heads...in terms of how to refresh it." Another characterized the polar bear cliché as "out of date—though useful on fundraising" (Climate Outreach 2016).

photographs in the category of whether the viewer felt motivated to change his or her own behavior. One survey participant commented, "In choosing photos to test it's a good idea to have an idea of what emotions you want to evoke" (Climate Outreach 2016).

If it seems elementary that people would rather look at pictures of smiling children than desperate polar bears or drowning men, it follows that they would rather think about—and work toward—a positive future than dwell upon the fears of a negative one. Futerra's argument on positive messaging is both simple and intuitive: if the vision of the future that climate change communicators want people to work toward isn't more desirable than the present, why bother working toward it? (Futerra 2009). Yet a future of climate-changed Earth that is less desirable than the present is exactly what most climate change messaging asks its audiences to embrace, at significant cost. Even fully achieving the optimistic goal of the Paris Accords of holding warming to 1.5 °C by 2100 comes with considerable negatives, for example, "committed sea level rise" that is projected to occur regardless of whatever CO_2 mitigation actions are taken now (Rhodium 2014). While this is undeniably a reality of the climate change situation, it creates a difficult selling dilemma: let's sacrifice today to live in a world that's significantly worse than the one we have now, but one that is *less* bad than the one we'll have if we do *not* sacrifice. This is like asking a child who does not like vegetables to give up a favorite toy today for the privilege of eating Brussels sprouts for dinner tomorrow, in order to avoid being forced to eat mud. The choice between Brussels sprouts and mud is a rational one, but commands no emotional cachet—the whole transaction is a net negative. When viewed this way, it is unsurprising that messaging attempts that try to motivate behavior change by "turning up the dial" on depicting apocalyptic futures make no real difference to audiences, and sometimes even backfire (Futerra 2009).

The alternative is to communicate climate change in a context that motivates stakeholders to make a choice to embrace a positive future, and to coalesce behind making it happen. In our analogy with the child who dislikes vegetables, the transaction flips completely if you convince the child to give up his favorite toy in favor of a new and different toy that's better, shinier and more gratifying. Will the child eat Brussels sprouts tomorrow night as the price of receiving the new, better toy today? Even a selfish child could be easily persuaded to accept this bargain, especially when the even more negative consequence of refusing—eating mud— can be avoided. In this iteration, the rational choice is merely an additional finger on the scale for an emotionally-invested choice with a positive outcome. The "selling" job that needs to be done here is also much easier and can be done with much more enthusiasm: sell the child on the positives of the shiny new toy she's going to get in exchange for the dirty old one, instead of trying to convince her how much worse mud is than Brussels sprouts.

A groundswell of literature on climate change communication bears out this simple analysis. "When communicators help people envision solutions to climate change," counseled the Center for Research on Environmental Decisions, "they provide a positive vision of what the future could be like." Leading with positive solutions—rather than another description of the problem—even helps with what might otherwise be considered threshold questions, such as whether climate change exists in the first place (CRED 2014). Psychologists and social scientists have argued that climate change communications should propose individual behavior change as part of a coordinated global strategy to transition the world's economy away from fossil fuel dependence: in other words, a grand plan for constructing a better and cleaner world (Corner and Groves 2014). They have also argued that, because of the psychological nature of how human beings evaluate potential loss versus potential gain, "shifting the policy conversation from the potentially negative future consequences of not acting (losses) on climate change to the positive benefits (gains) of immediate action is likely to increase public support." Furthermore, comparisons of negatively-framed scenarios versus positively-framed ones have shown that positive messages increase support for mitigation and adaptation efforts (Van der Linden et al. 2015).

Two interlocking issues should be considered next: first, which specific audiences should climate change communicators target with a proactive and positive message, and second, exactly which messages should they use? This paper seeks to raise the conceptual issue of positive messaging, but actually "market-testing" specific messages for discrete audiences is a task that is beyond the scope of my research, which has principally been to recognize general trends in developing literature. Nevertheless, "market-testing" and specific message campaigns could be considered by organizations, preferably with an international reach, interested in climate change communication. Climate Outreach (UK) and the Center for Research on Environmental Decisions (Columbia University, USA) have laid the groundwork for this sort of research. It should be broadened and applied on a larger scale, across national boundaries and involving international media, perhaps by NGOs or other organizations whose traditional focus has been on communicating climate change facts and science—in other words, the messengers whose messages could benefit most from a shift. Much more work needs to be done in this area.

4 Communicating Extreme Events: How Do You Make a Disaster "Positive"?

It is one thing to recognize the benefits of a positive messaging strategy on climate change in the abstract. It is quite another to fit extreme weather events into a more positive context. How does one "spin" events like the South Asia floods of 2017, or Hurricane Maria, into a forward-looking strategy to induce stakeholders to coalesce behind the efforts needed to build a better and cleaner world?

One suggestion is to deploy extreme events strategically to support other communication objectives. Many people feel that climate change is something distant, that's happening somewhere else and affecting someone else—like the polar bears in the ubiquitous photo—and is not a local and immediate threat.

Effective climate change communication should seek to narrow this gap by making people see local and personal impacts of climate change (CRED 2014). Extreme weather events offer the advantage of providing many opportunities for personal stories of how climate change is affecting real people.

One example is the story of Jayden F., a 13-year-old girl from Rayne, Louisiana, who filed a declaration with a U.S. federal district court in the groundbreaking climate change case of *Juliana v. U.S.* Jayden described the storms of August 2016 that caused rising floodwaters in her hometown. Awakened by her siblings at 5:00 AM, Jayden testified:

> "I noticed there was water coming from under the door to my room…When I stepped out of my bed, I stepped in water that came up to my ankles. I stepped right in the middle of climate change….All day, floodwater continued to pour into our home…Our toilets, sinks, and bathtubs began to overflow with awful smelling sewage because our town's sewer system also flooded. Soon the sewage was everywhere. We had a stream of sewage and water running through our house." (Declaration of Jayden F. 2016)

Though certainly a "dispatch from Hell," Jayden's declaration also contains a statement of steadfast hope for the future: "*I am scared. But I will not back down. We will conquer climate change.*" To hear these words from a 13-year-old girl— who will likely be alive to experience the better world that climate change communicators should be describing to their audiences—injects a positive emotional note into what otherwise could be a relentlessly depressing story. This message certainly resonates more than the typical "I told you so" or "This is the future" narratives.

Though extreme weather events are by definition negative, people are magnetically drawn to the positive stories they generate. Stories focusing on kindness or altruism have a tendency to "go viral," especially on social media platforms. For example, after Hurricane Sandy in 2012, numerous photos circulated on the Internet depicting random people in the New York area hanging power strips from fences and gates for strangers to charge their cell phones, homemade signs advertising free food ("*Free Tacos @ Tacombi, 267 Elizabeth, today*"), or messages thanking first responders posted at Union Square in Manhattan. One such collection was shared tens of thousands of times (Haberman 2012). Such displays invariably follow disasters with high media presence; Hurricane Irma's version was the tale of Ramon Santiago, an Orlando resident who gave a total stranger he met at a Lowe's home improvement store a power generator in advance of the hurricane (Fantozzi 2017). This kind of behavior is not as random or unusual as it might seem given the special attention often paid to such acts. While popular conception holds that the default reaction to a disaster situation is irrationality, selfish behavior and panic, scientific study has shown that these responses are comparatively rare (Cocking and Drury 2014). Indeed, witnessing a major disaster, especially one with an environmental component, seems to increase the psychological tendency toward altruism and cooperative behavior (Li et al. 2013).

Extreme weather events, therefore, clearly have the potential to motivate positive, cooperative and determined responses. Climate change communicators focused on positive messaging should always try to generate support for the choice to build a better world—a cleaner economy, with more (and better) jobs, enhanced national security, greater economic opportunity, technological responsibility and environmental justice—in response to climate change. Extreme weather events are incongruous with this vision, but they should be framed as challenges that must be overcome, rather than, as is so often depicted, punishments that must be endured for failing to heed prior warnings. Extreme weather events can also be used to enhance understanding of the enormous human capacity for cooperation, altruism and positive response. *"Look how people came together during this event. Can't we all do that, on a societal level, in response to climate change?"* In any event, communicators discussing extreme weather events should, above all else, avoid making the problems of climate change communication worse by reinforcing uniformly destructive, dystopian and depressing narratives that are clearly counterproductive.

5   Conclusions

Communication about extreme weather events, such as the dramatic and destructive events of 2017, tends toward the sensational, the spectacular and the depressing. Discourse in the media about weather events tends to follow negative patterns: "I told you so" or "This is the future" narratives frame extreme events either as punishments earned by ignoring previous climate change predictions, or warnings to prepare for an unpleasant future. Studies by psychologists and social scientists have shown that messages such as these increase audiences' feelings of helplessness, futility or motivation to deny the science or factual basis of climate change.

Research on a small scale has shown that messaging emphasizing positive possibilities rather than negative outcomes, however, engages audiences, increases their motivation to change their own behavior and to support meaningful action to address climate change. The key point is to ask potential stakeholders to make a choice for a positive future from which they will benefit—a cleaner, stronger world that we can build together—rather than trying to motivate them simply to avoid a parade of future horribles.

While fitting extreme weather events into a positive, proactive context is challenging, potential themes that could be employed include the use of extreme events to personalize and localize climate change, as well as leveraging audiences' interest in themes of altruism and cooperation in the face of disaster. More research and market-testing concerning which audiences to target, and which specific messages to use, should to be done to flesh out these basic concepts.

Acknowledgements *Special thanks to the Rose Law Firm, especially Adam Rose, Brian Riske and Cynthia Johnson.*

References

Barker A (2017) South Asia floods: estimated 40 million across India, Bangladesh, Nepal affected. ABC News, Available at Online http://www.abc.net.au/news/2017-09-08/40-million-forced-to-rebuild-lives-after-south-asia-floods/8886264 (Accessed 23 Oct 2017)

Berwyn B (2017) Simultaneous widespread global heat waves are the new norm. Pacific Standard, Available at Online https://psmag.com/environment/so-yeah-these-heat-waves-are-basically-the-new-normal (Accessed 23 Oct 2017)

Carpenter Z (2016) Is This 'Game Over' for the Planet? The Nation, Available at Online https://www.thenation.com/article/is-this-game-over-for-the-planet/(Accessed 23 Oct 2017)

CRED (Center for Research on Environmental Decisions) (2009) The psychology of climate change communication. Available at Online http://guide.cred.columbia.edu/ (Accessed 23 Oct 2017)

CRED (Center for Research on Environmental Decisions) (2014) Connecting on climate: a guide to effective climate change communication. Available at Online http://ecoamerica.org/wp-content/uploads/2014/12/ecoAmerica-CRED-2014-Connecting-on-Climate.pdf (Accessed 23 Oct 2017)

Climate Outreach (2016) Resource: climate visuals—7 key principles for visual climate change communication. Appendix, Available at Online https://climateoutreach.org/resources/visual-climate-change-communication/ (Accessed 23 Oct 2017)

Cocking C, Drury J (2014) Talking about Hillsborough: 'panic' as discourse in survivors' accounts of the 1989 football stadium disaster. J Commun Appl Psych 24:86–99

Corner A, Groves C (2014) Breaking the climate change communication deadlock. Natu Clim Change 4:743–745

Davis M (2006) Ecology of fear: Los Angeles and the imagination of disaster. Metropolitan Books, USA

Declaration of Jayden F. in Support of Plaintiffs' Opposition to Defendants' Motions to Dismiss (2016). Available at Online https://static1.squarespace.com/static/571d109b04426270152feb e0/t/57d0fac03e00be689aac4a09/1473313478990/JaydenDeclaration.pdf (Accessed 23 Oct 2017)

Fantozzi J (2017) A photo is going viral of a man who gave his generator to a stranger after Lowe's ran out ahead of Hurricane Irma. Business Insider, Available at Online http://www.businessinsider.com/photo-man-generator-stranger-lowes-hurricane-irma-2017-9 (Accessed Oct 23 2017)

Futerra (2009) Sell the sizzle. Available at Online https://www.wearefuterra.com/wp-content/uploads/2015/10/Sellthesizzle.pdf (Accessed 23 Oct 2017)

Haberman S (2012) 15 Amazing acts of kindness during sandy. Mashable.com, Available at Online http://mashable.com/2012/11/01/acts-of-kindness-sandy/ (Accessed 23 Oct 2017)

Hunt E (2017) Paris climate agreement: world reacts as trump pulls out of global accord. The Guardian, Available at Online https://www.theguardian.com/environment/live/2017/jun/01/donald-trump-paris-climate-agreement-live-news (Accessed 23 Oct 2017)

Hurricane Maria Updates: In Puerto Rico, the Storm 'Destroyed Us' (2017). New York Times, Available at Online https://www.nytimes.com/2017/09/21/us/hurricane-maria-puerto-rico.html (Accessed 23 Oct 2017)

Jervis R (2017) It's like the end of the world' inside San Juan's steaming airport. USA Today, Available at Online https://www.usatoday.com/story/news/2017/09/25/thousands-weary-travelers-stuck-steamy-san-juan-airport/699581001/# (Accessed 23 Oct 2017)

Karnad R (2017) Floods in drought season: is this the future for parts of India? The Guardian, Available at Online https://www.theguardian.com/commentisfree/2017/sep/04/floods-drought-season-mumbai-india-extreme-weather (Accessed 23 Oct 2017)

Klein N (2017) Harvey didn't come out of the blue. Now is the time to talk about climate change. TheIntercept.com, Available at Online https://theintercept.com/2017/08/28/harvey-didnt-come-out-of-the-blue-now-is-the-time-to-talk-about-climate-change/ (Accessed 23 Oct 2017)

Li Y, Li H, Decety J, Lee K (2013) Experiencing a natural disaster alters children's altruistic giving. Psychol Sci 24(9):1686–1695

Mann M (2017) What can we say about the role of climate change in the unprecedented disaster? Facebook, Available at Online https://www.facebook.com/MichaelMannScientist/posts/1515449771844553 (Accessed 23 Oct 2017)

Marshall G (2014) Don't even think about it: why our brains are wired to ignore climate change. Bloomsbury, USA

Masters J (2017) Summary of the Great Southwest U.S. heat wave of 2017. Weather Underground, Available at Online https://www.wunderground.com/cat6/summary-great-southwest-us-heat-wave-2017 (Accessed 23 Oct 2017)

Munger S (2017) Global warming summer: making sense of the disasters of 2017. Centric Law Blog, Available at online http://centriclaw.com/global-warming-summer-making-sense-of-the-disasters-of-2017/ (Accessed 23 Oct 2017)

NASA (2017) The consequences of climate change. Available at Online https://climate.nasa.gov/effects/ (Accessed 23 Oct 2017)

Nordhaus T, Shellenberger M (2009) Apocalypse fatigue: losing the public on climate change. Yale Environment 360, Available at Online http://e360.yale.edu/features/apocalypse_fatigue_losing_the_public_on_climate_change (Accessed 23 Oct 2017)

Rhodium Group (2014) American climate prospectus: economic risks in the United States. Available at Online https://gspp.berkeley.edu/assets/uploads/research/pdf/American_Climate_Prospectus.pdf (Accessed 23 Oct 2017)

Rich F (2016) Getting to green: saving nature, a bipartisan solution. W.W. Norton & Company, USA

Sandman P (2009) Climate change risk communication: the problem of psychological denial. Available at Online http://www.psandman.com/col/climate.htm (Accessed 23 Oct 2017)

Skerritt R (2017) UNGA 72nd session statement. Available at Online https://gadebate.un.org/sites/default/files/gastatements/72/dm_en.pdf (Accessed 23 Oct 2017)

Snider A (2017) Why America still hasn't learned the lessons of Katrina. Politico.com, Available at Online http://www.politico.com/magazine/story/2017/08/27/hurricane-harvey-katrina-lessons-louisiana-215543 (Accessed 23 Oct 2017)

Steffen A (2016) There will never be a better time to save the planet. Medium.com, Available at Online https://medium.com/@AlexSteffen/there-has-never-been-a-better-time-to-save-the-planet-d290132eaee (Accessed 23 Oct 2017)

Van der Linden S, Maibach E, Leiserowitz A (2015) Improving public engagement with climate change. Perspect Psychol Sci 10(6):758–763

Wallace-Wells D (2017) The uninhabitable earth. New York Magazine, Available at Online http://nymag.com/daily/intelligencer/2017/07/climate-change-earth-too-hot-for-humans.html (Accessed 23 Oct 2017)

Blogging Climate Change: A Case Study

Erangu Purath Mohankumar Sajeev, Kian Mintz-Woo,
Matthias Damert, Lukas Brunner and Jessica Eise

Abstract Public perception of the magnitude of challenges associated with climate change is still lower than that of the majority of scientists. The societal relevance of climate change has raised the need for a more direct communication between scientists and the public. However, peer-reviewed scientific articles are not well-suited to engaging a wider audience. This begets a need to explore other avenues for communicating climate change. Social media is a vibrant source for information exchange among the masses. Blogs in particular are a promising tool for disseminating complex findings on topics such as climate change, as they are easier to comprehend and are targeted at a broader audience compared to scientific publications. This chapter discusses the usefulness of blogs in communicating climate change, using our blog *Climate Footnotes* (climatefootnotes.com) as a case study. Drawing from communication theory and our experiences with *Climate Footnotes*, we identify and describe elements such as message framing, translation of scientific data, role of language, and interactivity in aiding climate change communication. The insights outlined herein help understand the nature and impact of online climate change communication. The chapter may also serve as a useful blueprint for scientists interested in utilizing blogs to communicate climate change.

1 Introduction

While the impacts of anthropogenic climate change are constantly increasing (Fischer and Knutti 2015), the same cannot be said for the public's concern. Climate change is a slow phenomenon. Manifesting over years and decades, it lacks

E. P. M. Sajeev (✉)
Rothamsted Research, AL5 2JQ Harpenden, UK
e-mail: em.sajeev@rothamsted.ac.uk

E. P. M. Sajeev · K. Mintz-Woo · M. Damert · L. Brunner
DK Climate Change, University of Graz, Graz 8010, Austria

J. Eise
Brian Lamb School of Communication, Purdue University, West Lafayette, IN 47907, USA

© The Author(s) 2019
W. Leal Filho et al. (eds.), *Addressing the Challenges in Communicating Climate Change Across Various Audiences*, Climate Change Management,
https://doi.org/10.1007/978-3-319-98294-6_9

129

the timeliness and punch of so many of today's headliners and hot-button issues of concern. It is a uniquely difficult challenge insofar as it requires people to act with urgency to prevent future scenarios that many find unimaginable or even unbelievable in their present condition. As such, human interest in climate change has been slow to garner and difficult to maintain.

Yet despite this challenge, public interest in and perception of climate change is critical. Without an informed public, governments, scientists and organizations will lack the social license necessary to act decisively to mitigate the impacts of this phenomenon. What is more, individuals who may wish to take action in their personal lives to lower their carbon footprint may lack the resources to do so or even the understanding that doing so is a decision that they could make.

Public perception of the magnitude of challenges associated with climate change is not where it needs to be. In a 2016 study conducted in a sample of countries in Europe, 16% of Germans and 14% of British were sceptical about human activity as a cause for climate change or did not believe that climate change occurs, and on an open-ended question less than 3% in both countries listed "climate change" as one of the toughest challenges their country will face in the next 20 years (Steentjes et al. 2017). In the United States, over half of Americans do not believe that climate change is due to human activity. A 2016 Pew Research study reports that roughly three-in-ten say it is due to natural causes (31%) and another fifth say there is no solid evidence of warming (20%) (Funk and Kennedy 2016). However, the same 2016 Pew Research study finds that a large majority (67%) want climate scientists to have a major role in climate policy—the most supported group considered—and a larger majority (78%) trust climate scientists "some" or "a lot" to provide accurate climate information. The public believes climate scientists have roles both in policy *and* in communication. As such, engaged participants in climate change—and scientists in particular—might assume responsibility for conducting outreach around the issue themselves. What is more, when it comes to issues with high societal relevance, researchers have an obligation to help inform a more general audience about their findings and associated implications (Leshner 2003).

2 Blogs as a New Medium of Climate Change Communication

The primary method of communication and dissemination of scientific findings has been and continues to be peer-reviewed scientific articles. While this method is uniquely suited to communicating to an academic audience, ensuring rigor and accuracy of scientific findings, it is less appropriate for a wider audience. Many aspects of peer-reviewed scientific articles make them a poor choice for communicating with a broader audience, aspects ranging from paywalls and jargon to (frequently) dense writing styles.

These are not the only challenges associated with the peer-reviewed scientific model of communication. While the scientific community has recognized the importance of disseminating information to the public, many still operate under the assumption that the critical barrier to communication is the lack of information that exists among the public. Hence a lot of effort around science communication is focussed on providing more information on the basis that once the public understands the science; they would be more inclined to agree with science and to endorse action. However, communication theory suggests otherwise and lists factors such as interactivity, inclusivity, trust and social identity as having stronger impacts than providing information alone (Bubela et al. 2009). Hence, merely amplifying exactly what is in the peer-reviewed scientific literature as many times as possible with the assumption that people will eventually understand and take desired action is too simplistic.

These communication challenges beget the need to explore new avenues for exchanging information. With the advent and progress of the technological era, there has been a shift from traditional modes of communication that employ a one-to-many communication strategy to a more interactive form of communication that uses a many-to-more strategy. Today's online communication environment, Web 2.0, is "a collaborative medium that allows users to communicate, work together and share and publish their ideas and thoughts" (Rollett et al. 2007). Much of that communication and collaboration occur across social media, a communication avenue that has emerged as a vibrant source for information exchange in society. One among the many promising new communication avenues that have emerged are blogs. Unlike traditional communication channels, blogs offer a platform to the users where information is interactive and inclusive irrespective of geography and socio-economic standing (O'Neill and Boykoff 2011).

According to the internet use statistics for 2017, almost 52% of the world population are internet users with a growth rate of 976% in the last 7 years.[1] According to the Pew Research Center,[2] a survey conducted in 2016 reveals that around 38% of Americans get their news online (via mobile or desktop), easily dwarfing legacy communication channels such as the radio (25%) and print media (20%). A breakdown by age group indicates online sources as the dominant medium of consumption of news for people below 50 years of age. As of 2012, there were 50 million blogs and this is expected to rise as numbers of users and blog posts on popular blogging platforms such as Tumblr, Wordpress and Blogger are increasing. Regarding science blogs, earlier estimates were in the range of 1000–1200 (Trench 2012). A review study by Schäfer (2012) estimates the number of science blogs to be about 1900 of which 1400 were "climate" blogs with only 323 actually categorized as "climate science" blogs. Even accounting for the existence of psuedoscience blogs folded into that tally, 1900 blogs compared to the 50 million that currently exist makes that number paltry indeed. As a form of new media, blogs

---

[1]http://www.internetworldstats.com/stats.htm.
[2]http://www.journalism.org/2016/07/07/the-modern-news-consumer/.

are part of the worldwide communication network and thus accessible "to diverse geographical and socio-economic populations, which provides new opportunities for engaging individuals with climate change" (O'Neill and Boykoff 2011). Given the potential of the internet and social media, the resources of Web 2.0 are tools that ought not to be ignored in the quest for wider outreach.

One of main functions of Web 2.0 is as 'places to learn' for formal and non-formal education as well as informal learning (Schugurensky 2000). Adult learners often benefit by optimizing these web-based spaces as alternative environments for informal learning when they navigate through information, network with other people, or produce wanted identities (Selwyn 2007). Informal learning is any activity involving the pursuit of understanding, knowledge or skill that happens without externally imposed curricula (Livingstone 2001). Informal learning is important for adults because it highlights the learner as an agent of learning and also since it happens in everyday life, it expands the conventional meaning of learning (Livingstone 2001; Marsick and Watkins 2001). With this perspective of online spaces, social media is not just a space for entertainment but also informal learning. Adult learners can determine their roles and degrees of engagement in the activities found in these web-based spaces. More interactive ways of using Web 2.0 may guarantee not only more diverse, but also better quality learning (Heo and Lee 2013).

Blogs in particular are a promising tool for disseminating complex findings on topics such as climate change for a number of reasons. Web 2.0 is a medium that allows users to share and make public their ideas and thoughts with no barrier to entry beyond an internet connection and literacy. Researchers are often hesitant to put forward their findings as they are concerned that they may not be properly validated or posited. Yet in most online spaces, no one else shares those qualms. Blogs are a good way for researchers to get over this and join the fray. Blogs necessitate neither the brevity nor the pith of a 140-character tweet, yet remain an informal and easily-accessible mode of communication for the broader public. They make for good training wheels, so to speak. Researchers can use blogs as a platform to engage with a broader audience allowing for collaboration and networking due to the quick and timely discussions that can take place. As far as young researchers are concerned, blogs serve as a testing ground to improve their writing skills, build an online identity and provide a learning opportunity by exposing them to new perspectives and the latest advances in technology and research (Putnam 2011). What is more, science to public communication is viewed as helpful for the process of conducting science as it can assist in generating ideas and disseminating policy proposals (Blanchard 2011). Academic journals, which traditionally used to communicate only via peer-reviewed articles, have also taken a step in the direction of new media. Examples include blogs by powerhouses of academic publishing such as *Nature*, *Science*, and *Cell* (Blanchard 2011).

In addition, topics around climate change are complex. This makes blogs an appropriate medium over other social media options. Blogs allow for depth and range. A blog can be as short as a couple hundred words or as long as several thousand (although best practices would suggest you should cap your length to around 400–700 words per entry). While videos are compelling, it is unreasonable to expect researchers to become videographers. It is less unreasonable to expect that researchers write. In addition, a blog may be contributed to as a 'guest' writer and could be a periodic activity in which a scientist might engage, not something he or she is forced to maintain on a daily basis such as Twitter or Instagram. Perhaps more to the point, however, blogs have the potential to reach a broad audience beyond the confines of academia and research communities. They allow for timely publication on issues as there is no review lag time, making it more interesting and engaging for a public readership.

The direct and rapid outreach of scientific information through new communication channels such as blogs creates a means for engagement and participation for readers and writers alike. For these reasons, we examine the use of blogs to disseminate scientific information on the complex phenomenon of climate change. In this chapter, we shed light on effective strategies, approaches and limitations using our own blog *Climate Footnotes* (climatefootnotes.com)[3] as a case study. We believe these insights may help in understanding how information related to climate change can be disseminated through online platforms such as blogs. This type of understanding of public engagement helps build the knowledge base necessary for a holistic approach encouraging the interplay between the scientists and the public in addressing topics related to climate change.

3 Our Blog—*Climate Footnotes*

While there are climate blogs that aim to provide timely climate scientific context (e.g., www.RealClimate.org) and projects that aim to provide attribution analyses for particular extreme weather events (e.g., www.climatecentral.org/), we created our blog motivated by the desire to increase the ranks of graduate students in scientific blogging (Wilkins 2008). *Climate Footnotes* serves as a space to discuss our scientific ideas and achievements as well as address personal and political topics. Different members of the Doctoral College Climate Change (DKCC)[4] at the University of Graz participated on a voluntary basis, with a strong majority of the students contributing at least once. As of October 2017 more than 70 posts have been published on the blog on a more or less regular basis. We ultimately settled on a posting schedule of twice per month with a rotating schedule of students, which proved to be a reasonable expectation measured against the pressures of other academic work. We also periodically invited guest bloggers to participate.

An important component of our success is an automated reminder system that allows contributors to publish on their own time schedule with only a small role for editors. Initially, editors spent a lot of energy contacting people and arranging for

---

[3]Throughout, when referring to particular blogposts on *Climate Footnotes*, we simply put the blogpost title in quotation marks.

[4]http://dk-climate-change.uni-graz.at/en/.

comments. This proved untenable and ultimately the amount of editing needed was limited. We ultimately settled on "editing as optional" although there is constant transparency and editing access (an email address is dedicated to blogposts in progress). Everyone has individual accounts which they can use to self-publish. We elected to publish in a private forum—i.e. not on the university system—as it afforded more control and no restriction in terms of content and we paid for the domain registration privately. Moreover, modern publishing platforms make setting up a blog rather easy, for *Climate Footnotes* we used a basic WordPress (https:// wordpress.org/) blog template, which provides all of the required functionality.

Before starting with *Climate Footnotes*, the team behind the project decided not to target a specific audience or limit contributions to specific aspects of climate change. Behind this decision were two main considerations, one theoretical and one functional. First, as explained above, blogs are theoretically accessible for a large amount of people around the globe and although its impacts may vary, climate change represents a challenge that is not restrained to specific regions or groups in society. Second, in order to benefit from the interdisciplinary composition of the team of authors, we did not want to define a specific scope of topics. We view our blog as an outlet that provides us with the opportunity to creatively experiment with different formats of texts and writing styles to convey scientific messages. The underlying rationale was the wish to understand how different formats appeal to readers and why. From our point of view, this openness and potential for creative freedom would have been impeded by limiting the scope of the blog. The different contributors provided a breadth of styles, which benefitted the blog without making it feel fragmented.

4 Our Experience with Blogging Climate Change

In terms of geographical coverage, our website statistics illustrate that people reading our blog are primarily from German or English-speaking countries (see Fig. 1). Almost 43% of the readers were from Austria. This is probably due to the fact that the team of authors is mainly based in Austria, which facilitates word-of-mouth advertising in this region. Users from Germany also comprised around 8% of the readership due to a few German blog posts and bloggers from Germany. However, due to the primary language of published articles being English readers from English speaking countries such as the United States, Canada and United Kingdom or countries that operate in English academically like India and South Korea were also prevalent.

In the following sections, we critically discuss our experiences with regard to drivers and barriers to successful climate communication via weblogs. Based on the website statistics of *Climate Footnotes* and discussions with colleagues and readers, we have identified the following main factors that can further help blogs in effective communication:

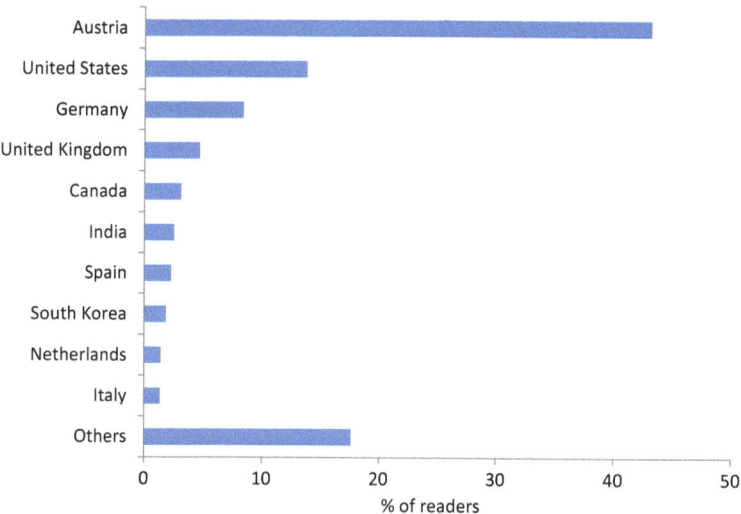

Fig. 1 Geographical distribution of readers

1. Breaking down complex scientific results into commonly understandable language (Sect. 4.1)
2. Responding to recent major events or articles that have recently been published in popular newspapers or magazines and addressing topics that are controversially discussed in a climate change context (Sect. 4.2)
3. Relating posts to people's everyday life, e.g., by means of storytelling or referring to specific regional and local contexts (Sects. 4.3 and 4.5)
4. Including interactive contents (Sect. 4.4)
5. Discussing weblog ethics and the possibility of controversial or dubious research (Sect. 4.6)
6. Promoting the blog through social media (Sect. 4.7).

4.1 Translating Scientific Data into Simple Terms

Results of scientific research are often solely published in peer-reviewed journals aimed at a small sub-group of the scientific community. But keeping in touch with a broader audience can help scientists reflect on their work and objectives (Dudo and Besley 2016). Also, in publicly funded research, an argument can be made for an obligation to communicate the results to the society (Lubchenco 1998), and when it comes to topics with a high societal relevance such as climate change and its implications, communication and interaction with society becomes even more crucial (Bonetta 2007).

In our blog, several posts break down topics connected to climate change, ranging from very broad questions like "Who measures global temperature?" to quite specific ones, like "Cheese versus Meat—Which Carbon footprint is higher?" The challenge of communicating a counterintuitive scientific concept also surfaces at times, such as with the post "Why is the pressure in meter?" This post originated from a long discussion of the author with the University's public relations department. In meteorology, upper-air pressure systems are frequently depicted as height anomalies (and are hence associated with the unit of length); however, for a layman these units of measurement could create confusion since pressure is usually measured in pascal or bar. In situations like this figures proved to be a powerful addition to mere text in explaining the underlying concepts and reasons. A blog provides a favourable frame for such an explanation which is uncommon in peer-reviewed scientific journal and would be at best available only when being entirely comprehensive, e.g., in a textbook.

4.2   Providing Perspectives on Current Events

One important advantage for blogs is that they can quickly amplify and respond to points in the news. One example which received a large amount of international public attention was the 2016 US Presidential Election. So, for instance, while it was fairly unknown what Secretary Clinton's climate policies were in the general election process, one of us wrote about what she had promised ("Clinton and Climate Change"). After the election, when Trump won unexpectedly, there was a lot of demand for discussion of potential policy implications and there was a new post on that topic ("The International Climate Effects of Trump"). By responding to potential policy implications of current events, the blog is able to participate in an interesting and important broader discussion. It is also worth introducing the science to help explain major events. For instance, after the devastating hurricane Matthew in 2016, a post addressed the question "Does global warming bring Matthew?" and also discussed the possibility of humans ever being able to stop such hurricanes.

4.3   Role of Language

For non-natives, the English language—much like jargon—can pose a barrier in understanding scientific text. However, even research focusing on regional and local issues is often published in international (and therefore English-language) journals making it harder for the regional (or local) community to become informed. In this case, breaking down the results in a blog post written in, for instance, German can reach more people. Some of the topics presented in our blog have a quite strong local focus (e.g., "The Raab Valley—A Catchment Description from a Hydrological View Point") and as Fig. 1 shows more than 50% of our readers come from German-speaking countries. The post "The pressure is on!— About the European cold spell in spring 2016"/"Nur nicht zu viel Druck! – Der Kälteeinbruch im Frühling 2016" was simultaneously published in English as well as German. The German version had almost 50% more views, highlighting the potential of communicating local and regional results in the respective native language.

Another possibility of making scientific knowledge more comprehensible for a broad audience is storytelling. Dahlstrom (2014) points out that "research suggests that narratives are easier to comprehend and audiences find them more engaging than traditional logical-scientific communication". Embedding scientific information into a story based on personal experience or pure fiction can help readers to identify themselves better with the issue and increase feelings of empathy. This especially applies to narratives that involve certain individuals to which a reader can emotionally connect; a concept called "personification" (Dahlstrom 2014). Our blog statistics confirm this claim. One of the more popular blog posts titled "Climate conferences and flying" addresses a paradox commonly experienced in a climate scientist's everyday life: climate scientists fly more frequently than the average population and thus have a large carbon footprint. Engaging with this paradox with reference to personal experience increased the narrative value and interest. Indeed, our website statistics show that blog posts that relate to personal experiences attract more readers and we would therefore suggest to carefully use storytelling for science-to-public communication on weblogs as, e.g., in the post "Diagnosis 'Cartastrophe': depressed cars are damaging the environment". This post uses cars as protagonists in order to recontextualize the counterproductive way that cars are treated.

4.4 Interactive Communication

The blogging platform is designed in a way which is conducive to textual inter-action, collaboration and social networking. The ability to organize information with the use of tags, categories etc. along with the use of embedded hyperlinks and even videos lends a dynamic nature and provides a greater depth to texts bringing in a diversity of voices to the fray. This allows for better access and ease of infor-mation to the readers compared to journal articles and print media (Davies and Merchant 2007). This new interactive model characterized by openness and short response time leads to the generation of a multitude of perspectives (O'Neill and Boykoff 2011). Such feedback and critiques allows for rapid development of ideas in contrast to academic journals where responses are usually incorporated towards the end stage of the publication process (Bubela et al. 2009). A survey on indicators of interactivity in academic blogs show that the level of interaction found in online blogs is similar to those at scientific conferences and meetings (Luzón 2011). Thus academic blogs can serve as an online meeting place facilitating information exchange and collaboration without the rigors of organizational issues related to conferences and meetings (Wulff et al. 2009). A particular post we would like to highlight in this regard was an online interactive quiz to test and spread awareness and information on climate change ("Take the Ultimate Climate Change Quiz!"). The post was created using a freely available quizzing platform called Qzzr. The quiz had ten multiple-choice questions related to climate change basics, news and politics. The quiz was the most popular post on the blog. One of the reasons for the popularity was most likely the interactive aspect, but it was also easy to share it with readers on other social media platforms which also encouraged participation.

4.5   Focussing on Topics Close to Home

Communication of climate change is easier when reflecting local experience (Shome et al. 2009). The cold spell which hit parts of Europe in late spring 2016 was coincidently strongly connected to one of the authors' research fields. A scientific article (Brunner et al. 2017) on the more general development of spring cold spells had been published only a few months earlier. The blog served as an ideal platform to communicate the specific meteorological condition which led to this cold spell based on the publication to a broader audience ("The pressure is on! —About the European cold spell in spring 2016").

Policy and local extreme weather events can even be connected such as in a post on Canadian forest fires ("Discussing Climate Disasters—Canada's Forest Fires"). Like Australia, Canada's economy is highly dependent on energy sources which contribute to climate change. But when forest fires hit the towns which are disproportionately responsible for contributing to the problem, there is an irony which is hard to miss. Another blogpost detailed the participation of one of the authors at the European Geosciences Union's General Assembly in Vienna ("EGU 2016 or Vienna stuffed with over 12,000 scientists"). This gave readers a first-hand narrative of a major European conference and the experience of a participating researcher. Such initiatives give the public access and a chance to appreciate the social workings of science, which is exclusive otherwise.

4.6   Weblog Ethics

Blogs are basically unregulated and non-peer-reviewed outlets and as such do not have strict underlying quality control. Yet, as is the case when doing research and presenting it at conferences and in scientific journals, researchers should ensure the correctness and transparency of what they publish on weblogs. Blood (2002) provides useful suggestions for universal weblog ethics and summarizes them in the following six points:

(1) one should publish as fact only that one believes to be true;
(2) if the material one refers to exists online, link to it;
(3) one should publicly correct any misinformation;
(4) one should write each article as if it could not be changed; one can add to it, but should not rewrite or delete, any entry;
(5) one should disclose any conflict of interest; and
(6) state questionable or biased sources.

When looking at these guidelines, it becomes obvious that most of them are also relevant as recommendations for best-practice research. In contrast to scientific publications, however, one major difference is worth highlighting: researchers might want to deliberately discuss biased or even dubious sources in their blog posts. Although there is danger of misleading communication if such sources are not acknowledged, a blog post can serve as a valuable platform for confronting questionable studies and their results with a researcher's personal opinion or other (reliable and transparent) scholarly work.

This especially applies to the climate change context. In order to tackle climate denial, for example, recent research argues that "normal academic response does not work since it presupposes that both parties follow basic rules of rational argumentation such as looking at the evidence as a whole" and that "denialism cannot be defeated by just providing more information" (Björnberg et al. 2017). In contrast to academic journals, which are mostly subject to these limitations, blogs can provide researchers with the opportunity to openly reflect on such issues and react to current developments with only minimal delay (see Sect. 4.2). Due to the more liberal format, blogs can thus also be used to transfer a (purely scholarly) discussion from the literature to a broader audience.

4.7   Readership and Promotion

To increase readership and target a wide range of audiences, we tried to promote our blog through different communication channels, such as Facebook, university homepages, academic conferences or mailing lists. Interestingly, however, most readers seem to have accidentally "stumbled upon" our blog while searching the internet for specific terms. Our website statistics show that search engines were the most common referrers (82% of clicks), indicating that is important to include relevant keywords in blog posts and use search engine optimization for webpages.

Facebook proved to be the second most common referrer (6% of clicks), which shows that sharing new blog posts on social networks with friends and acquaintances is an important factor for knowledge dissemination. It should be noted that Twitter might be another useful communication outlet. The integration of Twitter into our communication strategy, however, is still in the planning phase and at this time little can be said about its potential benefits.

In general, the challenge of spreading word of our blog was one of the most difficult issues we faced. While a few blog posts gained traction, the most popular was only seen by a few hundred people, sobering statistics for the amount of time invested (despite being a number far greater than the average academic journal article). While it does take time for blogs to gain momentum and traction (search engine optimization can take years, audiences time to acclimate to a new source, differentiation with competition is timely) and therefore success remains undefined for some time, in the end the primary motivation has been what little communication we *can* facilitate, a sense of obligation as scientists to contribute to a public debate and the challenge of stretching our capacities and improving our writing skills.

5 Conclusions

Social media is a viable and effective medium to disseminate information about climate change. However, the field would benefit from substantially more research on the use and implications of social media concerning climate change. Consequently, analysis of the various facets related to online climate change communication is relevant and requisite. This study describes the potential of one such medium of online communication—blogs—using our own experience with *Climate Footnotes* as a case study. Through our experience and a review of relevant literature, we highlight the usefulness of blogs in providing scientific education, furthering scientific discussions and encouraging participatory science. Furthermore, drawing from communication theory and our experience, we also identify key factors that may help researchers utilize blogs to communicate issues associated with climate change effectively across a wider audience such as:

- Making information memorable and impactful by reducing scientific complexity
- Quickly addressing and responding to current events that are in the news
- Using experiential and emotional frames like stories and personal experiences
- Tapping into the interactive potential of blogs using hyperlinks, comments and other features
- Fostering interest of the public using a regional or local message framing
- Ensuring proper research conduct and transparency
- Using other social media platforms to share and market the blog.

We believe that the ideas outlined here could be helpful for researchers who are trying or considering trying blogs as a medium to communicate climate change science. However, while the potential of scientific blogging is apparent, it has not yet been fully realized and our blogging experiences are not necessarily representative. It would be beneficial for other scientific blogging practitioners to continue to publish their experiences so that we can collectively determine best practices. This calls for further research and initiatives to deepen our understanding of the nature and impact of online climate change communication outlets, including blogs. This will enable the scientific community to explore how we might leverage our voice in the conversation around the pressing global challenge of climate change.

Acknowledgements The authors want to thank all other colleagues who also were part of the blogging team for valuable contributions and discussions, namely Arijit Paul, Sungmin O, Eike Düvel, Johannes Haas, Clara Hohmann, Philipp Babcicky, Javier Lopez Prol, Vincent Heß, Micheal Kriechbaum and Daniel Petz. This work was funded by the Austrian Science Fund (FWF) under research grant W 1256-G15 (Doctoral Programme Climate Change—Uncertainties, Thresholds and Coping Strategies). L. Brunner was financially supported by a Marietta Blau Grant by the Austrian Exchange Service (OeAD), financed by funds from the Austrian Federal Ministry of Science, Research and Economy (BMWFW).

References

Björnberg KE, Karlsson M, Gilek M, Hansson SO (2017) Climate and environmental science denial: a review of the scientific literature published in 1990–2015. J Clean Prod 167:229–241

Blanchard A (2011) Science blogs in research and popularization of science: why, how and for whom? In: Cockell M, Billotte J, Darbellay F, Waldvogel F (eds) Common knowledge: the challenge of transdisciplinarity. EPFL, Lausanne, pp 219–232

Blood R (2002) "Weblog ethics", the weblog handbook: practical advice on creating and maintaining your blog. Perseus Publishing, Cambridge, pp 114–121

Bonetta L (2007) Scientists enter the blogosphere. Cell 129(3):443–445

Brunner L, Hegerl GC, Steiner AK (2017) Connecting atmospheric blocking to European temperature extremes in spring. J Clim 30(2):585–594

Bubela T, Nisbet MC, Borchelt R, Brunger F, Critchley C, Einsiedel E et al (2009) Science communication reconsidered. Nat Biotechnol 27(6):514–518

Dahlstrom MF (2014) Using narratives and storytelling to communicate science with nonexpert audiences. Proc Natl Acad Sci 111(Supplement 4):13614–13620

Davies J, Merchant G (2007) Looking from the inside out: academic blogging as new literacy. A New Literacies Sampler pp 167–197

Dudo A, Besley JC (2016) Scientists' prioritization of communication objectives for public engagement. PLoS ONE 11(2):e0148867

Fischer EM, Knutti R (2015) Anthropogenic contribution to global occurrence of heavy-precipitation and high-temperature extremes. Nat Clim Change 5(6):560–564

Funk C, Kennedy B (2016) Public views on climate change and climate scientists. Retrieved 08 Sept, 2017. From http://www.pewinternet.org/2016/10/04/public-views-on-climate-change-and-climate-scientists/

Heo GM, Lee R (2013) Blogs and social network sites as activity systems: exploring adult informal learning process through activity theory framework. Educ Technol Soc 16(4):133–145

Leshner AI (2003) Public engagement with science. Science 299(5609):977

Livingstone DW (2001). Adults' informal learning: definitions, findings, gaps, and future research. NALL Working Paper# 21

Lubchenco J (1998) Entering the century of the environment: a new social contract for science. Science 279(5350):491–497

Luzón MJ (2011) 'Interesting post, but I disagree': social presence and antisocial behaviour in academic weblogs. Appl Linguist 32(5):517–540

Marsick VJ, Watkins KE (2001) Informal and incidental learning. New Dir for Adult Contin Educ 2001(89):25–34

O'Neill S, Boykoff M (2011) The role of new media in engaging the public with climate change. In: Whitmarsh L, O'Neill S, Lorenzoni I (eds) Engaging the public with climate change: behaviour change and communication. Routledge, London, pp 233–251

Putnam L (2011) The changing role of blogs in science information dissemination. Issues Sci Technol Librariansh 65(4)

Rollett H, Lux M, Strohmaier M, Dosinger G, Tochtermann K (2007) The Web 2.0 way of learning with technologies. Int J Learn Technol 3(1):87–107

Schäfer MS (2012) Online communication on climate change and climate politics: a literature review. Wiley Interdiscip Rev: Clim Change 3(6):527–543

Schugurensky D (2000) The forms of informal learning: Towards a conceptualization of the field. NALL Working Paper #19, 2000. Cardiff: Cardiff University. From https://tspace.library. utoronto.ca/bitstream/1807/2733/2/19formsofinformal.pdf

Selwyn N (2007). Web 2.0 applications as alternative environments for informal learning—a critical review. Paper presented at the OECD-KERIS expert meeting—session 6—alternative learning environments in practice: using ICT to change impact and outcomes. From http:// www.oecd.org/dataoecd/32/3/39458556.pdf

Shome D, Marx S, Appelt K, Arora P, Balstad R, Broad K et al (2009) The psychology of climate change communication: a guide for scientists, journalists, educators, political aides, and the interested public

Steentjes K, Pidgeon NF, Poortinga W, Corner AJ, Arnold A, Böhm G et al (2017). European perceptions of climate change (EPCC): topline findings of a survey conducted in four European countries in 2016

Trench B (2012) Scientists' blogs: Glimpses behind the scenes. In: Rödder S, Franzen M, Weingart P (eds) The sciences' media connection—public communication and its repercussions. Springer, Heidelberg, pp 273–289

Wilkins JS (2008) The roles, reasons and restrictions of science blogs. Trends Evol Ecol 23(8): 411–413

Wulff S, Swales JM, Keller K (2009) "We have about seven minutes for questions": the discussion sessions from a specialized conference. Engl Specif Purp 28(2):79–92

Creative Collaborations: Museums Engaging with Communities and Climate Change

Jennifer Newell

Abstract When museums and communities collaborate on fostering public engagement in climate change the results are powerful. Coming together around collections, co-producing and sharing knowledge, and creating outreach programs, members of specific cultural or interest groups can catalyze changes in perceptions within broad audiences. Museums are now considering the human and cultural dimensions of climate change more fully than in previous, more science-focused, phases. Museums have been gradually exploring more imaginative ways to captivate audiences and communicate views from people on the "front line". This paper employs views from inside two particular institutions and their relationships with Pacific Islanders to explore the dynamics and challenges of collaborating creatively and effectively around climate change communication.

1 Introduction

The Australian Museum launched an innovative, community-driven program of arts and activism in March 2018: *Oceania Rising: Climate Change in our Region.* At the launch Joseph Zane Sikulu, Sydney-based leader of 350.org Pacific, and a Tongan man, spoke about how the Australian Museum makes time and space for him and the 350.org Climate Warriors, whether it is for one of their Talanoa for Climate, or a program for Pacific youth (Sikulu 2018). He then said:

> Climate change conversation and climate change activism at the museum always makes so much sense to us because we always say that if we don't do anything about climate change then one of the few places we are going to be able to experience our culture is in the museum.

J. Newell (✉)
Pacific and International Collections, Australian Museum,
1 William Street, Sydney, NSW 2010, Australia
e-mail: Jenny.Newell@austmus.gov.au

© Springer Nature Switzerland AG 2019
W. Leal Filho et al. (eds.), *Addressing the Challenges in Communicating Climate Change Across Various Audiences*, Climate Change Management,
https://doi.org/10.1007/978-3-319-98294-6_10

If you've had the chance to walk through the archives here and see the Pacific collection, you'll understand how valuable our culture is, and how much of it there is. (Sikulu 2018)

These are valid points. We at the Australian Museum hope, however, that Joseph and the Climate Warriors will feel, through our work together, that museums are not only useful as storage houses of things now gone from lived lives, and more than sentinels of what can be lost if humans fail to get their act together. We hope Joseph and his colleagues will also see that museums have the capacity to be change agents; powerful allies in advancing climate change conversations and galvanizing action (Cameron and Neilson 2018; Janes 2010; Newell et al. 2016).

Museums are effective meeting places for people and perspectives from across boundaries. They are conscious of reaching out to all, regardless of background and ability, and of offering relatively safe spaces for difficult conversations. They are supportive, non-judgmental places for making enriching connections and exploring beyond comfort zones. As Esme Ward, Director of the Manchester Museum, stated at the opening of the International Symposium for Climate and Museums (Manchester Museum, April 2018), museums are *places of care*.

Museums are also fundamentally creative. They are well-structured for creating and sharing ideas and co-producing knowledge. They create entrance points for people of a great diversity of backgrounds to broaden their horizons and see new ways of navigating their worlds. This paper explores the potency of museums when they work closely with communities, with a focus on responding to climate change. Here we will demonstrate that working across boundaries and with communities beyond the institution is highly valuable and challenging for participants on either side.

There are three main points this paper makes:

1. Creating trust between a museum and a group that is not of the museum's dominant culture is hard work.
2. This work is worth it. What results from this trusting collaboration is revelatory, and carries great potential for creating conceptual shifts for broad audiences.
3. These relationships can be established on any scale. A substantial team, plenty of resources and a management supportive of community work and climate change projects makes the work easier, but there are still very effective engagements that can be activated by a team of one, with minimal resources.

This paper employs views from inside two particular institutions and their relationships with Pacific Islanders on two different scales. The aim is to highlight some of the dynamics, potentials and challenges in developing collaborative, cultural initiatives around climate change. This paper is not a detailed or sustained analysis of the factors at work in the nexus between museums, communities and climate change, but is rather an ideas piece, aspiring to share insights into that nexus based on specific examples. It draws on the author's personal experiences over eighteen years of working as a museum curator on projects with Pacific communities in Canberra, London, New York and Sydney, and, over the past six years, working to foster climate change-engaged citizens through museums. Ideally, the learnings gleaned from these precious, mutually rewarding collaborations may be a useful resource for others.

Initiatives to bolster engagement through museums are increasingly being informed by current scholarship on climate change communication. Researchers in this new field, notably those at the Yale Program for Climate Change Communication, the Tyndall Research Centre, Climate Outreach, and the Australian National University Climate Institute work from within diverse disciplinary approaches including anthropology, sociology, psychology, media studies and political science. The complex dynamics of a rapidly intensifying degree of change in global systems, divergent approaches to consumption and custodianship within and between societies, and the social and psychological dimensions of facing —and not facing—this wicked problem all make for communication requirements that are anything but easy. As the Yale Program notes, climate change communication is not a matter of a message being sent out to a receiver, but a great complex of interactions whereby "individuals, communities, and societies come to understand, care, and act on climate change through their communication with other people" (climatecommunication.yale.edu) (Fig. 1). There is a need for more research into cross-cultural communication to support effective conceptual shifts within industrialised societies. The shift that is needed, urgently, is the foundational one of moving from a broadly consumptive relationship to the natural world to a sustainable relationship, a shift from imagining humans as separate from nature to understanding humans, plants and animals as "components of each other's environments" (Ingold 2011: 87).

Whether global, national or more localized, studies of attitudes to climate change are revealing the importance of variables such as access to basic education and the

Fig. 1 A group from Kiribati at the Australian Museum, including (centre) Maria Tiimon Chi-Fang and Pelenise Alofa, director of KiriCAN, Kiribati Climate Action Network, and other participants in the Resilient Leaders program, Edmund Rice Centre and Pacific Calling Partnership, Sydney, March 2018. *Photo* Author

availability of information about local impacts. People's own observations about local environmental change can be comprehensive, but without hearing that these alterations fit into a worldwide phenomenon (as is more likely to be the case in remote, non-industrialised regions), changes can be interpreted as temporary and specific to the place. Indeed, global surveys in 2017 demonstrated that 40% of adults have never heard of climate change (Leiserowitz and Howe 2015). International research collaborations have been producing longitudinal, comparative analyses of the predictors of attitudes across many countries. A study by Tien Ming Lee, et al., published in *Nature Climate Change* of attitudes within 119 nations establishing that some variables such as education can be identified as broadly applicable predictors of awareness and attitudes across countries, with other predictors, such as socio-economic status, being highly localized (Lee 2015). As Lee states, the study reveals the need for climate change communication that is tailored to individual countries, "and even for areas within the same country" (Dennehy 2015). Co-author Anthony Leiserowitz notes that their results "also indicate that improving basic education, climate literacy and public understanding of the local dimensions of climate change are vital for public engagement and support for climate action." (Dennehy 2015).

Awareness and acceptance of climate change as a phenomenon, while an important first step, does not lead directly to engagement because it is often seen as a "distant threat" (Scannell and Gifford 2013: 62). As George Marshall (founder, Climate Action) states, a key challenge for achieving engagement is that people (primarily in the West) perceive climate change as happening far away and to someone else (Marshall 2015: Chap. 13).

Leading from this, of particular use for museums are the studies of the characteristics of effective climate change communication (such as Scannell and Gifford). One of the more comprehensive psychological studies, by Sander van der Linden et al., concludes that rather than framing climate change as "a future, distant, global, nonpersonal, and analytical risk", which is an "overt loss for society", those engaging in climate change policy and communication can most effectively adopt five key approaches:

(a) emphasize climate change as a present, local, and personal risk; (b) facilitate more affective and experiential engagement; (c) leverage relevant social group norms; (d) frame policy solutions in terms of what can be gained from immediate action; and (e) appeal to intrinsically valued long-term environmental goals and outcomes. (van der Linden et al. 2015)

Emphasising the local, social and personal dimensions of climate change, enabling emotional and creative engagement, and the positive outcomes of prompt action: these are the kinds of insights that can ensure effective education and public programs, exhibitions and other forms of outreach with communities. The factor that surfaces repeatedly throughout the literature is keeping the local firmly in view when tackling this global challenge.

2 Cultural Dynamics of Climate Change

Climate Change is cultural. Its causes are cultural, and it has cultural impacts. Consequently, responding effectively to climate change requires deep conceptual shifts—paradigm shifts—to limit and cope with it; changes in cultures of relating to the natural world, to personal and group responsibility for others in the living world, and to consumption. Those in Western, industrialised cultures can potentially learn much from those in cultures with stronger traditions of sustainable modes of living; primarily Indigenous/First Nations cultures.

From the first climate exhibition in 1992: *Global Warming: Understanding the Forecast*, American Museum of Natural History (which attracted 700,000 people and a curatorial award (Novacek 2016; Zelig and Pfirman 1993)) and for some sixteen years thereafter, museums focused on explaining the *science* of climate change. In 2008, the phenomenon of global warming had clearly become familiar enough in many regions for there to be three exhibitions in that single year that dealt in a sustained way with *social* and cultural dynamics of the phenomenon (in Rotterdam, Melbourne and New York) (https://mccnetwork.org/exhibitions/). Museums are now in a phase of being much more comfortable with creating multi-dimensional, creative, emotive presentations considering impacts of the climate crisis on societies and their ecologies (Newell et al. 2016: 7).

Co-producing exhibitions with community members about the climate crisis is still quite rare, but signs indicate this will increase, in line with general democratizing, decolonizing trends in the sector (Newell et al. 2016). During the five years this author was Pacific curator at the American Museum of Natural History (AMNH) in New York, it was common for the talks presented with Pacific colleagues about climate change and impacts on Pacific communities to draw a shocked response from audience members. People would come up after a talk, looking stricken, saying "I had no idea". Their next statement was usually "what can I do?". A substantial part of this author's collaborative research and curatorial work that has followed these encounters has been directed at finding ways to more adequately answer this question, and assist more people to get to the point of asking it.

The American Museum of Natural History's pioneering climate change exhibitions of 1992 and 2008 (*Global Warming* exhibit and *Climate Change: The Threat to Life and a New Energy Future*) have meant that for the period since 2008 until at least 2016, key decision makers determining the temporary exhibition program felt the museum had already "done" the subject for the time being. Excellent climate content was provided instead by the Education department for online and onsite courses. The AMNH is now, in 2018, developing new climate content for the redesign of the Hall of Planet Earth.

While at the AMNH, one of the ways that this author worked to enable community connections to collections was to run "Open House" sessions from time to time so members of the Pacific diaspora in New York could discover, in their new home, treasures made by their ancestors. Discussions afterwards over afternoon tea were opportunities to learn about diverse ways of experiencing and responding to climate change. The Open Houses were an important first step in establishing a way into the museum and into conversations for those who previously had not felt particularly welcome. The Hall of Pacific Peoples at the AMNH is Margaret Mead's 1970s hall, and the displays give the general impression of cultures being dead and gone. Pacific people can find this representation of their cultures to international audiences quite upsetting. When Kristina Stege, a Marshall Islander, anthropologist and activist living in New York, saw the Hall she decided to never return to the museum. Fortunately, a friend encouraged her to come to an Open House in 2013, and she afterwards decided to participate in a community connections project around the impacts of increasingly severe weather that the AMNH ran with the Museum of Samoa (Principal Officer Lumepa Apelu). The collaboration established has been generating effective climate change projects, education programs, publications and research grants ever since.

Thus, something as simple and low-cost as opening up the collections for an afternoon from time to time, in a way that is sensitive and respectful, can help build trust and great outcomes. Trust can also be fostered when those on the museum side understand something of the tug-of-war that members of source communities usually feel on entering collections. It can be deeply painful to see things of home (which are often not only "things" but kin, in a very real way), divorced from the life of home, detained on shelves in colonial institutions. At the same time there is the often awe-inspiring experience of being in the presence of beautiful, astounding things made by ancestors, of being reunited with kin and their treasured works. These special visits also help visitors, those outside the dominant culture of the institution, to gradually feel that the place is as much for them and as relevant to them as anyone else, and that their relationships to the treasures in the storerooms are valued. These entry points help break down fears about museums allowing the forbidding, paternalistic Ghost of Museums Past to start to give way to the rather more open-armed, receptive and lively sprite of Museums present.

None of this trust-building is easy; there are expectations on both sides that are opaque to the other until a boundary of politeness or protocol is unwittingly crossed. Sometimes bridges have to be rebuilt. Community members can be unsure about what is acceptable to ask, not sure if it is all right to lay hands on an object, to sing in the storeroom, to bring in green leaves. Museum staff facilitating the visit are often uneasy about what requests might be made, whether the head of collections will allow the visiting group to touch an ancestral treasure, and how to defuse tensions between all parties (Lythberg et al. 2015: 199–200). There are always relationships to balance and protocols to get right, but it is standard to see patience, good will and humour on both sides. Carla Krmpotich and Laura Peers' book *This is Our Life* offers a useful set of reflections on exactly these dynamics, with contributions from everyone involved in a Haida Gwai collection reconnection project at the Pitt Rivers and British Museums (Krmpotich and Peers 2013).

3 Recognising a Changing Pacific

Collaborating allows for the co-creation of new understandings, and access to alternative sources of knowledge and experience that would not otherwise be possible. The jointly-designed, jointly-carried out work that this author has been involved in, the learning has been deep on all sides; whether it has been interviewing people with Kristina Stege in the Marshall Islands about their methods for dealing with the environmental changes they have been experiencing (Niarchos 2016), or bringing community groups in Samoa and New York together to compare personal impacts of hurricanes (AAM 2018), or creating a video based on interviews with Larry Raigetal, traditional navigator of Lamotrek (Federated States of Micronesia) about the new challenges to reading the sea, and his activism. Raigetal is sending a traditional woven canoe sail, signed by all on the island, to museums and other public venues around the world to raise awareness of the place they are losing (Fig. 2). The Pacific is one of the regions in the abominable situation of having contributed least and suffering most from global warming. In the face of deep and devastating losses, much of what shines through in Pacific responses is profound strength, creativity, and persistence, neatly summed up in the 350.org Climate Warriors' "Not Drowning, Fighting" slogan.

Through the collaborative projects in the Pacific our museum teams have learnt particularly about practices of care in different cultures. For instance, in the outer islands of the Marshall Islands, when people are asked what methods they have for dealing with food plants dying after being flooded repeatedly by seawater, and how they manage the empty water tanks that come with lengthening droughts, they point to a cultural resource: the Marshallese practice of *laledron*; caring for each other. Neighbours will share what food, water and shelter they have. There is a sense of cultural resilience in the responses from Marshallese respondents referred to as being a key method for managing climate change impacts. There is a real sense of shared strength, often pointing to traditions of self-support such as canoe building and weaving, enshrining many key cultural values, crucial to pass on to future generations. As Noah Luther of Namdrik said in an interview about what helps in dealing with changes to the environment:

> The skills we need are: canoe-making, so that we don't depend on fuel. And handicraft-making like mat weaving, because it's our tradition. We need to preserve our skills, our custom, our language, our traditional food. I want our next generation to know all these things. (12 August 2016, interview, Niarchos project)

In places across the Pacific there are usually local words for "caring for each other", and "working with each other". These foundations of care are at the heart of so much of the operation of everyday life, as well as picking up again after disaster— after another one of the increasingly intense cyclones has hit, a storm surge washes right across an atoll, another stretch of ancestral land is lost to rising waters, a drought stretches even longer, all the bananas and breadfruit die off, or a reef fails to bounce back after bleaching. These practices of care typically extend to the land, sea

Fig. 2 First display of Lamotrek's woven pandanus leaf sail, at the University of Guam. The sail was sailed into Guam harbour at the start of the Pacific Arts Festival 2016, and has been displayed at the UN Headquarters in New York, Hawaii, Germany, and Sydney in 2018. Photo: Sandra Okada

and living things beyond people and ensure the maximum amount of resilience in the face of both the slow and fast violence of deep environmental change (Nixon 2013).

Sharing these learnings—philosophical and practical—through the museum's public programs, publications and exhibitions is increasingly important, as the necessity to find ways to cope with unpredictable conditions and increasing losses is rapidly becoming universal. Talks and story-telling sessions beyond the museum —at schools, community centres, and public events—as well as sharing presentations and podcasts online—are simple methods. More complex is finding ways to incorporate climate change experiences and responses into exhibitions. Museum managers can be uneasy about tackling what is seen to be a "bad news" topic and nervous about possibly getting government supporters and big funders off-side. However, there are opportunities that curators can pursue—pitches can be made for small scale exhibitions or single cases focusing on a particular angle and can be framed in a way that includes positive approaches and actions, all of which helps to enable audiences to remain hopeful and take action of their own. Small-scale exhibitions will not make as big a splash as a major exhibition, but they are more readily managed with limited resources, and have greater chances of getting off the ground. These exhibitions can be more nimble and experimental than major exhibitions, can be more readily co-curated with communities, and travel beyond the museum more easily, yet still provide clear, engaging and important content.

In 2016 several staff at the AMNH, with a team of Columbia University students, worked with several collaborators from the Marshall Islands, Tonga and Aotearoa/New Zealand to bring in their insights—and their cultural objects—to enliven the Hall of Pacific People. With their advice we were able to create "A Changing Pacific", a single case that was divested of its Margaret Mead content and given a set of themes to do with the Anthropocene in the Pacific: globalization, migration, connections to nature, and climate change. The case features poetry by Kathy Jetnil-Kijiner, one of the world's most potent climate activists. Addressing the increasing need for moving away from ever-more challenging home islands, there are vibrant works in feathers and plastic by Tongan artists living in the diaspora in Aotearoa New Zealand and Sydney, as well as a range of contemporary works drawn from the Museum's collection, highlighting our global interconnectedness: a *bilum* net bag by a young woman in Papua New Guinea, hand-knotted with a Toyota logo; and a print by a young Maori artist combining a traditional *tiki* with a barcode. Kristina (Tina) Stege donated a Marshallese *jaki-ed* fine mat for the case, and selected from the collection a *meto* "stick" navigational chart, to allow her to speak of the ways Marshallese are navigating challenging, rising seas. There is no chance that a visitor to the Hall now could miss the message that Pacific cultures are as vivid and dynamic as ever. And they can also get the message there, simply and easily, that climate change is having a big impact on people's lives. Not at some future point; now.

4   Collaborating in Sydney

At the Australian Museum, a team is engaging with Pacific communities and climate change on a larger scale, but with many of the same principles for building working relationships as in New York. The crucial difference is the Pacific collections team is not one but six people, and half of the team are Pacific Islanders, including Dr. Michael Mel, co-manager of the Pacific and International Collection. Trust with the communities in Sydney has already been established over many years, with many joint projects, the Museum's staff welcoming groups into the collection and for Pacific-centric events on-site, as well as working off-site, running youth reconnection projects at community centres and in the juvenile justice system, and taking an Australian Museum booth to community events and festivals, presenting what is in the collection, and making it clear that it is open for access. Sydney, particularly Western Sydney, is home to the majority of Australia's 340,000-strong Pacific Islander and Maori population (Batley 2017). While it is a substantial commitment of time and money for community members to come from Western Sydney to the Museum in the CBD (a three hour round-trip), groups visit the storerooms about once or twice a month, and for some events the Museum provides a bus to bring in groups.

There are a range of people within the Sydney Pacific community who are frustrated by the very large Australian apathy about the growing climate change crisis and want to work with the Australian Museum to try to reach out to the general public. They also see the Museum as a means to captivate sections of their own communities. For instance, some Pacific elders note that many of their youth are giving little serious thought to the issues, and complain that the only messages they listen to are in rap. It is fortunate, therefore, that Fijian rap artist and youth worker, Thelma Thomas (*a.k.a.* MC Trey), is part of the Australian Museum's Pacific team.

We are confident that Pacific communities in Sydney, and out in the Pacific, will continue to ask the Museum to help with a wide variety of ways of enacting the power of objects in the storeroom, enabling powerful conversations and amplifying voices.

Being open to the direction and drive the Pacific diaspora provides means we can convey their "up close and personal" perspectives on what is happening not only in home islands but also in Sydney; issues that the mainstream population tends to keep in their blind spot. Orators, artists, and others bring "ground truthing", a real kick to messaging, and effective, emotive, bearing of witness. Angela Tiatia, a video artist of Samoan-Australian heritage says of her video work "Tuvalu":

> The Pacific is the canary in the global mine. We can see a glimpse of the challenges we may face in our collective future by looking at life in some of our smallest nations. The remoteness of these island nations mean many of these warnings are going unwitnessed. This work bears witness to, and laments, what we are losing to climate change. (Tiatia 2018)

Angela's *Tuvalu* (Fig. 3) is the centerpiece of the program *Oceania Rising Climate Change in our Region*. It is part of the Australian Museum's growing climate change initiative, headed up by Director Kim McKay. This initiative includes the development of a permanent display in one of the galleries, as well as a pop-up touring exhibition, which participants in the Symposium on Climate Change and Museums in Manchester contributed ideas to. Programming includes hosting a

Fig. 3 Angela Tiatia's work, *Tuvalu*, 3 channel video. 20:34 min (2016), on display at the Australian Museum, 2018. Edition 1 of 3, Australian Museum. *Photo* Author

lecture series with high caliber local and international speakers, *HumanNature,* designed by a consortium of four universities in Sydney. The Australian Museum is also developing additional education programs and updated website content. We have been talking to Angela Tiatia, Joseph Sikulu, artist Latai Taumopeau, and others who share our desire to ramp up climate awareness and engagement in Sydney. The Museum is partnering with two art centres in Western Sydney; Casula Powerhouse Art Centre and Blacktown Arts who have large Pacific constituencies. We are developing the program together, sharing the reins. The video work *Tuvalu* is showing at the Museum first, and it will then travel to the partner institutions, where it will be surrounded by other artworks, talks, workshops, performances and film showings (Fig. 4).

At the launch event (at which Joseph Sikulu spoke), we placed pens and a page on seats in our auditorium, asking: "What are your key questions about climate change?" and "What question would you like us to answer together?". Most of the 120 people in the audience responded. Distilling these down into four key question types, the most common question was "what can we do?". The second most common was "how can we make the Australian government act", and thirdly "how can we communicate better on climate change?" The fourth type of question was one seeking specific information about the workings of climate change.

Part of what we will be doing to seek good answers to these questions is holding talanoa (a Fijian model for supportive, equitable group discussions) around each of those questions. We will also be having "story parties", some in the storeroom, eliciting stories about environmental and cultural change, and also about actions and activism that people have tried, and things that others might like to try. Stories about environmental loss (for instance noting the shells that appear in a woven

Fig. 4 Community kick-off of the *Oceania Rising: Climate Change in Our Region* program, featuring the Matavai dance group and Angela Tiatia's *Tuvalu* video artwork. *Photo* Author

Tongan bag can no longer be found on beaches. Figure 5) can help audiences see what is already changing. But we should not leave it there. We need to make sure the stories can reach further, grab hearts and minds and lead in easy steps to something concrete like joining one of the activist groups already doing important work. There are many groups who leverage the power of collaboration, in Sydney and in the Pacific, all whom are very accessible online. There are groups addressing a wide range of issues and sub-sections of society. For youth, for instance, there is the Youth Climate Coalition in Australia; groups in the Pacific such as Kathy Jetnil-Kijiner's Jo-jikum youth group; and Miranda Massie at The Climate Museum in New York has just established a Youth Climate Advisory Panel, a useful model that could be replicated elsewhere. These many groups are, with resilience and creativity, creating *communal* action. This is what we, too, hope to achieve with *Oceania Rising.*

Learning ways to tell powerful stories is one thing we hope to enact for as many people as possible through the course of the year. As one of my favourite historians, William Cronon says, "the stories we tell change the way we act in the world" (Cronon 1992: 1375). This is the impetus behind the workshops on "Social Networking and Activism" that Angela Tiatia will run as part of *Oceania Rising,* with participants using iPhones, green screens and footage to create short video works that will get them seen and heard on social media.

We expect the outcomes from *Oceania Rising* to be new artworks, all those potent stories, and many answers to not just the "what can I do?" question, but more powerfully, "what can *we* do?". We are looking for answers that work well in Sydney, for the great diversity of communities there, and we expect to unsettle the apathy of a wide range of Sydneysiders.

Fig. 5 Lady Fielakepa draws attention to the diminishing lopa seeds & sea shells that were once widely used throughout Tongan material culture. Mrs Snell collection documentation project. *Photo* Yvonne Carrillo-Huffman. ©Australian Museum

5 Conclusions

Community members are increasingly recognizing museums as a place for making things happen. Museums will continue to provide a platform for talking between otherwise disparate communities. Members of communities are seeing that far from being a set of closed doors, museums are often warmly open spaces; still regulation bound, but willing to make accommodations. Museums are places of which community members can make requests, use resources, use things from home countries in new ways, to speak of tradition and history, of now and the future.

The relationships forged between people within and without the museum are always carefully negotiated, rarely straight forward, often outside comfort zones. For those outside, working with a monolithic institution can be daunting, and it can often feel that relationships, reputations within one's own community, and any chance of a neatly resolved project, are all at risk. Making it work is exhilarating; but like any complex project, collaborations are rarely a complete success, with all expectations met. This is, however, all right. The nexus between community, collections and our changing environment will always be rich, layered, experimental, and guaranteed to be impactful. Bridging gaps, enabling powerful conversations to be held, in the museum space and beyond, museums and communities working together is a creative, dynamic force that can unleash capacities to initiate the paradigm shifts that we urgently need. These collaborations across cultures can inspire re-imagined practices of care, enabling us to create ourselves as intimately involved, responsible members of the world.

Acknowledgements The author acknowledges the Gadigal People of the Eora Nation, the traditional owners and custodians of the land the Australian Museum is sited and where the work for this piece was carried out. She also gratefully acknowledges the many Pacific scholars, artists, and community members who have generously shared their wisdom and care over the years. Newell's cited research projects were funded by the Museums Connect grant program of the American Alliance of Museums, part of the US State Department's Bureau of Educational and Cultural Affairs (2014); the Constantine Niarchos Foundation ("Analysing the dynamics shaping community responses to climate change in the Republic of the Marshall Islands", 2016); and Sydney Environment Institute ("Living with a Changing Sea: navigating through climate change in the Pacific", 2016). The author wishes to thank Walter Leal for his constructive comments.

References

American Museum of Natural History (AMNH) and American Alliance of Museums (AAM) (2018) Rethinking home: climate change in New York and Samoa. https://www.amnh.org/our-research/anthropology/projects/rethinking-home/. Last accessed 1 May 2018

Batley J (2017) What does the 2016 census reveal about Pacific Islands communities in Australia? State, Society and Governance in Melanesia Program, Australian National University, series: In Brief, 2017/23

Cameron F, Neilson B (eds) (2018) Museum change and museum futures. Series: Routledge Research in Museum Studies (8). Routledge, London and New York

Cronon W (1992) A place for stories: nature, history, and narrative. J Am Hist 78(4):1347–1376

Dennehy K (2015) Study reveals what the world thinks of climate change. Yale News, 27 July 2015. (https://news.yale.edu/2015/07/27/study-reveals-what-world-thinks-climate-change)

Ingold T (2011) The perception of the environment: essays on livelihood, dwelling, and skill. Routledge, London and New York

Janes R (2010) Museums in a troubled world: renewal, irrelevance or collapse?. Routledge, London and New York

Krmpotich C, Peers L (2013) This is our life: Haida material heritage and changing museum practice. University of British Columbia Press, Vancouver

Lee TM, Markowitz EM, Howe PD, Ko C-Y, Leiserowitz AA (2015) Predictors of public climate change awareness and risk perception around the world. Nat Clim Change 5:1014–1020

Leiserowitz A, Howe P (2015) Climate change awareness and concern in 119 countries. Yale Project for Climate Change Communication

Lythberg B, Newell J, Ngata W (2015) House of stories: the whale rider at the American Museum of Natural History. Mus Soc 13:195–202

Marshall G (2015) Don't even think about it: why our brains are wired to ignore climate change. Bloomsbury USA

Museums and Climate Change Network (2018) Chronology of exhibitions relating to climate change and/or anthropocene. https://mccnetwork.org/exhibitions/. Last accessed 20 May 2018

Newell J, Robin L, Wehner K (eds) (2016) Curating the future: museums, communities and climate change. Routledge, London and New York

Niarchos Foundation project (2016) Principal investigators: Jennifer Newell and Kristina Stege, summary of research findings (video). www.amnh.org/explore/amnh.tv/(watch)/research-and-collections/climate-change-in-the-marshall-islands/(category)/131059. Last accessed 18 May 2018

Nixon R (2013) Slow violence and the environmentalism of the poor. Harvard University Press, Cambridge, MA

Novacek M (2016) Foreword. In: Newell J, Robin L, Wehner K (eds) Curating the future: museums, communities and climate change. Routledge, London and New York

Scannell L, Gifford R (2013) Personally relevant climate change. The role of place attachment and local versus global message framing in engagement. Environ Behav 45:60–85

Sikulu JZ (2018) Presentation for "oceania rising: climate change in our region" program, Australian Museum, Sydney, 7 March 2018

Tiatia A (2018) Artist's statement accompanying a presentation of *Tuvalu*, 3-channel video, 20:32 mins (2016), at the Australian Museum, Sydney, March 2018

Van der Linden S, Maibach E, Leiserowitz A (2015) Improving public engagement with climate change: five "best practice" insights from psychological science. Perspect Psychol Sci 1–6

Yale Program on Climate Change Communication (2018) What is climate change communication, webpage: http://climatecommunication.yale.edu/about/what-is-climate-change-communication/. Last accessed 20 May 2018

Zelig E, Pfirman SL, (1993) Handling a hot topic—global warming: understanding the forecast. Curator: The Mus J 36:256–271

Climate Changes Cities—A Project to Enhance Students' Evaluation and Action Competencies Concerning Climate Change Impacts on Cities

Katharina Feja, Svenja Lütje, Lena Neumann, Leif Mönter, Karl-Heinz Otto and Alexander Siegmund

Abstract Concerning the social acceptance and realization of adaptation strategies, a raising awareness on the impacts of climate change among the population is indispensable, especially among young people as future decision makers. In this context, the article presents the structure and implementation of the environmental education project "Klimawandel findet Stadt" (*Climate changes cities*), carried out in cooperation between the universities of Bochum, Heidelberg and Trier. The project shall facilitate the development of students' evaluation and action competencies with regard to climate change consequences and sustainable adaptation strategies by using a new educational concept of climate change communication. It implies the design of learning modules with an emphasis on health and risk prevention, urban climate and planning, and urban ecology and biodiversity. External stakeholders, e.g. biological stations, environmental departments and municipal

---

Author contributions KF, SL and LN collaboratively conceptualized the paper and wrote the general chapters. KF presents her accompanying research on combined learning spheres as motivational triggers in Sect. 4.1, LN describes her research on the development of environmental action skills in Sect. 4.2 and SL outlines her research on behavior-based environmental attitudes in Sect. 4.3. LM, KHO and AS developed the three-step approach and conceptualized the framework of the project.

K. Feja · K.-H. Otto
Department of Didactics of Geography, Ruhr-University, Universitätsstr. 104, 44799 Bochum, Germany
e-mail: katharina.feja@rub.de

S. Lütje · L. Mönter
Trier University, Geography and Its Didactics, Behringstr. 21, 54296 Trier, Germany
e-mail: luetje@uni-trier.de

L. Neumann (✉) · A. Siegmund
Research Group for Earth Observation (rgeo), Department of Geography,
Heidelberg University of Education, Czernyring 22/11-12, 69115 Heidelberg, Germany
e-mail: lena.neumann@ph-heidelberg.de

A. Siegmund
Heidelberg Center for the Environment and Institute for Geography, Heidelberg University, Berliner Str. 48, 69120 Heidelberg, Germany

© Springer Nature Switzerland AG 2019
W. Leal Filho et al. (Eds.), *Addressing the Challenges in Communicating Climate Change Across Various Audiences*, Climate Change Management,
https://doi.org/10.1007/978-3-319-98294-6_11

159

offices, are involved in the planning and implementation of the above mentioned modules leading to an enhanced cross-sectoral cooperation of institutions. The methodical approach of the project is based on the dialectical intertwining of a three-step approach of spheres, namely observation sphere, laboratory sphere and sphere of action. In the spheres, students are confronted cognitively and affectively with climate change. Thereby, students shall be enabled and motivated to understand, evaluate and communicate climate adaptation strategies. As the concept is so far only normatively justified, there is need for empirical evidence of its effectiveness. In order to address this need, three efficacy studies are designed. If the methodical-didactical concept proves to be efficient, it could be implemented as a new form of climate change communication in educational institutions.

1 Introduction

Anthropogenically induced climate change is one of the biggest social, ecological, economical, and political challenges nowadays (cp. Federal Cabinet 2011; cp. IPCC 2014a, b). In addition, results of a study conducted by the Pew Research Center (2017) revealed that climate change is perceived as one of the most serious security threats for the respondents from 38 countries. Urban spaces are particularly affected by climate change impacts due to their high population and infrastructure density, which leads to a unique microclimate and heat island effects (cp. EEA 2016).

In Germany, a three-quarters majority of the population lives in densely and intermediately populated urban spaces—with a rising tendency (DESTATIS 2016). The 21st century is said to be the century of urban areas, the latter of which will become the central organizational form of humankind (WBGU 2016). In the context of the transitory century it is in the cities where the success or failure of approaches to the *Great Transformation*[1] (WBGU 2011) will become visible. In this respect, urban spaces serve a dual function. They offer the potential to adapt to climate change impacts. Moreover, they constitute the immediate environment of the majority of people, making the perception of climate-related phenomena possible and the opportunity to learn about climate change impacts more tangible. Nevertheless, awareness of the impacts of climate change needs to raise among the population in order to induce climate-conscious behavior (cp. BfN 2006), especially among young people as multipliers and *change agents* (UNESCO 2014) of the future. Therefore, the above-mentioned complexities should be didactically reduced for students[2] to ensure a fact-based assessment of the risks of climate change. The impacts, especially those experienced on a regional level, could raise the awareness

---

[1]The term *Great Transformation* stands for the global "structural transition [...] into a sustainable society, which must inevitably proceed within the planetary guard rails of sustainability" (WBGU 2011).

[2]In the context of the project, "students" refers to pupils, year 6 to year 10, aged approx. 12–17.

of the problem. This in turn enhances the insight of the need to develop appropriate measures in order to cope with climate change impacts.

This development underlines that—in addition to the broad field of climate mitigation—research on, and implementation of, climate adaptation strategies needs to receive increased attention (cp. UBA 2015). In 2008 the Federal Cabinet launched the *German Strategy for Adaptation to Climate Change* (Deutsche Anpassungs Strategie, DAS), followed by concrete measures introduced in the *Adaptation Action Plan of the German Strategy for Adaptation to Climate Change* in 2011 (Federal Cabinet 2011). These documents build the medium-term theoretical basis for integrating climate change adaptation into the political and municipal landscape in Germany. Curriculum analysis of three federal states, namely Baden-Wuerttemberg (BW), North Rhine-Westphalia (NRW) and Rhineland-Palatinate (RLP), reveal that only the latest versions of BW and RLP curricula include climate adaptation as a relevant topic (MWWK-RLP 2016; KM-BW 2016).

There are several studies identifying a gap between students' knowledge and application of that knowledge into action (Bogner and Kaiser 2012; Lehmann 1999). These results outline that the topic of climate change adaptation—in which the capacity and will to act play a major role—needs a new methodical-didactical approach. This paper is outlining an approach to communicate consequences of and possibilities to adapt to climate change to students. The joint project *Climate changeS cities* of the universities of Bochum, Trier and Heidelberg attempts to be a promising link between recently introduced curriculum contents and their implementation in educational practice. It promotes the communication between education researchers and education practitioners by developing learning modules at the universities and refining them in cooperation with schools.[3] Furthermore, external stakeholders like biological stations, environmental departments and municipal offices are involved in the development of the learning modules, which allows cross-sectoral exchange and new ways of communication. The project attempts to make scientific insights relevant for personal life on the part of the students by including cognitive as well as affective elements. This shall be achieved by focussing on climate change on a regional level and dealing with target-group relevant topics. The project aims to develop students' evaluation and action competencies with regard to climate change consequences and sustainable adaptation strategies in urban spaces. To do so, a new methodical-didactical approach was developed (Chap. 3). In case its effectiveness can be empirically supported (Chap. 4), the new approach could be permanently implemented in educational institutions. This in turn could be a new way of climate change communication, leading to an active engagement concerning the implementation of adaptation strategies.

---

[3]In the context of the project, "school" refers to secondary schools, namely comprehensive schools (German *Gesamtschule*) and grammar schools (German *Gymnasium*).

2 Climate Change Impacts and Climate Adaptation in German Cities

The analysis of annual mean values of the air temperature reveals that an increase of 0.5–1.5 °C is expected for Germany during the time period 2021–2050 (Federal Cabinet 2008). However, there is no consistent increase in temperature over the decades—regional and seasonal fluctuations of rising temperatures are characteristic (UBA 2015). Since 1951, the number of 'hot days'[4] increased in average from three days up to eight days per year (UBA 2015). Hot days may especially affect urban spaces due to their distinctive climate conditions and the resulting urban heat island (UHI) effect. Furthermore, a rising tendency of weather extremes and related weather phenomena such as heat waves, heavy precipitations, storms, and tornados have been observed in recent years (Federal Cabinet 2008; UBA 2015). Consequences of climate change include health risks, reduced labor productivity, safety risks, transport route blockage, shifting of the phenological seasons and drying of roadside greenery (cp. EEA 2016; cp. Wittig et al. 2012). As all people contribute to climate change and have to deal with its effects the one way or the other, climate-conscious behavior is a challenge to be addressed by society as a whole. Mitigation and adaptation have to be brought together in order to prevent a further increase of carbon dioxide concentrations as well as to deal with already unavoidable climate change consequences in a sustainable way (cp. EC 2015a, b). Exemplary spheres of activity for climate adaptation are the creation of retention areas for rainwater, the building of heat and flood resistant infrastructure and the planning of adaptive urban greenery (EEA 2016). The aim of climate adaptation strategies is to increase the adaptability of natural and societal systems in order to be prepared for a future in times of climate change (UBA 2015), i.e. to increase the resilience and to minimize the vulnerability of rural and urban spaces (UN 2017). Yet, implementing the integrated approach of adaptation strategies may lead to spatial conflicts (cp. Federal Cabinet 2011): the densification of cities with the aim to reduce further consumption of land competes with the creation of green spaces for UHI compensation and supply of fresh air (Milner et al. 2012).

As shown above, anthropogenically induced climate change is characterized by a high system complexity (cp. Jacobeit 2007). Consequently, the DAS (definition see Chap. 1) is comprised of 13 specific fields of action, which in turn consist of 55 impact-indicators and 42 response-indicators (UBA 2015). Additionally, two cross-cutting issues, namely spatial, regional, and urban land-use-planning as well as civil protection, were designated (UBA 2015). The strategy implies the diversity of climate change impacts and their dependence on the region being focused on (cp. EEA 2017). Cities constitute a specific case, as they are not only influenced by the regional macroclimate, but also by microclimatic factors depending on land

---

[4]Maximum temperature of at least 30 °C.

development, surface sealing and release rates of waste heat within the city (Wittig et al. 2012). Regarding its contents, the DAS constitutes the framework of the project specified below.

3 About the Project

The project *Climate ChangeS Cities*[5] is funded by the German Federal Environmental Foundation, which promotes innovative environmental projects and supports environmental communication. It started in June 2016 and runs until February 2019. Figure 1 shows the project structure. The central project aim is to design nine learning modules, dealing with various aspects of climate adaptation. The cooperating universities are the Ruhr-University in Bochum, Heidelberg University of Education and Trier University (Fig. 1)—particularly the Didactics of Geography led by Prof. Karl-Heinz Otto, Prof. Alexander Siegmund and Prof. Leif Mönter. Each university location offers a network of different faculty departments, participating schools and external stakeholders. The latter function as experts in the fields of biodiversity, climatology, environmental protection and/or urban planning. In addition, trainings for teachers regarding methodology and use of media will be implemented in the course of the project. Trainees of regional partners and trainee teachers will be involved in the testing of project modules. The project approach is interdisciplinary in order to address the complexity of climate change.

The central part of the learning modules (laboratory sphere) (Fig. 1) takes place in the student laboratories, which can be defined as open learning spaces that offer possibilities to engage in self-regulated learning through experiments and own research. The participating student labs are the Alfried-Krupp laboratory at the Ruhr-University Bochum, the Geco-Lab at the Heidelberg University of Education and the BioGeoLab at Trier University. The student labs are members of the German Federal Association of student labs (LernortLabor).

3.1 Methodical-Didactical Approach

From a didactical perspective, the combination of three different spheres (observation, laboratory, and action), lays the foundation for the design of learning modules (Fig. 1). Each university is responsible for three modules dealing with different aspects of the same subject area (Fig. 1). The observation sphere, i.e. the subject of observation, can be defined as the living environment ("Lebenswelt") of the students, which includes various social settings like school/work, family and

---

[5]Further information on the project can be found on the homepage http://klimawandel-findet-stadt. de.

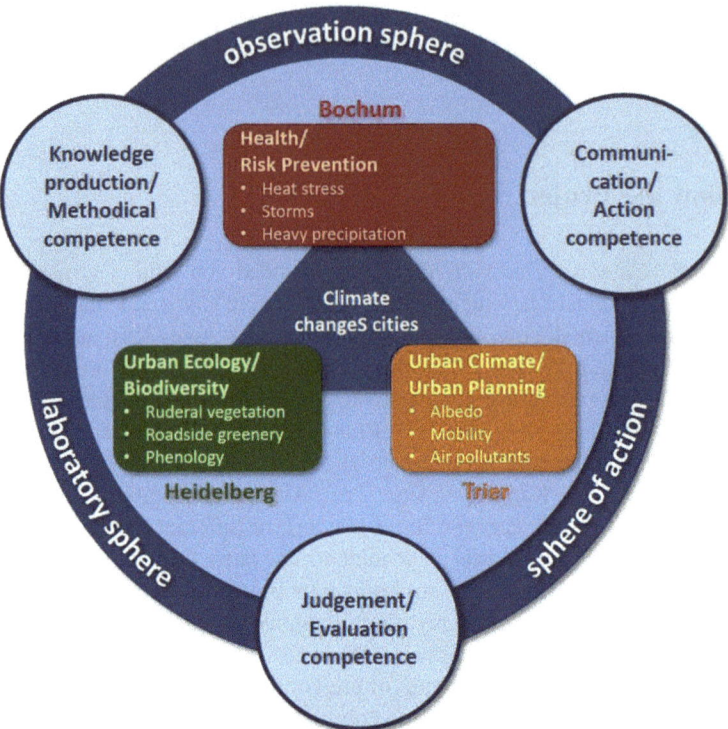

Fig. 1 Graphic describing the structure of the project *Climate changeS cities*

private life. The laboratory sphere corresponds to the student labs of the universities, where scientific insights can be discovered through various experimental methods ranging from the use of simplified models up to more complex experiments. The sphere of action contributes to the application of knowledge about sustainable development, ranging from theoretical strategies to activities like planting trees or altering unused urban spaces. Activities and tasks carried out within the spheres contribute to the fostering of several competencies, in particular action, evaluation, and methodical competence (Fig. 1). The normatively justified three-step approach is going to be empirically tested through accompanying research initiatives (Chap. 4).

The primary learning objective of the project is the enhancement of cross-cutting competencies, e.g. evaluation and action skills. It is expected that the three-step approach enables students to develop their competencies stepwise while passing through the different spheres. Against this background, the observation of climate-related phenomena in urban spaces serves the purpose of providing knowledge to students. For example, the mapping of inner-city structures informs students about traffic routes, urban soils or vegetation. In the student labs, the focus is on the application of scientific methods to explore a variety of adaptation strategies and evaluate the results of the investigations. In the sphere of action, students implement sustainable strategies or communicate their acquired knowledge to others. By integrating the action competence, which in turn builds on the other competencies (DGfG 2014), students shall be enabled to act responsibly and considerately in their everyday life—especially in the context of climate change impacts. For a better understanding of the dialectical intertwining of spheres, an example module is described in Sect. 3.2.

Furthermore, *exploratory learning* and *inquiry-based learning* (cp. Ausubel 1962; cp. Bruner 1961) are the main didactical concepts for the design and implementation of the learning modules. Inquiry-based methods include investigations that are self-directed by students and enable them to learn about specific topics (Lazonder and Harmsen 2016). The investigations are governed by one or more research question(s) that are ideally proposed by the students in the observation sphere. In the labs, students adhere (more or less) to the stages of the experimental algorithm and learn about (scientific) regularities through experiments, models, or the creation and usage of databases. The most important practical implication is that inquiry-based teaching procedures should employ guidance to assist students in accomplishing the tasks (Lazonder and Harmsen 2016). This is also true for exploratory learning, even though the focus of this concept is more on cooperative learning in small groups (Otto and Schuler 2012). In the project, students are offered possibilities to work in small groups of up to four, allowing for active participation in the tasks for all of them.

3.2 *Example Module:* Urban Mobility in Times of Climate Change

The module *urban mobility in times of climate change* encompasses the following: (1) to uncover mobility-related critical issues in the direct vicinity of schools, (2) urban planning processes with the help of a computer simulation, and (3) restructuring measures for urban spaces. Within the observation sphere, students shall identify critical issues in urban mobility. They shall be encouraged to analyze their own interests regarding the field of mobility. Therefore, results are not predictable and the use of appropriate (geographical) methods is flexible. According to *inquiry-based learning*, students frame their own hypotheses to be investigated. This is done by collecting data on the basis of the problems that the students detect, e.g. by the use of sound level meters, traffic counting or questioning of passers-by. The data collection is carried out in small groups and in a study area defined by the students' research questions. To learn more about the complexity of urban planning processes and spatial conflicts that may arise, a computer simulation game called *Mobility*[6] is used in the laboratory sphere. *Mobility* is a city building simulation game that gives

---

[6]The simulation game *Mobility* can be downloaded under the link www.mobility-online.de/de/.

direct response to the structures being built. This function provides the opportunity to try different types of solutions for critical issues that students have detected in the observation sphere.

On the basis of the theoretical input students shall be enabled to propose solutions to the problems they identified in their living environment. In the sphere of action the results are entered on a website.[7] Here, other students, teachers, and parents can acquire information about specific mobility-related problems and how to solve these, e.g. by adjusting one's own mobility behavior. Students also have the opportunity to discuss their results and proposals with cooperation partners from the urban planning authority in a face-to-face manner. This procedure reflects the high relevance of climate-adapted spatial and land-use planning (cp. IPCC 2014a, b). Furthermore, the module includes main features of an effective climate change communication between various local actors.

3.3 Rethinking Climate Change Communication

Climate change and climate adaptation are highly relevant topics in the context of Education for Sustainable Development (cp. UNESCO 2010). The current trend in the United States underlines the relevance of both a factual and demonstrative climate change education. Due to their topicality and spatial relatedness, the topics offer the potential to make use of new communication-related concepts to teach students sustainability in a more practice-oriented manner. Climate change is traditionally taught on a global scale which provokes psychological distance (Chiari et al. 2016; Moser 2014), whereas in the project it is communicated on a regional scale. This allows for more concrete spatial references and a sense of being affected by climate change impacts. The first-hand experience with climate change impacts and adaptation strategies is facilitated by knowledge acquisition in the living environment of the students, which is also thought to operate as a motivational trigger (Chiari et al. 2016; Moser 2010). Furthermore, exploratory and inquiry-based learning supports self-reliant learning and communication among students, while in traditional settings teacher-student communication prevails, following the well-known question-response sequence (cp. Li 2017).

The project does not only aim to invent a new kind of methodical-didactical approach for climate change communication, but also intends to establish a more intense cooperation between schools and universities. It often takes years until didactical concepts are transferred into school practice (cp. Schön 2011). Teaching staff are involved in the project as they are integrated in the development of the learning modules and offered teacher trainings. This leads to an immediate implementation of the learning modules in schools.

---

[7]The website containing students' results for the module on mobility can be viewed under the link www.klimawandel-findet-stadt.de.tl.

On a regional level, actors in climate change from different sectors are connected. The cooperation of policy makers, scientists, and school practitioners provokes a new kind of discourse. It is expected that the discursive practice involving experts from multiple sectors produces several valuable approaches to sensitise students toward climate change consequences. This might result in the development of truly innovative adaptation strategies.

4 Accompanying Scientific Research

The project is accompanied by three field studies aimed towards examining the efficacy of the—so far—normatively justified three-step approach. The research is focused on different psychological and didactical aspects namely motivation, action competence and environmental attitude. It focuses on students as respondents, particularly the students participating in the project. As the three-step approach has already been the core piece in a previous project,[8] it is essential to prove its effectiveness for the potential implementation of the approach in the didactics of Geography and educational practice in general.

4.1 Combined Learning Spheres as Motivational Triggers?

Motivation is part of every person's psychological condition and a relevant aspect of the learning process when it comes to the design of meaningful and sustainable learning situations. Motivation is defined as the activating focus of a current state towards positively assessed goals (Rheinberg 2004). Motivational phenomena can be understood only when the interaction between a person (*traits*) and a situation (*state*) is considered (cp. Lewin 1936; cp. Murray 1938). It is further assumed that people act according to their motives. There are three main motives or motive systems that can be defined as temporally stable features, namely achievement, power and affiliation (McClelland 1987). Motives are not always activated; they are context-dependent and vary according to the individual (Vollmeyer 2005). They are stimulated by situation criteria. Situation criteria that match a certain motive are called incentives (Rheinberg 2004). The interaction of motive and incentive(s) results in the current motivation of a person that in turn influences the behavior (Rheinberg 2004). Consequently, current motivation functions as a kind of mediator between person, situation and behavior.

To find out whether the combination of the spheres, conceptualized as a three-step approach, is more effective than using a single learning sphere, a

---

[8]"Regionalen Klimawandel beurteilen lernen (ReKli:B)" (*Learn to evaluate regional climate change*).

standardized questionnaire to measure personal and motivational conditions based on the Questionnaire on Current Motivation (QCM) (Rheinberg et al. 2001) and the short scale on intrinsic motivation (KIM) (Wilde et al. 2009) will be used. Items will be chosen to measure motivational factors in (1) field learning and laboratory learning and (2) laboratory learning only. The field learning situation is characterized by a high degree of self-regulated learning opportunities in various learning spheres namely local park or wood and inner-city areas. The motivational factors to be measured are, among others, probability of success, interest, and challenge (Rheinberg et al. 2001). It is assumed that the three-step approach as well as the didactical realization in terms of self-regulated learning will result in a high current motivation. A high motivation in turn might predict more sustainable learning outcomes.

4.2 Intervention-Based Efficacy Study on the Development of Action Competence

Even though the need of environmental education resulting in responsible engagement of students is commonly accepted, few concepts and studies exist on how to encourage the enhancement of action skills (Chiari et al. 2016; Bogner and Kaiser 2012). The three-step approach aims to support the development of students' action competencies concerning the handling of climate change consequences. The objective of the described study is firstly to evaluate whether the methodical-didactical concept is suitable to strengthen students' action competencies, and secondly, to investigate if there is an interdependence between the nature of action and the increase of action competencies. To answer the second research question, there was made a distinction between three natures of action:

1. Concrete realization of action—Students implement adaptation measures into practice
2. Structuring action on a planning level—Students develop a concept for the realization of adaptation measures
3. Informative action—Students serve as multipliers and inform different target groups about climate change consequences and sustainable adaptation strategies.

Figure 2 visualizes the study design. After a pre-test the whole sample group passes through the observation and the laboratory sphere. For the sphere of action, the sample group is subdivided into four groups: three intervention groups and one control group. While the three intervention groups perform different types of action, listed in Fig. 2, the fourth group serves as a control and deepens the acquired knowledge by learning with geographical models. The design is completed by a post-test following the intervention.

Another distinction is made concerning the subsections of action competence. According to the DGfG (2014), four areas of action competence exist: the

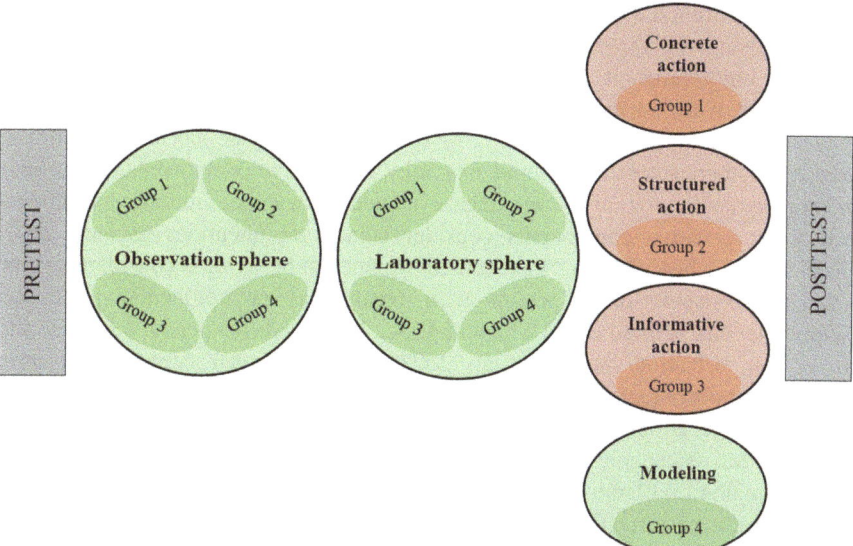

Fig. 2 Design of the efficacy-study dealing with impacts of the three-step approach on students' action skills

knowledge of information and strategies relevant to action (A1), the motivation and interest in geographical fields of action (A2), the willingness to take specific action in geographically relevant situations (A3) and the ability to reflect upon actions with regard to their effects on natural and social spaces (A4). The competencies A1 and A4 concern the ability to act appropriately, while A2 and A3 concern the motivation to act. Both elements are essential to fulfill an action. It is assumed that the concrete realization of action enhances the subsequent motivation to act, whereas the structuring of an action on a planning level predominantly supports the ability to act responsibly. The study aims to obtain further information about the dependence of the ability and motivation to act as well as the influence of the realized nature of action on both sub-areas (ability to act & motivation to act). Furthermore, follow-up interviews are planned, providing information about the relation of enhanced action skills and the concrete realization of climate-conscious behavior.

4.3 Behavior-Based Environmental Attitude

The research project examines the effectiveness of the three-step approach on behavior-based environmental attitude. The aim is to establish the three-step approach as a missing link between students' (theoretical) concepts of environmental behavior and their actual environmental performance (Wilhelmi 2011).

Therefore, within the project a method to enhance environmental attitude and behavior using the three-step approach is developed. Environmental attitude is equal to environmental awareness (cp. Katzenstein 1995; cp. Spada 1990). Environmental awareness is defined as a combination of values, attitude, and behavior towards the environment. A distinction is made between cognitive, affective, and conative components: examples of affective components are feelings and consternation; cognitive components include knowledge, thinking and rational evaluation; conative components consider behavioral intentions (Schick 2001; Spada 1990; Fishbein 1997). However, environmental behavior is considered as the actual performance of a person with regards to environmental protection (Winiwarter and Schmid 2008) which implies a link between a person's attitude and his or her behavior (Kaiser et al. 2007). In educational contexts the term *Environmental Competence* is more commonly used. Environmental competence is composed of (1) environmental knowledge such as coherent knowledge, knowledge of options for sustainable action and of potential impacts of action, (2) relationship with nature and (3) "Umwelthandlungskompetenz" (literally "environmental action competence") meaning the competence to transfer this knowledge to observable action (cp. Kaiser and Wilson 2004; cp. Kaiser et al. 2008). Therefore, the three-step approach aims at improving the environmental attitude in order to encourage sustainable environmental action.

Kaiser et al. (2007) calibrated the General Ecological Behavior (GEB) scale for adolescents for the purpose to measure attitudes as behavioral dispositions. The original GEB scale was developed for adults by Kaiser and Wilson (Kaiser 1998; Kaiser and Wilson 2004). The GEB scale for adolescents consists of six subscales (energy conservation, mobility, waste prevention, recycling, consumer behavior and indirect environmental behavior) with a total of 40 items, of which 20 were adopted from Kaiser and Wilson (2004). The study carried out in this project is based on the GEB scale and uses a pre-post-follow-up-design. The questionnaire software *Enterprise Feedback Suite Survey* hosted by QuestBack is used for data collection.

5 First Insights

The implementation of modules started in May 2017. In the context of the compulsory evaluation concept consisting of short questionnaires for students and teachers, first results showed that the majority of participants had a positive view on the contents and methodology of the respective modules. According to the questionnaire, students highly valued the supplied materials/worksheets, the establishment of a relation between the topic and one's own living environment and the opportunities for self-regulated learning in the student labs. Some tasks were considered too difficult, i.e. not easy to comprehend. An analysis of the supplied materials collected so far revealed that many students had difficulties concerning the experimental algorithm, i.e. the structured sequence from generating appropriate hypotheses to interpreting achieved results. One reason for that may be the low frequency of experiments in the (Geography) classroom (Hemmer and Hemmer 2010). It has been shown that the application of the theoretical knowledge at the end of the project day was particularly useful for the students, e.g. posters displaying a redesigned, climate-adapted school building planned by the students revealed the awareness of required modifications in times of climate change. Although modules can be carried out either in form of a project day or in several double periods, difficulties that occurred from the first implementations were mainly due to problems of coordination and time on the side of the teachers, who are tied to school procedures.

6 Outlook

In the paper, a new didactical-methodical approach to communicate climate change effects and adaptation measures in schools was outlined—a topic, which is highly relevant but new to the school curricula and therefore in need of a methodical and didactical processing. The concept combines cognitive and affective learning by focussing on a regional level of climate change following the approach of exploratory and inquiry-based learning. Within three spheres the competences of the students shall be developed stepwise resulting in the will and ability to realize sustainable climate change adaptation measures largely self-reliantly. As the concept is only normatively justified by now, investigations concerning the efficacy of the approach are necessary. Even if the studies prove the effectiveness of the concept, some limitations still remain concerning the transferability of the results: The learning modules are embedded in the German educational system only and designed for students attending specific forms of secondary schools. Due to the limited timeframe of the project, the results cannot provide statements about long-term effects of the three-step approach. The current focus is on the implementation and completion of the learning modules. To ensure their high quality, they are continuously revised and edited when necessary. Furthermore, teacher trainings focusing on media and methods being used are on the agenda. They aim at training the participants' methodical-didactical know-how as well as theoretical knowledge on climate adaptation and mitigation. By offering these trainings, it shall be ensured that key structures and outcomes of the project remain in the schools beyond the end of the project. To guarantee the implementation of the modules across all three university locations, a lab-rotation is planned for 2018. Consequently, the regional focus of each module needs to be shifted in order to be valuable for the other regions. The closing conference in 2019 and further events aim at giving participating students the possibility to display their results and adaptation strategies to a wider public in the form of posters, advice on websites, and implemented strategies like plantings on wasteland areas. In conclusion, the project *Climate changeS cities* aims to contribute to an effective climate change communication by modifying learning spheres, tasks and teaching processes.

References

Ausubel DP (1962) Learning by discovery. Educ Leadersh 20(2). Washington: The Association
    for Supervision and Curriculum Development, pp 113–117
BfN—Federal Agency for Nature Conservation (ed) (2006) Biologische Vielfalt und Klimawandel
    – Gefahren, Chancen, Handlungsoptionen (BfN-Skripten 148), Bonn. https://www.bfn.de/
    fileadmin/MDB/documents/service/skript148.pdf. Accessed 8 Aug 2017
Bogner FX, Kaiser FG (2012) Umweltbewusstsein, ökologisches Verhalten und Umweltwissen:
    Modell einer Kompetenzstruktur für die Umweltbildung, In: Bayrhuber H, Harms U,
    Muszynski B, Ralle B, Rothgangel M, Schön L-H, Vollmer HJ, Weigand H-G (ed) Formate
    Fachdidaktischer Forschung: Empirische Projekte – historische Analysen – theoretische
    Grundlegungen (Fachdidaktische Forschungen, vol. 2), Münster: Waxmann, pp 163–182
Bruner JS (1961) The act of discovery. Harvard Educ Rev 31(1). Cambridge, Mass.: Harvard
    University Press, pp 21–32
Chiari S, Völler S, Mandl S (2016) Wie lassen sich Jugendliche für Klimathemen begeistern?
    Chancen und Hürden in der Klimakommunikation, GW-Unterricht 141(1). Wien: Forum
    Wirtschaftserziehung, pp 5–18
DESTATIS—Federal Statistical Office of Germany (ed) (2016) Statistisches Jahrbuch
    Deutschland und Internationales 2016, Wiesbaden. https://www.destatis.de/DE/
    Publikationen/StatistischesJahrbuch/StatistischesJahrbuch2016.pdf?__blob=publicationFile.
    Accessed 8 Aug 2017
DGfG—German Society for Geography (ed) (2014) Bildungsstandards im Fach Geographie für
    den Mittleren Schulabschluss mit Aufgabenbeispielen, 8th revised and updated version. DGfG,
    Bonn
EC—European Commission (2015a) LIFE and climate change mitigation. Luxembourg
EC—European Commission (2015b) LIFE and climate change adaptation. Luxembourg
EEA—European Environment Agency (2016) Urban adaptation to climate change in Europe 2016
    —Transforming cities in a changing climate. Copenhagen
EEA—European Environment Agency (2017) Climate change, impacts and vulnerability in
    Europe 2016. An indicator-based report. Copenhagen
Federal Cabinet of Germany (2008) German Strategy for Adaptation to Climate Change adopted
    by the Federal Cabinet on 17 December 2008, Executive Summary. http://www.bmub.bund.de/
    fileadmin/bmu-import/files/pdfs/allgemein/application/pdf/das_gesamt_bf.pdf. Accessed 8
    Aug 2017
Federal Cabinet of Germany (2011) Adaptation action plan of the German strategy for adaptation
    to climate change, adopted by the German Federal Cabinet on 31st August 2011. http://
    www.bmub.bund.de/fileadmin/bmu-import/files/pdfs/allgemein/application/pdf/aktionsplan_
    anpassung_klimawandel_en_bf.pdf. Accessed 8 Aug 2017
Fishbein M (1997) Einstellung und die Vorhersage des Verhaltens. In: Hormuth SE
    (ed) Sozialpsychologie der Einstellung. Athenäum Hain Scriptor Hanstein, Königstein, Ts
Hemmer I, Hemmer M (eds) (2010) Schülerinteresse an Themen, Regionen und Arbeitsweisen des
    Geographieunterrichts. Ergebnisse der empirischen Forschung und deren Konsequenzen für die
    Unterrichtspraxis (Geographiedidaktische Forschungen 46). Hochschulverband für Geographie
    und ihre Didaktik e.V, Weingarten
IPCC—International Panel on Climate Change (2014a) Climate Change 2014: Mitigation of
    Climate Change. Contribution of Working Group III to the Fifth Assessment Report of the
    Intergovernmental Panel on Climate Change [Edenhofer O, Pichs-Madruga R, Sokona Y,
    Farahani E, Kadner S, Seyboth K, Adler A, Baum I, Brunner S, Eickemeier P, Kriemann B,
    Savolainen J, Schlömer, S, von Stechow C, Zwickel T, Minx J C (ed)], IPCC, Cambridge, New
    York
IPCC—International Panel on Climate Change (2014b) Synthesis report. Contribution of working
    groups I, II and III to the fifth assessment report of the intergovernmental panel on climate
    change [Core Writing Team, Pachauri RK, Meyer LA (ed)]. IPCC, Geneva

Jacobeit J (2007) Zusammenhänge und Wechselwirkungen im Klimasystem. In: Endlicher W, Gerstengarbe F-W (eds) Der Klimawandel – Einblicke. Rückblicke und Ausblicke, Berlin, Potsdam, pp 1–16

Kaiser FG (1998) A general measure of ecological behavior. J Appl Soc Psychol 28(5). Washington: Wiley-Blackwell, pp 395–422

Kaiser FG, Wilson M (2004) Goal-directed conservation behavior: the specific composition of a general performance. Personal Individ Differ 36(7). Oxford: Pergamon Press, pp 1531–1544

Kaiser FG, Oerke B, Bogner FX (2007) Behavior-based environmental attitude: development of an instrument for adolescents. J Environ Psychol 27(3). Amsterdam, London: Elsevier, pp 242–251

Kaiser FG, Roczen N, Bogner FX (2008) Competence formation in environmental education: advancing ecology-specific rather than general abilities 2(12). Lengerich: Pabst Science Publishers, pp 56–70

Katzenstein H (1995) Umweltbewusstsein und Umweltverhalten. Fernuniversität-Gesamthochschule, Hagen

KM-BW—Ministry of Education, Youth and Sports Baden-Wuerttemberg (2016) Gemeinsamer Bildungsplan der Sekundarstufe I. Bildungsplan 2016. Geographie. Neckar Verlag, Stuttgart

Lazonder, AW, Harmsen R (2016) Meta-analysis of inquiry-based learning: effects of guidance. Rev Educ Res 86(3). Washington: Thousand Oaks, pp 681–718

Lehmann J (1999) Befunde empirischer Forschung zu Umweltbildung und Umweltbewusstsein. Leske + Budrich, Opladen

Lewin K (1936) Principles of topological psychology. McGraw-Hill, New York

Li M (2017) Zweitsprachförderung im frühen naturwissenschaftlichen Lernen. Linguistisch hochwertige Formate und interaktive Elemente der Unterrichtskommunikation. Beltz, Weinheim, Basel

McClelland DC (1987) Human motivation. Cambridge University Press, Cambridge, New York

Milner J, Davies M, Wilkinson P (2012) Urban energy, carbon management (low carbon cities) and co-benefits for human health. Curr Opin Environ Sustain 4(4). Amsterdam, London: Elsevier, pp 398–404

Moser SC (2010) Communicating climate change: history, challenges, process and future directions, WIREs climate change, vol 1. Wiley-Blackwell, pp 31–53

Moser SC (2014) Communicating adaption to climate change: the art and science of public engagement when climate change comes home, WIREs climate change 5(3). Wiley-Blackwell, pp 337–358

Murray HA (1938) Explorations in personality. Oxford University Press, New York

MWWK-RLP—Ministry of Education, Science, Further Education and Culture of Rhineland-Palatinate (2016) Lehrplan für die gesellschaftswissenschaftlichen Fächer. Erdkunde, Geschichte, Sozialkunde. MF-Druckservice, Mainz

Otto K-H, Schuler S (2012) Pädagogisch-psychologische Ansätze. In: Haversath J-B (moderator), Geographiedidaktik, Theorie - Themen - Forschung, Westermann, Braunschweig, pp 133–164

Pew Research Center (2017) Globally, People Point to ISIS and Climate Change as Leading Security Threats. Concern about cyberattacks, world economy also widespread. http://www.pewglobal.org/2017/08/01/globally-people-point-to-isis-and-climate-change-as-leading-security-threats/. Accessed 8 Aug 2017

Rheinberg F, Vollmeyer R, Burns BD (2001) FAM: Ein Fragebogen zur Erfassung aktueller Motivation in Lern- und Leistungssituationen (Langversion, 2001). Diagnostica, 47(2) Göttingen, Bern: Hogrefe, pp 57–66

Rheinberg F (2004) Motivationsdiagnostik. Hogrefe, Göttingen

Schick A (2001) Umweltbewusstsein. In: Schulz WF (ed) Lexikon Nachhaltiges Wirtschaften. Oldenbourg, München

Schön L-H (2011) Vorbemerkungen. Empirische Fundierung in den Fachdidaktiken. Ergebnisse einer Fachtagung der Gesellschaft für Fachdidaktik. In: Bayrhuber H, Harms U, Muszynski B, Ralle B, Rothgangel M, Schön, L-H, Vollmer HJ, Weigand H-G (ed) Empirische Fundierung in den Fachdidaktiken (Fachdidaktische Forschungen, vol. 1), Waxmann, Münster, p 7

Spada H (1990) Umweltbewusstsein: Einstellung und Verhalten. In: Kruse L, Grauman Friedrich C, Lantermann DE (ed) Ökologische Psychologie. Psychologie Verlags.-Union, München

UBA—Federal Environmental Agency (2015) Monitoringbericht 2015 zur Deutschen Anpassungsstrategie an den Klimawandel. Bericht der Interministeriellen Arbeitsgruppe Anpassungsstrategie der Bundesregierung. Dessau-Roßlau. https://www.umweltbundesamt. de/sites/default/files/medien/376/publikationen/monitoringbericht_2015_zur_deutschen_anpas sungsstrategie_an_den_klimawandel.pdf. Accessed 8 Aug 2017

UN—United Nations (2017) New urban agenda. Quito: UN. http://habitat3.org/wp-content/ uploads/NUA-English.pdf. Accessed 8 Sept 2017

UNESCO—United Nations Educational, Scientific and Cultural Organization (2010) Climate change education for sustainable development, Paris: UNESCO. http://unesdoc.unesco.org/ images/0019/001901/190101E.pdf. Accessed 6 Sept 2017

UNESCO—United Nations Educational, Scientific and Cultural Organization (2014) UNESCO roadmap for implementing the global action programme on education for sustainable development. UNESCO, Paris

Vollmeyer R (2005) Einführung: Ein Orientierungsschema zur Integration verschiedener Motivationskomponenten. In: Vollmeyer R, Brunstein J (eds) Motivationspsychologie und ihre Anwendung. Kohlhammer, Stuttgart, pp 9–19

WBGU—German Advisory Council on Global Change (2011) World in transition—A social contract for sustainability, flagship report 2011. WBGU, Berlin

WBGU—German Advisory Council on Global Change (2016) Humanity on the move: unlocking the transformative power of cities, flagship report 2016. WBGU, Berlin

Wilde M, Bätz K, Kovaleva A, Urhahne D (2009) Überprüfung einer Kurzskala intrinsischer Motivation (KIM), Testing a short scale of motivation, Zeitschrift für Didaktik der Naturwissenschaften, vol 15. Springer Spektrum, Wiesbaden, pp 31–45

Wilhelmi V (2011) Geographische Umweltbildung weiterdenken, Praxis Geographie, vol 2. Braunschweig, Westermann, pp 4–8

Winiwarter V, Schmid M (2008) Umweltgeschichte als Untersuchung sozionaturaler Schauplätze? Ein Versuch, Johannes Colers "Oeconomia" umwelthistorisch zu interpretieren. In: Knopf T (ed) Umweltverhalten in Geschichte und Gegenwart, Vergleichende Ansätze. Attempto, Tübingen, pp 158–173

Wittig R, Kuttler W, Tackenberg O (2012) Urban-industrielle Lebensräume. In: Mosbrugger V, Brasseur G, Schaller M, Stribrny B (eds) Klimawandel und Biodiversität: Folgen für Deutschland. Darmstadt, Wissenschaftliche Buchgesellschaft, pp 290–307

Degree Programs on Climate Change in Philippine Universities: Factors that Favor Institutionalization

Jocelyn Cartas Cuaresma

Abstract The Philippine Climate Change Act of 2009 (RA 9729) mandates the integration of climate change in school curricula, and in capacity building, research and extension programs. Encouraged by the desire to know the policies and strategies that academic institutions are taking to address climate issues, this research seeks to examine efforts of universities in integrating climate change issues in academic endeavors and the extent to which these are sustained or institutionalized. The experiences of 10 universities covered in the study show that efforts have to be well coordinated, require a mandate at the highest level, partnerships can boost efforts, and success requires the participation of many disciplines and sectors. Some universities are more deeply involved than the others, but otherwise each has its own unique strategies to sustain decisions. The universities studied have developed their respective academic area of expertise to serve the knowledge and skills needs of the community and region where they operate. The paper hopes to contribute to a greater understanding of institutionalizing climate action by putting together information on climate change-related degree programs and initiatives of HEIs.

1 Introduction

The road towards the acceptance, adoption and institutionalization of climate change as an academic framework is not simple. At the global level, it took 23 years from the 1992 Rio Earth Summit to the 2015 Paris Climate Conference (Meakin 1992; UN 2015) for countries to agree to commit themselves to adopt concrete targets in reducing greenhouse gas emissions. As of October 5, 2016, only 169 out of 197 Parties have ratified the Paris Agreement (UNFCCC December 12 2015). At the country level, the Philippine Government has actively passed and

J. C. Cuaresma (✉)
National College of Public Administration and Governance,
University of the Philippines, Diliman Campus, 1101 Quezon City, Philippines
e-mail: jccuaresma@up.edu.ph; joycepcc@yahoo.com

© Springer Nature Switzerland AG 2019
W. Leal Filho et al. (eds.), *Addressing the Challenges in Communicating Climate Change Across Various Audiences*, Climate Change Management,
https://doi.org/10.1007/978-3-319-98294-6_12

formulated laws and plans on or related to climate change and mandated every agency of government to craft their respective policies and programs to integrate climate change in government undertakings. Although the language of the law is rather weak in terms of directly mandating universities and the Commission on Higher Education (CHED) to be among those agencies with clear and direct responsibilities in integrating climate change in higher education curricula (Cuaresma 2017: 78), universities have actually undertaken programs and projects that directly address climate change issues and challenges.

The task of integrating climate change in primary and secondary education curricula has been ordered in Republic Act (RA) 9279 or the Climate Change Act of 2009. The Department of Education, but not the CHED, as member of the Climate Change Commission (CCC) has been tasked to do so (Section 15(a), RA 9729). Whether or not this is an oversight, there is no doubt that higher education institutions (HEIs) as well as the CHED must similarly be tasked. In fact, Section 15(u) of RA 9729 ordered the inclusion of a "Representative of the academe" in the membership of the CCC.

2   Research Objectives

Universities as a sector of society have to show their commitment to the common struggle against climate change. Philippine universities have actively initiated programs and projects to raise awareness about climate change and to directly address climate change issues, but these are not widely documented and disseminated. Filho (2010, 2017) called attention to the remarkable initiatives of universities worldwide in search of solutions to climate change, but efforts reach limited audience and lacked a broad perspective.

The objective of this paper is to identify the factors that reflect the strategies of universities to integrate climate change into academic activities, and assess their level of diffusion and institutionalization of the climate change framework in university governance. It seeks to contribute to a better understanding of university governance and how universities are institutionalizing the climate change agenda. It aims to contribute to raising the level of awareness about the degree programs offered by public universities in the Philippines on or related to climate change, and the support activities the universities engaged themselves with to give substance to the fulfillment of their mandates.

This research is encouraged by the growing attention in daily conversations of things such as adaptation, mitigation, greenhouse gases emissions, carbon dioxide, causes and effects of climate change, and changing policies and popular attitudes, promoting awareness and knowledge about climate change, and impact of human activity on the environment. Moreover, the interest on universities is inspired by the acceptance that universities as institutions of society are established precisely to develop scientific knowledge for the sustainability of the planet and the survival of humankind (Benayas et al. 2010: 48)

3 Methodology and Scope

This paper seeks to examine the policies and programs that selected universities have instituted as their strategies for undertaking their mandate on climate change and assess university performance using Roger's (1962) concept of diffusion of innovations into organizations. Diffusion (of innovation, in this case climate change as an innovative framework of university governance), can go through five stages from knowledge, to persuasion, decision (reject or accept), implementation and confirmation. Diffusion comprises the key elements of innovation (climate change), adopters (universities, officials and professors, students and the academic community), communication channels (degree programs, research and extension), time, and social system (government, citizens, media). The paper also utilized Filho's (2009, cited in Brandli et al. 2015: 64) levels of implementation of sustainability as complementary assessment tool for the rapid assessment of university website information on climate-related programs and activities. Accordingly, universities may be in any of three stages of evolution:

Stage 1. The principles of sustainable development (SD) are not widely understood. No significant efforts exist towards promoting sustainability in university operations. There are no systematic projects meant to promote sustainability.

Stage 2. The principles of SD are widely understood. Efforts towards promoting sustainability at university operations are significant. Projects, research and extension activities are present to promote sustainability in the university as a whole.

Stage 3. Universities have fulfilled the requirements at stage 2 and committed to sustainability on a long-term basis. Evidences include certification from quality assurance organization (e.g., ISO certification) or accreditation bodies, existence of senior staff members, and centrally-funded continuing projects.

Successful diffusion leads to institutionalization, and institutionalization is characterized by sustainability and regularization (Rogers 1983: 175). Institutionalization is achieved when the innovation (e.g., of climate change) becomes a regularized part of operations. A policy, program or strategy is institutionalized when it becomes a regularized activity, supported by written policies, and applicable over many years of operation. A mandate, in this case on climate change, is institutionalized, where policies, vision and mission, and plans are issued to the effect, and programs become part of the organization's regular activities. To institutionalize and sustain climate change in university agenda, change must occur in five dimensions: (1) mission, vision or school philosophy; (2) academic programs, research & extension; (3) faculty expertise and student enrolment; (4) facilities; and (5) awards and recognition.

The study similarly crudely applied Mulvey et al' s (2016) climate accountability scorecard to make a basic assessment of the performance of HEIs in communicating climate change. A simple rating of activities was used to assess the extent of communicating climate change in terms of selected criteria to measure the extent of

Table 1 Rating of performance of universities

Advance	Good	Fair	Poor	Egregious
The HEI demonstrates best practices	The HEI meets emerging expectations	The HEI performance is neither positive nor negative	The HEI falls short of emerging societal expectations	The HEI acts very irresponsibly

Source Adapted from Mulvey et al. (2016: 2)

integration of climate change and environmental issues in academic programs and activities (Table 1). The criteria include the vision/mission statements on climate change, offering of degree programs, conduct of research and extension activities, establishment of support facilities, and awards and recognition received. Other criteria such as faculty resources and partnerships may be later included. The highest favorable rating is ADVANCE, which means that the university is showing high level of commitment and best practices to counter climate change.

Data gathering and analysis was done in two levels. First, a rapid survey and assessment of website information of 80 (75 out of 112 state universities, and 5 private universities) was done to establish web presence and gather information on degree program offerings. This search yielded the information that SUCs actually offer degree programs on Environmental Science/Studies/Management, Resource Management, Environmental Governance, Biodiversity, Ecotourism, Solid Waste Management), Agriculture and related disciplines (Forestry, Fishery, Food Science, Crop Science, Animal Science, Marine Science, Farming Systems, etc.), and Engineering (Agricultural Engineering, Mining Engineering, Petroleum Engineering, and Sanitary Engineering).

Then, 10 universities were purposively selected for the second level of analysis, focusing on the adoption of climate change as a university framework of governance and the corresponding programs adopted to institutionalize the integration of climate change in university governance. The selection of SUCs was further guided by the amount budget subsidy from the national government (DBM 2016 GAA; Cuaresma 2017).

Six of the 10 universities were visited and interviews conducted with the faculty/person in charge of the degree programs. HEI partnership with the Department of Energy was used as additional criteria in the selection of universities for the in-depth study, since it indicates active promotion of clean and renewable energy. The DOE-HEI partnership also led to the establishment of Affiliated Renewable Energy Centers, which suggests the development of expertise in renewable energy research and extension services. The 10 universities[1] selected for the in-depth analysis are as follows:

---

[1]UP Los Baños and UP Diliman are counted separately.

1. University of the Philippines Los Baños-School of Environmental Science and Management (UPLB-SESAM)
2. University of the Philippines Diliman-College of Science (UPD-CS)

   a. Marine Science Institute (MSI)
   b. Institute of Environmental Science and Meteorology (IESM)

3. Mariano Marcos State University (MMSU)
4. Isabela State University (ISU)
5. Central Luzon State University (CLSU)
6. Cavite State University (CvSU)
7. Western Philippines University (WPU)
8. Central Bicol State University of Agriculture (CBSUA)
9. Bicol University (BU)
10. Miriam College (MC, private university).

Two colleges/institutes of the University of the Philippines (one in UP Los Baños and one in UP Diliman) were chosen in view of the highly developed academic programs related to climate change that these academic units have to offer. Choosing UP Diliman is also expedient. The Miriam College was added because of its environmental orientation and programs related to climate and the environment. During the visits and interviews in 6 universities, peer and personal observations of physical facilities were noted to gain greater validity and accuracy in findings. In many ways, the research is limited by the scope and content of data presented in HEI websites. The interviews and ocular inspection of facilities confirm website data recency and completeness. Data on HEIs not visited were gathered from published documents and university websites.

4 Findings

4.1 The Philippine Universities

As of September 2014, the Philippines had a total of 112^2 state universities and colleges (SUCs) counting only the main campuses (CHED 2014; cited also in Cuaresma 2017: 73). Aside from the top 25 SUCs in the country in terms of budget subsidy, the remaining 87 SUCs are rather small with an average annual budget subsidy of Php215 million (DBM 2016; see also Cuaresma 2017: 76). Given the amount of their budgets, complying with the provisions of RA 9729 is thus a

---

[2]The Department of Budget and Management lists 115 SUCs in the 2016 General Appropriations Act.

daunting challenge to SUCs. Some 24 HEIs have partnered with the Department of Energy (DOE) since the 1990s to promote and develop renewable energy as alternative energy source (DOE 2001: 10). The partnership brought opportunities for the partner universities to acquire additional funds for research, and the challenge to develop strength in renewable energy research and extension, and create the Affiliated Renewable Energy Center (AREC) as an extension arm of the DOE.

The study is particularly interested in covering SUCs and one private university that actually offer degree programs on climate change. The rapid assessment of the website presence of SUCs and the courses they offer became the basis for the selection of SUCs for a more in-depth assessment of climate change diffusion and institutionalization. The rapid assessment of 80 (75 SUCs and 5 private) university websites generated the information that at least 32 SUCs and 2 private universities actually are strong in course offering on Agriculture, Agricultural Engineering, Fishery, Forestry, Animal Science, Food Science and cognate programs, which can easily be considered program intervention towards climate change adaptation. This finding is in keeping with the natural environment where Philippine universities operate.

A second significant finding arising from the rapid survey of website information on academic offerings of 80 universities is that 10 SUCs and 2 private universities have institutionalized policies and programs to address climate issues more strongly and more systematically, and 40 HEIs have efforts to incorporate climate change in education, research and extension. This is evidenced by the offering of degree programs on Environment Science, Environmental Studies, Environmental Management, Environment and Resource Management in at least 40 SUCs at the Ph.D., Master and undergraduate levels. The rest (28 SUCs) have not yet systematically regularized climate change concerns into university policy, programs and projects. Using Filho's (2009, cited in Brandli et al. 2015: 64; see discussion below) levels of implementation of sustainability, the performance of 80 universities in the implementation of climate-related policies and programs are rated as shown in Table 2.

It is further observed that 2 SUCs have no websites, but have Facebook accounts. Three websites were not accessible during the time of research. Fifteen (15) SUCs with websites have little or no information on degree program offerings.

The following discussion is based on data gathered from 10 HEIs (9 SUCs and 1 private university). The performance of HEIs are assessed in terms of their vision/mission statements (VM), degree program offering, research and extension services, facilities, partnerships, awards and recognition, and related strategies.

Table 2 Rank of 80 HEIs using Leal Filho's stages of implementation of sustainability

Stages	Stage 1	Stage 2	Stage 3	Total
No. of HEIs	28	40	12	80

4.2 HEI Vision and Mission Statements

The vision and mission statements of the HEIs mirror the general direction of governance of the universities. At a glance (Table 3), the UP units, CLSU, and Miriam College have more explicitly stated their commitment to environmental sustainability and climate change mitigation and adaptation. Miriam College's commitment to environmental protection in academic and non-academic endeavors is inspiring considering its Catholic tradition and women orientation.

Seven (7) VM statements showcase the universities' direction in environmental governance. For instance, the CLSU, envisions itself to become a world-class

Table 3 Statements of Vision and Mission of 10 HEIs

HEI	Vision	Mission	Score
UPLB-SESAM	A center of excellence in environmental science and management … in a manner that fully respects the limits of nature	#1 Mission: … To develop/ offer higher academic degrees on environmental science and management	Advance
UPD-CS[a]	MSI: #1 Objective: To generate basic information for the optimal and sustained utilization, management, and conservation of the marine environment and its resources	MSI: Pursue research, teaching, and extension work in marine biology, marine chemistry, physical oceanography, marine geology, and related disciplines	Advance
	IESM: A center of excellence in environmental and atmospheric sciences in the Asia-Pacific	IESM: Serve as a center of excellence in the environmental and atmospheric sciences …, investigating natural phenomena (including human interactions with nature), …	Advance
CvSU	Excellence in the development of globally competitive and morally upright individuals	Provide excellent, equitable and relevant education opportunities in the arts, sciences and technology through quality instruction and responsive research and development activities	Good
CBSUA	A leading university of agriculture in the Philippines by 2018 and in the ASEAN Region by 2024	Produce globally competitive graduates to develop viable agri-industrial technologies and to help build resilient and sustainable communities	Advance

(continued)

Table 3 (continued)

HEI	Vision	Mission	Score
MMSU	A world-class university dedicated to the development of virtuous human resources and innovations for inclusive growth	Develop globally competitive professionals and industry-ready graduates ...	Good
ISU	A leading, vibrant, comprehensive and research university in the country and the ASEAN region	#2 Mission: Generate innovative and cutting-edge knowledge and technologies for people empowerment and sustainable development	Advance
CLSU	A world-class National Research University for science and technology in agriculture and allied fields	#2 Mission: Generate, disseminate, and apply knowledge and technologies for poverty alleviation, environmental protection, and sustainable development	Advance
WPU	The leading knowledge center for sustainable development of West Philippines and beyond	Develop quality human resource and green technologies for a dynamic economy and sustainable development through relevant instruction, research and extension services	Advance
BU	A world-class university producing leaders and change agents for social transformation and development	Give professional and technical training, and provide advanced and specialized instruction in literature, philosophy, the sciences and arts, besides providing for the promotion of technological researches	Good
MC (Private HEI)	A premier Filipino Catholic institution of learning that forms leaders in service who combine competence with caring, are rooted in Filipino culture and Asian tradition, and yet are citizens of the world	Provide quality and relevant Christian education that prepares students to become effective leaders, lifelong learners, and productive citizens; pursues the core values of truth, justice, peace and integrity of creation	Advance

Source of data Data gathered from interviews and university websites
[a]The College of Science was created in October 1983. At present, it is composed of 9 institutes and 2 academic programs. Two institutes (MSI and IESM) are included in this study. (www.science. upd.edu.ph). Other colleges within the UP System also offer climate-related degree programs which are not covered in this study

National Research University for science and technology in agriculture and allied fields. Recently, it formulated a University Comprehensive Environmental Management Plan and the CLSU Strategic Plan 2016–2040. The CLSU has pioneered in agricultural education and technology since 1907. Among its current program strategies are its advocacy for a paradigm shift or environmental cultural evolution in the various aspects of environmental management, and integration of environmental education in all courses in the university. It pursued the development of clean technologies. It adopted the "polluter pays principle" (UNESCO 1992: 4, 7), where the polluter shall shoulder the cost of pollution that it produces and should see to it that damage to the environment is prevented. In its Strategic Plan, the CLSU sees itself to become a comprehensive research university by 2035 particularly in the field of agriculture and natural resources, and by 2040 to become a seamlessly integrated academic university fully connected and engaged to the outside world. (https://clsu.edu.ph).

Miriam College is a non-profit non-stock Catholic educational institution that was created in 1926. MC has 3 colleges and 3 institutes including the College of Arts and Sciences and the Environmental Management Institute. Steeped in the Catholic tradition, MC believes in the integrity of creation and commits itself to "care for the earth and to practice a lifestyle that sustains the health of the planet on which all life depends". MC presents itself since the 1970s as a Dark Green School, a steward of the environment, with course offerings in environmental education. It adopted its Seven Environmental Principles into all its academic programs from preschool to the graduate level.

4.3   Degree Programs Related to Climate Change

The degree programs related to climate change are interestingly diverse, and offered not only by the colleges of engineering and sciences but also by College of Agriculture and the Graduate School. Courses are also offered in distance mode. The MMSU and CLSU are among the few universities offering a graduate program on Renewable Energy Engineering. The MMSU's Professional Science Master's in Renewable Energy Engineering (PSM-REE) is a new addition to MMSU's roster of degree programs. Launched in August 2016, information on the program in the university website is still limited to the news of its launching (MMSU News, August 11 2016). Established with the support of the Research Triangle Institute[3] and Energy Development Corporation, the Master's program offering was encouraged by the presence of big commercial wind farms in Ilocos Norte Province (Table 4).

---

[3]RTI is the main implementing organization of the USAID-funded Science, Technology, Research and Innovation for Development (STRIDE) Project.

Table 4 Degree programs on climate change/DRRM/environmental science

HEI	Degree programs	Host unit	Score
UPLB-SESAM	MS in Environmental Science, since 1996, 32 units Ph.D. in Environmental Science, since 1996; Areas of Specialization: (1) Environmental Security and Management; (2) Protected Areas Planning, Development and Management; and (3) Social Theory and Environment.	SESAM	Advance
UPD-CS	MSI: M.Sc. and Ph.D. degrees in Marine Science, major in Marine Biology, Marine Physical Science or Marine Biotechnology Professional Masters in Tropical Marine Ecosystems Management (PM-TMEM)	College of Science	Advance
	IESM: Ph.D./Master in Environmental Science Ph.D./Master in Meteorology	College of Science	Advance
CvSU	BS in Environmental Science MS Agricultural Engineering Ph.D./MS in Agriculture	College of Agriculture, Environment and Natural Resources	Advance
CBSUA	BS Environmental Sciences MS in Resource Management, major in • Cooperative Management • Entrepreneurship and Environmental Management Diploma & MS in Disaster-Risk Management (Ladderized)	College of Agriculture & Natural Resources Graduate School	Advance
MMSU	BS Environmental Science Professional Science Masters in Renewable Energy Engineering (PSM-REE), since 2016	College of Agriculture, Food and Sustainable development	Good
ISU	BS Environmental Science	College of Arts & Sciences	Good
CLSU	Master & Doctor in Renewable Energy Systems, with Case Study Master in Environmental Management Certificate Programs: • Basic Environmental Impact Assessment • Diploma in Land Use Planning	Open University, College of Engineering, and Environmental Management Institute	Advance

(continued)

Table 4 (continued)

HEI	Degree programs	Host unit	Score
WPU	BS Environmental Management	College of Agriculture, Forestry and Environmental Sciences	Good
BU	Doctor of Medicine-Master of Public Administration (double) Program, major in Health Emergency & Disaster Management, since 2013	Graduate School and College of Medicine	Advance
MC	BS in Environmental Planning and Management, major in Urban Planning and Green Architecture, and Corporate Environmental Management MS/Ph.D. in Environmental Studies Ph.D. in Environmental Education Certificate courses (18 units each): Environmental Education, Environmental Studies, Environmental Management	Environmental Management Institute	Advance

Source of data data gathered from interviews and university websites

The UP-MS offers M.Sc. and Ph.D. degrees in Marine Science, with specializations in Marine Biology, Marine Physical Sciences, and Marine Biotechnology, and Professional Master in Tropical Marine Ecosystem Management. Established in 1974, the MSI is UP's coordinating base for research, teaching, and extension work in marine biology, marine chemistry, physical oceanography, marine geology, and related discipline (www.msi.upd.edu.ph).

The BU-Graduate School's master's degree, major in Health Emergency and Disaster Management is one of its kind. In 2012, the College of Medicine and the Graduate School agreed to incorporate the MPA-HEDM in the Doctor of Medicine Program. Graduates of the MD-MPA degree are expected to have competencies as medical doctors trained to work in various settings including in emergency and disaster situations. The MD-MPA-HEDM has attracted students from the government and nongovernment offices engaged in disaster response and rescue such as policemen, firemen, doctors, nurses. The BU-EDEN (Extension Disaster Education Network) partnership influenced the offering of the MPA-HEDM.

Environmental Science/Studies/Management is a common program offering in 10 HEIs CBSUA offers a master's level degree in DRM. First offered in 2008, enrolment and graduation in CBSUA's Diploma and MS-DRM has been sustained (Table 5).

At the BU, the first batch of 60 students enrolled in 2013–2014, and 54 students from the first batch will graduate in 2018, with a double degree in Doctor of

Table 5 Data on enrolment and number of graduates. MS-DRM, CBSUA

	Total number of enrollees in diploma		Total no. of graduates	Total number of enrollees in the master		Total no. of graduates
	FS	SS		FS	SS	
2010–11	43	18	0	33	45	2
2011–12	42	23	15	28	21	10
2012–13	24	29	13	41	18	2
2013–14	29	14	10	12	15	6
2014–15	31	23	10	16	24	5
2015–16	14	29	14	19	30	11
2016–17	37	21	19	18	35	14

Source of data Interview at CBSUA

Table 6 Enrolment in BU-DM-HEDM double program, Bicol University

Academic year	Enrolment
2013–14	60
2014–15	55
2015–16	55
2016–17	60

Medicine and Master in Public Administration (interview data). All students are on scholarship from the BU and study full time (Table 6).

Website data on faculty resources of the universities/colleges studied are rather limited. Table 7 counts only the number of faculty members teaching in the college where the CC-related courses are offered. As a whole, the faculty resources of colleges/units offering the CC-related programs can be beefed up from other units.

Table 7 Faculty resources in six specific colleges/units offering CC-related degree programs

HEI	Faculty resources
UPLB-SESAM	10 FT; 1 Prof. Emeritus; 2 Professorial Lecturers; 8 Researchers
UPD-CS	MSI: 19 regular faculty; 4 professor emeriti; 5 adjunct professors
	IESM: 11 FT, all Ph.D.s
CBSUA-Graduate School	19 FT; 20 AF; 1 visiting professor
ISU	4 in environmental science & forestry; 22 faculty in crop science; 6 in agribusiness; 11 in animal husbandry
CLSU-OU	9 FT; 6 AF
BU-Graduate School	19 faculty members

Notes FT—full time faculty; AF—affiliate faculty. Data are gathered from interviews and university websites

4.4 Research and Extension Programs

The search for research and extension program accomplishments of HEIs yielded an enormous amount of information, which cannot be included here for lack of space. All 10 HEIs were given a rating of ADVANCE in Research and Extension. Table 8 lists up to 3 examples of research and extension programs of the HEIs. In a nutshell, the HEIs researches and extension programs have been varied, ranging from bioethanol research, assistance to local governments in solid waste management plan formulation, ecosystem assessment, organic and resilient agriculture research, environmental impact assessment, and livelihood education to build

Table 8 Samples of research and extension (R&E) programs of HEIs

HEI	Research and extension programs	Rating
UPLB-SESAM	• Ten years after the millennium ecosystem assessment of Laguna de Bay: towards a sustainable future • Social network analysis of selected community-based forest management projects • 2nd international conference on climate change research and development 2017, and the 6th national climate research conference 2017, September 27–28, 2017	Advance
UPD-CS	MSI: post-doctoral seminars Bolinao marine laboratory training & extension program (BML-TEP)	Advance
	IESM: Benham rise potential productivity research project Assessment of hydrologic carrying capacity of island watersheds Emission factors from motor vehicle pollution sources using a wind tunnel	Advance
MMSU	• Bioethanol research • Climate-resilient agriculture • Sagip (or save) Quiaoat River	Advance
ISU	Integrated R&D program on climate change resiliency and environmental protection Environmental quality assessment: biophysical, socio-economic and cultural interrelationships in a protected national park	Advance
CvSU	Coffee research Urban agriculture research Biogas research	Advance
CBSUA	CCA measures Environmental management Food security, livelihood security and organic agriculture	Advance
CLSU	Ecological solid waste management program Dark green school program Biodiversity conservation and management program	Advance

(continued)

Table 8 (continued)

HEI	Research and extension programs	Rating
WPU	Climate change adaptation and risk reduction program Mangrove rehabilitation and solid waste management projects Coastal resource assessment of Turtle Bay, Puerto Princesa City	Advance
BU	• Adaptive Agriculture, Forestry & Nature Management in Response to Climate Change • Adopt-a-School-Sagip Buhay (Rescue Lives) Program: Integrating Community-Managed DRR Consideration in Local Development Management Process • Hazard Reduction and Vulnerability Analysis of Sagumayon River	Advance
MC-ESI	Radyo Kalikasan [radio environment] Southern Sierra Madre Wildlife Conservation center services 75-hectare Dona Remedios Trinidad reforestation project	Advance

Source of data Data gathered from interviews and university websites

resilient communities. The HEIs have also produced huge amount of research presented in local and international conferences and published in ISI publications and school journals.

The MMSU has been engaged in Coastal Resource Management. Its *Coastal Clean-up and Scubasurero* project aims to generate awareness of the local community about the present state of the marine ecosystem. Collection of non-biodegradable garbage along the shore and supervised underwater garbage collection is done with the community. In the *Sagip* (Rescue or Save) Quiaoat River project, the MMSU provides technical assistance to the City Government of Batac, Ilocos Norte in the rehabilitation, restoration of the Quiaoit River including its waterways and watershed. The *Pagrang-ayen ni Mangngalap* program aims to help partner-fishermen groups improve the quality of life of fishing households through extension activities and enhance public awareness on the importance of conserving fishery and aquatic resources.

Two of the 5 research agenda of CBSUA zero-in on climate change measures, and environmental management. A third agenda is about food security, livelihood security and organic agriculture. CCA measures include the development and dissemination of climate smart agriculture systems, indigenous knowledge systems, and crop-weather relations and climate predictive models. Under environmental management, CBSUA engages in sustainability assessment of socio-ecological systems, alternative livelihoods in environmentally critical areas, green and clean industries, and climate proofing ecosystems, among others.

The WPU is a recognized leader in sustainable development in Western Philippines and advocate of environmental protection and green technology. Three of its centers work on the pursuit of environment and sustainable development activities. The WPU-AREC promotes the utilization and commercialization of RE technologies in the Province of Palawan. WPU's partnership with the DOE enabled it to develop expertise in solar and hydro power technology, and community organizing and social preparation processes towards the promotion of the use of RE. It designed a hydropower system that can energize 100 households in Region IV-B. It was tapped by the United Nations Development Programme in 2005 to develop the El Nido solar power plant project.

BU's important collaborative extension program is the Bicol-EDEN, an acronym for Extension Disaster Network (EDEN). EDEN is a multi-stakeholder collaboration of various agencies across the Bicol Region and with the United States of America to moderate the impact of disasters and improve the delivery of services to citizens affected by disasters. Through EDEN, the BU engages stakeholders from local governments, other state and private universities, and nongovernment organizations in research and education programs on disaster mitigation, preparation, response and recovery.

Miriam College's College of Arts and Sciences offers Environmental Planning and Management programs in collaboration with the Environmental Management Institute (EMI). The EMI is one of MC's advocacy centers, focused on achieving MC's core values of Peace, Justice and Integrity of Creation. The EMI is described as an active player in the environmental movement in the country and a center of excellence in environmental education. It was awarded a reforestation area in Bulacan, the Dona Remedios Trinidad Reforestation Project, to become its laboratory for academic programs. Miriam College has adopted a number of approaches from learning to discovery, creation, exploration, innovation, transforming ideas, conduct of leadership forums to showcase its programs all in commune with the natural environment.

The UPLB-SESAM adopts a research and development framework, which highlights the environmental problems of the country and locates the application of its various activities in the natural environment. SESAM's research, training and extension activities have been extensive, for which it gained recognition from the CHED for accomplishments in the area of environmental science. The UPLB features SESAM as a platform for convergence and consolidation of university-wide efforts to develop and offer higher academic degrees on environmental science and management. SESAM hosts the UPLB-Climate Change and Disaster Risks Studies Center, created in 2007, to develop climate change adaptation and mitigation strategies as well as coordinate climate change initiatives in the University. SESAM was established in 1977 to address environmental degradation through instruction, research and public service. The School has produced 245 MS and 92 Ph.D. graduates.

4.5 HEI Support Facilities

Support facilities are herein understood to refer to centers, institutes, laboratories and similar facilities established in support of degree programs of HEIs some of which are cited here. All HEIs have support facilities dedicated to climate change as well as in support of the overall function of the universities. The UPLB grouped its 18 interdisciplinary studies centers into four clusters (Agriculture, Technology, Environment and Development), to make its research, development and extension programs more holistic, inclusive and responsive to society's needs. Six to eight of these centers are dedicated to climate-related research activities.

For the UP-Marine Science Institute (MSI), one important facility is the 5-hectare Bolinao Marine Laboratory, the research base for MSI's faculty and staff's scientific activities (Table 9).

The UP-IESM was conceived in 2003 to be a center of excellence in environmental and atmospheric sciences in the Asia-Pacific. The Institute is fully supported with laboratory and physical facilities in support of its research and teaching functions. Its faculty resources are highly qualified and well-researched.

The MMSU has actively pursued climate-related programs and activities recently with the announcement of the implementation of 3 big projects that include the National Renewable Energy Research and Innovation Center, created in support of its offering in 2016 of the Professional Science Master on Renewable Energy Engineering course. The project received Php150 million budget support from the national government to develop a laboratory for the development of *nipa* bioethanol feedstock, biofuel being the flagship project of MMSU. The establishment of the bioethanol facility had the support of the USAID-STRIDE project, DOE, Ethanol Producers Association of the Philippines, Pangasinan State University, Municipality of Bugallon, and Province of Pangasinan. The DOE supports the bioethanol project of MMSU considering its promising contribution to the objective of the Bioethanol Act of 2006 of developing indigenous, renewable, and clean energy sources to reduce dependence on imported oil (MMSU News, September 28 2015). The Village-Scale Bioethanol Facility, together with the MMSU-DOST Food Innovation Center, were launched and commissioned in September 2016.

MMSU's Center for Climate Resilient Agriculture (CCRA) is the Regional Research and Training Center for climate Change in Region 1. The CCRA was established in 2012 to study the impacts of climate change on agriculture. CLSU's ICCEM aims to provide quality education, research and extension services on the causes and consequences of climate change and provide environmental management. Miriam College's 75-hectare Reforestation area serves as laboratory for environmental courses.

4.6   Awards and Recognition

Almost all HEIs in this study have achieved a certain level of academic excellence in their respective areas of expertise. At least five HEIs have received recognition from the CHED as Centers of Excellence in various fields.

The MSI faculty include two National Scientists, several academicians/members of the National Academy of Science and Technology (NAST), several Outstanding

Table 9 HEI support facilities

HEI	Centers and support facilities	Rating
UPLB-SESAM	• UPLB Climate Risk Studies Center • Center for Integrated Natural Resources and Environment Management • Interdisciplinary Biofuels Research Studies Center	Advance
UPD-CS	MSI: Research laboratories, herbarium • Pilot seaweed processing plant and laboratories for marine colloids • 5-hectare Bolinao Marine Laboratory in Pangasinan; other laboratories	Advance
	IESM: Biogeography, Environment, Evolution & Climate Laboratory • EarthSMaP (Earth Surface Materials & Processes) Laboratory • Environmental Pollution Studies Laboratory	Advance
CvSU	• Affiliated Renewable Energy Center • National Coffee Research Development and Extension Center • Season Long Training on Organic Vegetable Production	Advance
CBSUA	• Disaster Risk Reduction Management • Organic Agriculture • Renewable energy	Advance
MMSU	• Center for Climate-Resilient Agriculture, since 2012 • Village-Scale Bioethanol Refinery Facility, since 2015 • GIS and Environment Monitoring Laboratory	Advance
ISU	• Climate Change Center-Education, Research and Development • Affiliated Renewable Energy Center (AREC) • Cagayan Valley Agriculture, Aquatics and Natural Resources Research and Development Consortium	Advance
CLSU	• Institute for Climate Change and Environmental Management (ICCEM), since 1997 • Biodiversity Center & the Regional Integrated Coastal Resources Management Learning Center (RIC), established at ICCEM in 2012 • Ramon Magsaysay-Center for Agricultural Resources & Environmental Studies • Affiliated Renewable Energy Center	Advance
WPU	• Affiliated Renewable Energy Center • National University for Agriculture and Fisheries	Advance
BU	• BU-EDEN	Advance
MC	• Environmental Studies Institute • Southern Sierra Madre Wildlife Conservation Center • 75-hectare Dona Remedios Trinidad Reforestation Project	Advance

Source of data Data gathered from interviews and university websites

Young Scientists, a NAST-Third World Academy of Science Awardee, several Hugh Greenwood Environmental Awardees (NAST), among other awards.

The CLSU is recognized as lead agency of the Muñoz Science Community. It is the seat of the Regional Research & Development Center in Central Luzon, and one of the premier institutions of agriculture in Southeast Asia, with breakthrough researches in aquatic culture, ruminants, crops, orchard, and water management. It is the first comprehensive state university to undergo institutional accreditation. It is recognized by the CHED as a National University College of Agriculture, and a National University College of Fisheries, and Center of Excellence in various fields including Agriculture, Agricultural Engineering, and Fisheries. The Department of Environment and Natural Resources designated it as Regional Integrated Coastal Resources Management Center. It is also a Model Agro-Tourism Site for Luzon (Table 10).

WPU is recognized as a National University for Agriculture and Fisheries, belongs to the top ten universities in agriculture, and among the top five in fisheries. Bicol University is ISO-certified (ISO 9001:2008). It has grown to become a premier state university in the Bicol Region with 13 colleges and institutes. It is CHED's Center for the Development of Fisheries Education, and Higher Education Research, among others. BU is also recognized as base agency for the Bicol Consortium for Health Research and Development and the Bicol Consortium for Agriculture Resources Research and Development.

Miriam College's environmental orientation is highly evident it its programs. It banners eco-friendly policies in academic programs as well as in administration, outreach and finance. It pursued campus planning, pedestrianization, no smoking policy, no balloons policy, anti-smoke belching, water and power conservation, and solid waste management (www.mc.edu.ph). The MC academic campus exudes in its green spaces that includes a mini-forest park and rock garden maintained to achieve perfect harmony with nature. Led by the ESI, the MC's Environment Week is an awareness-building activity on climate change adaptation and disaster risk reduction among members of the MC community. Fun activities include the launch of simple concepts and ideas such as casual car sharing, serving organic food, tumbler campaign, pedometer challenge, grab a bag sale, triathlon, create your own emergency kit preparedness plan, craft making from recyclables, and survival relay. The formal part of the activity involves the conduct of a DRRM seminar, scanning of strategies for the formulation of MC's climate change program, talk on hazards, Project NOAH,[4] lessons on environmental principles, online carbon footprint simulation, online simulation game on survival and resiliency, and mapping for resiliency workshop on Open Street Map.

---

[4]Nationwide Operational Assessment of Hazards.

Table 10 Awards and recognition received by HEIs

HEI	Awards & recognition	Rating
UPLB-SESAM	2016 CHED Center of Excellence Award in Environmental Science; publishes the Journal of Environmental Science and Management (a semi-annual ISI-accredited scientific journal)	Advance
UPD-CS	MSI: 1994, Designated "National Center of Excellence in Marine Sciences" by virtue of Presidential Proclamation No. 518 • 1998, CHED-designated First Center of Excellence for Marine Sciences • Regional Center of Excellence of the Global Environment Facility-Coral Reef Targeted Research Programme for Southeast Asia • July 2008, Co-awardee, Calouste Gulbenkian International Prize on Biodiversity and Defense of the Environment • 2017 BML-TEP Best Extension Program Award	Advance
	IESM: Various faculty scientist's achievement awards	Advance
ISU	CHED-designated Center of Excellence in Agriculture, ICT, and Teacher Education	Good
CvSU	AREC officials are accredited by DOE as Certified Renewable Energy Engineers	Good
CLSU	Fulbright fellowship award; Eco-friendly and sustainable School	Advance
WPU	• Recognition as National University for Agriculture and Fisheries • AREC officials are certified by the DOE as Certified Renewable Energy Engineers	Good
BU	ISO 9001:200 Certification valid for 2017–2018 CHED Center of Excellence in Development for Fisheries Education, Development for Nursing, and Higher Education Research	Advance
MC	Citation as "Dark Green School" from the Environmental Education Network of the Philippines "Garbology Master" Gold Certificate from the Quezon City The Miriam Public Education and Awareness Campaign for the Environment (Miriam-P.E.A.C.E.), received a Likas Yaman award from the DENR for its pioneering efforts in environmental education	Advance

Source of data Data gathered from interviews and university websites

5 Conclusion

This review of HEI activities to integrate climate change into academic activities shows that:

- Some HEIs have made more explicit commitments to pursue programs and projects on climate change. Notable are the CLSU's formulation of the University Comprehensive Environment and Management Plan 2016–2040 and

University Strategic Plan; and UPLB-SESAM's adoption of a R&E framework for environmental conservation, protection and restoration/rehabilitation.

- Some HEIs have developed a niche in certain academic fields such as in Renewable Energy (MMSU, CLSU), and DRRM (BU), offered these as degree programs in the undergraduate, master and doctoral levels.
- UPD-MSI's Bolinao Marine Laboratory has served as research base for the Institutes research activities. MC's Doña Remedios Trinidad Reforestation Project has the multipurpose of serving as a community project, a laboratory for environmental projects of students, and MC's contribution to the government's reforestation program.
- Field visits to HEI campuses—UP Diliman, UPLB, CLSU and MC—show their commitment to the practice of environmental management through keeping a green campus (MC and CLSU banner as Deep Green Schools).
- Research outputs and academic programs of many HEIs have received recognition from the CHED and other agencies and organizations (Table 11).
- Some degree programs require multi-perspectives from the hard sciences, agriculture, and management to the social sciences, suggesting that CC actions must have the concerted effort of highly divergent disciplines.
- Evidently, some degree programs were instituted through partnerships with other universities, government agencies, civil society organizations and the private sector.

Overall, the HEIs are demonstrating best practices in climate change education. Having said this, the question of factors that promote the diffusion and institutionalization of climate change in university framework of governance remains to

Table 11 Summary of scoring of performance of HEIs using Mulvey's climate accountability scorecard

HEI	VM	Degree program	R&E	Support facilities	Awards & recognition	Overall rating
UPLB-SESAM	Advance	Advance	Advance	Advance	Advance	Advance
UPD-CS: MSI	Advance	Advance	Advance	Advance	Advance	Advance
IESM	Advance	Advance	Advance	Advance	Advance	Advance
CvSU	Good	Advance	Advance	Advance	–	Advance
CBSUA	Advance	Advance	Advance	Advance	–	Advance
MMSU	Good	Good	Advance	Advance	–	Good
ISU	Advance	Good	Advance	Advance	Good	Good
CLSU	Advance	Advance	Advance	Advance	Advance	Advance
WPU	Advance	Good	Advance	Advance	Good	Good
BU	Good	Advance	Advance	Advance	Advance	Advance
MC	Advance	Advance	Advance	Advance	Advance	Advance

be answered. Based on data gathered, it is safe to say that, in general, the 10 universities studied have institutionalized climate change into university framework, programs and activities. The formulation of climate and environment-oriented mission and vision, offering of relevant degree programs, and putting in place the necessary facilities and resources have helped HEIs develop their respective areas of expertise in addressing climate change. Climate-related programs have been instituted together with support facilities, research and extension work.

6 Way Forward

Much of HEI activities related to climate change have not been accounted for in this research. A short visit and interviews would not suffice to gather the most relevant information about HEI activities on climate change, although effort is exerted to gather such information as much as possible. Much remains to be covered in terms of the integration of climate change in management, policy and social sciences degree programs such as public administration, and the numerous possibilities of integrating CC in almost any academic programs. This research also did not discuss mitigation vis-a-vis adaptation measures; measurement of HEI's reduction of GHG emissions; HEI's carbon sinks; and extent of use of RE in HEI communities. HEIs found to offer climate change-related programs that could not be covered in this study could be the subject of future research.

HEIs may adopt their own climate accountability scorecard (Mulvey et al 2016; Filho 2009) to measure their progress and contributions to GHG reduction, and to environmental sustainability. At the least, HEIs should monitor their contribution to GHG reduction, and disclose primarily in their websites climate-relevant information including their own energy utilization, waste management and disposal policies and practices, and natural resources conservation. Another direction in climate education is to directly link adaptation and mitigation strategies to home improvement, self-improvement, green spaces, and community spaces, and make climate change a home issue that people can act on a daily basis such as energy conservation and waste management.

Acknowledgements The support of the University of the Philippines-National College of Public Administration and Governance as well as the assistance of professors/heads of universities (BU, CBSUA, CLSU, CvSU, ISU, and Miriam College) interviewed are gratefully acknowledged.

References

Benayas J, Alonso I, Alba D, Pertierra L (2010) The impact of universities on the climate change process. In: Filho WL (ed) Universities and climate change: introducing climate change to university programmes. Springer, Berlin, pp 47–65. https://doi.org/10.1007/978-3-642-10751-1_1

Brandli LL, Filho WL, Frandoloso MAL, Korf EP, Daris D (2015) The environmental sustainability of Brazilian universities: barriers and pre-conditions. In: Filho WL, Azeiteiro U, Caeiro S, Alves F (eds) Integrating sustainability thinking in science and engineering curricula. World Sustainability Series. Springer, Cham

Commission on Higher Education (CHED) (2014) State Universities and Colleges Statistical Bulletin, Academic year 2013–2014, 103 pages. Link: http://www.ched.gov.ph/wp-content/uploads/2014/temp/10-03/home/State%20Universities%20and%20Colleges%20Statistical%20Bulletin.pdf

Cuaresma JC (2017) Philippine Higher Education Institutions' responses to climate change. In: Filho WL (ed) Climate change research at universities: addressing the mitigation and adaptation challenges. Springer, Hamburg, Germany, pp. 69–93

Department of Budget and Management (2016) General Appropriations Act

Department of Energy (2001) DOE guidebook for developing sustainable rural renewable energy services, 39 pages

Filho WL (2009) Sustainability at universities: opportunities, challenges and trends. Peter Lang Scientific Publishers, Frankfurt

Filho WL (2010) Climate change at universities: results of a world survey. Chap. 1. In: Filho WL (ed) Universities and climate change: introducing climate change to university programmes. https://doi.org/10.1007/978-3-642-10751-1_1, ©Springer, Berlin, Heidelberg, pp. 1–19

Filho WL (ed) (2017) Climate change research at universities: addressing the mitigation and adaptation challenges. Springer, Hamburg, Germany, eBook ISBN 978-3-319-58214-6

http://www.bicol-u.edu.ph/

https://clsu.edu.ph/

http://iccem.clsu.edu.ph/linkages.php

http://www.isu.edu.ph/

http://www.mc.edu.ph

http://www.mmsu.edu.ph/

http://iesm.upd.edu.ph/

http://www.msi.upd.edu.ph/about-msi

https://sesam.uplb.edu.ph; https://uplb.edu.ph/sesam

https://idsc.uplb.edu.ph

http://www.wpu.edu.ph

Meakin S (November 1992) The Rio Earth Summit: Summary of the United Nations Conference on Environment and Development. Science and Technology Division. Link: http://publications.gc.ca/Collection-R/LoPBdP/BP/bp317-e.htm

MMSU News, www.mmsu.edu.ph. PSM program launched today. August 11. 2016. Link: http://www.mmsu.edu.ph/news/details/read/psm_program_launched_today

MMSU News. 3 big dev't projects in MMSU seen. September 28, 2015. Link: http://www.mmsu.edu.ph/news/details/read/3_big_dev_t_projects_in_mmsu_seen

Mulvey K, Piepenburg J, Goldman G, Frumhoff PC (October 2016) The climate accountability scorecard: ranking major fossil fuel companies on climate deception, disclosure, and action. Union of concerned scientists. Link: http://www.ucsusa.org/sites/default/files/attach/2016/10/climate-accountability-scorecard-full-report.pdf

Republic Act No. 9729 or the Climate Change Act of 2009, dated October 23, 2009

Rogers EM (1983) Diffusion of innovation, 3rd edn. Free Press of Glencoe, New York

UNESCO (1992) The Rio declaration on environment and development, 19 pages. Link: http://www.unesco.org/education/pdf/RIO_E.PDF

UNFCCC December 12, 2015. Paris Agreement—status of ratification. Link: http://unfccc.int/
paris_agreement/items/9444.php
United Nations (UN) (2015) Paris Agreement, 27 pages. Link: http://unfccc.int/files/essential_
background/convention/application/pdf/english_paris_agreement.pdf

Associate Professor Jocelyn Cartas Cuaresma is graduated in Economics, has a master's degree
in Public Administration, major in Fiscal Administration, a second master's degree in
Administrative Sciences at the Post-Graduate School of Administrative Sciences in Speyer,
Germany, Doctor of Public Administration at the National College of Public Administration in the
University of the Philippines Diliman. Her current research interests include climate change
education, climate financing, and partnerships in climate change capacity building.

Climate Change Communication to Safeguard Cultural Heritage

Rosmarie de Wit, Mohammad Ravankhah, Dimitrios G. Kogias,
Maja Žuvela-Aloise, Ivonne Anders, Brigitta Hollósi, Angelika Höfler,
Jörn Birkmann, Charalampos Patrikakis, Vanni Resta and Silvia Boi

Abstract The protection and conservation of cultural heritage is important for our society, not only in order to preserve cultural identity, but also because it acts as a wealth creator, bringing tourism-related opportunities on which many communities depend. However, Europe's heritage assets are highly exposed to extreme weather events and natural hazards, which may be exacerbated as a result of climate change. The goal of the STORM (Safeguarding Cultural Heritage through Technical and Organisational Resources Management) project is to provide critical decision-making as well as technical tools to multiple sectors and stakeholders engaged in the protection of cultural heritage from extreme events and climate change. Here, the STORM framework will be presented, focusing on the communication of extreme events and climate change information through the risk assessment procedure, which serves as an important basis of the decision-making tools. Using the climate change communication methodology outlined here, the effect of climate change on cultural heritage specific natural hazards can be quantified and subsequently used in the risk assessment procedure, hereby resulting in an increased understanding of climate change risks on cultural heritage.

R. de Wit (✉) · M. Žuvela-Aloise · I. Anders · B. Hollósi · A. Höfler
Zentralanstalt für Meteorologie und Geodynamik (ZAMG), Hohe Warte 38,
1190 Vienna, Austria
e-mail: rosmarie.dewit@zamg.ac.at

M. Ravankhah · J. Birkmann
Institute of Spatial and Regional Planning, University of Stuttgart, Pfaffenwaldring 7,
70569 Stuttgart, Germany

D. G. Kogias · C. Patrikakis
University of West Attica, 250 Thivon & P. Ralli Str., GR12244 Egaleo, Greece

V. Resta
Kpeople Ltd., Chase Side House, 42 Chase Side, Enfield, UK

S. Boi
Engineering Ingegneria Informatica S.p.A, Via San Martino della Battaglia 56,
00185 Rome, Italy

© Springer Nature Switzerland AG 2019   199
W. Leal Filho et al. (eds.), *Addressing the Challenges in Communicating Climate
Change Across Various Audiences*, Climate Change Management,
https://doi.org/10.1007/978-3-319-98294-6_13

In addition, STORM-developed technical solutions that aid information transfer during and after extreme events will be highlighted.

1 Introduction

Cultural heritage is an irreplaceable source of inspiration and plays a crucial role in defining cultural identity (UNESCO 2007a). On top of that, it provides tourism-related opportunities to communities, thereby acting to improve economic outlook. In order to conserve these assets for the current and future generations, the protection of cultural heritage is of utmost importance. However, cultural heritage is threatened by both natural and anthropogenic hazards, and in recent years the impact of climate change on cultural heritage has been recognized as one of these threats (UNESCO 2007a). Examples of these climate change and extreme events related threats are (increases in the frequency of) heavy rainfall episodes, which could cause damage due to faulty or inadequate water disposal systems or the inability of historic rainwater goods to handle heavy rain, or the deterioration of facades due to thermal stress (UNESCO 2007b). This realization led the International Council on Monuments and Sites (ICOMOS) to adopt a resolution recommending "that climate change adaptation strategies for cultural heritage should be mainstreamed into the existing methodologies for preservation and conservation of sites" (ICOMOS 2007).

The goal of the Horizon 2020 funded Safeguarding Cultural Heritage through Technical and Organisational Resources Management (STORM) project is to provide critical decision-making tools to stakeholders engaged in the protection of European cultural heritage from extreme events and climate change, which ties in directly with this recommendation. The project aims to do this by considering the prevention, intervention, and the improvement of policies, planning & processes phases in an integrated fashion, and by developing a technological framework providing a collaborative platform to support stakeholders to act more efficiently during the prevention (in order to adapt to the effects of climate change) and intervention (when an extreme event occurs) phases. To test the developed framework, pilot site studies at five different cultural heritage locations (see Table 1), all with unique risk profiles, are performed.

Table 1 STORM cultural heritage pilot sites and their location

Pilot site name	Location
The Diocletian Baths	Rome, Italy
Roman Ruins of Tróia	Grândola, Portugal
Mellor Heritage Project	Greater Manchester area, UK
Fortezza Fortress	Rethymno, Greece
Ephesus Great Theatre	Izmir, Turkey

In terms of communication, information regarding future climate change and extreme events is key during the prevention phase, whereas real-time information and resilient communication is essential during the intervention phase. Hence, information and communication on two different timescales are considered within the project. Climate projections addressing changes in means and extremes are used to perform a risk assessment, which can in turn be used as input by cultural heritage stakeholders and site managers to define adaptation strategies. Within STORM, a climate change assessment and communication methodology, customised for cultural heritage sites, was developed and applied for the five pilot sites listed in Table 1. This method, relying on open-access data and software so it can not only be applied within the STORM context but also by other European sites, is used to quantify the climate change effects on natural hazards that are of importance to cultural heritage sites. The risk assessment approach, which relies on this climate information, is outlined in Sect. 2.1, with a focus on the climate analysis methodology and the translation and communication of this information to the disaster risk reduction and cultural heritage community in Sect. 2.2. In Sect. 3, real-time resilient communication during and after extreme events is discussed, and the technical solution developed within the STORM project is highlighted.

2 Communication on Long Timescales: Risk Assessment

2.1 General Risk Assessment Concept

The STORM project aims to provide cultural heritage properties with a systematic and integrated methodology for risk assessment and management in response to adverse effects of natural hazards and climate change-related events. Risk assessment is "a methodology to determine the nature and extent of risk by analysing potential hazards and evaluating existing conditions of vulnerability that together could potentially harm exposed people, property, services, livelihoods and the environment on which they depend" (UNISDR 2015). It is a central component of the risk management procedure to determine the nature and level of risks based on which risk reduction strategies should be developed. Moreover, "the process of producing a risk assessment will enable both public authorities and businesses, NGOs, and the general public to reach a common understanding of the risks faced as a community and help fostering an inclusive debate about the relative priority of possible prevention and mitigation measures" (EC 2010).

The STORM risk assessment procedure was developed based on the current risk assessment standards and guidelines, including ISO 31000: Risk management—Principles and guidelines (ISO 2009), the United Nations Office for Disaster Risk Reduction (UNISDR) framework for disaster risk management (UNISDR 2015), and the Intergovernmental Panel on Climate Change (IPCC) Managing the Risks of Extreme Events and Disasters to Advance Climate Change Adaptation (SREX) framework (IPCC 2012). Furthermore, current risk assessment and management approaches and methods specific to cultural heritage, such as the US Federal Emergency Management Agency's (FEMA) procedure for hazard mitigation planning (FEMA 2005) and the United Nations Educational, Scientific and Cultural Organization's (UNESCO) disaster risk management cycle (UNESCO World Heritage Centre 2010) have been considered. The risk assessment procedure developed as part of the STORM project comprises the following major steps:

- Identification and analysis of extreme events and climate change threats;
- Assessing the value of heritage properties exposed to the hazards;
- Analysing the vulnerability of the heritage properties to the identified hazards;
- Identifying and analysing the risks.

To adequately address short and long-term effects of natural hazards on heritage sites, both hazards leading to sudden-onset disasters (e.g. storms, flooding, wildfires) and threats leading to slow-onset disasters (e.g. change in freeze-thaw events, heat waves, and prolonged wet or dry periods) have been incorporated in the assessment procedure. A semi-quantitative ranking (adapted from HAZUS-MH; FEMA 2004) is applied to analyse the potential hazards affecting the pilot sites and to determine which hazards or threats need to be integrated into the further risk assessment procedure.

With respect to the exposure assessment, movable and immovable heritage assets, their associated intangible attributes and their values are considered as elements at risk. Therefore, exposure assessment is mainly focused on the analysis of the significance of heritage assets at the pilot sites.

Apart from hazard and exposure, a critical component of risk that significantly contributes to the overall loss is 'vulnerability'. IPCC (2014) defines vulnerability as "The propensity or predisposition to be adversely affected.". The concept of vulnerability, within the IPCC's climate change adaptation agenda, encompasses two key elements of sensitivity to harm, and lack of capacity to cope and adapt. Accordingly, the STORM project adapts the concept of vulnerability in the WorldRiskIndex (Birkmann and Welle 2015) based on which two components of susceptibility and coping and adaptive capacity are assessed to derive structural and non-structural vulnerability of the heritage sites to the identified hazards.

Following the assessment of the risk elements, potential impacts of the hazards on each pilot site are identified. Subsequently, the risks are analysed to measure the level of the risks while applying the Risk Index methodology. To adequately visualise the level of the risks in different parts of the pilot sites, relative risk maps are generated using ArcGIS tools. The output of the risk assessment, through the risk evaluation, will support the decision-making process to understand for which risks mitigation, adaptation and preparedness strategies need to be determined.

In the next section (Sect. 2.2) the methodology to evaluate climate change and extreme event related hazards is explained in more detail, with a special focus on tailoring this specialized knowledge to the use in the risk assessment and the communication of the results to the cultural heritage stakeholders.

2.2 Tailored Climate Change and Extreme Event Information

In order to provide relevant climate analyses, to be used outside the climate research community and that are of interest to cultural heritage site managers, knowledge needs to be shared between the cultural heritage, disaster risk reduction and climate change communities. Based on the pilot sites' needs a climate change analysis methodology was developed to present climate data in such a way they can easily be integrated in the risk assessment procedure and understood by cultural heritage stakeholders. The methodology developed in STORM consists of five steps, which are outlined below. It is interesting to note that the developed strategy only relies on open-access data and software. Therefore, it can be used by any European cultural heritage site, and the use of the method is not limited to the STORM participants.

Step 1: Assign relevant climate indices to the hazards

First of all, a general understanding of what climate change hazards and extreme events are threatening the cultural heritage site has to be established. In STORM, pilot site meetings were conducted during the initial project phase, which provided the opportunity to obtain this information. In the next phase, these hazards and threats were defined in a uniform matter for all sites. To this end, tables identifying hazards and threats leading to sudden-onset and slow-onset disaster, developed as part of the risk assessment procedure and filled out by the pilot site managers highlighting the relevant risks per pilot site, were used as a baseline. Standardized climate change indices as defined by the Expert Team on Climate Change Detection and Indices (ETCCDI: e.g. Karl et al. 1999; Peterson et al. 2001; Peterson 2005; Zhang et al. 2011; Sillmann et al. 2013) were assigned to these hazards. An example of this process is shown in Table 2 for the slow-onset hazard 'precipitation'. Here, it can be seen that the indices 'rr—precipitation sum', 'r10 mm—heavy precipitation days', 'r20 mm—very heavy precipitation days', 'rx1 day—maximum 1-day precipitation amount' and 'rx5 day—maximum 5-day precipitation amount' were assigned to the hazard 'intense rainfall', 'consecutive wet days' to the hazard 'prolonged wet periods' and the index 'consecutive dry days' to 'prolonged dry periods'. A definition of the indices is provided in the final column.

Step 2: Obtain meteorological data to define current climatic conditions

To obtain information regarding a change in climate or climate extremes, a baseline reflecting current climatic conditions needs to be established in a second step. In STORM, observations from meteorological stations in the vicinity of the pilot sites were used. To this end, data from the European Climate Assessment & Dataset (ECA&D) were used. This dataset provides a large number of quality-controlled time series of meteorological parameters for all European countries. The data can be accessed online (see www.ecad.eu), are all offered in the same standardized data format and are free for use in non-commercial research and education (Klok and Klein Tank 2009). Using data provided by ECA&D and

Table 2 Assignment of ETCCDI indices to hazards illustrated based on the example of the hazard 'precipitation'

Hazards leading to SLOW-ONSET disasters

Hazard		Index	Definition
Precipitation	Intense rainfall	rr[a]	Precipitation sum: total precipitation sum per time period
		r10 mm	Heavy precipitation days: number of days per time period with daily precipitation equal to or greater than 10 mm
		r20 mm	Very heavy precipitation days: number of days per time period with daily precipitation equal to or greater than 20 mm
		rx1 day	Highest one day precipitation amount: maximum of one day precipitation amount in a given time period
		rx5 day	Highest 5-day precipitation amount: highest precipitation amount for a 5-day interval
	Prolonged wet periods	cwd	Consecutive wet days: number of consecutive days per time period with daily precipitation amount of at least 1 mm
	Prolonged dry periods	cdd	Consecutive dry days: number of consecutive days per time period with daily precipitation amount below 1 mm

[a]Not an official ETCCDI index

selecting a station in the vicinity of the site of interest, any European site can obtain the data necessary to analyse the current climatic conditions.

Step 3: Climate projections

In step 3, climate projections are considered based on global model runs forced with the representative concentration pathway for the 8.5 W/m^2 radiative forcing (RCP8.5) towards the end of the 21st century scenario (IPCC 2013). In the description of the climate change analysis methodology provided to the pilot sites it is specifically highlighted that these projections depend on the radiative forcing *scenario* used, and that these scenarios are based on assumptions concerning future developments that may or may not be realized (IPCC 2014) and that as a result, it is important to note that the model results are climate *projections* and describe a *possible* future state based on the forcing assumptions.

As the spatial resolution of global models is too coarse for a detailed climate analysis at pilot site level, downscaling techniques have to be used in order to obtain local or regional scale climate projections. To this end, dynamical and statistical models are used to assess possible future climatic conditions for the pilot sites. For a local assessment of the future climatic conditions at the pilot sites a statistical downscaling of global climate model results to the specific pilot locations is performed, resulting in daily time series of temperature and precipitation parameters. Using four global climate model runs, provided by the Coupled Model Intercomparison Project Phase 5 (CMIP5, Taylor et al. 2012), as well as the observations outlined in Step 2 a statistical downscaling for air temperature as well as daily precipitation sum using the analog method (e.g. Zorita and von Storch 1999; Benestad et al. 2008) was performed. The basic idea of this method consists of picking observation data from a date in the past where the large-scale atmospheric circulation regime corresponds most closely to the atmospheric state simulated by the global model for the day for which the prediction is made, and gives an optimized climate projection for the station location. This method was applied as it is capable of producing a realistic distribution of the data in the past, which is an important criterion for the subsequent calculation of climate indices as well as the assessment of extreme events.

To obtain regional scale information, a dynamical downscaling of the global model results with regional climate models (RCMs) is used. With this technique, information about future climatic conditions in a region around the pilot site (maps) can be determined at a higher resolution than available from the global model. By averaging the grid boxes in the vicinity of the pilot site this information can also be shown as a time series. The regional information of future climate conditions under the RCP8.5 scenario is provided within the extensive database of regional climate model results from the Coordinated Downscaling Experiment—European Domain (EURO-CORDEX) project. The EURO-CORDEX project is the European branch of the international CORDEX initiative, which aims to provide regional climate change projections for climate change impact, adaptation and mitigation studies (e.g. Jacob et al. 2014). Within the EURO-CORDEX project, an ensemble of RCM runs forced by multiple global climate models from CMIP5 was conducted. The data are freely available through www.euro-cordex.net/ as gridded datasets with a spatial resolution of 0.11° (about 12 km, providing regional maps) and as daily output of air temperature (minimum, maximum, mean), precipitation sum, daily mean wind and daily mean relative humidity (the latter two for a subset of the models) for the time period 1951–2100, dependent on the climate scenario. In the STORM project, an ensemble of 9 different RCM model runs was selected for the evaluation. The large number of model ensemble members is used for statistical analysis and is intended to account for uncertainties in the individual climate model simulations. The use of multi-model ensembles provides an improved 'best estimate' projection, as the mean of the ensemble can be expected to outperform individual ensemble members under the assumption that simulation errors in different models are independent (IPCC 2007).

The evaluation time period 1971–2000 for historical simulations was selected as reference climate period and the time period 2036–2065 was analysed for the future climate scenarios for both the statistical and dynamical downscaling results.

Step 4: Determination of climate indices and initial analysis

Using the open-source climate data operator (cdo) software (Schulzweida 2017), provided by the Max Planck Institute for Meteorology, the climate indices were determined based on the observations, the statistical downscaling time series, and the RCM results for all five pilot sites. In case of the modelling data sources, the climate indices were calculated for each model separately and then evaluated as an ensemble (multi-model mean and spread). The indices time series are then analysed and displayed in a number of ways (climatologies, frequency distributions, Walter-Lieth diagrams, and time series) to aid the interpretation of current climatic conditions as well as projected changes.

Step 5: Communication of results

In order for the climate analysis to be incorporated in the risk assessment procedure and be easy to understand for cultural heritage stakeholders and site managers, these results have to be summarized in a qualitative as well as quantitative manner. To this end, the tables developed in Step 1 are expanded to contain this information (illustrated in Table 3 for the example of 'Extreme temperatures - heat

Table 3 Summary of the quantitative and qualitative analysis of the "extreme temperatures—heat waves and cold waves" hazard for the Roman Ruins of Tróia[a]

Hazards leading to slow-onset disaster		Observations/ historical run	Statistical downscaling (SD)	RCM		TRÓIA
Hazard	Index	1971 -2000 baseline	Change 2036-2065 relative to 1971-2000	Change 2036-2065 relative to 1971-2000	Comments	Qualitative classification of change
Extreme temperature — Heat waves	tx[b]	21.0°C 21.1±0.0°C 19.7±1.2°C	+2.0±0.0°C	+1.7±0.2°C	Cold bias in RCMs ~1.5-2.0 °C increase	
	su[c]	89.7 d/yr 94.5±0.3 d/yr 75.6±29.9 d/yr	+42.2±13 d/yr	+28.3±19 d/yr	-5% overestimate in SD, ~15% underestimate in RCMs -35-45% increase	
	csu[d]	18.3 d/yr 29 d/30-yr period n/a 24.2±13.1 d/yr 54.6±23.7 d/30-yr period		+17.5±10.4 d/yr +29.2±23.6 d/30-yr period	High model variability and overestimate in RCMs ~70% increase Extreme values: ~50% increase	
Extreme temperature — Cold waves	cfd[e]	0 d/30-yr period n/a 2.3±1.6 d/30-yr period		-0.2±2.1 d/30-yr period	Slight cold bias in RCMs and large model variability No cold waves are to be expected under current and future climatic conditions	

[a]Quantitative description of the current climate baseline derived from observations (no shading), statistical downscaling (light grey shading) and RCMs (dark grey shading) in the column labelled "1971–2000 baseline" as well as the change for the period 2036–2065 relative to the 1971–2000 baseline based shown for the statistical downscaling (abbreviated "SD" in the table) and RCM ensemble (ensemble mean ± spread). See text for more information

[b]Maximum temperature

[c]Number of summer days

[d]Number of consecutive summer days

[e]Number of consecutive frost days. Due to the nature of the analog method, "csu" and "cfd" cannot meaningfully be determined from the statistical downscaling results

waves and cold waves' for the Roman Ruins of Tróia). These tables provide an assessment of the current hazard level (for the 1971–2000 period, based on observations (where available) or the historical RCM ensemble) as well as the projected hazard level (2036–2065). Results based on the historical runs of the statistical downscaling and RCMs are included here as well in order to estimate how well the models replicate past conditions. The projected levels are determined based on statistical as well as dynamical downscaling results (indicated with 'Statistical downscaling' and 'RCM' in the tables).

The change for the period 2036–2065 relative to the baseline period is determined by a comparison of the climate projections under the RCP8.5 scenario to the historical model runs. First, the difference between the projected 30-year average and historical simulated 30-year average have been determined. Subsequently, the ensemble mean difference and spread (standard deviation) have been considered.

In the 'Comments' column, an assessment of the ensemble performance, based on the models' ability to replicate the observations and on the ensemble spread, is given. An increase has been colour coded red, a decrease blue to aid a quick assessment. The 'Qualitative classification of change' column translates projected change from a quantitative to a qualitative scale, where red indicates a very high, orange a high, yellow a medium, light green a low and dark green a very low increase. The colour codes are assigned based on a quantitative scale that assesses the projected change either based on absolute increase (in this example maximum air temperature: "tx", and number of summer days: "su") or a relative increase (consecutive number of summer days/frost days: "csu/cfd") (procedure not shown).

The climate tables contain information related to all hazards to which a climate index could be applied in order to provide a consistent analysis for all sites, making the results comparable for the risk analysis process. However, as a result, the tables also contain information on hazards that are not directly relevant for each individual pilot site. Therefore, short 'executive summaries' regarding current climatic conditions as well as projected changes are written for all pilot sites, highlighting only the most relevant results and figures for the cultural heritage stakeholders and site managers.

Using the 5-step method outlined above, climate change and extreme event related hazards are analysed and presented both in a quantitative and qualitative manner. The analyses are tailored for use in the risk assessment procedure as well as customized to be understood by cultural heritage site managers, and can be incorporated in the prevention step, helping cultural heritage site managers make informed decisions on how to efficiently adapt to climate change. Moreover, as open-access and open-source data and software have been selected, the use of this methodology is not limited to the STORM project. Any European cultural heritage site can obtain information regarding natural hazards and extreme events as well as changes therein as a result of climate change by following the outlined approach, and incorporate this in their risk assessment procedure.

3 Communication on Short Timescales: Resilient Communication During and After Extreme Events

Communicating the effects of climate change on long time scales is important in order to be able to manage the risks associated with those hazards. However, when facing the actual hazard and potentially associated communication infrastructure loss, resilient communication on short time scales is essential. Therefore, the resilience in communication, especially during or after extreme natural events or disasters, has been under constant investigation from the research community due to the crucial role that they can play in such critical situations. This role has been enhanced, with the recent advances in technology and the rapid growth of the ecosystem of the Internet of Things (IoT). In IoT-based solutions, the human can act as a sensor and provide information through text or multimedia files (e.g., photos, videos) in the system that could be used by first-responders or authorities during emergencies and extreme events to better prepare and deal with the task at hand. Therefore, the resilience in the communications will be critical for the efficient management of such solutions.

In Mauthe et al. (2016), network resilience is defined as "the ability of a network to maintain the same level of functionality during internal changes or/and external disturbances as a result of large-scale disasters". To this end, the solutions that were proposed include the enhancement of the hardware's redundancy in an effort to provide extra links to several network points, opting to offer diverse routes, of the same quality, that will be chosen when the optimal choice is not available anymore. In addition, the network should be continuously monitored for any anomaly detection that can point to possible failures in the network that will trigger a reaction before an aggravation of the network's possibly harmful condition. Ad hoc solutions have also been studied in an effort to provide basic communications, usually among the first-responders, in crisis situation where the main communication infrastructure has been damaged and cannot be used.

In STORM, a solution to deal with the resilience in the communications under extreme conditions is studied based on the fact that Wi-Fi is present in almost every mobile phone, tablet and smart watch. The purpose is to use a black box as a special device, whose Wi-Fi capabilities will allow it to send encrypted messages to the mobile phones of the people that are involved in a possible hazardous situation during an extreme event. To this end, the proposed black box can act as a Wireless Access Point and be used to broadcast its Service Set Identifier (SSID), which can contain predefined instructions, in case of an emergency. This way, the people that receive the message to their mobile (or other smart) devices will be guided to predefined spots where help is easier to arrive and where the danger levels are less high. This device is expected to perform well within the STORM context, since there is no need for any special infrastructure and will only use the Wi-Fi frequency (and, of course will be based to its popularity) in order to reach a larger audience for help under extreme conditions. The instructions that can be broadcasted, in case of emergency, can be changed anytime with the use of a web interface, as long as the device is not broadcasting. Using this black box technique, enabling the sending of information to mobile devices during disasters, communication during and after extreme events is guaranteed.

4 Discussion

In Sect. 2.2, the STORM-developed method to assess natural hazards relevant to cultural heritage, and projected changes therein as a result of climate change, was outlined. As the assessment is based on open-access data and open-source software this method can be applied by any interested cultural heritage site in Europe to perform a similar hazard assessment. However, it should be noted that although the method is straightforward and reproducible, the actual data gathering and analysis still requires expert knowledge. The same is true for the full risk analysis procedure as outlined in Sect. 2.1. As a result, cultural heritage experts would have to team up with experts from the disaster risk reduction and climate change communities to perform the analyses suggested in Sect. 2 for their site.

The use of the ETCCDI indices provides a solid base for the assessment of most hydrometeorological hazards, however, several hazards that are of interest to cultural heritage sites, such as for example wind-driven rain or rising damp, cannot be assessed using this approach. Furthermore, not all data needed for their derivation is provided by all data sources mentioned in Sect. 2. Temperature and precipitation data are widely available, however, for example wind gust data that are needed for the assessment of storms, are not. As a result, not all relevant hazards can be analysed in a quantitative manner. One solution to overcome this problem is to provide a qualitative hazard assessment based on literature research to provide an estimate for use in the risk analysis. Alternatively, data from nearby monitoring stations that are not necessarily freely available, could be acquired from local sources.

5 Conclusion and Outlook

The STORM project aims to provide cultural heritage stakeholders with a systematic and integrated methodology for the risk assessment and management in response to adverse effects of climate change-related events. Here, the project was outlined with a focus on the incorporation of climate change and extreme event information in the risk assessment procedure, as well as resilient communication during extreme events.

The STORM risk assessment procedure is developed based on current risk assessment standards and specifically tailored for the project by considering UNESCO as well as IPCC guidelines, thereby including the viewpoints from both the cultural heritage and climate change communities. A methodology to provide relevant climate information for use in the risk assessment and to cultural heritage stakeholders was designed using freely available data and open-source software, so that these methods can be used by a wide audience and are not limited to STORM. Following the developed approach, the hazards of importance to the cultural heritage site under consideration were identified by the site managers. Subsequently, climate indices were assigned to the identified hazards, and analysed using observational data provided by ECA&D, as well as CMIP5 and EURO-CORDEX climate projections. Finally, to ease the communication of climate change and extreme event information within the project, customized climate and extreme event information was provided to be incorporated in the risk assessment procedure using tables showing the current climate baseline as well as the projected changes in a quantitative (numeric) and qualitative (using a colour coding system) manner. Furthermore, site-specific information that can be used by the pilot site managers is communicated in general, short 'Executive Summaries'. Finally, a novel approach to resilient communication in case of extreme events as developed in STORM, using a black box that can be used to transmit warning or instruction messages to the mobile devices of those in the vicinity of the site, was highlighted.

After performing the full STORM-developed risk assessment described in Sect. 2.1, combining information related to the identified and analysed (climate change) hazards, the pilot site's exposure, and vulnerability, risk management and adaptation strategies will be designed. Firstly, a set of actions to be taken in case of an emergency will be devised. Secondly, guidelines for the application of cultural heritage 'first aid' will be developed. Together, these guidelines and strategies target the protection of the specific cultural heritage sites.

Acknowledgements This project has received funding from the European Research Council (ERC) under the European Union's Horizon 2020 research and innovation programme (grant agreement n° 700191). The authors would also like to thank the whole STORM-team for sharing their knowledge on a wide range of different topics.

References

Benestad RE, Hanssen-Bauer I, Chen D (2008) Empirical-statistical downscaling. World Scientific Publishing Co Inc., Singapore
Birkmann J, Welle T (2015) Assessing the risk of loss and damage: exposure, vulnerability and risk to climate-related hazards for different country classifications. Int J Glob Warm 8(2):191–212. https://doi.org/10.1504/IJGW.2015.071963
EC (2010) Risk assessment and mapping guidelines for disaster management SEC (2010) 1626 final. Brussels. https://ec.europa.eu/echo/files/about/COMM_PDF_SEC_2010_1626_F_staff_working_document_en.pdf. Accessed Nov 2016
FEMA (2004) Using HAZUS-MH for risk assessment: how-to-guide, Federal Emergency Management Agency. http://mitigationclearinghouse.nibs.org/content/fema-433-using-hazus-mh-risk-assessment-how-%C2%A0guide-2004

FEMA (2005) Integrating historic property and cultural resource considerations into hazard mitigation planning. How-to guide (FEMA 386-9). Washington, DC. http:// wyohomelandsecurity.state.wy.us/grants/hmpg/Integrating_Historic_Property_Cultural_ Resource_Considerations_into_hmplanning.pdf. Accessed Nov 2016

ICOMOS (2007) ICOMOS international workshop on impact of climate change on cultural heritage, New Delhi, 22 May 2007—resolution. http://www.icomos.org/climatechange/pdf/ New_Delhi_Resolution_EN.pdf. Accessed Oct 2017

IPCC (2007) Working group 1 report 'the physical science basis'. In: Solomon S, Qin D, Manning M, Chen Z, Marquis M, Averyt KB, Tignor M, Miller HL (eds) Contribution of working group I to the fourth assessment report of the intergovernmental panel on climate change, 2007. Cambridge University Press, Cambridge, United Kingdom and New York, NY, USA

IPCC (2012) Managing the risks of extreme events and disasters to advance climate change adaptation: glossary of terms. In: Field CB, Barros V, Stocker TF, Qin D, Dokken DJ, Ebi KL, Mastrandrea et al MD (eds) A special report of working groups I and II of the intergovernmental panel on climate change. Cambridge, UK, and New York, NY, USA: Cambridge University Press

IPCC (2013) Climate change 2013: the physical science basis. In: Stocker TF, Qin D, Plattner G-K, Tignor M, Allen SK, Boschung J, Nauels A, Xia Y, Bex V, Midgley. PM (eds) Working group I contribution to the fifth assessment report of the intergovernmental panel on climate change. Cambridge University Press, Cambridge, United Kingdom and New York, NY, USA

IPCC (2014) Climate change 2014: impacts, adaptation, and vulnerability. Part B: Regional Aspects. In: Barros VR, Field CB, Dokken DJ, Mastrandrea MD, Mach KJ, Bilir TE, Chatterjee et al M (eds)Contribution of Working Group II to the Fifth Assessment Report of the Intergovernmental Panel on Climate Change. Cambridge (UK & USA): Cambridge University Press

ISO (2009) ISO Guide 73/2009: Risk management vocabulary, International Organization for Standardization. https://www.iso.org/standard/44651.html

Jacob D, Petersen J, Eggert B et al (2014) EURO-CORDEX: new high-resolution climate change projections for European impact research. Reg Environ Change 14:563. https://doi.org/10. 1007/s10113-013-0499-2

Karl TR, Nicholls N, Ghazi A (1999) CLIVAR/GCOS/WMO workshop on indices and indicators for climate extremes: workshop summary. Clim Change 42:3–7

Klok EJ, Klein Tank AMG (2009) Updated and extended European dataset of daily climate observations. Int J Clim 29(8):1182–1191. https://doi.org/10.1002/joc.1779

Mauthe A, Hutchison D, Ceetinkaya EK, Ganchev I, Rak J, Sterbenz JPG, Gunkel M, Smith P, Gomes T (2016) Disaster-resilient communication networks: principles and best practices. In: Proceedings RNDM 2016—8th international workshop on resilient networks design and modeling, 13–15 Sept 2016, Halmstad, SE, pp. 1–10. https://doi.org/10.1109/rndm.2016. 7608262

Peterson TC et al (2001) Report on the activities of the working group on climate change detection and related rapporteurs 1998–2001. WMO, Rep. WCDMP-47, WMO-TD 1071, Geneve, Switzerland, 143pp

Peterson TC (2005) Climate change indices. WMO Bull 54(2):83–86

Sillmann J, Kharin VV, Zwiers FW, Zhang X, Bronaugh D (2013) Climate extremes indices in the CMIP5 multimodel ensemble: Part 2. Future climate projections. J Geophys Res Atmos 118:2473–2493. https://doi.org/10.1002/jgrd.50188

Schulzweida U (2017) CDO user guide—climate data operators, Version 1.8.1. https://code. mpimet.mpg.de/projects/cdo/embedded/cdo.pdf. Accessed Oct 2017

Taylor KE, Stouffer RJ, Meehl GA (2012) An overview of CMIP5 and the experiment design. Bull Amer Meteor Soc 93:485–498. https://doi.org/10.1175/BAMS-D-11-00094.1

UNESCO World Heritage Centre (2007a) Case studies on climate change and world heritage. http://whc.unesco.org/document/134011. Accessed Oct 2017

UNESCO World Heritage Centre (2007b) World Heritage Reports n°22. Climate change and world heritage: report on predicting and managing the impacts of climate change on world heritage and strategy to assist states parties to implement appropriate management responses. http://whc.unesco.org/documents/publi_wh_papers_22_en.pdf. Accessed Oct 2017

UNESCO World Heritage Centre, ICCROM, ICOMOS, and IUCN (2010) Managing disaster risks for world heritage. Paris: World Heritage Resource Manual, UNESCO. http://whc.unesco.org/uploads/activities/documents/activity-630-1.pdf. Accessed Nov 2016

UNISDR (2015) Proposed updated terminology on disaster risk reduction: a technical review: The United Nations Office for Disaster Risk Reduction. www.preventionweb.net/files/45462_backgoundpaperonterminologyaugust20.pdf. Accessed Nov 2016

Zhang X, Alexander L, Hegerl GC, Jones P, Klein Tank A, Peterson TC, Trewin B, Zwiers FW (2011) Indices for monitoring changes in extremes based on daily temperature and precipitation data. Wiley Interdiscip Rev: Clim Change 2(6):851–870. https://doi.org/10.1002/wcc.147

Zorita E, von Storch H (1999) The analog method as a simple statistical downscaling technique: comparison with more complicated methods. J Clim 12:2474–2489. https://doi.org/10.1175/1520-0442(1999)012%3c2474:TAMAAS%3e2.0.CO;2

Capacity Development to Support Planning and Decision Making for Climate Change Response in Kenya

Sheila Shefo Mbiru

Abstract Kenya's Climate Change Act 2016, obligates national and sub-national governments to mainstream climate change responses into development planning, decision-making and implementation. To enhance the capacity of the public service to comprehensively address climate change challenges, the Ministry of Environment and Natural Resources and relevant stakeholders developed a training program on *"Climate Change Policy, Planning and Budgeting at National and County Level"*. The program targets middle level managers and technical government officers involved in policy formulation, planning, budgeting and implementation of programs in sectors vulnerable to the effects of climate change. The paper presents the design and development process of the training program which was an iterative, multi-stakeholder consultative process. This includes the development of the curriculum and training manual and describes the inaugural training program where twenty seven national and sub-national government officers were successfully trained. The training program enhanced the capacities of national and sub-national government officers to understand climate change, its impacts and response actions and will enable them support planning and decision making for climate change response actions. This will form a significant contribution towards implementation of the Climate Change Act (2016) and Kenya's Nationally Determined Contribution (NDC). Experiences and lessons learnt will inform efforts to develop similar targeted climate change training program and contribute to capacity building efforts to improve climate change response.

S. S. Mbiru (✉)
The Low Emission and Climate Resilient Development (LECRD) Project,
Project Officer—Knowledge Management and Capacity Development,
Nairobi, Kenya
e-mail: shefombiru@gmail.com

© Springer Nature Switzerland AG 2019
W. Leal Filho et al. (eds.), *Addressing the Challenges in Communicating Climate Change Across Various Audiences*, Climate Change Management,
https://doi.org/10.1007/978-3-319-98294-6_14

213

1 Introduction

One of the major challenges currently facing Kenya is coping with the unprecedented threats posed by climate change. These include an increase in the frequency
and intensity of extreme climate events like droughts and floods. There is also
evidence of rising temperatures and increased variability in the rainfall patterns. The
impacts of climate change have adversely affected Kenya's economy and pose a
threat towards the realisation of Vision 2030, Kenya's economic blueprint, which
seeks to create a competitive and prosperous nation.

Kenya recognises the need for concerted global effort to comprehensively
address climate change. It is in this respect that Kenya ratified the Paris Agreement
in December 2016, signalling its resolve to make its contribution towards meeting
the global adaptation and mitigation goals. Further, Kenya's Nationally Determined
Contribution (NDC) signals the country's deliberate resolve to address climate
change mitigation and adaptation on equal footing for national good. The country
has put in place a number of measures to address climate change that include the
National Climate Change Action Plan (NCCAP 2013–2017 under review) that
charts a low carbon climate resilient development pathway, and the National
Adaptation Plan (2017–2030) that addresses adaptation across all sectors of our
economy. Further, the country now has a Climate Change Act that commenced in
May 2016 following Presidential assent. The Act is in tandem with the Constitution
of Kenya and Vision 2030 and was developed through an inclusive and consultative
process that guides response to climate change at national and county levels.[1]

The Kenya Climate Change Act 2016[2] provides for a regulatory framework for
the development, management, implementation and regulation of mechanisms to
enhance climate change resilience and low carbon development for the sustainable
development of Kenya.

This Act shall be applied in all sectors of Kenya's economy by the national and
county (sub national) governments to among other things;

- Mainstream climate change responses into development planning, decision
  making and implementation,
- Build resilience and enhance adaptive capacity to the impacts of climate change,
- Facilitate capacity development for public participation in climate change
  responses through awareness creation, consultation, representation and access to
  information and
- Provide mechanisms for and facilitate climate change research and development, training and capacity building.

---

[1]Ministry of Environment and Natural Resource (MENR) 2017. Climate Change Policy, Planning
and Budgeting at National and County Level: Facilitators Manual.
[2]Republic of Kenya (13th May 2016) Climate Change Act. Nairobi. Government of Kenya.

1.1 Rationale

The transition to green, low emission and climate resilient development requires unprecedented levels of awareness, knowledge and skills. Policy makers, technical experts, government officers are faced with the challenge of enhancing their knowledge and competencies to integrate adaptation and mitigation into sectorial policy cycles and implementation, develop skills to integrate climate resilience in city planning and draft sound project proposal to access climate finance funding among others. Capacity assessments in various countries confirmed that human resource and skills gaps constitute a major bottleneck to effectively address climate change (UNITAR 2013).

This is true in Kenya where many government officers lack the basic climate change knowledge and skills necessary to be effective at their jobs. The Sixth Schedule of the Constitution mandates the National Government to enhance capacities of sub-national governments to govern and provide services ensuring that workers acquire the necessary skills, through training and capacity-building.

Notwithstanding the efforts put in place to address the capacity needs of the national and sub-national governments, there still exist capacity gaps in the public service to comprehensively address challenges related to climate change.

To bridge the capacity gap, a climate change training program was designed to equip government officers with relevant knowledge, skills and attitudes to mainstream climate change into national and sub-national policy, planning and budgetary process given the high priority that the government has placed in addressing climate change and honouring its international obligations to the United Nations Framework Convention on Climate Change (UNFCCC). The program will form a significant contribution towards implementation of the Climate Change Act (2016) and Kenya's Nationally Determined Contribution (NDC). This will go a long way to climate proof national and sub-national county plans and ensure national priorities across all sectors, are climate smart.

The *"Climate Change Policy, Planning and Budgeting at National and County Level"* training program was designed and developed by the Ministry of Environment and Natural Resources, through the Climate Change Directorate,[3] in collaboration with the Kenya School of Government,[4] Kenya Institute of

---

[3]The *Climate Change Directorate (CCD)* established by the Climate Change Act 2016, is the lead agency of the Kenya Government on climate change plans and actions to deliver operational coordination. It is under the Ministry of Environment and Natural Resources.
[4]The *Kenya School of Government (KSG)* established by the KSG Act (No. 9 of 2012), is a State Corporation established to offer management training, research, consultancy and advisory services to the public sector. KSG offers services to both National and County (sub national) governments, private sector players as well as those from the Non-Governmental Organizations (NGOs) http://www.ksg.ac.ke/.

Curriculum Development,[5] the Council of Governors[6] and other stakeholders. This training program will enhance the capacities of national and sub-national governments, build capacity of government officers and enable them deliver on their mandates to serve citizens efficiently and effectively and form a significant contribution towards implementation of Kenya's national and international climate change obligations.

This paper presents the training program design and development process which was an iterative, multi-stakeholder consultative process. This includes the development of the curriculum and training manual and describes the inaugural training program where 27 national and sub-national officers were successfully trained. Experiences and lessons learnt will inform efforts to develop similar targeted climate change training program and contribute to capacity building efforts to improve climate change response.

2 Capacity Development for Climate Change Response

Dagnet et al. (2015) noted that in order for the Paris Agreement[7]; the international climate agreement to be universally effective, capacity building[8] is vital for enabling developing countries to contribute to the global effort to reduce emissions and adapt to climate change. However, countries are not all at the same stage of development, nor do they have the same levels of capabilities. This reality must be taken into account in building a low-carbon and climate-resilient world in an equitable way.

According to Dagnet et al. (2015), capacity building for climate action can be understood as *"the process through which individuals, organizations and societies*

---

[5]The *Kenya Institute of Curriculum Development (KICD)* is an Institute established through the KICD Act No. 4 of 2013 of the laws of Kenya to conduct research and develop curricular for all levels of education. The Institute also develops print and electronic curriculum support materials, initiates and conducts curriculum based research, organizing and conducting in-service and orientation programmes for curriculum implementers https://www.kicd.ac.ke/.

[6]The *Council of Governors (COG)* is composed of the Governors of the forty-seven counties and its main functions are the promotion of visionary leadership; sharing of best practices and; offer a collective voice on policy issues; promote inter—county consultations; encourage and initiate information sharing on the performance of County Governments with regard to the execution of their functions; collective consultation on matters of interest to County Governments http://www. cog.go.ke/.

[7]UNFCCC 2015. Paris Agreement https://unfccc.int/files/meetings/paris_nov_2015/application/ pdf/paris_agreement_english_pdf.

[8]Capacity building and capacity development are two phrases often used interchangeably and understandably assumed to mean the same thing. Literature however argues that there is infact a clear distinction between the two with capacity building implying there is no capacity to begin with and capacity development acknowledging existing capacities and focusing on strengthening what is already there. Capacity development is a more holistic, collaborative approach that encourages ownership (Freeman 2010).

obtain, strengthen and maintain the capabilities to achieve the ability to mitigate and adapt to climate change over time." Building capacity is not a simple process of merely imparting knowledge or experience to individuals in isolation. Capacity is systemic, and so to build and then sustain individual capacity, efforts must also address the other two dimensions of a country's capacity system: its organizations and institutional arrangements. Capacity building efforts must therefore result in capacity built at all three levels simultaneously and in a synergistic manner that is appropriate for each national context.

Action for Climate Empowerment (ACE) is a term adopted by the United Nations Framework Convention on Climate Change (UNFCCC). It refers to Article 6 of the Convention's original text (1992), focusing on six priority areas: education, training, public awareness, public participation, public access to information, and international cooperation on these issues. The implementation of all six areas has been identified in recent years as the pivotal factor for everyone to understand and participate in solving the complex challenges presented by climate change. The importance of ACE is reflected in other international frameworks such as the Sustainable Development Goals (SDGs, 2015); the Global Action Programme for Education for Sustainable Development (GAP on ESD, 2014); the Aarhus Convention (2011); and the Bali Guidelines (2010).[9]

Article 6 of the UNFCCC on Education, Training and Public Awareness[10] calls on governments to develop and implement education and training programmes, including the strengthening of national institutions, training of scientific, technical and managerial personnel, as well as implementing public awareness programmes on climate change and its effects in order to improve the capacity to implement mitigation and adaptation actions. Article 6 seeks to reduce the impact of climate change by enabling society to be a part of the solution.

Broadly all Sustainable Development Goals (SDGs) have a bearing on climate change. However Goal number 13 specifically focuses on the need for urgent global action to combat climate change and its impacts and Target 13.3 on capacity development.[11]

In addition and more recently, the need for capacity strengthening at both global and local level is also considered so vital by government representatives that this requirement forms a key pillar of the Paris Agreement Article 11.1 (Bickersteth et al. 2017)

> Article 11.1 Capacity-building under this Agreement should enhance the capacity and ability of developing country Parties, in particular countries with the least capacity, such as the least developed countries, and those that are particularly vulnerable to the adverse

---

[9]UNESCO and UNFCCC 2016. Action for EMPOWERMENT Guidelines for accelerating solutions through education, training and public awareness.

[10]UNFCC 1992 http://unfccc.int/files/essential_background/background_publications_htmlpdf/application/pdf/conveng.pdf#page=17.

[11]SGD Target 13.3 Improve education, awareness-raising and human and institutional capacity on climate change mitigation, adaptation, impact reduction and early warning https://sustainabledevelopment.un.org/sdg13.

effects of climate change, such as small island developing States, to take effective climate change action, including, inter alia, to implement adaptation and mitigation actions, and should facilitate technology development, dissemination and deployment, access to climate finance, relevant aspects of education, training and public awareness, and the transparent, timely and accurate communication of information. (Paris Agreement, December 2015)

According to Bickersteth et al. 2017, the skills base for transitioning to a more climate-resilient, low-carbon economy is weak in both developing and developed countries. The number of trained personnel with the understanding to 'climate proof' investments for future climate-resilience and to design and shepherd low-carbon policies through implementation is low and is clustered in a few government departments.

Dagnet et al. (2015) noted that a sufficient level of capacity is needed to undertake climate action. There needs to be sufficient personnel dedicated to climate issues in the main organization responsible for climate as well as other key agencies, ministries, research organisations, academia, private sector, non-governmental organisations and other stakeholders.

In Kenya, climate change capacity building has remained a central theme in national documents. The National Climate Change Action Plan, outlined a comprehensive national capacity development and knowledge management strategy. One of the recommendations includes integrating climate change in national education curricula at all levels. In addition, the Climate Change Act 2016 mandates national and county governments to facilitate capacity development for public participation in climate change responses through awareness creation, consultation, representation and access to information and provide mechanisms for and facilitate climate change research and development, training and capacity building. Further, the Act obligates the national and sub-national governments to mainstream climate change responses into development planning, decision-making and implementation.

3 Training Program Development Process

Climate change is difficult to communicate by its very nature. Citizens, governments, private sector and all stakeholders cannot factor climate change in their decisions without a reasonably accurate understanding of the climate change problem. To make informed decisions, people must have at least a basic knowledge of the causes, likelihood and severity of the impacts and the range, cost and efficacy of different options to mitigate and adapt to climate impacts. The countries decision makers; current and future need to be enabled to make informed decisions (NRC 2010).

Education and training programs are powerful tools to bring complex issues such as climate change to public attention, understanding and action. Training is widely recognized to be a critical component of Parties' to the UNFCC and other relevant stakeholders' efforts to address climate change. The target group for such training is mainly relevant stakeholders with specific roles to play in tackling climate change.

Training programmes can develop skills of government officers, empower citizens as agents of change, enhance public participation in decision making processes and mobilize solutions to climate change.

Training is any planned activity to transfer or modify knowledge, skills and attitudes through learning experiences that can have an immediate practical application. Training is learning by doing. Personnel may require training for a variety of reasons including the need to maintain levels of competence, respond to the demand of changing circumstances and new approaches and technologies.[12]

This thinking informed the development of the *Climate Change Policy, Planning and Budgeting at National and County Level Training Program*. The development of the training program was supported by the Low Emission and Climate Resilient Development (LECRD) Project,[13] which is funded by the United States Agency for International Development (USAID) through the United Nations Development Program (UNDP). The Ministry of Environment and Natural Resources through its State Department of Environment provided leadership and guidance in development of the training program including the curriculum, facilitator's manual and planning and implementation of the program roll out. The Climate Change Directorate and the LECRD Project provided overall coordination and quality assurance. The Training Program was developed with technical guidance from the Kenya School of Government and Kenya Institute of Curriculum Development.

3.1   Curriculum Development

As climate change is a cross-sectoral issue, an integrated response was required. The climate change curriculum was developed through a multi-stakeholder, participatory and iterative process, led by curriculum development experts at the Kenya School of Government and the Kenya Institute of Curriculum Development and involving a wide range of experts in climate science, policy, planning, finance and knowledge management from National and County government, CBOs and academia. The various government ministries and institutions represented included; Ministry of Environment and Natural Resources, Ministry of Devolution and Planning, The National Treasury, Kenya Meteorological Department, University of Nairobi, National Environment Management Authority, National Green Growth Secretariat, Care International among others.

---

[12]https://www.msh.org/sites/msh.org/files/mds3-ch52-training-mar2012.pdf.
[13]The Low Emission and Climate Resilient Development (LECRD) Project funded by USAID through UNDP and implemented through The Ministry of Environment and Natural Resources aims to support Kenya's efforts to pursue long-term, transformative development as well as accelerate sustainable climate resilient economic growth, while slowing the growth of greenhouse gas emissions.

The program used local technical experts to design and develop the program, transferring their knowledge, expertise and skill. The program took into consideration international, national and subnational climate change processes, policy and legal frameworks, plans and strategies and will contribute to achieving national and global climate change priorities.

Relevant topics were selected and prioritized and the curriculum outline developed and refined during several curriculum development workshops. The adopted approach provided ground to learn from the subject matter experts, resulting in a robust and relevant curriculum. A modular approach was selected.

3.2   Facilitators Manual Development

Based on the curriculum, a comprehensive facilitator's manual was developed by subject matter experts through an iterative and highly interactive peer and technical review process.

With technical guidance from the Kenya School of Government and the Climate Change Directorate during several workshops; the experts were divided into teams based on technical expertise to develop the facilitator's manual. For each of the four modules the following was developed; objectives, technical content, definition of technical terms, learning activities, discussion topics, assignments and additional reading and supplementary material was identified. During development the experts were instructed to; avoid using complicated scientific terms, use images and experiential tools such as videos and use real world analogies with local context examples.

After the draft manual was developed, it was circulated several times among module teams and technical experts for technical editing. An external technical and editorial expert then reviewed and typeset the manual before final production and printing. The manual will be used to deliver and guide the training process.

All the contributions of institutions and individuals was credited and acknowledged in the Facilitator's Manual.

3.3   Training of Facilitators (ToF)

To prepare facilitators to deliver the climate change training program, the Kenya School of Government held a 5 day *Training of Facilitators (ToF)* course. The 25 trainees were selected mainly from a pool of subject matter experts in climate change and related relevant subject areas.

The objective of the course was to ensure that facilitators are well trained on various adult training methodology approaches for proper facilitation of the training program. The following topics were covered; adult training approach, effective presentation skills, evaluating training programs and an overview of the four modules of the training program. In addition the participants were introduced to the *World Climate Simulation Tool*.[14] The tool was developed by Climate Interactive and Massachusetts Institute of Technology and is an integral part of the 10 day training program.

The ToF course was facilitated by four Kenya School of Government trainers who employed both interactive and participatory methods of training, which are best suited for the training of adult learners. The methodologies used included; short lectures, plenary discussions and sharing of experiences, use of simulation exercises, critique, group discussions, interactive group presentations and group projects. The facilitators were also trained on appropriate use and development of visual aids. Pre and post assessments were undertaken to determine the percentage level of increase in knowledge and assess the training program's content, the facilitators, and the general learning facilities.

The facilitators also; interrogated the climate change curriculum, refined learning objectives making them clear, concise and relevant and arranged the topics in logical order taking into account that adult learners prefer learning in easy and progressive stages.

All the 25 participants successfully completed the ToF Course and were awarded certificates. The facilitators qualified to deliver climate change training programs which will be offered through the Kenya School of Government.

4   Climate Change Training Program Description

4.1   Program Introduction

This program is premised on the notion of mainstreaming climate change into national and sub-national policy, planning and budgeting processes. This will enhance decision making processes and will enable national and county officers better understand climate change and enhance their capacities to climate proof all sectors and integrate climate change adaptation and mitigation in national and sub national development plans.

The 10 day training program is learner-centred with emphasis on case studies, experience sharing, field excursions and other practical learning approaches including role play and simulation. The training will be facilitated by leading technical experts certified by the Kenya School of Government. The training program details are presented in Table 1 outlining the targets, entry requirements, objectives, learning outcomes and duration.

---

[14]The World Climate Simulation is a role playing exercise of the UN climate change negotiations for groups. It is unique in that it uses an interactive computer model to rapidly analyse the results of the mock-negotiations during the event. https://www.climateinteractive.org/programs/world-climate/.

Table 1 Climate change training program details

Target group	Middle level managers and technical cadres from National and sub-national governments involved in; • Policy formulation, • Planning, • Budgeting and • Implementation of programs In sectors vulnerable to the effects of climate change
Entry requirement	Professional public servants who should have served in the public service for at least 3–5 years. In particular officers serving in planning, budgeting and finance positions will be considered Officers working in environment, natural resources and agriculture will be given preference
Program objectives	By the end of the program the participants should be able to; • Recognize climate change and its impacts, • Appreciate sub-national, national and international processes supporting the climate change agenda, • Mainstream climate change adaptation and mitigation actions into planning and budgeting processes, • Develop strategies to mobilize climate financing and • Monitor, evaluate and report on implementation of climate change actions
Learning outcomes	At the end of the program participants will have acquired knowledge and skills to; • Integrate climate change actions and considerations when designing, planning and budgeting for programs, projects and activities, • Recognize the importance of mainstreaming climate change actions within their jurisdiction of work, • Ensure that policies and strategies include aspects of climate change, • Ensure that all plans at national and sub-national level are climate proofed and • Develop proposals to attract funding for climate change actions
Program duration	The program will run for 10 days (60 training contact hours)

The Training program is structured into four (4) modules and a session on climate change information and knowledge management. This program content summary is presented in Fig. 1.

5 Training Program Rollout—Inaugural Training

The capacity of the national training institute for public officers, the Kenya School of Government (KSG) has been strengthened to design, develop and deliver climate change programs. To ensure sustainability, the *Climate Change Policy, Planning and Budgeting at National and County level* training program will now be offered by Kenya School of Government to eligible public officers through self-sponsorship or support by their institutions.

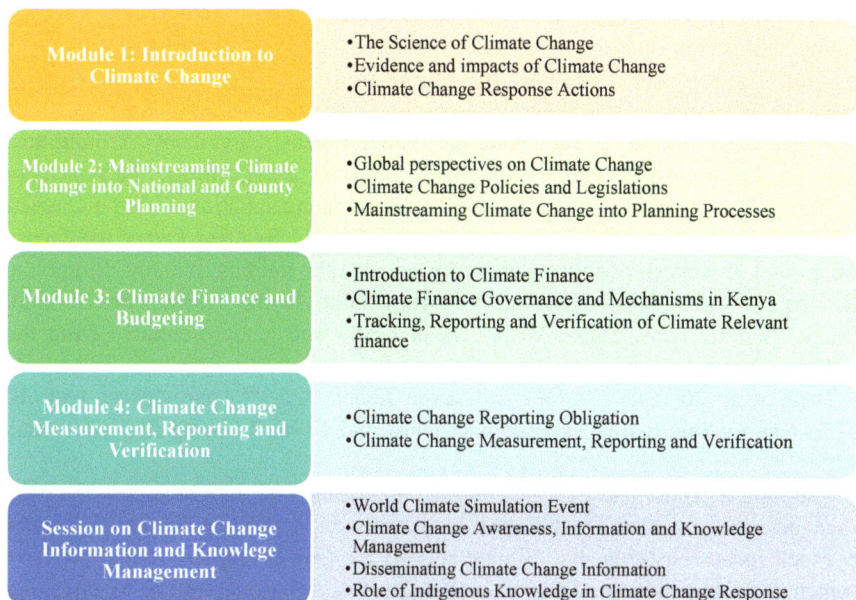

Fig. 1 Climate change training program content summary. *Source* Author

To kick off the training program rollout, Kenya School of Government (KSG) in collaboration with the Ministry of Environment and Natural Resources and the Council of Governors (COG) held the inaugural 10 day training program in Nakuru County in June 2017.

Nakuru County was selected as the location for the 10 day training as it was the county selected to host the World Environment Day (WED) celebrations on 6th June 2017. The participants would benefit from being a part of the celebration and get involved in the WED activities thus enhancing and broadening their learning experience. A training venue was selected with a conducive and comfortable learning environment and a checklist developed to ensure that all logistical and training venue preparations were well handled prior to and during the training.

5.1 Participant Selection

Participants were selected from both national and county governments targeting middle level managers and technical cadres involved in policy formulation, planning, budgeting and implementation of programs in sectors vulnerable to the effects of climate change. Invitation letters to nominate participants to the training program were sent to relevant Ministries, Departments and Agencies for national government and Council of Governors (COG) for county government indicating the eligibility criteria for the program.

Twenty seven 27 Senior Government Officers were selected; 10 (ten) representing Ministries, Departments and Agencies (MDAs) and 17 from selected counties. The representatives from the MDAs were; Ministries of Devolution and Planning, Environment and Natural Resources, Energy, Transport, and Agriculture, The National Treasury, The National Environment Complaints Committee, National Environment Trust Fund and National Environment Management Authority. The twenty five counties were selected in consultation with the Council of Governors to get an equitable representation of the 47 counties. Out of the twenty five invited, seventeen counties were represented at the training, namely; Bungoma, Kakamega, Nandi, Kitui, Garissa, Migori, Bomet, Kilifi, Transnzoia, Baringo, Vihiga, Narok, Nyeri, Siaya, Wajir, Uasin Gishu and a representative from the COG Secretariat.

There were 20 male (74.1%) and seven female (25.9%) participants with 10 (37%) being youth; that is 35 years and below and 17 (63%) being above 35 years old.

Adopting a gender and intergenerational approach is an important consideration when designing Action for Climate Empowerment (ACE) activities. A gender approach means ensuring that climate actions are gender-responsive and promote women's participation in decision-making. Intergenerational refers to engaging people of all ages in finding solutions for climate change, taking into special consideration the vulnerabilities of youth and the elderly, who have a reduced capacity to cope independently. Future generations are likely to be the most vulnerable to the impacts of climate change, yet they are also the least represented in current decisions on climate action (UNESCO and UNFCCC 2016).

5.2 Facilitation Selection and Training Material Preparation

Thirteen (13) facilitators were selected from the pool of trained and certified Kenya School of Government trained facilitators and were advised to use both participatory and interactive methods of training best suited for training of adult learners. These included; lectures (presentation slides, flip charts, audio-visual aids and hand-outs), plenary and group discussions, case studies, sharing of experiences, role play and simulation, field excursions, individual and group assignments.

Facilitators were asked to develop and provide relevant training materials for the participants and use a blended learning approach taking into cognizance the different learning styles namely; visual, auditory and kinaesthetic.[15] The facilitators were also encouraged to avoid using jargon and complicated scientific terminology.

Pre and post assessment tools were developed to assess increase in knowledge as a result of the training, collect feedback on the inaugural training program and how

---

[15]http://open.lib.umn.edu/humanresourcemanagement/chapter/8-4-designing-a-training-program/.

satisfied they were with the training experience and assess the facilitators. This would inform the need for additional training and adjustment/revision of the training program methodology and materials

Participant assessment was continuous and based on their ability to respond to practical and theoretical tasks given individually and in groups.

At the end of each session the facilitators were evaluated on specific issues namely; punctuality, presentation flow, handling questions, active participation of learners, use of visual aids, relevance of subject to their work place, use of relevant examples, knowledge of subject, treats participants with dignity and respect and finally on variety and appropriateness of training methods used.

5.3   Training Sessions

The participants were taken through the four modules namely; Introduction to Climate Change, Mainstreaming Climate Change into National and County (sub-national) Planning Processes, Climate Change Financing and Budgeting and Climate Measurement, Reporting and Verification. In addition to the four modules there was an interactive and vibrant *World Climate Simulation*[16] event and a session on Climate Change Awareness, Information and Knowledge Management, Disseminating Climate Change Information and the Role of Indigenous Knowledge in Climate Change Response.

The participants had a field trip to the Nakuru Metrological Department where they had an opportunity to see manual and digital weather instruments, interact with meteorologists and understand weather and climate monitoring. In addition the participants participated in the World Environment Day activities at the Egerton University, Njoro, Nakuru County.

One of the major highlights of the World Environment Day celebrations was the launch of the Inaugural Training Program on Climate Change Policy, Planning and Budgeting at National and County levels by the Cabinet Secretary, Ministry of Environment and Natural Resources (Fig. 2). The Cabinet Secretary noted that, "*the program, the first of its kind will form a significant contribution towards the implementation of the Climate Change Act 2016, as well as the Kenya's Nationally Determined Contribution (NDC)*". The launch was also attended by the Principal Secretary, State Department of Environment among other top government officials. The launch was featured on the Ministry of Environment and Natural Resources website[17] and several tweets sent out.

At the end of the training program, each participant developed an action plan to enhance transfer of knowledge, skills and competencies to the work place. This

---

[16]https://www.climateinteractive.org/programs/world-climate.
[17]http://www.environment.go.ke/?p=3672.

Fig. 2 Launch of the inaugural climate change training program by the Cabinet Secretary, Ministry of Environment and Natural Resources

included; key learning areas, proposed activities, responsibilities and areas of further training and development.

All participants dubbed the *"Climate Change Champions"* successfully completed the training program and were awarded certificates. The *"Climate Change Champions"* were challenged to put into practice what they have learnt by; climate proofing their County (sub national) Integrated development Plans (CIDPs),[18] ensuring national priorities across all sectors are climate smart, mainstreaming adaptation and mitigation actions into policy formulation, planning and budgeting processes at national and sub-national levels and developing proposals to attract funding for climate change response actions.

A social networking group was established with membership of all the *"Climate Change Champions"* and key facilitators to continue exchanging information and opportunities on climate change.

---

[18]The County Integrated Development Plan (CIDP) is a 5 year plan that shall inform the county's annual budget and shall reflect the strategic midterm priorities of the county governments. The CIDP will contain specific goals and objectives, a costed implementation plan, provisions for monitoring and evaluation and clear reporting mechanisms. It will contain information on investments, projects, development initiatives, maps, statistics, and a resource mobilization framework.

5.4 Limitations

Political—The timing of the program may have affected the participation of county (sub national) government officers as it was two months before the general elections. The number of county representatives was fewer than those invited; only 17 out of the 25 invited, attended the training.

Financial—Due to limited funding, only a small number of national and government officers could be trained during the inaugural training session.

5.5 Lessons Learnt

After the successful inaugural climate change training program, the following lessons were learnt from the design, development and implementation process;

- Engagement and involvement of stakeholders at all stages of the training program design, development and implementation enhanced the ownership and success of the program and its quality, delivery and outcome,
- The use of local subject matter experts to develop and deliver the training made it relevant and contextualised local climate circumstances,
- Interaction between the national and sub-national government officers enriched the discussions, sharing of experiences and helped bridge the understanding between different levels of government,
- With regard to future participation; closer consultations will be done with the relevant stakeholders to reduce the disparity in gender and youth representation and though there was a fairly good representation of youth in the training, this representation can be increased,
- The *World Climate Simulation Event* was highly rated as a learning method as it gave the participants an opportunity to apply the knowledge and skills in a real life situation. Most participants indicated they wanted more field trips and experiential learning opportunities,
- Though there was some media coverage of the training program launch, there should have been more involvement of media during the training to give the program more visibility and raise public awareness about the program,
- A pre and post assessment was undertaken and one of the findings indicated that most participants rated highly their level of knowledge, skills and abilities in the learning objectives after the training. In addition 56% participants stated the needs for the training were met and 25% participants said their needs were met and surpassed,
- Most facilitators were ranked as excellent so the facilitators training was vital, and
- Need to review training program duration to rationalise time allocated for various modules and topics. Some trainees felt that the time allocated for some key topics was not adequate.

5.6 Recommendations

The following recommendations are proposed for further training programs

- Develop training case studies from selected counties to showcase best practice and results-driven examples of how county governments and other key stakeholders are addressing climate change,
- Organise knowledge exchange visits and provide knowledge sharing opportunities for participants to exchange lessons learnt and good practices to build and strengthen existing skills and capacities,
- Develop E-learning modules which will increase the reach of the training program and introduce new target groups,
- Lessons learnt from this process can be fed into efforts of Kenya Institute of Curriculum Development (KICD) to integrate climate change into national education curricula at all levels as per the Climate Change Act 2016,
- Development of relevant and specialised training programs for specific target groups such as, *Coding and Tracking of Climate Finance* for Finance Officers,
- Sensitization of political and economic decision makers to including policy makers, financial institutions, private sectors and other relevant stakeholders to enhance climate change awareness across sectors and governance levels to promote and implement mitigation and adaptation actions,
- Future training programs should take into consideration gender and intergenerational concerns and increase the participation of women and youth. Supporting women's empowerment and drawing on their experiences, knowledge and skills and mobilizing youth participation in climate change processes will make climate change responses more effective,
- Follow-up after to monitor outcomes and access the impact of training and continuing performance of the participants in the workplace and
- Review feedback from the evaluations to improve on the technical content and training methods to suit participants' needs and remain relevant.

6 Conclusions

Building and sustaining individual and institutional capacity and providing an enabling environment is fundamental to climate change response both at global and national level. The *Climate Change for Policy, Planning and Budgeting at National and County Level* training program responded to Kenya's need to mainstream climate change into national and sub national policy, planning and budgeting process and will contribute to achieving national and international climate change priorities. Engagement of key stakeholders throughout the design, development and implementation process ensured; ownership, commitment to, and success of the climate change training program.

The first cohort of successfully trained government officers is expected to contribute to planning, budgeting, decision making and implementation of climate change response actions in Kenya. However, this is only a fraction of the number of national and county government officers who need training and require their capacity to be built and consequently more training programs need to be held.

Running future training programs may require additional financial support including reviewing the existing curriculum and developing other relevant and targeted programs. In addition, there are discussions between Kenya School of Government and relevant stakeholders on approaches to build the capacity of non-state actors involved in climate change related activities.

The *Climate Change Policy, Planning and Budgeting at National and County Level* training program will contribute to Kenya's efforts to mainstream climate change responses into development planning, decision-making and implementation and will contribute to developing capacity at individual and organisational level to adequately respond to climate change and meet national and global climate obligations.

Acknowledgements The financial support from United States Agency for International Development (USAID) and United Nations Development Program (UNDP) through the Low Emission and Climate Resilient Development (LECRD) Project and all the organisations and subject matter experts, too many to mention, are highly acknowledged for the successful design, development and implementation of the *Climate Change Policy, Planning and Budgeting at National and County Level* training program.

References

Bickersteth S, Dupar M, Espinosa C, Huhtala A, Maxwell S, Pacha MJ, Sheikh AT, Wesselink C (2017) Mainstreaming climate compatible development. Climate and Development Knowledge Network, London. ISBN 1-909464-96

Dagnet Y, Northrop E, Tirpak D (2015) How to strengthen the institutional architecture for capacity building to support the Post-2020 climate regime. Working Paper. Washington, DC: World Resources Institute. Available online at http://www.wri.org/publications/capacitybuilding

Freeman K (2010) Capacity development theory and practice: lessons learnt from CORD and KITWOBEE in Northern Uganda. http://architecture.brookes.ac.uk/research/cendep/dissertations/KatherineFreeman.pdf

Government of Kenya (2015) Addressing climate change: success stories from Kenya. Ministry of Environment and Natural Resources (MENR), Nairobi, Kenya

Government of Kenya (2017) Climate change policy, planning and budgeting at national and county level: facilitators manual. Ministry of Environment and Natural Resources, Nairobi, Kenya

National Research Council (NRC) (2010) Informing an effective response to climate change. The National Academies Press, Washington, DC. http://nap.edu/12784. ISBN 978-0-309-14594-7

Republic of Kenya (13 May 2016) Climate change act. Nairobi: Government of Kenya. http://kenyalaw.org/kl/fileadmin/pdfdownloads/Acts/ClimateChangeActNo11of2016.pdf

UNESCO (2010) Climate change education for sustainable development. The UNESCO climate change initiative

UNESCO and UNFCCC (2016) Action for EMPOWERMENT Guidelines for accelerating solutions through education, training and public awareness. http://unesdoc.unesco.org/images/0024/002464/246435e.pdf

UNFCCC (2015) Paris Agreement. https://unfccc.int/files/meetings/paris_nov_2015/application/pdf/paris_agreement_english_pdf

United Nations Framework Convention on Climate Change (1992) http://unfccc.int/files/essential_background/background_publications_htmlpdf/application/pdf/conveng.pdf#page=17

United Nations Institute for Training and research (UNITAR) (2013) Guidance note for developing a national climate change learning strategy. The one UN Climate Chane Learning Partnership. UN CC: Learn

Climate Change Litigation: A Powerful Strategy for Enhancing Climate Change Communication

Paola Villavicencio Calzadilla

Abstract In a context in which national and international policy-making have been inadequate and insufficient for dealing with climate change, more recently courts have become a critical forum in which the climate crisis and the consequences of inaction are under debate. Climate change litigation (CCL) is emerging as a valuable strategy to hold governments and private entities accountable for their lack of action and to advance policy and regulation in both, mitigation and adaptation. In addition, CCL also appears to be a powerful tool for communicating the urgency of climate change. Win or lose, climate-related cases can help to promote a better understanding of climate change, raise awareness and enhance dialogue and public engagement in the debate over the actions needed to confront the challenges linked to it. Against the backdrop of the climate change communication discourse, this paper explores how CCL assists in communicating climate change issues. By looking at the experience of some of the most significant climate-related cases that have set important precedents in recent years, this paper highlights that CCL contributes to the public understanding of the causes, risks and consequences of climate change, as well as the adaptation needs, by bringing its realities closer to people—within and outside courtrooms—and by presenting complex related issues in a clear and easy-to-understand manner. Thus, as climate-related cases are reported in a variety of sources gaining national and international attention, they help to increase the public's understanding of climate issues, raise public and politic awareness and inspire action.

1 Introduction

Climate change communication has played an important role in raising awareness on and fostering public engagement towards climate change. As a result of important efforts done to improve communication, significant developments have

P. Villavicencio Calzadilla (✉)
North-West University, Potchefstroom, South Africa
e-mail: p_villavicencio@hotmail.com

© Springer Nature Switzerland AG 2019
W. Leal Filho et al. (eds.), *Addressing the Challenges in Communicating Climate Change Across Various Audiences*, Climate Change Management,
https://doi.org/10.1007/978-3-319-98294-6_15

231

been achieved and concerns about climate change have grown in many parts of the world in recent years (Capstick et al. 2015). Yet, due to the complex nature of climate change and persistent failures in communicating climate change issues, the adequate understanding of it continues to be limited. Indeed, the understanding of the causes and impacts of climate change is still superficial and incomplete and for many audiences climate change is still perceived as an ambiguous, invisible, abstract, complex and distant—both spatially and temporally—issue. Climate change is susceptible to be overwhelmed by more immediate threats and doubt and scepticism still linger in various sectors of society (Moser 2010; Moser and Dilling 2011; Wolf and Moser 2011).

In the light of the urgency of climate change, more remains to be done in order to foster communication of climate change and catalyse public involvement (Leal Filho 2009). Thereby, it is essential to re-examine existing communication strategies, but also consider alternatives and complementary tools that can be used to connect climate change issues with different audiences. One of these is precisely climate change litigation (CCL).

While it is not a new phenomenon, CCL has gained momentum over the past decade as an important strategy for raising the profile of climate change, to bypass slow political progresses in mitigation and adaptation to climate change impacts, and to regulate how states respond to climate change at the global, regional and local level (Sprinz and Hefele 2017; Lin 2012). By applying a wide range of different legal theories and creative forms of argument, carefully selected, climate mitigation and adaptation cases are being filed by individuals, groups of citizens and non-governmental organizations (NGOs) in different adjudicative forums (supra-national or domestic) around the world, in order to press legislators and policymakers to be more ambitious in their approach to climate change, to escalate the development of existing regulatory regimen, to fill the gaps left by legislative and regulatory inaction, to force changes in industrial practices and corporative behaviour (for example, in terms of GHG emission reductions or assessment of the impacts of those emissions), or to obtain recognition and seek redress for climate-related injuries, among others (UNEP 2017; Peel and Osofsky 2015; Abate 2016; Faure and Peeters 2011; Burns and Osofsky 2009; Averill 2009). Thus, as Osofsky (2010) points out, "courts have become a critical forum in which the future of GHG-emissions regulation and responsibility are debated" (p. 4).

However, the positive implications of CCL are well beyond the courtrooms and the litigation's specific claims and results of an individual case. As Averill (2008) noted "[a]rguments and decisions in climate cases may have repercussions far beyond the narrow interest of the litigants themselves...[and] decisions...also have impacts that extend beyond the legal system itself" (p. 916). Indeed, the acts of preparing, announcing, filing, advocating and forcing a response in a climate-related case have positive effects, for example, in advancing climate change communication (Hunter 2007). CCL plays an important role in promoting the communication of the causes and impacts of climate change, raising awareness and understanding and in stimulating dialogue and public engagement in the debate over the necessary actions to tackle the challenges of climate change. Whether successful or not, by presenting complex material in an accessible and understandable manner, climate-related cases can help to explain specific issues linked with climate change, make climate change more visible to a range of audiences, mobilise public sentiment on climate change issues, legitimise scientific evidence and affect public perceptions about climate science, reconfiguring therefore the public discourse (Averill 2007; Osofsky 2010; Lin 2012; Peel and Osofsky 2015). Thus, while CCL cannot be considered as a substitute for other approaches to address the challenges of climate change, climate-related cases provide valuable material that can be used by others to promote the public understanding of the causes and consequences of climate change and to communicate the urgency of addressing it while fostering social action (Averill 2007).

By focusing on some of the most significant climate mitigation and adaptation cases that have been filed in different jurisdictions in recent years, this paper considers the evolving importance of CCL as a communicative and awareness-raising tool in the face of climate change and explores its contribution to overcoming the challenges in communicating climate change across various audiences. Because of the increasing volume of climate-related cases filed in the last decade, the scope of the review of these cases is limited, and does not include every significant climate case. Nevertheless, the analysis of the selected cases will provide a framework to highlight the role of CCL in communicating climate change issues.

First, the paper provides a brief overview of three climate-related lawsuits filed in three different countries, which have led to different outcomes beyond the courtrooms: *Urgenda* (the Netherlands), *Leghari* (Pakistan) and *Lliuya* (Germany/Peru). Looking at the experience of these cases, the paper then moves to discuss how CCL, in general, and these three cases, in particular, help to increase the visibility and understanding of climate change by bringing it closer to the people and by presenting complex issues in a comprehensible language, so people—within and outside courtrooms—can understand the urgency of climate change. As climate-related cases are reported in a variety of sources, gaining national and international attention, as has been the case of the three lawsuits analysed, the paper also explains the value of CCL to improve education, increase public and political awareness and foster debate on climate change. Finally, some concluding remarks are made.

2 Significant Climate Change Cases

During the last decades, hundreds of climate-related lawsuits have been brought in several jurisdictions around the world (Fig. 1) and many more are expected to be raised in the context of the Paris Climate Agreement of 2015 (UNEP 2017). However, while CCL is not a new phenomenon, some of the climate-related cases filed in the last years are particularly noteworthy not only for their policy and regulatory repercussions, but also for their implications in promoting communication of climate change mitigation and adaptation issues. Three of these cases are briefly summarised below.

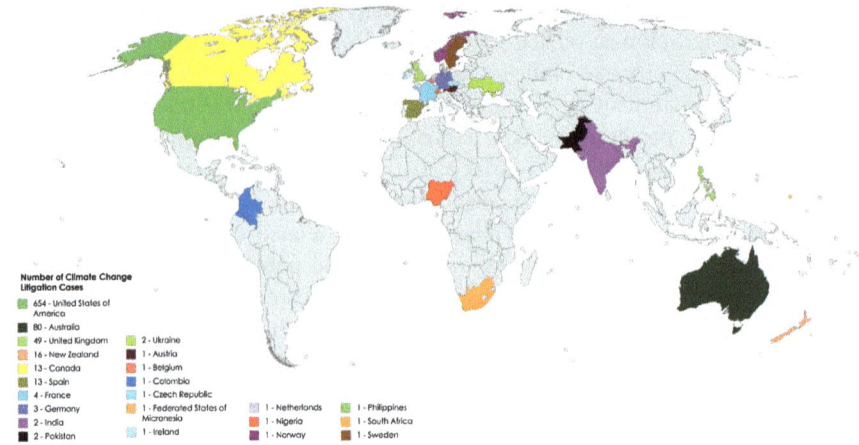

Fig. 1 Climate change cases filed—March 2017 (UNEP 2017, pp. 12–13)

2.1   Urgenda v. The Netherlands

In November 2013, the Dutch *Urgenda* Foundation, together with 866 concerned Dutch citizens, filed a case against the Dutch Government.[1] In an extended sub-poena of 144 pages, *Urgenda* requested the District Court of The Hague to declare, *inter alia*, that the current annual GHG emissions' level in the Netherlands was unlawful and that the Dutch State was liable for its role in causing dangerous global climate change. On this basis, *Urgenda* asked the court to order the State to reduce the joint volume of annual GHG emission within its territory by 40%, or in any case at least by 25%, compared with 1990 levels, by the end of 2020.

In its defence, the government claimed that it had not acted unlawfully in addressing climate change, given that it had committed to preventing dangerous climate change and, therefore, had adopted an adequate climate policy that includes mitigation and adaptation measures. By arguing that, among others, there was no legal duty under the national or international law for the Netherlands to adapt its climate policy and to take measures to achieve the reduction claimed by the plaintiffs, the Dutch State considered that the claim should be rejected.

In June 2015, the District Court of the Hague agreed with the petitioners that the Dutch government should take more steps to combat climate change and, in a landmark decision setting a powerful precedent, ordered the Dutch government to reduce its GHG emissions by at least 25% by the end of 2020 (Court of The Hague 2015). In September 2015, the government appealed the case and when this paper was written it had not yet been resolved. This was the first time that a court determined that states have an independent obligation toward their citizens and that

---

[1]For information about the case, see http://www.urgenda.nl/en/climate-case/.

a government was found liable for a climate policy that is substandard according to international norms (Loth 2016).

2.2 Leghari v. Pakistan

In another landmark case, in August 2015, *Ashgar Leghari*, a Pakistani farmer, sued the government of Pakistan for its delay in implementing the 2012 National Climate Policy and Framework and for failing to apply priority actions and make no effort to develop adaptive capacity and resilience to climate change.[2] The plaintiff accused the Federal Government and the Government of the Punjab region, where his home was situated, of failing to ensure water, food and energy security in the face of climate change. In his petition, he argued that the government was obligated to implement the national policy and other measures to help people to adapt to the effects of climate change. Thus, he alleged that the inaction, delay and lack of seriousness of the government to address the challenges and to meet the vulnerabilities associated with climate change violated his constitutional rights to life and dignity. The plaintiff did not demand a compensation for damages; instead his claim was aimed at generating governmental action.

In September 2015, the Lahore High Court Green Bench agreed with the case and determined that the delay and lethargy of the government in implementing the national policy and framework offends the fundamental rights of the citizens that need to be safeguarded (Lahore High Court 2015). Thus, concluding that climate change is the most serious threat faced by Pakistan and that its impacts are already felt across Pakistan and the plaintiff's region, the court ordered the relevant government agencies to each nominate "a climate change focal person" to ensure the implementation of the Framework and assist the court in overseeing actions undertaken. In addition, the court ordered the creation of a Climate Change Commission, consisting of 21 individuals representatives of key ministries, NGOs and technical experts, to push forward the actions promised by the government. This commission is also called to file interim reports as and when directed by the Court. In a nutshell, this case highlights "a simple but fundamental truth—individuals can and do make a difference" (Mehra 2015).

2.3 Lliuya v. RWE

In November 2015, *Saúl Luciano Lliuya*, a Peruvian farmer, supported by the environmental organisation Germanwatch, filed a lawsuit against the German

---

[2]For information about the case, see http://www.lse.ac.uk/GranthamInstitute/litigation/ashgar-leghari-v-federation-of-pakistan-lahore-high-court-green-bench-2015/.

energy company RWE in a court of Essen, Germany, where the company is based.[3] Lliuya's suit alleged that his home town in Peru (Huaraz) was acutely threatened by a potential flood wave from a glacial lake located above his town, which had substantially increased its volume as a result of the melting of nearby mountain glaciers due to climate change. The plaintiff alleged that RWE, Germany's largest electricity producer and owner of a large number of coal-fired power plants, was one of the major polluters and contributors to climate change that is causing glacial retreat in the Peruvian Andes, increasing the risk of flooding in his home town. *Saúl Lliuya* demanded RWE to accept its responsibility for its emissions and contribution to climate change and help to reduce the risk of flooding by bearing part of the resulting cost of adaptive measures that needed to be taken in the lake. In concrete, he asked the German court to order RWE to provide a compensation of 17.000 Euros for its pollution. The demanded amount represented the 0.47% of the estimated repair total cost in case of flooding, the same percentage than the estimated company's share of global emissions from 1751 to 2010. In its response, RWE basically claimed that climate change is a complex issue and that individual companies cannot be linked to specific impacts, like flooding in Huaraz.

In December 2016, the Essen Regional Court dismissed the lawsuit pointing to a lack of legal causality. It basically argued that due to the complexity of climate change it was not possible to prove an individual link between the emissions of individual emitters and specific local impacts of climate change. According to the court, as countless emitters of carbon dioxide exist worldwide, the share of an individual emitter, like RWE, is irrelevant for the causation of climate change impacts (Germanwatch 2017). In January 2017, the plaintiffs appealed against the negative ruling before the Higher Regional Court Hamm. At the time of writing this paper, the appeal is still pending resolution. Despite the unsuccessful outcome, the case, the first of its kind in Europe, forms part of a larger trajectory of legal initiatives that demand action and "could be a precursor to additional transnational litigation against large emitters for adaptation measures or compensation" (Banda and Fulton 2017).

3 Telling Stories in CCL and Promoting the Visibility of Climate Change

Storytelling, as a powerful tool for organising and transmitting information, is considered a way to effectively communicate information about climate change issues (Jones 2014). This is due to the fact that stories are inherently interesting, can provide concrete examples of abstract concepts, can be crafted to speak to people's lives and experiences, can be used to put large-scale issues into a familiar context, and can be applied across different audiences, among others (Kearney 1994). By

---

[3]For information about the case, see http://germanwatch.org/de/der-fall-huaraz/.

bringing home and making tangible the invisible and often abstract concept of climate change and connecting discussion about climate-related issues to people's everyday lives, storytelling can promote a better understanding of the meaning and scope of climate change and can be used as an important tool for engaging audiences and motivating action (Kearney 1994; Moser 2010; Corner 2013).

CCL can help to communicate climate change issues as storytelling is precisely an important and valuable tool in this kind of litigation. As in other complex litigations that involve highly intricate legal and scientific issues, in CCL facts and central themes related to climate change are used to craft and present stories that appeal to the community's shared sense of values (Kanner and Nagy 2006). Climate-related cases tell stories about climate change victims—people, communities and ecosystems—who are suffering or are going to suffer the impacts of climate change in the future, making them more visible to the public and promoting a better understanding of their situation. These cases make climate injures claims come alive and, by identifying the winners and losers of climate change, highlight which voices should be heard and which groups should be protected by society. For example, the *Urgenda* case stressed the consequences of the increase in global temperature for the Netherlands, especially for the most vulnerable groups of society, including the elder, people in social isolation, the homeless and children, grandchildren and generations of Dutch citizens after them. Similarly, in *Leghari v. Pakistan*, the case highlighted the vulnerability of Pakistani people, especially small-scale farmers from Punjab region, who are suffering the consequences of unpredictable weather. Even if a climate-related case does not succeed, it can tell human stories linked to the causes and impacts of climate change. The case of *Lliuya v. RWE* revealed, *inter alia*, the threats of accelerated melting of glaciers over people living in the Peruvian Andes (Table 1).

Hunter (2007) notes "[t]he story-telling quality of [climate-related cases] makes climate change more tangible and more immediate, which significantly changes the tone of the climate debate" (p. 4). In international negotiations and previous national climate policy debates the focus has primarily been on the GHG concentrations and the global impacts of climate change, *inter alia*, rise in average temperatures and sea levels, changes in precipitation patterns or glacial melt. However, in CCL the discussion moves beyond the scientific debate concerning the reality and implications of climate change. While emphasising on human stories, climate-related cases increase the saliency of other issues, like social and ethical concerns (Averill 2009). In this cases the focus shifts from abstraction to more specific injures, linking climate change with the lives of ordinary people and showing how harms caused by it know no territorial or generational boundaries (Hunter 2007). Thus, although in the last years the situation of vulnerable people and communities threatened by climate change—specially from the Artic region and pacific islands—has gained important visibility as they have been able to tell their stories to the climate community, in CCL those stories are given names and faces reaching a greater level of specificity and particularity, attracting public attention on them (Averill 2009). This is important because people who are not

Table 1 Climate-related cases: Implications on climate change communication

Climate change case	Focus	Status	Climate change issues communicated
Urgenda v. The Netherlands	Mitigation	Approved (under appeal)	• Impact of temperature rise on people (present and future) in the Netherlands • Relying on reports of the IPCC, etc. • Climate change impacts and risks: heat stress, coastal floods, sea level rise, droughts, etc. • Responsibility of the government for effectively controlling the Dutch emission levels • Well reported: http://www.urgenda.nl/en/climate-case/
Leghari v. Pakistan	Adaptation	Approved	• Impact of climate change on people from Pakistan, especially small-scale famers • Relying on reports of the IPCC, etc. • Climate change impacts and risks: floods, heavy precipitations, droughts, etc. • Responsibility of government to promote resilience and adaptation measures • Well reported (media coverage), e.g.: https://www.nytimes.com/2016/05/11/science/climate-change-citizen-lawsuits.html
Lliuya v. RWE	Adaptation	Dismissed (under appeal)	• Threats of melting glaciers over poor people in the Peruvian Andes • Relying on scientific studies on climate change: Carbon Majors Report • Climate change impacts and risks: melting glaciers, flooding, etc. • Responsibility of polluting companies for climate change and transboundary impacts • Well reported: http://germanwatch.org/de/der-fall-huaraz

scientists are not necessarily interested on scientific explanations about climate change, rather in what it means for them and for their family at present. In *Urgenda* the plaintiffs stressed how the effects of climate change, including heat stress, coastal floods, overflowing rivers, sea level rise, droughts, soil salinity, among others, could negatively affect the production of food and the availability and supply of water, jeopardising the life, health and well-being of Dutch citizens, especially of the most vulnerable that will be hit first and hardest. Likewise, in *Laghari*, the plaintiff who relies on farming for a living highlighted the impacts of climate change on his life and dignity. He argued that water scarcity and temperatures shift in the Punjab region, where he lives, are stressing crops and, therefore, making impossible for him, his family and other small-scale farmers around him to continue to make a living there. In the same way, the Peruvian case revealed how glacier retreat in the mountains of the Peruvian Andes due to human-induced climate change has led to an increased risk of flooding for the plaintiff and for people living in Huaraz, who face the imminent threat of losing their properties, culture and lives in a massive glacial lake outburst flood (Table 1).

Thus, as the impacts of climate change are more understandable when a specific person or group expresses the implications of such impacts for their lives (Hunter 2007, p. 4), climate cases that tell stories about specific climate-related injuries and showcase the urgency of climate change can help to increase the public's understanding of climate issues, raise awareness on the plight of victims of climate change and press for political action to mitigate climate change or to support adaptation measures (Lin 2012).

4 Visualising and Understanding the Science of Climate Change Through CCL

The complex nature of climate change has limited people's understanding and awareness of climate issues (Moser and Dilling 2004). Despite the improved efforts to communicate climate change science, many people remain confused and do not realize, or do not accept, climate scientists' findings. While the strong consensus among scientists is not widely appreciated, "[t]here is a disturbing large gulf between the research community's knowledge and the general public's perception" (Somerville and Joy Hassol 2011, p. 49). Although the complexities linked to the science of climate change are of greatest interest to scientists, they are hard to comprehend and mostly uninteresting or esoteric for lay people (Moser and Dilling 2004, 2011). Thus, communicating climate change in terms of natural scientific indicators, such as global mean temperatures limits, fails to resonate with lay audiences (Shaw 2014). However, laypeople have important roles to play in mitigating and adapting to climate change; individually or collectively, they can not only respond to climate change through changes in their lifestyle, but can also influence and participate in climate science and policy dialogue (Wibeck 2014). Providing non-scientists with tangible, concrete and understandable complex science-based messages and information about climate change is therefore important to overcome the barriers to public communication and public engagement in order to allow a more informed debate.

In light of the above, CCL can make the science of climate change more accessible to the public, promoting a better understanding of the multidimensional nature and effects of it and affecting public perceptions. In CCL, courts facilitate the connection between climate science and society by providing a forum in which scientific information on climate issues must be presented in a way that is understandable for laypeople. As Averill (2009) notes, "[l]awsuits bring complex climate science to an understandable level and explain how causes such as [GHG] emissions can result in injures to human beings and ecosystems far away in time and space" (p. 144). In order to enable non-expert courts and judges, who often have little training in science supporting climate change claims, to understand the facts alleged in a case, litigants (plaintiffs and defendants) and their attorneys collect, synthetise, communicate and present climate scientific information—statements and findings—in a language that is accessible and understandable by non-scientists (Averill 2009). By doing so, CCL makes scientific information about climate change more accessible to a wider audience and help to raise awareness not only on the scientific issues, but also on the other implications of climate change (Hunter 2007; Averill 2008).

In *Urgenda v. The Netherlands*, the plaintiffs' claim relied on scientific studies on climate change, in particular on the reports of the Intergovernmental Panel on Climate Change (IPCC), to support their claim (Table 1). Beyond complexities, they stressed how climate change requires an urgent response from the Dutch government to avoid irreversible consequences. For its part, after reviewing the scientific evidence, the court concluded that the Dutch reduction target was below the standard deemed necessary by climate science and the international climate policy. Thus, in order to prevent a dangerous temperature rise of 2 °C or more, the court ordered the State to reduce its emissions by 25% by 2020 as an absolute minimum. Although the Court accepted being confronted with complex issues and lacking specialized knowledge, it were able to address such difficulties sending a message that lowers the scientific complexities: given that the internationally agreed GHG reduction targets of States are insufficient to limit warming to below 2 °C and to avoid some of the worst climate impacts, States have the obligation to take measures in their own territory to prevent climate change.

Similarly, in *Leghari* the petitioner emphasised the strengthened scientific consensus around climate change and, based on that, underlined the impacts of climate change on Pakistan, especially on vulnerable farmers. In its ruling the court also included references to climate science, mainly the conclusions of the IPCC's reports, and observed that extreme weather events such as floods, heavy precipitations and drought, all increased by climate change, are threatening water and food security in the country (Table 1). Thus, the court underscored that climate change appears to be the most serious threat facing Pakistan.

In *Lliuya v. RWE*, the plaintiff supported his claim with the research findings of the Carbon Majors Report. For the first time, this report quantified and traced industrial GHG emissions and linked them to the world's largest multinational and state-owned *producers* of crude oil, natural gas, coal and cement. Using this report, *Saúl Luciano Lliuya* alleged that the 0.47% of all carbon dioxide accumulated in the atmosphere could be traced back to RWE. Precisely, based on this finding, he asked the German court to order RWE to cover the 0.47% of the total cost of protecting his town from the outburst of the glacial lake. Although the Carbon Majors Report includes complex and debated calculations, its use in this climate case helped to spread the information of who may be identified as the major fossil fuel producers that contributed significantly to climate change and how their behaviour is putting vulnerable populations in remote areas at risk of suffering the worst consequences of climate change (Table 1).

Therefore, as climate-related cases rely heavily on scientific findings and the testimony of scientific experts, they can affect the public perceptions of the credibility, salience and legitimacy of the scientific information related to climate change (Averill 2007). Although the judicial treatment of climate science in courts can be different, depending on the nature of such courts, putting climate science in a forum that is perceived as credible, legitimate, less politicised and impartial provides the opportunity to influence the public perception of climate change. This is due to the fact that "[c]onvencing arguments are best received if they come from highly credible and legitimate sources" (Moser and Dilling 2004, p. 41). As Hunter (2007) points out "[w]hen courts and other highly credible institutions validate the basic science of climate change, the general public's perception of the climate debate shifts from *whether* climate change is occurring to what the appropriate remedies should be. For the public, Judicial decisions can move the debate [over climate science] from an esoteric one among scientists to an issue *decided* by impartial judges whose job...is to resolve such matters" (p. 6). Thus, the manners in which litigants present climate science and the ways in which courts, as social actors, respond by taking a specific stance on scientific issues, can affect public perceptions of the relevance, credibility and legitimacy of climate experts and climate science (Averill 2008; Peel and Osofsky 2015).

5 Educating People and Stimulating Public Debate: Two Additional Positive Effects of CCL

By informing about the complexity of climate change to courts, judges, litigants, the legal and other communities, governmental and non-governmental organisations and civil society in general and by bringing the scientific debate to a level that allows participation by laypeople, CCL and related cases have the potential to serve as an exercise of civil education over the science and impacts of climate change, affecting people's perception, triggering public debate about the appropriate responses and, ultimately, promoting social action (Averill 2007).

In litigation over climate (mitigation or adaptation) measures, while judges can teach society by writing decisions which include climate change issues, courtrooms and other quasi-judicial forums became important public stages for dialogue and debates over those issues. Furthermore, climate-related cases for themselves can also facilitate people's understanding of climate change and promote engagement in debates on a number of issues, including the causes, impacts, vulnerabilities or appropriate responses to climate change, among others (Peel and Osofsky 2015). This is because climate cases, win or lose, are reported in a variety of sources gaining national and international attention (Table 1).

Media coverage of these cases, for example, can increase awareness and set the agenda for public debate about climate change topics (Averill 2009). Indeed, as stressed by Kaminakaité-Salters (2011) "one of the key aims of climate change litigation is likely to be awareness-raising through media exposure" (p. 170). Likewise, Smith and Sherman (2006) argue that "[climate change-related] lawsuits, whether successful or not, focus public attention upon key issues through media exposure and can be effective in influencing governmental and corporate policies" (p. 12). Thus, while judges with different levels of scientific training, knowledge and willingness to engage with complex issues of climate science have, for example, the opportunity to learn from the stories told in courtrooms, civil society and the public at large can also learn from these stories as they are reported by the media.

Furthermore, as media practitioners, who have to compete for the attention of their audiences, routinely rely on stories, anecdotes, and other narrative formats to communicate and resonate with their audiences (Dahlstrom 2014, p. 13614), climate cases reported by the media using these formats can attract a broader public who can learn from exciting and unforgettable climate stories, enhancing thus their understanding of climate change issues (Green 2004). For example, the three climate cases described in this paper have been reported in the media, attracting wide attention. The stories of the *Urgenda* Foundation, *Ashgar Leghari* and *Saúl Luciano Lliuya* appeared in a number of newspapers and online news portals, including the New York Times, The Guardian, International Business Times, BBC News, VICE News, among others.[4]

In addition, organisations and their lawyers who launched climate cases can hold press conferences or other public communications to announce milestones in the course of litigation, gaining coverage in newspapers, magazines, blogs and on TV news broadcasts. For example, Roger Cox—a Dutch environmental lawyer and the legal counsel for the plaintiffs in *Urgenda* proceeding—gave interviews to the media, participated in conferences and gave lectures and TED Talks about the case, given it an even greater publicity and visibility.[5] These organisations and their lawyers can also maintain very accessible and understandable websites and other channels of communication with the public for a better knowledge, awareness and understanding of the cases and related issues (Lin 2012). For example, *Urgenda* Foundation and Germanwatch both offer very accessible, easy and understandable information of the cases through their websites.[6]

Moreover, lawyers and academics can publish analyses of climate cases in a wide variety of academic and professional journals, magazines or books. The *Urgenda* case is particularly exemplary in this sense. While the legal reasoning of the case is explained in an exclusive book: "Revolution Justified"[7]; the lawsuit and especially the judgement, that was unusually translated into English, received a fair amount of scholarly attention since the beginning, appearing in several legal and

---

[4]See a summary of the media cover, specially of the *Urgenda* and *Leghari* cases, in http://www.urgenda.nl/en/climate-case/ and http://germanwatch.org/de/der-fall-huaraz). See also https://www.theguardian.com/environment/2017/may/23/climate-change-government-court-cases-study.

[5]See a summary of Cox's presentations in http://www.revolutionjustified.org/rj-media.

[6]See http://www.urgenda.nl/en/climate-case and http://germanwatch.org/de/der-fall-huaraz.

[7]Cox (2012).

multidisciplinary academic journals, including the Journal of Environmental Law,[8] Climate Law,[9] Journal of Energy & Natural Resources Law,[10] Transnational Environmental Law,[11] Nature[12] and Science,[13] just to name a few. The ruling also generated blog commentaries,[14] and was discussed in several conferences and academic meeting.[15]

In addition to the above, teachers also discuss climate cases in classrooms, environmental reporting services provide reports on cases fillings and court decisions, and advocacy groups use information about court cases to promote their points of view and encourage social action, for example, through newsletters, websites, and other outlets (Averill 2007). Thus, climate-related cases, arguments and opinions that are well communicated by media or any other source, as described above, can help to improve public understanding of the different dimensions of climate change, raise awareness on its causes and impacts, and serve as vehicles for high-profile public debate on a wide variety of climate-related issues, keeping those in the public eye and on the political agenda.

6 Conclusion

In the light of the urgent need to address climate change challenges, this paper has explored the role of CCL in communicating climate change issues across various audiences. It showed how CCL, beyond being an important strategy to exert pressure on the executive and legislative branches of governments to act on climate change, can help communicate the risks of inaction on climate change, not in the abstract sense, but in concrete and real terms, as well as the feasible solutions.

The three climate-related cases showcased in this paper—*Urgenda*, *Leghari* and *Lliuya*—reveal that every climate lawsuit, won or lost, provides important material to facilitate the communication on the urgency of addressing climate change, while fostering social action. In particular, climate change cases allow a greater understanding of the meaning and scope of climate change by telling stories about specific victims and injuries in concrete places, making visible the current and

---

[8]De Graaf and Jans (2015).

[9]Lin (2015).

[10]Cox (2015) and Roy and Woerdman (2016).

[11]Van Zeben (2015) and Galvão Ferreira (2016).

[12]Schiermeier (2015).

[13]Enserink (2015) and Science (2015).

[14]Just to mention some: Verschuuren (2015) and Warnock (2015). See also information about the three cases in Climate Law Blog http://blogs.law.columbia.edu/climatechange/.

[15]For example, in the 14th Annual Colloquium of the IUCN Academy of Environmental Law (http://iucnael2016.no/wp-content/uploads/2015/05/Final-Programme-23.06.pdf) and the 2nd International Forum on Environmental Justice (*II Foro Internacional de Justicia Ambiental*) (http://www.tribunalambiental.cl/ii-foro-internacional-de-justicia-ambiental/#programa).

future consequences of climate change and bringing them closer to people's lives and homes. In addition, by presenting complex scientific information on climate change in a language that is understandable by non-experts, within and outside courtrooms, these cases make climate science more accessible to the public. Thus, as climate-related cases are reported in a variety of sources, especially national and international media, gaining wide attention, they foster education over climate change issues, including possible solutions, and help raise awareness and enhance public debate, motivating social and political action. The three climate lawsuits analysed above are just representative examples of the enormous communicative potential of CCL. As they appear to have encouraged a new wave of claims, further analysis of other existing cases in different jurisdictions, but also of those that may arise in the context of the Paris Agreement, can be useful to explore this indirect implication of CCL and its contribution to the field of climate change communication. In any case, CCL is emerging as a novel tool for communicating the urgency of climate change and for promoting action.

Funding Acknowledgement The generous financial support of the South Africa National Research Foundation (NRF) in the form of a 'travel grant for individual researcher' is acknowledged with appreciation.

References

Abate R (ed) (2016) Climate justice: case studies in global and regional governance challenges. Environmental Law Institute, Washington, DC, USA

Averill M (2007) Climate litigation: Shaping public policy and stimulating debate. In: Moser S, Dilling L (eds) Creating a climate for change: communicating climate change and facilitating social change. Cambridge University Press, New York, USA, pp 462–474

Averill M (2008) Climate litigation: ethical implications and societal impacts. Denver Univ Law Rev 85:899–918

Averill M (2009) Linking climate litigation and human rights. Rev Eur Comp Int Environ Law 18 (2):139–147

Banda M, Fulton S (2017) Litigating climate change in national courts: recent trends and developments in global climate law. Environ Law Rep 47(2):10121–10434

Burns W, Osofsky H (eds) (2009) Adjudicating climate change: state, national, and international approaches. Cambridge University Press, New York, USA

Capstick S, Whitmarsh L, Poortinga W, Pidgeon N, Upham P (2015) International trends in public perceptions of climate change over the past quarter century. WIREs Clim Change 6(1):35–61

Corner A (2013) The 'art' of climate change communication. The Guardian. https://www.theguardian.com/sustainable-business/art-climate-change-communication. Accessed 25 Sept 2017

Cox R (2012) Revolution justified. Planet Prosperity Foundation, The Netherlands

Cox R (2015) A climate change litigation precedent: Urgenda Foundation v The State of the Netherlands. J Energy Nat Resour Law 34(2):143–163

Dahlstrom M (2014) Using narratives and storytelling to communicate science with nonexpert audiences. PNAS J 111:13614–13620

De Graaf K, Jans J (2015) The Urgenda decision: Netherlands Liable for role in causing dangerous global climate change. J Environ Law 27(3):517–527

Enserink M (2015) In surprise, Dutch court orders government to do more to fight climate change. Science, June 24. http://www.sciencemag.org/news/2015/06/surprise-dutch-court-orders-government-do-more-fight-climate-change. Accessed 25 Oct 2017

Faure M, Peeters M (eds) (2011) Climate change liability. Edward Elgar Publishing, Cheltenham, UK

Galvão Ferreira P (2016) Common But Differentiated Responsibilities' in the National Courts: lessons from Urgenda v. The Netherlands. Trans Environ Law 5(2):329–351

Germanwatch (2017) First Climate Lawsuit Against Energy Company Before German Courts. http://germanwatch.org/en/14191. Accessed 25 Sept 2017

Green M (2004) Storytelling in Teaching. The Association for Psychological Science 17(4). https://www.psychologicalscience.org/observer/storytelling-in-teaching. Accessed 28 Sept 2017

Hunter D (2007) The implications of climate change litigation for international environmental law-making. Washington College of Law Research Paper 14:1–19. https://papers.ssrn.com/sol3/papers.cfm?abstract_id=1005345. Accessed 20 Sept 2017

Jones MD (2014) Communicating climate change: are stories better than "just the facts"? The Policy Stud J 42(4):644–673

Kaminakaité-Salters G (2011) Climate change litigation in the UK: its feasibility and prospects. In: Faure M, Peeters M (eds) Climate change liability. Edward Elgar Publishing, Cheltenham, UK

Kanner A, Nagy T (2006) Legal strategy, storytelling and complex litigation. Am J Trial Advocacy 30(1):1–26

Kearney A (1994) Understanding global change: a cognitive perspective on communicating through stories. Clim Change 27:419–441

Leal Filho W (2009) Communicating climate change: challenges ahead and action needed. Int J Clim Change Strateg Manag 1(1):6–18

Lin J (2012) Climate change and the courts. Legal Stud 32(1):35–57

Lin J (2015) The first successful climate change negligence case: a Comment on Urgenda Foundation v. The State of the Netherlands (Ministry of Infrastructure and the Environment). Clim Law 5(1):65–81

Loth M (2016) Climate change liability after all: a Dutch Landmark case. Tilburg Law Rev 21:5–30

Mehra M (2015) Pakistan ordered to enforce climate law by Lahore court. Climate Home News. http://www.climatechangenews.com/2015/09/20/pakistan-ordered-to-enforce-climate-law-by-lahore-court/. Accessed 15 Oct 2017

Moser S (2010) Communicating climate change: history, challenges, process and future directions. Wiley Interdiscip Rev: Clim Change 1(1):31–53

Moser S, Dilling L (2004) Making climate HOT: communicating the urgency and challenge of global climate change. Environ: Sci Policy Sustain Dev 46(10):32–46

Moser S, Dilling L (2011) Communicating climate change: closing the science-action gap. In: Dryzek JS, Norgaard RV, Schlosberg D (eds) The Oxford Handbook of climate change and society. Oxford University Press, Oxford, UK

Osofsky H (2010) The continuing importance of climate change litigation. Clim Law 1(1):3–29

Peel J, Osofsky H (2015) Climate change litigation: regulatory pathways to cleaner energy. Cambridge University Press, Cambridge, UK

Roy S, Woerdman E (2016) Situating Urgenda v the Netherlands within comparative climate change litigation. J Energy Nat Resour Law 34(2):165–189

Science (2015) Dutch court orders government to cut more CO_2. Science 349(6243):10

Schiermeier Q (2015) Landmark court ruling tells Dutch government to do more on climate change. Nat News. June 24, http://www.nature.com/news/landmark-court-ruling-tells-dutch-government-to-do-more-on-climate-change-1.17841. Accessed 25 Oct 2017

Shaw C (2014) Reframing climate risk to build public support for radical emission reductions: the role of deliberative democracy. Carbon Manag 5(4):349–360

Smith J, Sherman D (2006) Climate change litigation: analysing the law, scientific evidence and impacts on the environment, health and property. Presidian Legal Publications, Adelaide, South Australia

Somerville R, Joy Hassol S (2011) Communicating the science of climate change. Phys Today 64 (10):48–53

Sprinz D, Hefele P (2017) Compensating for climate change impacts? Priorities for research and public policy. German Konrad Adenauer Foundation (KAS) and the Potsdam Institute for Climate Impact Research (PIK) http://www.kas.de/recap/en/publications/50024/. Accessed 15 Sept 2017

UNEP (2017) The status of climate change litigation—A global review. UNEP. http://columbiaclimatelaw.com/files/2017/05/Burger-Gundlach-2017-05-UN-Envt-CC-Litigation.pdf. Accessed 10 Oct 2017

Van Zeben J (2015) Establishing a governmental duty of care for climate change mitigation: will Urgenda Turn the Tide? Trans Environ Law 4(2):339–357

Verschuuren J (2015) Spectacular Judgment by Dutch Court in Climate Change Case. Tilburg University Blog, June 25 https://blog.uvt.nl/environmentallaw/?p=109. Accessed 25 Oct 2017

Warnock C (2015) The Urgenda Decision: Balanced Constitutionalism in the Face of Climate Change?' Oxford University Press Blog, July 22. http://blog.oup.com/2015/07/urgenda-netherlands-climate-change. Accessed 25 Oct 2017

Wibeck V (2014) Enhancing learning, communication and public engagement about climate change—Some lessons from recent literature. Environ Educ Res 20(3):387–411

Wolf J, Moser S (2011) Individual understandings, perceptions, and engagement with climate change: insights from in-depth studies across the world. Wiley Interdiscip Rev: Clim Change 2:547–569

Court Decisions

Court of The Hague (2015) Urgenda v. Government of the Netherlands, 24 June, ECLI:NL: RBDHA:2015:7196, Rechtbank Den Haag, C/09/456689/HA ZA 13-1396 (English version). http://www.urgenda.nl/documents/VerdictDistrictCourt-UrgendavStaat-24.06.2015.pdf. Accessed 10 Sept 2017

Lahore High Court (2015) Leghari v. Federation of Pakistan, WP No. 25501/2015. http://edigest.elaw.org/sites/default/files/pk.leghari.090415.pdf. Accessed 3 Oct 2017

Paola Villavicencio Calzadilla is a Postdoctoral Research Fellow at the Faculty of Law of the North West University, Potchefstroom Campus (South Africa) and a Visiting Researcher at the Faculty of Law of the University of Groningen (The Netherlands). She holds a Master's Degree and Ph.D. in Environmental Law from the University Rovira i Virgili (Spain). She has been involved in different research projects and is author and co-author of several papers related to climate change law, environmental and climate justice, energy policy, sustainable development and rights of nature.

Transnational and Postcolonial Perspectives on Communicating Climate Change Through Theater

Nassim Winnie Balestrini

Abstract While theater cannot prevent climate change, it can engage with conveying knowledge and attitudes. This paper does not measure the impact of specific theater performances on test participants. It rather analyzes which artistic methods the authors of a specific corpus of texts use in order to communicate climate change. In this sense, this paper is not concerned with pragmatic suggestions regarding climate change per se; instead, the focus is on communicating climate change through theater. Integrating climate-change science into theatrical performances generates aesthetic challenges: how can dramatists represent a long-term global phenomenon within the spatiotemporal limits of a performance? How can drama convey scientifically sound information along with captivating characters and plots? How can performances elicit more nuanced viewer responses than panic in the face of impending disaster or apathy based on lacking concern? Taking transnational American Studies and postcolonial literary theory as points of departure, this paper will discuss English-language theatrical works linked to Climate Change Theatre Action (CCTA), an initiative originally launched by artists in the United States and Canada to publicize the 2015 Paris Climate Conference (COP21) and designed to occur every other year. This activism-oriented project translates issues related to global climate change into a transnational theater practice that experiments with innovative drama aesthetics and that fosters communication across boundaries between theater professionals and amateurs, climate-change specialists and the scientifically untrained general public as well as local action and international orientation. Despite continuing notions that science represents rational thinking whereas artistic depictions express or arouse predominantly fearful emotions, this body of very short performances and the online forum in which some of the same theater practitioners exchange ideas and experiences offer working models for effective collaboration that may support widespread activism.

N. W. Balestrini (✉)
American Studies, University of Graz, Attemsgasse 25/II, 8010 Graz, Austria
e-mail: nassim.balestrini@uni-graz.at

© Springer Nature Switzerland AG 2019
W. Leal Filho et al. (eds.), *Addressing the Challenges in Communicating Climate Change Across Various Audiences*, Climate Change Management,
https://doi.org/10.1007/978-3-319-98294-6_16

247

1 Introduction

The goal of this paper is, first, to highlight the dilemmas of communicating climate change as defined in selected manuals published for this very purpose and, second, to analyze a corpus of contemporary short plays which, in turn, were written and have been performed as part of theater practitioners' climate change awareness activism. While climate change communication specialists have not yet discovered the stage as a potent vehicle of conveying climate change issues to non-scientist audiences, playwrights and theater professionals have channeled their environmentalist agendas into performances. As research conducted from the disciplinary perspectives of literary and cultural studies, this essay enquires into the artistic means which playwrights and performers are currently using to express their understanding of climate change and to elicit emotional and intellectual responses from implied audiences.

For over a decade, climate change communication has been concerned with the perceived chasm between those who are conversant in the language of the natural sciences and those who are not. As a result, scientists, policy makers, and environmental activists have been struggling to devise methods that render climate change understandable to non-scientists and communicate scientific knowledge in a manner that triggers desirable forms of grassroots action. Publications such as *Americans and Climate Change: Closing the Gap Between Science and Action* by the Yale School of Forestry & Environmental Studies (Abbasi 2006) and *The Psychology of Climate Change Communication: A Guide for Scientists, Journalists, Educators, Political Aides, and the Interested Public* (hereafter: *PCCC*) by Columbia University's Center for Research on Environmental Decisions (Shome and Marx 2009) exemplify the trend to include various social contexts and institutions as well as academic disciplines in the discussion of how climate change communication should be packaged.

Oddly enough, these publications lack discussion of how the arts can participate in the same effort. That dramatists and performers have been active in conveying their views on general ecological issues and on climate change is unquestionably true.[1] Collaborations between scientists and theater artists have taken place, even if highly visible publications on climate change communication do not yet reflect such efforts. As the following discussion of climate change plays will illustrate, the techniques that dramatists use to awaken the activist potential in their audiences contain some of the concerns and recommendations voiced in climate change

---

[1]Theater scholar Una Chaudhuri, for instance, has been urging these matters since the 1990s. See, for instance, the special issue of *Theater* on "Ecology and Theatre" (1994) that she edited, her essay "The Silence of the Polar Bears: Performing (Climate) Change in the Theater of Species" (2012), and the volume she co-edited with Shonni Enelow *Research Theatre, Climate Change, and the Ecocide Project* (2014).

communication manuals. Thus, an expansion of collaborative efforts between climate change scientists and theater artists may be on the horizon, at least in contexts in which those who want to communicate climate change–related knowledge assume that artistic works can appeal both to the rational and emotional faculties of audience members.

In addition to not mentioning contemporary artist-activists, climate change communication manuals show little awareness of how literary and cultural studies contribute to unraveling the mechanisms of meaning-making related to climate change outside the scientific community. References to the potential use of narratives or of images remain devoid of scholarly contextualization. While the social sciences and (social) psychology are part of the outlook, the critical perspectives of schools of thought like postcolonial studies and transnational American Studies remain invisible.

Postcolonial theory scrutinizes how economic and social conditions of colonization impact cultural self-expression: while economic hegemony led colonizers to represent people in and from (former) colonies as 'Other,' implying that 'Otherness' equals lower status, the colonized have been asserting the empowering features of cultural "hybridity." This means that they merged the colonizers' and their own cultural traditions. From this in-between "third space" innovative artistic forms evolved that transcend hierarchical binaries of self and 'Other' (see Anzaldúa 1987; Bhabha 1994).

Transnational American Studies theory, a related approach which is also concerned with dismantling hierarchies and dichotomies in cultural discourse and practice, argues that cultural historians focusing on the United States must go against replicating 'exceptionalist' myths. The term exceptionalism has been used to characterize a type of national historiography which construes the United States as pursuing a divine mission whose fulfillment justifies all means. According to the transnational American Studies approach, United States literary and cultural history must take into account the diverse global patterns of human mobility and its cultural, economic, environmental, and political impacts. The depth of time and breadth of space which applying this approach requires should then be dealt with in a manner that neither privileges nor ignores the role of national boundaries, but which contextualizes them thoroughly. When jointly contemplating postcolonial and transnational arguments, it becomes clear that it is helpful to analyze artistic renderings of climate change–related histories of pollution from the perspective of economic and cultural hegemonies.[2]

---

[2]Ecocriticism—the study of environmental issues and representations in literature and culture—has adopted both a postcolonial and transnational outlook, thus acknowledging the necessity to go beyond an "Anglo-American focus" (Heise 2008, p. 387). Also see Cilano and DeLoughrey (2007) and Huggan and Tiffin (2010).

2 Diagnosing the Ills of Climate Change Communication

According to *PCCC*, the work of Columbia University's Center for Research on Environmental Decisions (CRED) integrates a "broad spectrum of disciplines [...]: psychology, anthropology, economics, history, environmental science and policy, and climate science" (Shome and Marx 2009, p. 2).[3] The absence of literary and cultural studies from this list notwithstanding, this CRED publication argues that people process information better when it is "actively communicated with appropriate language, metaphor, and analogy; combined with narrative storytelling; made vivid through visual imagery and experiential scenarios; balanced with scientific information; and delivered by trusted messengers in group settings" (p. 2). Although theater and performance do not seem to be part of the studies from which this set of maxims was derived, to a literary scholar it appears striking that the entire ground sketched in this list is covered by theater semiotics, i.e., a sign system that includes mostly audible verbal language (which can be figurative); visual features like stage sets, costumes, and lighting; characters and plots which represent experiences that may be based on real-life—and thus also scientific—knowledge and that are performed in theaters or other venues by actors towards whom audience members may have a positive attitude.

The same holds true for the foci of subsequent chapters of *The Psychology of Climate Change Communication*: acts of "[f]raming" and "context[ualizing]" in order "to achieve a desired interpretation or perspective" (p. 6) inadvertently exist in any context of performance, even if the desired effect may not be of the unequivocal and unambiguous kind described by Shome and Marx. Awareness of how the respective audience's set of beliefs and cultural self-definition affects their reaction to what they are presented with (p. 7) coheres with literary scholar Stanley Fish's theory of "interpretive communities"[4] which make certain readings of literary texts plausible to sets of people, a similarity that the authors' interest in mobilizing group action confirms (see Shome and Marx 2009, pp. 28–36). Based on related notions of local or otherwise social rootedness, the authors surmise that audiences will react more deeply to a story linking climate change to their own environment—a circumstance which then raises the problems of potentially unscientific ideas about causal relations between climate change and specific natural disasters, and of disregarding the global interdependence and scale of climate in general (pp. 9–10). This set of considerations is particularly relevant for the outlook offered by transnational American Studies because the trend that has led to refocusing the field encourages the study of "intricate interdependencies" (Radway 2002, p. 53)

---

[3]A similar approach can be found in Abbasi (2006). While communicating climate change is linked to the social sciences and to the necessity of transforming vast amounts of data into "narrative storytelling," the arts are not mentioned (p. 11). The societal domains included in the conference proceedings are "Science, News Media, Religion & Ethics, Politics, Entertainment & Advertising, Education, Business & Finance and Environmentalists & Civil Society" (pp. 17–18).

[4]For details, see Fish (1980).

that shed light on how even seemingly locally restricted cultural phenomena need to be studied contextually through the lens of long-term and geographically far-flung relations.[5]

Despite this overlap between the CRED authors' approach and the areas of theater practice and analysis from a cultural studies–informed literary studies perspective, one major quandary resides in the way in which Shome and Marx depict human beings as processing information in a bifurcated manner that is presented as a clear distinction between the communicative possibilities and practices of the natural sciences as opposed to those attributed to non-scientific discourse. In a table that lists characteristics of what the authors describe as the "analytic processing system" and the "experiential processing system" of the brain, scientific forms of representation ("numerical statistics in tables, figures, graphs, charts") contrast with implicitly non-scientific and, in part, art-related media and genres like "images or stories" which, as the authors generalize, are "emotionally charged and vivid" (p. 16).[6] The emotions listed in the table are "fear, dread, and anxiety" (p. 16). The table thus creates the impression that visuals and narratives are—by default— neither "logical" nor "deliberative" nor "analytic" (p. 16) but rather upsetting. While the authors eventually conclude that "[t]he most effective communication targets both processing systems of the human brain" (p. 18), it remains unclear which genres achieve this goal.[7]

As a practice-oriented manual, *PCCC* may use intentionally simple explanations, and not including the arts could be rationalized by the fact that the manual is being marketed to non-artists. Curiously enough, however, an article by playwright and climate change activist Chantal Bilodeau entitled "A Climate of Collaboration: Environmental Scientists Urge Artists to Humanize the Stories behind the Research" (2013a) reveals that Columbia University scientists participated in a 2009 conference that explicitly focused on bringing together scientists and artists. Bilodeau quotes one of CRED's co-directors as saying that "[e]ngaging people through the arts—not necessarily in ways that are alarmist and fear-provoking, but in ways that are thought-provoking—can remind them of their long-term goals. The arts can comment on abstract things—like the sustainability of planet Earth—so much better than statistics can" (Elke Weber qtd. in Bilodeau 2013, p. 13).[8] In other words, if the arts manage to be "thought-provoking" regarding climate change,

---

[5]Also see Fishkin 2011, Fluck 2007, Hornung 2011, Lenz 2011, Shu and Pease 2015.

[6]Similarly, Abbasi (2006) characterizes science as verbally oriented and "Society" as visually oriented, but casts journalism and television as solutions to science's non-inclusive use of language (pp. 107–108), particularly regarding the desideratum of giving climate change a human face (pp. 114–115) rather than focusing on landscapes and animals. In the same vein, people engaged in the area of "religion & ethics" are supposed to "harness visual media as well as traditional written and oral media" (p. 138).

[7]"Cli fi" or climate fiction has been around for decades or, some critics argue, much longer than that. While some authors certainly address horrifying disasters, others merge scientific knowledge and literary narrative. See Mayer and Weik von Mossner 2014, Leikam and Leyda 2017.

[8]Regarding such dialogs, see Ereaut and Segnit 2006, Moser and Dilling 2007, Norgaard 2011.

they can channel an experience-focused and affective appeal into a rational response that implies analysis and planning rather than activism born of sheer panic. Arguing from the perspective of American literary scholarship, Paweł Frelik stresses that works of "cli-fi" (short for "climate fiction") should—analogous to science fiction —be seen as "a tool for thinking" and as "invaluable cognitive and didactic tools" which encourage interdisciplinary exchange between science and art (Frelik in Leikam and Leyda 2017, p. 128) rather than being irrationally dystopian for the sake of shocking readers.

In addition to acknowledging that artistic works can elicit analytical *and* emotional responses, which in themselves are more complex than any schematic two-column table can convey, it is also helpful to distinguish between the dramatic representation or "scripted performance of emotions" (Tait 2015, p. 1504) and an audience member's emotional response to such a representation. The scriptedness of emotional representation in a play offers yet another pathway into the interdisciplinary study of climate change performance, which necessarily precedes research on audience responses—but this has not happened yet outside literary and cultural studies circles. As Philip Smith and Howe (2015) point out, "[d]espite good-faith attempts to imagine alternative possibilities, culture is largely understood in social-psychological terms as a handicap to full rationality" (p. 18). In their study, *Climate Change as Social Drama*, they thus harness narrative theory and Aristotelian concepts, anthropology-based performance theory, and theories of the social sciences in order to develop a "cultural–sociological approach" (p. 34) that is meant to bridge the gap between cultural expression and public reception. Smith and Howe argue that "climate change enters public life not as a set of scientific facts but rather as a set of collective representations replete with genre, plot, characters with particular attributes, and performative actions that are worthy or unworthy" (p. 49). Despite the drama metaphor in their theoretical approach, their study of climate change art focuses on film (as in the chapter on "From Awareness to Action: Taking into Consideration the Role of Emotions and Cognition for a Stage Toward a Better Communication of Climate Change") and the visual arts (as in the chapter on "Strengthening Personal Concern and the Willingness to Act Through Climate Change Communication"). Beyond those art forms, seeing public interaction and discourse regarding climate change through the lens of "social drama" implicitly raises the question as to how actual climate change plays and performances represent and possibly simulate such interaction and discourse.

Those who contemplate climate change communication grapple, as shown with the above examples, with defining central characteristics of and attitudes towards scientific and non-scientific discourse, with defining how human beings process and respond to information, and with imagining how social contexts affect whether communicated content will yield desired types of action. Non-scientific forms of discourse thus do not come as neutral vehicles without cultural-historical baggage and cannot simply 'be filled' with science-based information. Instead, it is time to contemplate how different genres and their contexts of public reception offer medium- and form-specific possibilities. In the following, I will discuss how several of the 100 short manuscripts of CCTA 2015 and 2017 negotiate the challenge of 'working' as pieces to be performed and to be understood as artistic interventions in the public debate on climate change.[9]

3   Climate Change Theatre Action 2015, 2017, and Beyond

Artists in Canada and the United States jointly initiated Climate Change Theatre Action (CCTA) in order to raise public awareness of the 2015 Paris Climate Conference (COP21) through events carried out between 1 November and 12 December 2015.[10] During that time period, more than 100 participants in 25 countries presented—be it in public or private places, in educational institutions, at festivals or other occasion—readings, performances, and adaptations selected from the 50 short pieces (plays, performance poems, songs) that artists from various countries provided for the purpose.[11] Between 1 October and 18 November 2017, CCTA 2017 events in 37 countries made use of pieces that can be performed in about five minutes, submitted by artists affiliated predominately with Canada and the United States, including indigenous peoples, but beyond North America also with Brazil, Ethiopia, Fiji, Greece, India, Iraq, Jamaica, Japan, Jordan, Kenya, Lebanon, New Zealand, the Philippines, Samoa, Scotland, Spain, Uganda, and Zimbabwe.[12]

The long-term and global characteristics of climate change heighten the challenge of writing a very short performance piece that is simultaneously condensed and complex. Transnational and postcolonial perspectives on culture share this challenge to think across long stretches of time and space, albeit not on the same scale as developments related to climate change. Also, economic inequality, political power, and individual agency figure prominently in studying both climate change and culture. In both areas of inquiry, researchers must keep local, national, borderlands, and global phenomena in mind. Not surprisingly, the CCTA corpus of plays includes works that focus on specific situations, for instance, of rural populations in developing countries, in order to address the suffering of relatively weak

---

[9]On climate change drama in a wider sense, i.e., including full-length plays and series of dramas, see Johns-Putra (2016), Balestrini in Leikam and Leyda (2017), Balestrini (2017a, b), Balestrini (2018).

[10]Regarding CCTA, see Bilodeau, "Arctic in Context" (2013b) and "As the Climate Change Threat Grows" (2016).

[11]"Climate Change Theatre Action 2015 Edition." See https://www.thearcticcycle.org/ccta-2015/ (Last Accessed 21 Sept 2017).

[12]See https://www.thearcticcycle.org/ccta-2017/ (Last Accessed 21 Sept 2017).

links in the political landscapes of specific locations.[13] Several plays demonstrate
the use of such intersections and their potency as dramatic material that may inspire
climate change awareness.[14]

The combination of postcolonial and transnational features resulted in several
striking short plays which jointly encourage viewers not to fall into the trap of
assuming that Western nations are in charge of defining climate change–related
actions not only for themselves but, first and foremost, for previously colonized
nations. The majority of the authors who penned these works self-identify as being
affiliated with former colonies (India, New Zealand, Uganda), or they list multiple
countries (including Canada, the US, the Czech Republic, and Iraq). While all of the
works in this group rely on verbal arguments, presented mostly in dialog format,
they bifurcate: they either focus on sarcasm and comedy, or they foreground
poignant physical sensations which jolt a character into understanding a point. In
the latter case, Elspeth Tilley's "Flotsam" (2015) presents a mother–daughter
conflict regarding laws and court decisions about deporting refugees. When the
mother, who is a judge, insists on following the law, her daughter puts salt into her
mother's tea in order to force her literally to taste the salt water that Kiribati
islanders must rely on when returning to their storm-tossed home threatened by
further tidal waves. Hassan Abdulrazzak's "American Nightmare" (2017) turns the
tables of dependence and oppression: an American male character imagines being
sexually humiliated by a Middle Eastern business partner. This nightmare mixes his
own sense of guilt at the long history of Western imperialism and economic
exploitation with a heavy dose of Orientalist clichés regarding sexual mores and
violent tendencies. Less harrowing than that, Elaine Ávila's "Portuguese Tomato"
(2015) pinpoints how some Americans romanticize poorer countries (for instance
by idealizing their products) and simultaneously expect the residents of these
countries to fight climate change, monoculture, and other issues on their own rather
than realizing that geographical remoteness must not preclude joint action.

In Ugandan playwright Deborah Asiimwe's "'Them' and 'Us'" (2015),
fast-paced argumentative dialog about climate change becomes effective drama.
A Western film maker considers his climate change documentary film both
objective and productive, but his female interlocutor reads his representation as
talking condescendingly to people from developing countries like her own. While
conversing, they are drinking and commenting on coffee: a beverage that Western
nations import from the global South and that has become a gourmet product.

---

[13]For instance, the 2017 corpus includes plays about tribal communities in India that are mistreated
by powerful corporations; about rural populations in Pakistan whose way of life is deemed less
valuable than certain kinds of modernization; about the clash between economic growth and
environmental protection in Ethiopia; and about the impact of corruption on waste management in
Lebanon. The CCTA plays are made available to participants but have not been published. In the
following, the year in parentheses indicates to which CCTA corpus a play belongs.

[14]It goes without saying that the entire corpus of CCTA plays in itself, with works by authors from
developing and industrial nations around the globe, embodies the principle of abolishing cultural
hierarchies based on economic and political power.

Cynthia confronts Mark directly with the hypocrisy of demanding change of others without facing the impact of one's own actions ("Why should Africa and Asia be put under pressure to tiptoe around their own industrial revolution because someone else has been and still is irresponsible?"); in mock-folktale fashion, she uses an analogy that emphasizes how closely nations affect one another: "Those who have destroyed more and are still in the process of destroying the little that is remaining are the ones that need to be preached to. Leave us alone to industrialize at our own pace and in our own way. Don't set your house on fire and expect to bully your neighbor into putting out the fire you have set." Thus, a woman from a developing country switches positions with a man who would prefer to be seen as charitable and fair towards people in need; she stresses that pretending that there is no "them" and "us" erases who has been doing what to whom. This terse dialog, which ends when Cynthia has drained her coffee cup, is not an alarmist lament. Instead, strong emotions are poured both into allegorical ways of making a point and into sharp-tongued statements based on cogent reasoning.

David Geary (Canada/New Zealand) also stresses transnational linkages and the notion of a surprisingly small world sharing a global climate, but his play "Morehu and Titi" (2015) primarily relies on comic effects. The main characters—a reptile and a bird with symbolic features rooted in Maori culture—are not simply shocked by the appearance of an iceberg in their part of the world, but both the iceberg and the Aurora Australis are represented by an actor who plays Al Gore dressed up as these natural phenomena. On top of that, the folksy anthropomorphic New Zealand animal heroes intone a parody of "Let It Go" from Disney's animated film *Frozen* (2013). The hilariously ironic staging of global interconnectedness is comple-mented by the extensive temporal perspective of a reptile whose looks evoke the era of dinosaurs, scenarios of extinction, and the dissolution of the boundary between animal and human existence.[15]

The scriptedness of the social drama—of disagreeing individuals or of similarly-thinking associates who criticize others—in these short dialogs fosters clarity and conciseness of argument. Not unexpectedly, fast-paced dialog can swiftly reveal and develop dramatic conflict and be engaging for viewers, and this may be the reason why one third of the 100 plays feature two characters—in about half of the cases two characters who are a couple or family members on the cusp of either going separate ways or jointly deciding to work together on preventing the further exacerbation of anthropogenic climate change. The two-character plays thus

---

[15]Comedy as a means of inspiring audiences to consider the possibility of extinction, based on scientific publications, also occurs in Chantal Bilodeau's "Homo Sapiens" (2017), which is set in the distant future and treats audience members as specimens of *homo sapiens* as perceived by the dramatic characters who hail from a more advanced human species. Animal characters also recur as commentators on human folly in plays that feature animals equipped with symbolic implica-tions: bald eagles in Elaine Ávila's "Brackendale" (2017), a play which cites scientific research, and Elspeth Tilley's "The Penguins" (2017), a play which alludes to Aristophanes's *The Birds* and is designed to provoke thoughts about post-humanism, as indicated in the author's note preceding the play.

present the social drama of climate change communication as intrinsic to the performance's fictional world.

A different communicative situation characterizes the monologs which constitute another third of the corpus. These single-actor plays (in some instances designated as performance poems) tend to directly address someone or, more often than not, an implied audience that either functions as a silent listener or as a potential interlocutor and activist-in-arms. Thus, the social drama moves from the performed words into the extrinsic realm of the performance space. In the two-character plays which present character constellations such as parent and child or close friends, the conflict or balance between rational argument and irrational emotion frequently relies on viewers' expectations regarding the dynamics of the given personal relationship, which then assumes an allegorical dimension through which the characters' interaction embodies their respective attitude and its resulting behavior regarding climate change.

Among the two-actor plays, Deborah Zoe Laufer's "The Cow Is Dead" (2015) juxtaposes a teenage daughter who—despite her impatience and irascible way of speaking—represents rational thought and wants to transform scientific knowledge into action (such as not eating meat and not driving) with a mother who has an unhealthy life style (she smokes heavily while being pregnant) and spouts fatalistic platitudes based on pseudo-reasons for not taking responsibility, mainly because individual action presumably makes no difference and only causes extra work on her part. In the one-character plays, the balance between argument and emotion becomes more precarious in the sense that viewers are drawn into the fray as implied interlocutors, which on the one hand risks alienating them but which on the other hand may lead to pledging commitment (at least momentarily).

Various plays attempt to forestall alienating viewers by focusing on the monolog speaker's personal sense of responsibility, by using non-human speakers, humor, or unequivocal emotional appeal. Dipika Guha's performance poem "A Song for the Sleepers" (2015) presents a speaker from India who alternately looks at the world through newspaper reporting in the US and in the UK, and through the eyes of his grandmother in India. The speaker concludes that the same problematic issues recur in different times and places. He fears that his empathy with Syrian refugees does not go deep enough because he may be too comfortable in this transnational well-educated context in which, for instance, drowning refugee children are images on the news rather than individuals with names. While the speaker's self-deprecating sarcasm may incite listeners to think about their own position of privilege, the poem does not accost listeners regarding their personal approach to confronting other people's suffering.[16]

---

[16]Similarly, Andrea Lepcio's "Alone" (2015) succeeds in depicting a woman representing the wealthy "1%"-ers. She lives on literally and symbolically high ground that saves her from flooding, within earshot of the voices of those further below. The rather dry representation of "1%"-er arguments regarding their sense of entitlement implicitly results in a plea for empathy, but without being maudlin or melodramatic.

Humor and sarcasm certainly work well in making clear argumentative points about environmental hazards and climate change, as also shown in Mindi Dickstein's "Starving to Death in Midtown" (2015), a densely-written monolog of a dying bee in New York City which alludes to American political and cultural history and offers incisive comments on faulty human behavior. But here, as in Darrah Cloud's "TESS Talk" (2015)—which parodies TED Talks by presenting a bartender's logically incoherent rejection of climate change by comparing it to the menopausal woman's hot flashes—one may wonder whether the humor will prevent serious engagement with the (parodied) argumentations.

At the other end of the range between irony and empathy, Chantal Bilodeau's "Mother" (2015) has Mother Earth address "you," that is, humankind. The personified globe gently chides her offspring for averting their gaze and concludes that everything will improve once she can see her own reflection in her children's eyes. While this monolog mentions scientific concepts such as the double helix and evolution, it argues that only an empathetic relation between the earth and her residents offers hope. Thus, the emotional appeal inherent in softly pressuring the viewers to lock eyes, to see eye to eye, and to acknowledge their interdependence is presented as a prerequisite for transforming scientific knowledge into action. Bilodeau's short play seeks to achieve the very ideal of merging the rational and emotional sides of the human brain—a strategy that other works in the CCTA corpus also pursue: in Catherine Banks's "The Project Hope" (2017), a character finds that hope cannot be nurtured solely through rational arguments but that it requires personal recollections of experiencing natural beauty, which are translated into theater semiotics through visualizing such memories as beautifully enshrined sources of light. Rather than externalizing something psychological, Clare Duffy ("The Blue Puzzle" [2017]) and Lynn Rosen ("Dot to Dot to Dot" [2017]) evoke organic designs and the embeddedness of earthly life in an extensive universe which needs to be conceptualized through vast depths of time and space.

Just as climate change communication manuals wish to close the gap between science and action, CCTA artists and activists seek to transform theater experiences into climate change–related actions. Thus, several plays contain components which break the fourth-wall illusion and which involve audience members as part of the performance. Kendra Fanconi's monolog "Finale" (2015) invites a silent audience member to dance with the actor throughout her/his metatheatrical musings about theater artists and audience members being friends. This ostensible friendship then becomes the basis for offstage collaboration. The context and concrete details of this collaboration remain unmentioned; thus, the play's impact will differ, depending on its framing of when, where, and for what kind of audience it is performed. Less intimate than dancing and more conventional in its approach, August Schulenburg's performance poem "The Reasons" (2015) lists ostensible reasons for (not) becoming active in preventing further climate change. The anaphoric sequence of lines beginning with "Because" concludes with allowing the audience to join in. As the poem progresses from unreasonable excuses for inaction to possible motivations for constructive action, audience members may be inspired to question their own passivity and contemplate personal change.

The fact that "social drama" underpins individuals' reasons for considering themselves climate change activists or not becomes clear in such participatory works as much as in the plays that represent climate change through easily recognizable character constellations. Uniting arguments and varied emotional responses, Anita Majumdar's "We (don't) Deserve Nice Things" (2017) powerfully connects the "social drama" of a wedding with a call to consequential awareness of the impact of human actions on climate-related environmental conditions. Letting audience members see the actors perform a symbolic act after the end of the play proper forcefully conveys the process of forging argument and emotion into action. By addressing wasteful conventions associated with wedding celebrations, Majumdar's drama pits an environmentally conscious bride against one of her tradition-oriented bridesmaids: among other things, the bride decides to use crumpled-up balls of newspaper instead of flowers at her wedding. While the bridesmaid vehemently rejects such an unconventional procedure, the "Flower Girl" arranges the paper balls into a pleasant design, which the agitated bridesmaid ruthlessly destroys. At the end, *the entire production team should work together to smooth out each scrunched paper ball on stage and place it into the appropriate paper recycling bin. This should take place as the audience leaves so it is seen as a public act.* This "public act" demonstrates that climate change theater transcends its own performative context and extends into the world at large. That the audience could carry out simple acts of conservation and recycling is thus made clear non-verbally, yet unmistakably.

4 Conclusion

Although the climate change communication manuals discussed in this essay do not mention theater as a vehicle for conveying scientific knowledge about climate change, they certainly address a broad range of communicative practices that—from the perspective of a literary scholar—belong to the repertoire of drama and performance. While the large corpus of CCTA plays discussed here contains a share of texts that drift towards unconvincingly simplistic transformations of climate change deniers into activists or that are rather too emotional or didactic, numerous works in this collection forcefully demonstrate how climate change drama can merge theater semiotics and scientific knowledge when arguing in favor of individual action to prevent further destruction of the planet. Covering themes ranging from the extinction of both human and non-human species and the ecological impact of the meat and fishing industries and of burning fossil fuels to water scarcity, floods, and the discrepancy between a class- and race-based sense of entitlement and despair at ongoing oppression, playwrights effectively link postcolonial predicaments and transnational interrelations in highly concentrated scenes whose inherent "social drama" balances science-based knowledge with emotional responses on the part of the dramatic characters. The predominance of dialogs (often between actual or potential loved ones within the play) and of monologs (often with an implied addressee outside the play) among the CCTA pieces confirms the focus of performed climate change communication on "social drama." Rational thought and fearful emotions are combined in either variant, often in a situation which is or could be a turning point in the characters' relationship or in the respective individual character's attitude towards climate change. Frequently, these performances use humor, even sarcasm, which appears to be meant to trigger insights or changed behavior, but which runs the risk of eluding serious engagement with the topic or of alienating audience members by offending them.

Further research needs to be done:

1. on the evolving international corpus of climate change plays and actual performances;
2. on how—beyond theaters and other performance venues—theater practitioners concerned with climate change communication have made their ideas known online. For instance, HowlRound.com, an open access website hosted by Emerson College, self-defines as a "knowledge commons by and for the theatre community"; the website features a series entitled "Theatre in the Age of Climate Change," which consists of essays by playwrights, actors, directors, dramaturgs, scenic designers, composers, librettists, and scholars.[17] The argumentative focus of numerous texts, quite a few of which were written by playwrights who contributed to the CCTA corpus, imparts the authors' goal of bringing their artistic and scholarly work to as large a public as possible, and to communicate across perceived boundaries of nations, languages, and disciplines. Ultimately, these contributions—mostly essays and blog entries—need to be discussed in conjunction with the semiotics of climate change dramas as well as their production, performance, and reception contexts.
3. on how collaboration between theater artists and scientists has been developing and on how such collaboration has impacted practices of explaining and conveying climate change both on and off the stage.

It stands to reason that theater will become increasingly prominent in the realm of climate change communication, particularly if more and more artists will strike a balance between rational argument and emotional impact, and between communicating scientific insights and social drama both on and off the stage. As is true for shared concerns of postcolonial and transnational American Studies theories, climate change communication must confront axes of time and place that belie all attempts at particularizing humanity into ostensibly separable and distinct groups. Climate change as a global problem requires of those who try to understand it a comparable willingness to wrap their brains around solutions which demand local actions and worldwide coordinated actions that will necessarily impact one another.

---

[17]http://howlround.com/journal-series-theatre-in-the-age-of-climate-change.

Works Cited

Abbasi DR (2006) Americans and climate change: closing the gap between science and action. Yale School of Forestry & Environmental Studies. http://environment.yale.edu/climate-communication-OFF/files/americans_and_climate_change.pdf. Last Accessed 14 Nov 2017

Anzaldúa G (1987) Borderlands/La Frontera: the New Mestiza. Aunt Lute Books, San Francisco

Balestrini N (2017a) Cli-Fi drama and performance. In: Leikam S, Leyda J (eds) Amerikastudien/American Studies, vol 62, no 1, pp 114–120

Balestrini N (2017b) Transnational space and anthropocentric time: theorizing Chantal Bilodeau's plays as climate change drama. In: Schmidt K, Aghoro N (eds) J Contemp Drama Engl Spec Issue Theatre Mobil 5(1):70–85

Balestrini N (2018) Climate change drama across time and space. Chantal Bilodeau's (2016) *Forward.* In: Löschnigg M, Braunecker M (eds) Green matters: reading literature as cultural ecology. Brill, Leiden

Bhabha H (1994) The location of culture. Routledge, London

Bilodeau C (2013a) A climate of collaboration: environmental scientists urge artists to humanize the stories behind the research. Am Theatre, 62–65

Bilodeau C (2013b) 'Arctic in context' climate change theater action. World Policy Institute. http://www.worldpolicy.org/blog/2016/02/10/climate-change-theater-action. Last Accessed 14 Nov 2017

Bilodeau C (2016) As the climate change threat grows, so does a theatrical response. Am Theatre, 30 March. http://www.americantheatre.org/2016/03/30/as-the-climate-change-threat-grows-so-does-a-theatrical-response/. Last Accessed 27 Aug 2016

Chaudhuri U (ed) (1994) Theatre, special issue on "ecology and theatre, vol 25, no 1

Chaudhuri U (2012) The silence of the polar bears: performing (climate) change in the theater of species. In: Arons W, May T (eds) Readings in performance and ecology. Palgrave Macmillan, New York, pp 45–57

Chaudhuri U, Enelow S (eds) (2014) Research theatre, climate change, and the ecocide project: a casebook. Palgrave Macmillan, New York, p viii

Cilano C, DeLoughrey E (2007) Against authenticity: global knowledges and postcolonial ecocriticism. ISLE Interdiscip Stud Lit Environ 14(1):71–86

Ereaut G, Segnit N (2006) Warm words: how are we telling the climate story and can we tell it better? Institute for Public Policy Research, London, UK. https://www.ippr.org/files/images/media/files/publication/2011/05/warm_words_1529.pdf. Last Accessed 27 Aug 2016

Fish S (1980) Is there a text in this class? The authority of interpretive communities. Harvard University Press, Cambridge

Fishkin S (2011) Redefinitions of citizenship and revisions of cosmopolitanism—transnational perspectives: a response and a proposal. J Transnatl Am Stud 3(1):1–11. https://escholarship.org/content/qt1qw5364p/qt1qw5364p.pdf. Last Accessed 14 Nov 2017

Fluck W (2007) Inside and outside: what kind of knowledge do we need? a response to the presidential address. Am Q 59(1):23–32

Heise U (2008) Ecocriticism and the transnational turn in American Studies. Am Lit Hist 20(1/2):381–404. https://academic.oup.com/alh/article-lookup/doi/10.1093/alh/ajm055. Last Accessed 11 Oct 2016

Hornung A (2011) 'Planetary citizenship' section of "symposium: redefinitions of citizenship and revisions of cosmopolitanism—transnational perspectives: a response and a proposal. Journal of Transnatl Am Stud 3(1):39–46. https://escholarship.org/uc/item/8n55g7q6. Last Accessed 14 Nov 2017

Huggan G, Tiffin H (eds) (2010) Postcolonial ecocriticism: literature, animals, environment. Routledge, Abingdon

Johns-Putra A (2016) Climate change in literature and literary studies: from Cli-Fi, climate change theater and ecopoetry to ecocriticism and climate change criticism. WIREs Clim Change. https://doi.org/10.1002/wcc.385. Last Accessed 3 Apr 2016

Leikam S, Leyda J (eds) (2017) 'What's in a name?': Cli-Fi and American Studies (extended forum). Amerikastudien/Am Stud 62(1):109–138

Lenz G (2011) 'Introduction' section of "symposium: redefinitions of citizenship and revisions of cosmopolitanism—transnational perspectives: a response and a proposal". J Transnatl Am Stud 3(1):4–17. https://escholarship.org/uc/item/8n55g7q6. Last Accessed 14 Nov 2017

Mayer S, Weik von Mossner A (eds) (2014) The anticipation of catastrophe: environmental risk in North American literature and culture. Winter, Heidelberg

Moser S, Dilling L (eds) (2007) Creating a climate for change: communicating climate change and facilitating social change. Cambridge University Press, Cambridge

Norgaard K (2011) Living in denial: climate change, emotions, and everyday life. MIT Press, Cambridge

Radway J (2002/1998) 'What's in a name?' Repr. In: Pease D, Wiegman R (eds) The futures of American Studies. Duke University Press, Durham, pp 45–75

Shome D, Marx S (2009) The psychology of climate change communication: a guide for scientists, journalists, educators, political aides, and the interested public. Center for Research on Environmental Decisions, Columbia University. http://guide.cred.columbia.edu/pdfs/CREDguide_full-res.pdf. Last Accessed 14 Nov 2017

Shu Y, Pease D (eds) (2015) American Studies as transnational practice: turning toward the transpacific. Dartmouth College Press, Hanover, New Hampshire

Smith P, Howe N (2015) Climate change as social drama: global warming in the public sphere. Cambridge University Press, Cambridge

Tait P (2015) Love, fear, and climate change: emotions in drama and performance. PMLA 130(5):1501–1505

Climate Change Communication: A Friendly for Users App

Constantina Skanavis, Aristea Kounani, Athanasios Koukoulis,
Georgios Maripas-Polymeris, Konstantinos Tsamopoulos
and Stavros Valkanas

Abstract Living in the era of technology and information, mobile devices, such as mobile phones, laptops, personal digital assistants (PDAs), tablet PCs, are becoming gradually popular and connected with people's daily lives. The conjunction of the intensification of online technologies and rising public awareness of the changing climate provides numerous opportunities and challenges for climate-change communication. This research concerns the establishment of an environmentally oriented application for mobile phones, focused on climate change, with the intention of raising the knowledge and altering the attitude and behavior towards this crucial environmental issue, based on internet support. The significance of the environmental problem at stake, classifies it as one that mandates immediate awareness. This paper contributes to Climate Change Communication by explaining the use of technological tools, as applications, that are youth friendly and fast, increasing quite effectively environmental awareness.

C. Skanavis · A. Kounani (✉) · A. Koukoulis · G. Maripas-Polymeris
K. Tsamopoulos · S. Valkanas
Department of Environment, University of the Aegean, University Hill,
Mytilene 81100, Greece
e-mail: akounani@yahoo.gr

C. Skanavis
e-mail: cskanav@aegean.gr

A. Koukoulis
e-mail: env12026@env.aegean.gr

G. Maripas-Polymeris
e-mail: env12045@env.aegean.gr

K. Tsamopoulos
e-mail: mar12074@marine.aegean.gr

S. Valkanas
e-mail: ct13009@ct.aegean.gr

© Springer Nature Switzerland AG 2019
W. Leal Filho et al. (eds.), *Addressing the Challenges in Communicating Climate Change Across Various Audiences*, Climate Change Management,
https://doi.org/10.1007/978-3-319-98294-6_17

1 Introduction

Climate change, as it is widely recognized, is a vital issue with implications registered not only globally, but also on national and local scales. It is a concern that seems too abstract to numerous people (Cohen et al. 1998). This fact poses a particularly challenging case of environmental communication because its foremost cause, greenhouse gas emissions, is invisible. Though there is unanimity within the scientists' world, towards the issue of the anthropogenic climate change, its impacts and the imperative need for mitigation and adaptation actions, remains in a low priority in public's insight. This is due to the multiple and conflicting messages about climate change, which are interpreted in several ways, as well (Schroth et al. 2014). The literature provides numerous recommendations for more efficacious communication of climate change, such as further research into communication technologies and the ethical use of visualizations; a focus on the communication of specific mitigation and adaptation measures going beyond general climate awareness, since awareness, information, and understanding are not enough to alter individuals' habits of mind and practice (Schroth et al. 2014; Moser 2010).

At the same time, in the current era, the importance of information is enhancing gradually, fact that has led the institutions to look for new methods in order to access the information and communicate it, as well. Some experts define the devices, especially internet, which plays a significant role in the procedure of transferring information, so as to implement different approaches in education. Nevertheless, these technologies were not initially produced for educational grounds or transferring information generally or communicating issues like climate change (Goksu and Atici 2013). In addition, during the last decade, it is observed a rapid increase at the number of mobile lines globally. In fact, there were more than 7.0 million users worldwide with a personal mobile line by end of 2015. The way people interact among them and the way they communicate with each other have evolved completely, incorporating the mobile gadgets and the mobile technologies as part of them (Briz-Ponce et al. 2017). The key to success lies in finding the appropriate points for integrating technology into a new pedagogical practice (constructivism) (Al Hamdani 2013).

This research is about the creation of an environmentally mobile application, focused on climate change, with the intention of raising the knowledge and altering the attitude and behavior towards this crucial environmental issue, based on support from internet. Also, it is exploring whether or not mobile learning could contribute to the communication of climate change effectively and whether a mobile application could increase knowledge and raise ultimately the environmental awareness of individuals. Total number of mobile phone users is rapidly increasing, reaching globally the number of 4.77 billion users in 2017 (STATISTA 2017), placing an enormous pressure on climate change communication to adapt to the modern era's requirements. This paper, based on an extended literature review, is presenting a technological tool that can contribute to a better understanding of climate change, using means that are more friendly, easy, pleasant and accessible to younger individuals. Additionally, this paper is presenting an innovative communicational tool, created by Skyros Team in support of environmental campaigns focusing on climate change.

1.1 Mobile Learning (m-Learning)

The deployment of communication and information technologies, their wide-range usage and their positive influences, have inspired, several innovative and strategic approaches. While e-learning is being included in the learning procedure, becoming extensive as a constituent part of traditional education, it has also initiated positive alterations in lieu of pedagogical, economic and technological perspectives. The necessity to access immediately and everywhere the given information has augmented the impacts of mobile learning and mobile technologies, and it has brought newfangled tactics to education processes (Goksu and Atici 2013).

Mobile technologies are defined as the electronic devices that are small enough to fit in a jacket pocket, including mobile phones, portable digital assistants (PDAs) and iPod while their functionality varies from moderately simple usage of SMS to the more advanced use for tutoring (Kaliisa and Picard 2017).

Today, mobile devices have become increasingly popular and connected with people's daily life style. Each new version of these devices brings innovative features that make them more convenient and affordable, while new applications frequently become accessible supporting individual's life convenience (Mannheimer-Zydney and Warner 2016). The recent results on the spread and subscription to mobile phone services, account a tremendous advance and penetration of mobile devices in both developed and developing countries' daily routines (Kaliisa and Picard 2017; Johnson et al. 2007).

Mobile learning (m-learning) is defined as the use of mobile technologies for educational purposes. These devices can provide learning potentialities that are: informal, portable, omnipresent, prevalent, unprompted, contiguous, and personal (Kukulska-Hulme et al. 2011). Therefore, learners are no longer the passive recipients of education, but are consumers that make choices in the learning market, take principal accountability and control of their learning process, as well as setting targets and estimating results (Rodriguez-Arancon et al. 2013; Pilling-Cormick and Garrison 2007). Consequently, as mobile devices are used in learning environments, they should be given much more significance. Predominantly, using mobile telephones more than the computers and their approachability to the common webpages such as Twitter, Instagram, Youtube, and Facebook specify that they have the impending to be used in learning environments. In modern times, the effectiveness of constructivism learning method, the transition from computer based learning to web based learning and the advance in technologies have made mobile learning as one of the most widely held learning styles (Goksu and Atici 2013).

The field of mobile learning comprises the innovative teaching and learning techniques and the wide-ranging variation of mobile applications. Till today, numerous scientists and practitioners have explored the delivery, methodology and practicability of mobile device usage in education context, technical support, the building of information technology infrastructure and other resources. The progress of information technologies has led to countless innovations in modern teaching and learning procedures (Bidin and Ziden 2013).

1.2 Climate Change Education (CCE)

Climate change is one of the most severe tasks humanity is being confronting with today (McCright et al. 2013; Marcinkowski 2009), an enormous peril to human development and a compromising agent that standstills the exertions to diminish extreme penury (Fernandez et al. 2014). Climate change presents a worldwide challenge the enormousness of which people have not come across formerly (Bangay and Blum 2010). Education is a critical component in the response to climate change that should not be disregarded (Bangay and Blum 2010) since it has a significant role to play in understanding, mitigating, and adapting to changing climate (UNESCO 2009). Climate change education (CCE) can help in planning and implementing adaptations with respect to current and future effects of climate change (Pruneau et al. 2013). Climate change education is about helping learners comprehend and address the impacts of global warming today, while simultaneously encouraging the alteration in attitudes and conducts required to put the world on a more sustainable path in the future (UNESCO 2009). Formal and non-formal educational approaches should be used together to supply rich, in-depth learning (Skanavis 2004). In order to address the intricate climate change problem, CCE curricula should be knowledgeable by dialogue between the academia and those who are anticipated to be influenced mostly by the impacts of climate change, in a transdisciplinary approach (Fernandez et al. 2014; McGregor 2010).

1.3 Environmental Education (EE) and m-Learning

Today, numerous researchers have conducted researches in order to investigate the potential of m-learning in Environmental Education. Over time, many environmental educators have deliberated the application of computer-based tools in EE as abstruse. The usage of the computer is conventionally perceived as contradictory to straight interaction with nature, particularly since it has so far kept the participant away from it (Shultis 2001). From this perspective, computer-mediated education may conduce to the alienation from nature, one of today's essential challenges for EE. Still, in search for novel methodologies to confront this challenge, environmental educators have become aware of the high interest in new technologies that can be observed specifically among youth. Consequently, there has been extra encouragement to add computer-based media to EE methodologies, based on virtual environments that cannot substitute traditional field trips in any way (Ruchter et al. 2010).

Mobile learning is presented to increase user's learning performance (Chang et al. 2011; Chen et al. 2009; Liu et al. 2009) and their ingenuity (Cavusa and Uzunboylu 2009). Research has shown that the mobile device as a mobile guide can help the user to increase his environmental knowledge and the motivation to engage in learning activities (Chang et al. 2011; Akkerman et al. 2009; Ruchter et al. 2010; Uzunboylu et al. 2009). Mobile devices hold a variety of features that can help to pair the benefits of computer—mediated learning with direct nature experience. Studies show that handhelds can accompany the learner into the field and can scaffold exploratory activities, for instance by enabling students to sense and record aspects of the local environment (Rogers et al. 2005), while at the same time they can assist the educator in guiding the participants and monitoring their progress (Ruchter et al. 2010; Abe et al. 2005).

Omnipresent computing and mobile technologies offer much scope for designing advanced learning experiences that can take place in a variety of outdoor and indoor settings (Rogers et al. 2005). Mobile learning (m-learning) is one achievable solution to the challenges of outdoor instruction because it has the handheld feature of portability, social interactivity, context sensitivity, connectivity and individuality (Chang et al. 2011).

1.4   Climate Change Communication (CCC)

Climate change communication intents to bring knowledge obtained through research into the public arena such that the outcomes are observable and relevant to daily life. Conversely, knowledge generated through scientific survey is invariably embedded in local social and political contexts. The crucial strains, within the social and political contexts that envelope scientific comprehension, are habitually crumpled into a restricted understanding of how the public obtains information (Grothmann et al. 2017).

Most of the current practices in communication of climate change adaptation have not been empirically assessed regarding their efficacy in creating awareness for climate change risks and stimulating adaptation action (Grothmann et al. 2017; Wirth et al. 2014). A communication framework that is entrenched in a local place can reveal the complex relations between environment, management, experience, politics and scientific knowledge in an integrated whole, with a fundamental aim of encouraging knowledge transfer, action and sustainable futures (Shrivastava and Kennelly 2013). Efficacious climate change communication is not just about informing the community about climate science, or enabling a public dialogue around its impending effects, but giving public the appropriate knowledge, implements and reflective spaces for developing a vision of how to influence and plan their own alteration in a place-based perspective (Dulic et al. 2016; Dieleman 2012).

Communication plays a vital part in erecting concepts of climate change and its association to the people within a society. The reproduction of pictures of climate change that reflect dominant social, political and cultural hierarchies and that insert a particular view of the world into much of daily life, represents a crucial feature of media coverage nowadays. Consequently, this fact decreases arguments towards the issue of climate change to restricted and possibly alienating speeches that diminish broader people's ownership of the issue, potentially discouraging citizen engagement in mitigation energies. Additionally, the occurrence of a narrow scientific vision of climate change, aimed at an hypothetically coherent and devious public, apportions a sense of issue-ownership to the few: those equipped with the necessary linguistic and literacy skills, along with a necessary sensation of empowerment. Appeals to an unreal audience that is considered to be both principally severed and opposed to change, earnestly restricts chances for communal decarburization. Subsequently, in order to avoid this, it is imperative that climate change communication tools recognize the public as a socially integrated, diverse and changeable public, which has wide implications for climate change mitigation efforts. Additionally, it is necessary the used means to provide a greatly nuanced comprehension of the structure and nature of public, while undoubtedly supply exciting prospects for engaging citizens in climate change debates and extenuation efforts in an approach that reflects their lifestyles, their daily routine and in their priorities what is most essential for them. This fact means that CCC should be adjusted in public's everyday practices and the cultural, material and social circumstances in which they are embedded, using a participatory approach that integrates key principles of deliberative democracy and could simplify new forms and levels of public engagement that integrate sincere dialogue regarding the risks and liabilities of climate change mitigation (Fox and Rau 2017).

As it can be obviously be observed from the aforementioned, the aim of climate change communication is an indispensable first thoughtfulness. There can be a multiplicity of scopes behind communication exertions, partially appointed by the preposition of the communicators, partially regulated by what is traditionally established. In the sequel, the first purpose is principally to enlighten and tutor individuals about climate change; the second basic purpose of communication efforts is to accomplish some type and level of social engagement and action; the third category of communication efforts targets even deeper by endeavoring to encourage not just political action or context-specific conduct modification, but to bring about alterations in social norms and cultural values that act more broadly (Moser 2010).

1.5 Climate Change Communication Using Mobile Applications

Taking into consideration the communicative paradigm of the so-called "digital natives" (Prensky 2001) and the hours youth spend using their mobile devices, it would be commonsense to pay more attention to these tools. Applications can help users to absorb new information more easily (Ouariachi et al. 2017). This lack of everyday interactions and exploration makes it difficult for people to develop an emotional *attachment for* or *interest in* ecology, thus limiting their enthusiasm for practicing environmental protection. Moreover, numerous studies of human interaction with real ecological environments have found that "emotion" is an imperative learning factor, but is commonly ignored. The significance of emotion in ecological environment education, and the problems that can arise when such emotion is absent, was pointed out by Reis and Roth in 2009 (Huang et al. 2016).

The commitment of the new for altering the world generation is vital and critical for evading the devastating impacts of global challenges such as climate change, but researches show that awareness of the issue is still restricted and a remarkable deficiency of commitment to espousing actions for mitigating and adapting to climate change can be perceived (Bofferding and Kloser 2015). Scholars acknowledge the restrictions of traditional education programs in supplying information and conventional media, while at the same time they call for innovative strategies to involve youth in climate change (Ouariachi et al. 2017).

Climate change is an issue that is probably studied and deliberated more than any other environmental issue in many social sectors. So far, education systems, through policy and practice, continue to contribute straight to the problem by offering their uncritical support of economic systems that are creating the climate crisis (Greenwood and Hougham 2015). In hunt of innovative approaches to increase climate change awareness among digital natives and applications are gaining currencies as new platforms for communication, education and social change (Ouariachi et al. 2017).

1.6   Raising Environmental Awareness Using Mobile Applications as Tools

Since, the most significant goal of environmental conservation is the impeding of destruction by premature recognition of potential perils; it is of great importance the comprehension of the complexities of the interrelationships among the natural environment and human activity, which is said to be an essential state for the preservation and upgrading of environmental quality. This is a target that can be achieved through the environmental education, one of the principal goals of which is to develop the responsible environmental behavior of human beings (Chang et al. 2011).

The role of education in understanding, preserving, and finding the key to the solutions of environmental issues has been commonly documented since the decade of '70s (Shobeiri et al. 2006). Though, attitudes and emotions are vital factors in influencing human conduct, and the relationship among attitudes toward the environment and human behavior has been frequently investigated, researchers found that outdoor experiences can influence individuals' attitudes as well (Uzunboylu et al. 2009).

The most significant point is that, to date; only a few researchers have explored the application of mobile learning for augmenting environmental awareness (Uzunboylu et al. 2009).

In 2009, Uzunboylu et al. made an effort to integrate mobile technologies, data services, and multimedia messaging systems in order to increase students' usage of mobile technologies and advance environmental awareness. The findings of this research revealed that the environmental awareness of participants has meaningfully increased, and attitudes concerning preserving natural environment and averting pollution enhanced (Chang et al. 2011).

Smart phones and tablet computers have recently emerged as mainstream devices for use in mobile learning, since the convenience and immediacy of mobile learning provides more learners with additional learning opportunities (Huang et al. 2016). Furthermore, mobile learning tolerates learning to be carried out in authentic outdoor learning environments, providing learners with a broader variety of opportunities to acquire knowledge. Besides authentic learning environments do more to evoke affective feeling than classroom learning environments (Huang et al. 2016; Gulikers et al. 2005).

Mobile devices provide a means to engage individuals to develop positive attitudes toward maintaining environments. Using mobile telephones in informal education is exciting and interesting for users, fact that is interpreted that mobile technologies should be used to increase environmental awareness of students. The broader use of ML and sophisticated technologies requires schools to establish core and elective courses that support environmental topics and other subjects. Students in many disciplines could be provided opportunities to engage in on-campus and outdoor activities that include ML and mobile technologies (Huang et al. 2016).

2 The Case of "Climapp"

"Climapp" is a mobile application that was created by some members of the research team of the Research Centre of the Environmental Education and Communication, Department of the Environment, University of the Aegean. It is an environmental communication approach, suggested by the "Skyros Team", a research team based at Skyros Island in Greece. The aim is to communicate the issue of climate change more easily and efficiently to the public and specifically to youth, through a trendy communicating tool. Skyros Project, a multi-awarded educational and communicational program, is based on the successful cooperation between two governmental organizations, the University of the Aegean and the Skyros Port Fund. The specific program aims to best implement a comprehensive environmental campaign focusing at disseminating information and raising environmental awareness. It includes environmental campaigns targeted at citizens, tourists, boat passengers and children. Various outdoor activities are organized in order to attract public's environmental interest on issues like climate change.

"Skyros Team", having an extensive background on innovative communication techniques tailored to the needs of the specific recipient groups, attempted to reach the young ones through applications that are part of their daily routine. There was a strong need to create a communicational application for climate change which was going to involve the Greek Youth in the universal efforts to halt climate change and related consequences.

The software used to create the application was Android Studio. It is able to function on all devices with an API (Application Programming Interface) level of 15 and higher, and there is no cross-platform support, meaning that the application can be installed and function only on Android devices. The application is available for every tablet or smartphone and is open source, as well.

The content of Climapp's menu is based on the appropriate information that must be communicated to public, according to retrieved literature, in order to be aware of the issue and raise users' awareness on it. In Fig. 1 is presented the topics that there are available in the application's menu, such as "Greenhouse effect", "Causes", "Impacts", "History of Climate Change", "Timeline", "Climate Change In Greece", "Solutions" and "Glossary". In other words, "Climapp" addresses the significant climate related themes. But the primary focus is on the causes of climate change (Fig. 2), naming specifically all the anthropogenic activities that contribute to the present scenery, starting from the industrial revolution era. "Climapp" has focused specifically on the impacts of climate change, which are presented in Fig. 3. "Climate change in Greece" is a topic that directs potential users to a local comprehension of how this vital issue affects them even at home.

"Solutions" is another major area of interest.

The various introduced terms can be found in the "Glossary" section. The specific application has chosen to explicitly define certain scientific or other concepts related to climate change in order to facilitate the user's understanding. Such concepts included greenhouse effect, renewable energy etc.

"Climapp" have incorporated the recommendations given by scholars on how to communicate climate change: (1) make it of local interest, (2) make it visual, and (3) make it connected. Moreover, it provides a local discourse and shows a general preference towards portraying a citizen with local scenarios, in which players make decisions at home or within their communities, such as how to reduce carbon emissions.

Focusing on solutions and encouraging actions, have been suggested as strategies, for raising awareness and enhancing engagement (Ouariachi et al. 2017). On the other hand, in order to motivate the users and alter their attitude and behavior into environmental friendly ones," Climapp" communicate them some messages, such as "Don't stay home and lock yourself in a closet. Get out in the world!", "You will make the difference", "Any futures are possible; do you dare to change the world? Take decisions today that will lead to a better tomorrow" "On the other hand, there is still room for improvement in various areas."

Fig. 1 Main menu of "Climapp"

Since the Industrial Revolution there has been an enormous raise in human activities, like the deforestation, the expand of agriculture fields, the burning of fossil fuels, that have affected the release and uptake of carbon dioxide (CO2). Carbon dioxide is being added to the atmosphere more rapidly than it can be removed by other parts of the carbon cycle.

The main activities that increase the amount of greenhouse gases in the atmosphere are the following:

• Power plant that utilize fossil fuels.

Fig. 2 Climate change causes as they appear in 'Climapp"

Current Impacts of climate change In past decades, climate change has had implications on natural and human systems on the whole planet, irrespective of its cause, indicating the sensitivity of natural and human systems to changing climate. Proof of the observed climate change effects is stoutest and most comprehensive for natural systems. In several regions, altering precipitation or melting snow and ice are changing hydrological systems, influencing water resources in terms of quantity and quality. Many marine, terrestrial, freshwater species have shifted their geographic ranges, seasonal activities, migration patterns, abundances and species interactions in response to continuing climate change. Some impacts on human systems have also been attributed to climate change, with a major or minor contribution of climate change distinguishable from other influence.

1. Climate Change effect on Forest wild fires

2. Impacts of climate change on Agriculture

Fig. 3 Climate change impacts as they appear in "Climapp"

Limitations of the study

The major limitation of this study is the lack of actual implementation of application in real time conditions.

3 Conclusions

In the current era, the demand for inserting the principles of lifelong learning in education and in wider development policies give the impression to be more imperative than ever before (Rodriguez et al. 2013). Disseminating information in fast and efficient ways is critical, especially in issues of global environmental concern. Integrating new technologies into environmental teaching gives educators the advantage of incorporating in their courses interactive multimedia material. A suitable standpoint toward such tools is mitigation and adaptation, while promising adaptations of these tools will allow learners to better comprehend the relation between people, natural environment and climate change (Greenwood and Hougham 2015).

Since schools have a good infrastructure of information technology, such as computer-classrooms, networks and teaching websites, they could use technological tools such as "Climapp" and integrate teaching settings to achieve creative ways of instructing students, thereby improving their learning abilities. Furthermore schools could build their curriculums based on communicational and educational technologies like "Climapp" in order to insert crucial environmental issues such as climate change and design the learning procedure and tasks as tools to assist them in instructing the courses. The goal is to promote the usage and learning behavior of e-learning towards climate change by incorporating e-learning into the students' courses. Many studies have provided teachers and students with learning system or environment to help them acquiring more learning experience and knowledge. There is great demand for developing m-learning applications and services for information provision, dissemination of material and communication supporting different user profiles and mobile platforms and devises (Economou et al. 2012). Mobile application "Climapp" is an effective assisting tool for both students and teachers, since it can help students to acquire more knowledge from extracurricular learning resources and foster diversified learning skills and can assist teachers in interacting with students frequently.

The main advantage of these applications is that they allow students to personalize the learning process according to their own needs and abilities. Learning process becomes learner-centered. An e-learning application could serve this goal by allowing learners to personalize their learning experience while it offers them the chance to discover and build the knowledge by themselves. Furthermore, knowledge is socially and individually constructed on the basis of experience. E-learning applications and especially m-learning applications facilitate students to networking and communication (Economou et al. 2012).

The usage of mobile applications as a means to communicate climate change in a fast, pleasant and efficient way, raising this way knowledge, altering attitudes and behaviors, is imperative. "Climapp" is a promising tool to communicate efficiently all the appropriate information that a young individual needs in order to comprehend better the complicated issue of climate change. Downloading information regarding Greece and how climate change is influencing community's life can facilitate promotion of active environmental citizenship. Empowering people at an early age and inspiring collective as well as individual actions is of critical importance in the environmental awareness campaigning. Change is more likely to happen at a local level when community perspectives are embedded in the residents' personal environmental agenda. Appropriate solutions and actions are more effective when based on experiences already familiar in a given community (Dulic et al. 2016).

Future research prospects should focus on the implementation of "Climapp" into young crowds, assessing the impact and effectiveness of this app on users learning and engagement. Also, it should be useful to conduct a qualitative survey, using a well-structured questionnaire in order to assess the knowledge and attitude towards the issue of climate change, before and after the usage of the specific application, in order to evaluate its efficacy. "Climapp" could be assessed in high school environments in order to assess its validity, as well.

References

Abe M, Yoshimura T, Yasukawa N, Koba K, Moriya K, Sakai T (2005) Development and evaluation of a support system for forest education. J For Res 10:43–50
Akkerman S, Admiraal W, Huizenga J (2009) Storification in history education: a mobile game in and about medieval Amsterdam. Comput Educ 52(2):449–459
Al Hamdani DS (2013) Mobile learning: a good practice. Procedia Soc Behav Sci 103:665–674
Bangay C, Blum N (2010) Education responses to climate change and quality: two parts of the same agenda? Int J Educ Dev 30(4):359–368
Bidin S, Ziden A (2013) Adoption and application of mobile learning in the education industry. Procedia Soc Behav Sci 90:720–729
Bofferding L, Kloser M (2015) Middle and high school students' conceptions of climate change mitigation and adaptation strategies. Environ Educ Res 21(2):275–294
Briz-Ponce L, Pereira A, Carvalho L, Juanes-Mendez JA, García-Penalvo JF (2017) Learning with mobile technologies—students' behavior. Comput Hum Behav 72:612–620
Cavusa N, Uzunboylu H (2009) Improving critical thinking skills in mobile learning. Procedia Soc Behav Sci 1:434–438
Chang CS, Chen TS, Hsu WH (2011) The study on integrating WebQuest with mobile learning for environmental Education. Comput Educ 57:1228–1239
Chen TS, Chang CS, Lin JS, Yu HL (2009) Context-aware writing in ubiquitous learning environments. Res Pract Technol Enhanc Learn 4(1):61–82
Cohen S, Demeritt D, Robinson J, Rothman D (1998) Climate change and sustainable development: towards dialogue. Glob Environ Change 8:341–371
Dieleman H (2012) Transdisciplinary artful doing in spaces of experimentation. Transdiscipl J Eng Sci 3:44–57
Dulic A, Angel J, Sheppard S (2016) Designing futures: inquiry in climate change communication. Futures 81:54–67

Economou D, Keable-Crouch A, Bouki V, Basukoski A, Getov V (2012) WMIN-MOBILE: a mobile learning platform for information and service provision. In: Venkatasubramanian N, Getov V, Steglich S (eds) Mobile wireless middleware, operating systems, and applications. MOBILWARE 2011. Lecture notes of the institute for computer sciences, social informatics and telecommunications engineering, vol 93. Springer, Berlin, pp 23–33

Fernandez G, Thi My Thi T, Shaw R (2014) Climate change education: recent trends and future prospects. In: Shaw R, Oikawa Y (eds) Education for sustainable development and disaster risk reduction. Disaster risk reduction (methods, approaches and practices). Springer, Tokyo, pp. 53–74

Fox E, Rau H (2017) Disengaging citizens? Climate change communication and public receptivity. Ir Polit Stud 32(2):224–246

Goksu I, Atici B (2013) Need for mobile learning: technologies and opportunities. Procedia Soc Behav Sci 103:685–694

Greenwood DA, Hougham RJ (2015) Mitigation and adaptation: critical perspectives toward digital technologies in place-conscious environmental education. Policy Futur Educ 13(1):97–116

Grothmann T, Leitner M, Glas N, Prutsch A (2017) A five-steps methodology to design communication formats that can contribute to behavior change: the example of communication for health-protective behavior among elderly during heat waves, SAGE Open, 1–15 (January–March 2017)

Gulikers JT, Bastiaens TJ, Martens RL (2005) The surplus value of an authentic learning environment. Comput Hum Behav 21(3):509–521

Huang TC, Chen CC, Chou YW (2016) Animating eco-education: to see, feel, and discover in an augmented reality-based experiential learning environment. Comput Educ 96:72–82

Johnson RB, Onwuegbuzie AJ, Turner LA (2007) Toward a definition of mixed methods research. J Mixed Methods Res 1(2):112–133

Kaliisa R, Picard M (2017) A systematic review on mobile learning in higher education: the African perspective, TOJET Turk Online J Educ Technol 16(1):1–19

Kukulska-Hulme A, Pettit J, Bradley L, Carvalho A, Herrington A, Kennedy D, Walker A (2011) Mature students using mobile devices in life and learning. Int J Mob Blended Learn 31(1):18–52

Liu TY, Tan TH, Chu YL (2009) Outdoor natural science learning with an RFID supported immersive ubiquitous learning environment. J Educ Technol Soc 12(4):161–175

Mannheimer-Zydney J, Warner Z (2016) Mobile apps for science learning: review of research. Comput Educ 94:1–17

Marcinkowski T (2009) Contemporary challenges and opportunities in environmental education: where are we headed and what deserves our attention? J Environ Educ 41(1):34–54

McCright A, O'Shea B, Sweeder R, Urquhart G, Zeleke A (2013) Promoting interdisciplinarity through climate change education. Nat Clim Change 3(8):713–716

McGregor S (2010) Education and climate change (book review). J Clean Prod 18(7):696–697

Moser SC (2010) Communicating climate change: history, challenges, process and future directions. Wiley Interdiscip Rev Clim Change 1(1):31–53

Ouariachi T, Olvera-Lobo MD, Gutierrez-Perez J (2017) Gaming climate change: assessing online climate change games targeting youth produced in Spanish. Procedia Soc Behav Sci 237:1053–1060

Pilling-Cormick J, Garrison DR (2007) Self-directed and self-regulated learning: conceptual links. Can J Univ Contin Educ 33(2):13–33

Prensky M (2001) Digital natives, digital immigrants. Horizon 9(5):1–6

Pruneau D, Kerry J, Blain S, Evichnevetski E, Deguire P, Barbier P, Freiman V, Therrien J, Langis J, Lang M (2013) Competencies demonstrated by municipal employees during adaptation to climate change: a pilot study. J Environ Educ 44(4):217–231

Reis G, Roth WM (2009) A feeling for the environment: emotion talk in/for the pedagogy of public environmental education. J Environ Educ 41(2):71–87

Rodriguez-Arancon P, Arus J, Calle C (2013) The use of current mobile learning applications in EFL. Procedia Soc Behav Sci 103:1189–1196

Rogers Y, Price S, Randell C, Fraser DS, Weal M, Fitzpatrick G (2005) Ubi-learning integrates indoor and outdoor experiences. Commun ACM 48(1):55–59

Ruchter M, Klar B, Geiger W (2010) Comparing the effects of mobile computers and traditional approaches in environmental education. Comput Educ 54:1054–1067

Schroth O, Angel J, Sheppard S, Dulic A (2014) Visual climate change communication: from iconography to locally framed 3D visualization. Environ Commun 8(4):413–432

Shobeiri SM, Omidvar B, Prahallada NN (2006) Influence of gender and type of school on environmental attitude of teachers in Iran and India. Int J Environ Sci Technol 3(4):351–357

Shrivastava P, Kennelly JJ (2013) Sustainability and place-based enterprise. Organ Environ 26 (1):83–101

Shultis J (2001) Consuming nature: the uneasy relationship between technology outdoor, recreation and protected areas. George Wright Forum 18(1):56–66

Skanavis C (2004) Environment and community, 1st edn. Kalidoskopio, Athens, (in Greek), p 246

STATISTA (The Statistics Portal) (2017) Number of mobile phone users worldwide from 2013 to 2019, https://www.statista.com/statistics/274774/forecast-of-mobile-phone-users-worldwide/. Last Accessed 30 Sept 2017

United Nations Educational, Scientific and Cultural Organization (UNESCO) (2009) Learning to mitigate and adapt to climate change: UNESCO and climate change education.http://climatefrontlines.org/ed_seminar_brochure.pdf. Last Accessed 1 Sept 2017

Uzunboylu H, Cavus N, Ercag E (2009) Using mobile learning to increase environmental awareness. Comput Educ 52:381–389

Wirth V, Prutsch A, Grothmann T (2014) Communicating climate change adaptation—state of the art and lessons learned from ten OECD countries. GAIA 23(1):30–39

Constantina Skanavis is a Professor in Environmental Communication and Education at the Department of Environment, University of the Aegean (Mytilene, Greece). She is also the Head of the Research Centre of Environmental Education and Communication. She joined the University of the Aegean 15 years ago. Before that she was a Professor at California State University, Los Angeles. She has developed several courses on issues of environmental health and education. She currently teaches environmental education, environmental communication and environmental interpretation courses in undergraduate and postgraduate levels. Professor Skanavis has numerous publications on a international basis and has given presentations all over the world.

Aristea Kounani is a Ph.D. Candidate at the Department of Environment of the University of the Aegean and a senior researcher of the Research Centre of Environmental Education and Communication. She has a Master Degree on "Agriculture and Environment", and a Bachelor Degree on the Agricultural Technology. Aristea Kounani is a researcher with interests in the area of climate change, sustainable development, environmental education and communication. She is currently conducting researches on the issue of refugees (environmental-climate refugees), Education for Sustainable Development and Climate Change Communication.

Athanasios Koukoulis is an undergraduate student of the Department of Environment, University of the Aegean, and a researcher at the Environmental Education and Communication. He has great passion with programming and technology. His research interests are in the area of renewable energies, the causes and effects of climate change, and the use of new technology tools for environmental education and communication.

Georgios Maripas-Polymeris is an undergraduate student of the Department of Environment and a researcher at the Research Centre of Environmental Education and Communication. His research interests beyond Environmental Awareness include climate change and environmental management.

Konstantinos Tsamopoulos is an undergraduate student in the Department of Marine Sciences of the University of Aegean and a researcher in the Centre of Environmental Education and Communication. His involvement with fine arts and technology lead him to research their application into environmental education and awareness. He is a person with broadened research interests, fact that can be seen on the various researches that he is involved.

Stavros Valkanas is an undergraduate student of the Department of Cultural Technology and Communication of the University of the Aegean. His academic interests include video editing, audiovisual production, and web and non-application development.

Linaria Port: An Interactive Tool for Climate Change Awareness in Greece

Constantina Skanavis, Kyriakos Antonopoulos, Valentina Plaka,
Stefania-Pagonitsa Pollaki, Evangelia Tsagaki-Rekleitou,
Georgia Koresi and Charikleia Oursouzidou

Abstract In Greece, the port of Skyros Island, Linaria, is a small multi-awarded public port. United Nations has characterized it as the "blue port with a shade of green", because Skyros Port Fund has adopted an environmentally sustainable agenda and has invested in innovations that make this port unique for its tourism and environmental high-end consideration. What has truly escalated the Linaria Port's global reputation is the highly spoken cooperation of the Port Authority of Skyros with the University of the Aegean's Department of Environment. The name of the above mentioned, academic collaboration is "SKYROS Project" and it has been in effect since 2015, mostly focusing on environmental campaigns that stimulate climate change awareness for locals and visitors. The SKYROS Project is generating data that are collected through the Tourist Observatory and Maritime Observatory that have been established at the Port. Furthermore a Guests' Book records the comments visitors make. As a result a holistic picture of the environmental and tourism consequences and/or good practices is gathered on yearly basis.

C. Skanavis · V. Plaka (✉) · S.-P. Pollaki · E. Tsagaki-Rekleitou
G. Koresi · C. Oursouzidou
University of the Aegean, Mytilene, Lesvos, Greece
e-mail: plaka@env.aegean.gr

C. Skanavis
e-mail: cskanavi@aegean.gr

S.-P. Pollaki
e-mail: env10080@env.aegean.gr

E. Tsagaki-Rekleitou
e-mail: mar13087@marine.aegean.gr

G. Koresi
e-mail: env13037@env.aegean.gr

C. Oursouzidou
e-mail: charisoursous@gmail.com

K. Antonopoulos
Skyros Port Fund, Linaria Port, 37007 Skyros Island, Greece
e-mail: litasky@otenet.gr

© Springer Nature Switzerland AG 2019   281
W. Leal Filho et al. (eds.), *Addressing the Challenges in Communicating Climate Change Across Various Audiences*, Climate Change Management,
https://doi.org/10.1007/978-3-319-98294-6_18

In an effort to interpret what should the next step be in the environmental awareness arena, the idea of an environmental camp for children came up. This study outlines how the SKYROS project has launched a national campaign teaching others how to implement climate change awareness through children's camps in various geographic locations.

1   Introduction

Climate change science can trace its origins back to the early 19th Century although interest really took off in the 1980s, when public interest and research activity proliferated as the potential negative effects of global warming became clear (Frost et al. 2017). It has become a major focus of attention because of its potential hazards and impacts on the environment, particularly in vulnerable systems like coasts (Sánchez-Arcilla et al. 2011). Climate change communication though is no longer the largely uncharted topic of 10 years, ago (Moser 2016). Communication has become an integral part of the climate change discourse as scientists, governments and civil society organizations have recognized the crucial role of an effective communication in raising awareness about the consequences of climate change on life (Harris 2014).

In Greece, Linaria Port of Skyros Island (Fig. 1) is famous for its innovative approaches of its Port Authority. A world wide environmental campaign, under the brand name "SKYROS Project", has launched there, presenting an innovative cooperation with the University of the Aegean and the Skyros Port Fund. Most environmental campaigns related to climate change are based on the principle that people need more information to behave pro-environmentally (Skanavis and Kounani 2017). However, this approach in terms of "information-deficit" has been widely criticized as being inadequate to promote behavioral change (Ockwell et al. Ockwell and WhitmarshL 2009; Skanavis and Kounani 2017). To tackle such concern, the SKYROS Project, taking off in 2015, established Linaria port as interactive lab which promotes environmental issues awareness through hands on experience.

2   Ports and Climate Change

Climate change reporting in particular has been at the center of a number of controversies related to the scientific independence (Williamson 2016; Nerlich 2010) and a robust. Transparent process for advice provision is therefore required that can mitigate against accusations of a lack of integrity or of bias in climate reporting (Frost et al. 2017).

In coastal areas, vulnerability assessments have focused mainly on the sea level rise (SLR) and its impact on coastal communities (Nicholls et al. 2011), addressing

Fig. 1 Skyros Island, Greece. *Source* Google maps

related effects on beaches (Sánchez-Arcilla et al. 2011; Monioudi et al. 2016), coastal defense structures (Burcharth et al. 2014), coastal ecosystems (Kane et al. 2015), or the flooding of urban areas (Paudel et al. 2015). In this way, ports are highly vulnerable to climate risks in terms of both their facilities and operations (Becker et al. 2012). Given the critical role that ports play in the global economy and supply chains (Ng and Liu 2014), their inability to adapt to such risks poses a significant contemporary problem (Yang et al. 2017). Thus, it is important to find effective ways for ports to adapt to risks posed by climate change. Further, policymakers and other port stakeholders must understand the potential risks to ports in order to develop appropriate adaptation planning and strategies (Yang et al. 2017).

In view of the above ignorance, the concept of green port (or low-carbon port) was officially proposed in the United Nations Climate Change Conference in 2009 (Wu and Ji 2013). On the basis of organic combination of port development, utilization of resources and environmental protection, green port refers to the one characterized by healthy ecological environment, reasonable utilization of resources, low energy consumption and low pollution (Chen 2009).

Today, the ports operate as mainstream business infrastructures and as a result they experience the typical environmental issues (Naniopoulos et al. 2004), like climate change. Smaller ports have set as their mission to safe guard their harbor operations and "protect the maritime area against the adverse effects of human activities so as to protect human health and conserve marine ecosystems" (DEFRA 2002). These ports seek a new discourse to manage their environmental impacts and adopt proactive port sustainability approaches (Kuznetsov et al. 2015). However mere investigation of sustainability management issues in smaller ports would fill us in with new knowledge, efficiency and awareness in order to proceed with the best possible practices (Dinwoodie et al. 2012).

As regards to the Greek ports' sector, their operation has multiple environmental impacts on respective coastal areas, due to transportation and marine environment impacts (Antonopoulos et al. 2016). Therefore, the implementation of an integrated environmental policy in Greek ports would gradually incorporate sustainable development principles, into the entire spectrum of financial and operational activities, taking place in the ports (Naniopoulos et al. 2004).

Looking at the Aegean Sea, the island of Skyros is right in the middle. Its harbor, Linaria, is a small ordinary public port, used for various purposes, carrying out successfully the tasks, for which, it has been charged by the regulating Authority (Antonopoulos et al. 2016, 2017a). Furthermore, the Linaria port has been highly competitive for securing high sustainability enforcement standards and operates in ways that complement the environmental quality (Antonopoulos et al. 2016). Since 2010, this small port constantly develops innovative ideas and implements strict regulations that assure environmental quality and the arrivals in the port was recorded as a 957% increase (Antonopoulos et al. 2017a).

3   Methodology

This research project is based on a case study approach through a quality assessment process. Linaria Port serves as an interactive lab and places the ones educated in real time conditions. University students, locals and visitorsbecome the decision makers for environmental issues taking in their hands environmental planning and management of the specific port area. This state of art educational approach has secured the brand name, SKYROS Project.

The above Project is a paradigmatic cooperation of two public sectors, the Research Center of Environmental Communication and Education of the Department of Environment of University of the Aegean with the Skyros Port Fund Authority. The enthusiastic students through their daily environmental investment in the area led the way to a permanently established remote training site that invests into environmental research and education practices (Antonopoulos et al. 2017a). Worthwhile accomplishments include the establishment of the tourism observatory and marine one, both located at Linaria Port of Skyros Island. Also a free of charge kids' camp was established at the same port.

The aim of this study is to outline how the SKYROS Project has launched a national campaign luring others on how to implement climate change awareness based on establishment of a children's camp that can be offered in various geographic locations.

4 Linaria Port—An Innovative Way to Communicate and Educate

There is a great word-wide interest to the small island tourist ports. This is based on the fact that small scale ports are characterized by diversity, hosting fishing interests while creating potential employment opportunities, while encouraging leisure functions (Antonopoulos et al. 2016). The question that is being addressed is pertinent to the potential outcome when a small port follows an environmental agenda.

The port of Linaria is reasonably attracting high tourist interest. Antonopoulos et al. (2016) characterized Linaria as an environmentally sustainable small port community, where an environmentally responsible behavior is promoted to both visitors and residents. Since 2010, this small port constantly develops innovative ideas, such as the construction of seadromes, the use of electric scooters, the PV panels, a gas station and the cooperation of the Port Authority of Skyros with the University of the Aegean's Department of Environment, which resulted in attracting the interest and respect of travelers.

4.1 The Guest Book of Linaria Port

A prior research (Antonopoulos et al. 2017b), analyzed the visitors' feedback from the guest book provided at the touristic port of Linaria-Skyros. Attention was given particularly on the visitors' comments related on how the quality of environment acts as an attractive factor. The guest book is a way for visitors to freely express themselves and in this way their objective impressions are presented. The comments analyzed in the guest book of the port covered the period of July 2012 up to September 2014. The results of the analysis revealed that tourists have shown more interest on the personal touch of the ones serving the Port than the actual services provided. According to the findings, the environmental actions taking place in the port's facilities appeared to be the main reasons of tourists' positive impressions, satisfaction, their willing to extend their stay and to visit the island again in the future. The warm welcome, the immediate assistance and the friendly attitude of the staff were among the major factors that led to their satisfactory stay at the island (Antonopoulos et al. 2017b). Thereby, it is possible to analyze the behavior and determine future improvement actions and programs based on the voice of the visitor (Arabatzis and Grigoroudis 2010). This Greek small public port, with few resources has succeeded into attracting tremendous publicity and has been recorded as the fuller and friendliest public port of Greece (Antonopoulos et al 2015). According to the results, the average stay was three days. Extended stays up to 10 days, primarily because of the ambience and the environment of Skyros were also reported. The positive influence of service quality on purchase intentions is greater when satisfaction is also greater (Taylor and Baker 1994). Researchers emphasized that this research, based on visitors' free expression, would prove to be even more useful if further research gets initiated based though on answers on specific questions.

4.2   Observatory of Sustainable Tourism

The Observatory of Sustainable Tourism is a movement of World Tourism Organization (UNWTO) with GOST Program (Global Observatories for Sustainable Tourism), to promote sustainable tourist destinations (Karavitakis and Chondromatidou 2016). In Linaria Port, based on another tedious research of "SKYROS Project", a Tourism Observatory at Skyros Island was established. The generated data were analyzed and a picture on the tourism aspect of the island was presented. An exploitation of the gathered information about the tourist product of Skyros was a valuable tool for the creation of sustainable tourist policies. The pertinent researchers created a tool of analysis and collection of useful information with the name of "Observatory of Sustainable Tourism", watching therefore the enrichment and the renewal of data on a continuous basis.

According to Karavitakis and Chondromatidou (2016), the research was conducted by questionnaires addressed to tourist establishments and other business owners as well as accommodation owners at Skyros Island. The purpose was to collect data on social, economic and environmental issues. The collected data are processed in specific databases, which would continuously monitor the tourist product of Skyros and its impacts on the entire island. It should be noted that the results and impacts depend on the specific characteristics of the destinations and are not the same for all destinations (Spilanis 2009).

It should be noted that the hospitality of the locals, the natural beauty and the innovations of the Port Fund Authority constitute the strong points of the island (Karavitakis and Chondromatidou 2016). It is important that tourism is not exerting pressure on the environment due its activity. Also not well planned construction in building facilities (Achilles, Molos) seem to have influenced the island's change of use and landscape ecology. The Observatory's operation concerns the recording of the major tourist resources and activities of the destination (tourist image creation) and the quantitative data on tourist supply, demand and return per island. In general, the results portrayed a weak tourism development. This is associated with factors such as the pertinent infrastructure, the lack of organization and the reduced visibility of the natural and cultural beauty of the island (Karavitakis and Chondromatidou 2016).

4.3 Marine Observatory

The researchers of SKYROS Project proceeded into also studying the marine tourism characteristics at Skyros Island. The Marine Observatory is based on a similar concept like the Observatory of the Sustainable Tourism. This type of observatory focuses on the economic, social and environmental development of the port, presenting specifically the tourist product of the port. Based on such data, Skyros Port Fund authority and local operators can improve their services in a sustainable way. A friendly and highly competitive tourism services' system has been enacted tailored to the needs of leisure crafts sector. The management team of the Port of Linaria, on one hand supports rapid developmental processes and on the other hand safeguards the strict standards set for the local environment and the residents' related rights (Antonopoulos et al. 2016).

Linaria Port Authority adopted a green approach through innovative environmental education projects disseminating respect for the marine, coastal, and community environments while supporting the economic growth of the area. This project was launched the summer of 2016 with an initial sample of 129 tourist vessels that berthed in the port (Antonopoulos et al. 2017a). Many of the participants in the research have strongly expressed their enthusiasm for such actions taking place in the port area of Skyros Island (Antonopoulos et al. 2017a).

When a port respects the environment and promotes a sustainable lifestyle then tourism rates increase. Deliverables promote environmentally responsible behavior for both visitors and residents. This type of port can be used as a learning and research environment that can promote sustainable development in the tourism sector (Antonopoulos et al. 2017a). In conclusion, the researchers claimed Linaria Port as an exemplary case of a small public port. Linaria port has enjoyed increased arrivals of leisure crafts up to 975%, since 2010.

5 Environmental Kid's Camp in Linaria Port

The environmental camp for children was offered to the local community children as well as the ones visiting the island for first time in the summer of 2015 (Apostolopoulou et al. 2016). Outdoor environments can enhance mental health of participating students, contribute to students' intellectual and emotional development, support their environmental awareness and can give them opportunities to play and get involved in creative activities as well as connect directly with nature (Plaka and Skanavis 2016).

A research group of SKYROS Project, specifically handling educational approaches for the young, created a well-prepared educational program, based on North American Association for Environmental Educators' (NAAEE) Guidelines for Excellence (NAAEE 2017) and the basic principles of Environmental Education (UNESCO 1977).

Having a high quality educational program tailored to the participants' interests and being inspired by a familiar surrounding, like the port, is very important for practicing theory in real time conditions. The objectives of the summer environmental camp at Skyros Island were related to the dissemination of environmental education to children and to the promotion of their responsible environmental behavior through theory and hands on experience in an outdoors set up (Skanavis and Kounani 2017). An climate change educational program, using the port as the focal point, provides opportunities for understanding the importance of our actions towards the protection of the environment. Furthermore, it develops critical analysis skills, through the realization of the consequences due to careless use of our resources. The program was composed of three sessions, covering "Terrestrial Ecosystems", "Aquatic Ecosystems" and "Human and Environment Interaction". Children were confronted with various regional and global environmental issues, such as, climate change, greenhouse effect global warming, biodiversity concerns, forest fires, natural disasters, floods, droughts, renewable energy, non-renewable energy, endangered species, etc. (Skanavis and Kounani 2017). But, the focus of the camp has been on climate change awareness as a result of human actions due to negligence or intention, through a port.

Participating kids were confronted with the topic of climate change, the causes and the impacts of this major global issue, and the necessary actions that should be taken in order to be environmentally active (Skanavis and Kounani 2017). The educational program takes action in a port area. The participants age ranges from 6 to 13 years. The port served both as an environmental stimulation and an open laboratory for practicing skills related to climate change concerns. Replicating of this environmental camp in other geographic locations is desirable in order to spread the climate change meaning across the country. Of prime importance is the appropriate training of the educators charged with the responsibility to inspire environmental concern to the young ones (Figs. 2 and 3).

Summer programs provide an ideal opportunity for environmental education in an interactive context (Larson 2008). Nature-based outdoor education programs are effectively improving environmental awareness and sensitivity (Okur-Berberoglu et al. 2014; Apostolopoulou et al. 2016). Most camp programs are considered to be part of a positive youth developmental movement, aiming to offer experiences that are not only safe and enjoyable but also facilitate children's progress towards adulthood (Hederson et al. 2007). The summer camps could be described as a process of participation, where individuals are empowered to follow their interests, while exploring nature and question human interaction (Bergman 2014).

On the other hand, summer environmental education programs expose children to unfamiliar environmental setups and concepts in an exciting context, which may induce interesting attitude changes (Larson 2008). Therefore, nature camps might

Fig. 2 Reality in environmental camp

Fig. 3 Environmental camp event

be more effective in promoting children's emotional affinity to nature, ecological beliefs and environmental behaviors than in-class environmental education programs (Collado et al. 2013). Understanding and cultivating children's environmental consciousness may be crucial to rectifying the environmental degradation and to mitigating the impacts of climate change (White 2004; Garner et al. 2015). Participating in nature-based summer camps raises familiarity with nature, increase environmental knowledge, awareness and behavior on environmental issues, such as climate change (Dresner and Gill 1994; Garner et al. 2015).

6 Discussions

Climate change is a problem of global scope, with significant consequences for coastal communities as well. Green ports' assessment requires a complex model involving many indexes including various concerns (Chengpeng et al. 2017). These indexes should be comprehensive, quantitative or qualitative. They should also reflect the real-time changes, capable of adapting to the needs of social development, while maintaining a relative stability for the evaluation of green port in certain periods (Ma et al. 2014).

Climate change communication is no longer the largely uncharted topic of 10 years, ago (Moser 2016). Communication has become an integral part of the climate change discourse as scientists, governments and civil society organizations have recognized the crucial role of an effective communication in raising awareness about the consequences of climate change on life (Harris 2014). The focus of the environmental educational program has been on climate change awareness as a result of human actions due to negligence or intention. Because of the vulnerability to climate changes, education programs should prepare children for future risks (Ebi and Paulson 2007). Communication about climate change should aim to achieve meaningful engagement in all three facets: understanding, emotion, and behavior (Ockwell et al. 2009).

As Bergman noticed (2014) an environmental camp can count as a success educating, empowering and even radicalizing both its participants and others in the broader environmental movement. This can be seen as transforming people from passive bystanders to citizen activists, which some see as the key to effecting the systemic changes needed for instant emissions' reduction (Skanavis and Kounani 2017). A well-structured and organized summer environmental program, such as this one at Skyros Island, which is staffed with experienced environmental educators, could be a key in communicating climate change, specifically to children. Obviously, a positive effect is observed on children's knowledge, attitudes and participation behavior on the issue of climate change, after their environmental summer camp completion. As the results of pre and post camp interviews showed, all of the children were happy doing things to help the environment (Skanavis and Kounani 2017).

The relationship with nature is not just a pleasant task, but it is also an essential component of the human wellbeing general goal (Plaka and Skanavis 2016). It is of paramount importance to understand children's perspectives, since children both now and in the future will influence and be influenced by environmental issues in many ways (Skanavis and Manolas 2015).

7 Conclusions

Linaria Port, "the blue port with a shade of green", as it has been labeled by United Nations, communicates about climate change, from an innovative environmental campaign in Greece, the Skyros Project. Through this campaign and its actions, the port has been serving as an environmental interactive tool. The SKYROS Project attempts connecting people and kids around the world based on a common vision, the protection of the environment and climate change abatement. Based on the success of the three consecutive years' operation of the children's environmental camp, the SKYROS Project has launched a national campaign teaching others how to implement climate change awareness through children's camps in various geographic locations. The program educates the youth into becoming active environmentally concerned citizens while it gives the opportunity to college students to get trained as potential environmental educators.

Linaria port has taken action in the battle against climate change, by promoting environmentally responsible behavior. Collected data on tourism related pressure is of enormous value when preventing environmental consequences in an area is of interest. Filling in comments from visitors in a guest book has proven to produce helpful information when analyzing the recorded data. Therefore, the marine observatory and the tourist observatory and the individual comments from tourists create a portfolio based on which environmental protection can be practiced.

The creation of a children's summer environmental camp has been complementing the above mentioned accomplishments, taking the interpretation of the initial analysis of collected data to the next level. All environmentally innovative activities are guided and supervised by the administrative and academic force of the SKYROS Project, the environmental awareness campaign program, which was delivered at the island of Skyros in Greece in summer 2015 (Skanavis and Kounani 2017).

Climate change is a problem of global scope, with significant consequences for coastal communities (Lieske and Wade 2014). The impact of climate change on the marine environment was receiving little attention at this time, but in recent years it has started to "catch up" both in terms of research activity and public and policy interest (Frost et al. 2017). Developing ways of communicating complex messages and implementing science-policy interface mechanisms are not ends in themselves. Collating, interpreting and disseminating information on climate impacts on marine systems has as a long-term goal to wisely use scientific information in policy and decision-making in order to plan and manage communities accordingly.

References

Antonopoulos K, Skanavis C, Plaka V (2015) Exploiting further potential of Linaria Port-Skyros: from vision to realization. In: Proceedings of the 1st Hellenic conference on tourist port, Marinas, Greece, Athens, pp 101–111

Antonopoulos K, Plaka V, Skanavis C (2016) Linaria port, Skyros: An environmentally friendly port community forleisure crafts. In: Proceedings of the 13th international conference on protection and restoration of the environment, Greece, Mykonos Island, pp 1054–1062

Antonopoulos K, Margariti A, Marini K, Skanavis C (2017a) The case of the port of Linaria-Skyros, Greece: the human factor as the main dimension of impression and other findings in the GUESTBOOK tourists' evaluation. J Psychol Res (in submission process)

Antonopoulos K, Plaka V, Mparmpakonstanti A, DimitriadouD, Skanavis C (2017b) The blue port with a shade of green: the case study of Skyros Island. In: Health and environment conference, UAE, Dubai, pp 175–187

Apostolopoulou S, Grigoroglou G, Karamperis M, Skanavis C, Kounani A (2016) Environmental summer camp in a Greek Island. In: 13th international conference on protection and restoration of the environment, Mykonos Island, Greece, 3–8 July 2016, pp 1024–1030

Arabatzis G, Grigoroudis E (2010) Visitors' satisfaction, perceptions and gap analysis: the case of Dadia–Lefkimi–Souflion National Park. For Policy Econ 12:163–172

Becker A, Inoue S, Fischer M, Schwegler B (2012) Climate change impacts on international seaports: knowledge, perceptions, and planning efforts among port administrators. Clim Change 110(1):5–29

Bergman N (2014) Climate camp and public discourse of climate change in the UK. Carb Manag 5 (4):339–348. https://doi.org/10.1080/17583004.2014.995407

Burcharth HF, Andersen TL, Lara JL (2014) Upgrade of coastal defence structures against increased loadings caused by climate change: a first methodological approach. Coast Eng 87:112–121

Chen YQ (2009) The development of fifth generation port. China Collectiv Econ 7:113–114 (Chinese version)

Chengpeng W, Di Zh, Xinping Y, Zaili Y (2017) A novel model for the quantitative evaluation of green port development—a case study of major ports in China. Transp Res Part D (in Press)

Collado S, Staats H, Corraliza AJ (2013) Experiencing nature in children's summer camps: affective, cognitive and behavioural consequences. J Environ Psychol 33:37–44

DEFRA (2002) Safe guarding our seas

Dinwoodie J, Tuck S, Knowles H, Benhin J, Sansom M (2012) Sustainable development of maritime operations in ports. Bus Strategy Environ 21(2):111–126

Dresner M, Gill M (1994) Environmental education at summer nature camp. J Environ Educ 25 (3):35–41

Ebi LK, Paulson AJ (2007) Climate change and children. Pediatr Clin North Am 54:213–226. https://doi.org/10.1016/j.pcl.2007.01.004

Frost M, Baxter J, Buckley P, Dye S, Stoker B (2017) Reporting marine climate change impacts: lessons from the science-policy interface. Environ Sci Policy 78:114–120

Garner MA, Doster-Taft E, Stevens LC (2015) Do children increase their environmental consciousness during summer camp? a comparison of two programs. J Outdoor Recreat Educ Leadersh 7(1):20–34. https://doi.org/10.7768/1948-5123.1238

Harris U (2014) Communicating climate change in the Pacific using a bottom-up approach. Pac J Rev 20(2):76–94

Henderson KA, Scheuler-Whitaker L, Bialeschki MD, Scanlin MM, Thurber C (2007) Summer camp experiences: parental perceptions of youth development outcomes. J Fam Issues 28 (8):987–1007. https://doi.org/10.1177/0192513X07301428

Kane HH, Fletcher CH, Frazer LN, Barbee MM (2015) Critical elevation levels for flooding due to sea-level rise in Hawaii. Reg Environ Change 15:1679–1687

Karavitakis L, Chondromatidou AM (2016) Tourist observatory as a tool for sustainable tourism development. The case of Skyros Island. Undergraduate thesis, University of the Aegean

Kuznetsov A, Dinwoodie J, Gibbs D, Sansom M, Knowles H (2015) Towards a sustainability management system for smaller ports. Mar Policy 54:59–68

Larson RL (2008) Environmental education and ethnicity: the impact of a summer education program on the environmental attitudes and awareness of minority children. Postgraduate thesis, Clemson University

Lieske JL, Wade T (2014) Climate change awareness and strategies for communicate. Estuar Coast Shelf Sci 140:83–94

Ma D, Ding Y, Yin H, Huang ZH, Wang HL (2014) Outlook and status of ships and ports emission control in China. Environ Sustain Dev 39(6):40–44 (Chinese version)

Monioudi IN, Karditsa A, Chatzipavlis A, Alexandrakis G, Andreadis OP, Velegrakis AF, Poulos SE, Ghionis G, Petrakis S, Sifnioti D, Hasiotis T, Lipakis M, Kampanis N, KarambasT Marinos E (2016) Assessment of vulnerability of the eastern Cretan beaches (Greece) to sea level rise. Reg Environ Change 16:1952–1962

Moser SC (2016) What more is there to say? Reflections on climate change communication research and practice in the second decade of the 21st century. WIREs—climate change (in Press) https://doi.org/10.1002/wcc.403

Naniopoulos A, Palantzas G, Nalmpantis D, Koutitas C, Makris D (2004) Sustainable environmental management of Greek ports. The case of the port of Thessaloniki. In: 2nd International congress on transport research in Greece, Hellenic Institute of Transportation Engineers & Greek Institute of Transport, Athens, Greece

Nerlich B (2010) 'Climategate': paradoxical metaphors and political paralysis. Environ Values 19 (4):419–442

Ng AKY, Liu JJ (2014) Port-focal logistics and global supply chains. Palgrave Macmillan, Basingstoke

Nicholls RJ, Marinova N, Lowe JA, Brown S, Gusmão D, Hinkel J, Tol RSJ (2011) Sea-level rise and its possible impacts given a 'beyond 4 °C world' in the twenty-first century. Philos Trans R Soc A Math Phys Eng Sci 369:161–181

North American Association for Environmental Education (NAAEE) (2017) Professional development of environmental educators: guidelines for excellence, Washington

Ockwell D, Whitmarsh L, O'Neill S (2009) Reorienting climate change communication for effective mitigation: forcing people to be green or fostering grass-roots engagement? Sci Commun 30(3):305–327

Okur-Berberoglu E, Ozdilek HG, Yalcin-Ozdilek S, Eryaman MY (2014) The short term effectiveness of an outdoor environmental education on environmental awareness and sensitivity of in-service teachers. Int Electron J Environ Educ 5(1):1–20

Paudel Y, Botzen WJW, Aerts JCJH (2015) Influence of climate change and socioeconomic development on catastrophe insurance: a case study of flood risk scenarios in the Netherlands. Reg Environ Change 15:1717–1729

Plaka V, Skanavis C (2016) The feasibility of school gardens as an educational approach in Greece: a survey of Greek schools. Int J Innov Sustain Dev 10(2):141–159

Sánchez-Arcilla A, Mösso C, Sierra JP, Mestres M, Harzallah A, Senouci M, El Raey M (2011) Climate drivers of potential hazards in Mediterranean coasts. Reg Environ Change 11:617–636

Skanavis C, Kounani A (2017) Children communicating on climate change: the case of a summer camp at a Greek Island. Leal-Filho W, Manolas E, Azul AM, Azeiteiro U (eds) Handbook of climate change communication, Elsevier, Amsterdam, (in Press)

Skanavis C, Manolas E (2015) School gardens and ecovillages: innovative civic ecology educational approaches at schools and universities. In: Transformative approaches to sustainable development at universities. World sustainability series. Springer International Publishing, Switzerland https://doi.org/10.1007/978-3-319-08837-2_37

Spilanis I (2009) Tourism and performance measurement—measuring results

Taylor S, Baker T (1994) An assessment of the relationship between service quality and customer satisfaction in the formation of consumers' purchase intentions. J Retail 70(2):163–178

UNESCO (1977) Final report: intergovernmental conference on environmental education. UNEP, Tbilisi

White R (2004) Young children's relationship with nature: its importance to children's development and the earth's future. White Hutchinson Leisure & Learning Group. http://www.whitehutchinson.com/children/articles/childrennature.shtml. Accessed 15 Sept 2017

Williamson P (2016) Take the time and effort to correct misinformation. Nature 8(540):171

294   C. Skanavis et al.

Wu XD, Ji L (2013) Research on the impact of climate change on port of China and the countermeasures. China Water Transp 13(10):116–118 (Chinese version)
Yang Z, Ng AKY, Lee PTW, Wang T, Qu Z, Rodrigues VS, Pettit S, Harris I, Zhang D, Lau Y (2017) Risk and cost evaluation of port adaptation measures to climate change impacts. Transp Res Part D (in Press)

Dr. Constantina Skanavis is a Professor in Environmental Communication and Education at the Department of Environment, University of the Aegean (Mytilene, Greece). She is also the Head of the Research Centre of Environmental Education and Communication. She joined the University of the Aegean 15 years ago. Before that she was a Professor at California State University, Los Angeles. She has developed several courses on issues of environmental health and education. She currently teaches environmental education, environmental communication and environmental interpretation courses in undergraduate and postgraduate levels. Professor Skanavis has numerous publications on a international basis and has given presentations all over the world.

Kyriakos Antonopoulos is the President of the Board of Skyros Port Fund. He has created a multi awarded small marina at Linaria Port of Skyros Island. This Port has been identified as the Blue Port with a Shade of Green by United Nations. He is the author of several papers and has given a lot of presentations on the environmental excellence expected in the ports and how to reach such a state.

Valentina Plaka is an environmental scientist, who holds a Master's Degree in Environmental Policy and Biodiversity Conservation from the University of the Aegean. Today, she is a Ph.D. candidate in Environmental Communication and Education in the University of the Aegean. She was awarded with Award of Excellence as a postgraduate student from the University of the Aegean. Also, Valentina is author of a list of researches and presentation on international and national conferences and articles in magazines and book chapters. She participates in environmental articles in an educational journal and in creating of educational kits about environmental issues. She aims to educate and communicate environmental issues to the society towards a new lifestyle, according to sustainable development.

Stefania-Pagonitsa Pollaki was graduated of the Department of Environment of the University of the Aegean. Today, she is a researcher—member of the Research Center of Environmental Communication and Education of the Department of the Environment with director Prof. Constantina Skanavis. The Active participation in environmental education through music and musico-kinetic therapy is her main interest.

Evangelia Tsagaki-Rekleitou is an undergraduate student in the Department of Marine Sciences in the Environmental School, University of Aegean. As a member of the Research Center of Environmental Communication and Education (R.C.E.C.E.), she did her internship in Skyros Project. Her thesis is about coastal waste in Mytilene, Lesvos. She is interested in creating educational material for environmental programs for children. In collaboration with other members of R.C.E.C.E, she is writing a kids' book about environmental issues and ways to protect our planet.

Georgia Koresi is an undergraduate student in the Department of Environmental Sciences in the Environmental school, University of Aegean. As a member of the Research Center of Environmental Communication and Education (R.C.E.C.E.), she did her internship in Skyros Island. She is interested in environmental education for children.

Charikleia Oursouzidou is an Environmental Scientist, researcher of Research Center of Environmental Communication and Education of the Department of Environment of the University of the Aegean. She has completed her thesis, titled as "Assessment of teachers' knowledge, training, perceptions and participation in environmental education at a Greek island", which was presented and published at the International Conference "Protection and Restoration of The Environment", which took place on 03/07/2016 on the island of Mykonos. She has also graduated from the first Greek Academy of Environmental Instructors which took place in the Island of Skyros and is a member of the multi awarded team "Skyros 2017". Finally, she was one of the creative members of the video titled "Can We?" that "Skyros 2017" created, and was awarded with the Peoples' Choice Award in the global video competition Film4Climate, during the COP22 in Marrakesh in 2016.

Communicating Sustainability: Promoting a Self-assessment Tool for Eco-villages

Georgios Antonopoulos, Emmanouil Avgerinos,
Kyriakos Antonopoulos and Constantina Skanavis

Abstract The German Advisory Council on Global Change (WBGU) has reasoned that climate change, resource scarcity and radical social change have been increasingly transforming the economies, lifestyles and communities work. Sustainable development has emerged as a way of shaping a new future, where environmental, social and economic structures are perceived as interconnected. Advancement in all three areas is needed to achieve a sustainable future. The Eco-Village movement has been pioneering into promoting sustainable development, by ways of bringing people together to create intentional communities where sustainability in all three areas constitutes their main drive and objective. Eco-Villages have been rather uncharted territory for academic research. There are though successful examples that have pioneered in communicating environmental issues and have disseminated knowledge towards local municipalities, whilst achieving their goals on sustainability. The object of this paper is to create a tool for evaluating the sustainability discourse of Eco-Villages by means of identifying emerging concepts, common patterns and occurring sustainability themes amongst them, while taking into account their individual characteristics, similarities and differences.

G. Antonopoulos · E. Avgerinos (✉) · C. Skanavis
Department of Environment, University of the Aegean,
University Hill, 81100 Mytilene, Greece
e-mail: mavgerinos@env.aegean.gr

G. Antonopoulos
e-mail: envm616004@env.aegean.gr

C. Skanavis
e-mail: cskanav@aegean.gr

K. Antonopoulos
President of the Board of Skyros Port Fund, Skyros Port Fund,
34007 Skyros, Greece
e-mail: litasky@otenet.gr

© Springer Nature Switzerland AG 2019   297
W. Leal Filho et al. (eds.), *Addressing the Challenges in Communicating Climate Change Across Various Audiences*, Climate Change Management,
https://doi.org/10.1007/978-3-319-98294-6_19

1 Introduction

Global warming and associated climate change, due to anthropogenic Greenhouse Gas emissions had been deemed "unequivocal" by the Intergovernmental Panel on Climate Change (IPPC), back in 2007. With the exponential growth in human populations and economies, the natural environment comes under increasing stresses that translate to significant concerns in resources depletion, deterioration of life-supporting ecosystems, waste increase, and pollution (Toros 2011; Banuri and Opschoor 2007). The German Advisory Council on Global Change (WBGU) recently predicted that, in the face of climate change, resource scarcity, and radical social change, we would see a profound and far-reaching "Great Transformation" in our economies, our lifestyles, and our communities (Andreas and Wagner 2012). The increasing environmental impact of human activities has created a universal movement towards a more sustainable future—economically, ecologically and socially.

Since the 1992 Rio Earth Summit, the concept of sustainable development has been central to both imagining and realizing this new future (Andreas and Wagner 2012). Sustainable development has a plethora of definitions, with the most frequently quoted one having derived from a United Nations report (Brundtland 1987) stating: "*Sustainable development is development that meets the needs of the present without compromising the ability of future generations to meet their own needs*". It should take full account of the three dimensions of sustainability—meaning environmental, social and economic—while including cross cutting, intangible as well as long-term considerations. In a climate perspective, development pathways need to be found that will make economies emit less GHGs, while reducing vulnerability to climate change impacts and sustain the growth momentum (Banuri and Opschoor 2007).

Some contend (Seyfang and Smith 2007; Boyer 2015) such pathways may emerge from community-based activities, create an ideological space for experimentation with alternative systems of production and consumption. One example of such activities has been the model of eco-villages. While multiple definitions of eco-villages exist (Dawson 2006; Gilman 1991), mostly all agree that eco-villages are permanent human settlements, with an ideological commitment to modelling positive solutions to global environmental crises (Boyer 2015). This concept has been implemented by groups spread around the globe, frequently depending on limited resources and with low support from institutions or governments (Bissolotti et al. 2006).

Their main contribution has been through being an example of a sustainable living, by successfully implementing alternative systems and thus demonstrating their potential of being replicated at higher scales of the broader community (Clarissa de Oliveira et al. 2013). But how sustainable are eco-villages? Currently, there is no universal framework for assessing eco-villages sustainability in all dimensions, and the majority of research has been concerned with their ecological dimension (Bissolotti et al. 2006). In order to meet this deficiency and assist eco-villages achieving their vision towards a culture of sustainability and thus being more efficient in defusing their paradigm and sharing their vision, eco-villages, specific tools to assist them are in need (Christian 2012).

This paper aims to review the available literature to identify eco-villages' most successful practices that can act as a guide towards their sustainability assessment and assist them in promoting their vision for change towards sustainable solutions. The currently available sustainability self-assessment and guiding tools developed for eco-villages will be reviewed for identifying their strong qualities and inefficiencies and propose ways for the development of a more comprehensive framework for sustainability assessment.

2   Sustainability in Eco-villages

Eco-villages have long served as experimental community models with the potential to help move society towards sustainability (Clarissa de Oliveira et al. 2013), aiming to create a culture that can be incorporated in the paradigm of sustainable development (Gesota 2008). While eco-villages have evolved from the diverse counterculture movements of the 60s and the 70s, they depart from their predecessors in many respects. It is evident that eco-villages have created forms of living that can be demonstration and education centres for people, organisations and governments for a gradual transition towards sustainability (Andriopoulos et al. 2016; Gesota 2008). The three sustainability levels that appear in communities that have established a harmonious living are ecological, social/communitarian and cultural/spiritual sustainability (Bissolotti et al. 2006). Each community has its base in one or several of these principles, not always being able to reach that goal, but they are in a constant struggle to evolve, improve and create ways to reach their objectives (Bissolotti et al. 2006). The eco-village movement is very diverse, as each project varies differently in terms of size (from small groups of people to whole neighbourhoods), vision, character and design and is dependant, among others, on location, culture and climate (Sevier 2008). Like other planned communities, the process of an eco-village starts with a shared vision that predicates common goals. Further planning, implementation and fundraising are some of them, employed in order to either expand the original concept or operationalize it in a step-by-step fashion (Toros 2011).

Typically, eco-villages built ecologically sustainable housing, grow a proportion of their own organic food, recycles waste products harmlessly, and, try to generate their own off-grid power, as well as offer a variety of tours and classes on sustainable living (Christian 2012; Andriopoulos et al. 2016). And while they are capable of demonstrating successful alternative systems (Boyer 2015), they frequently lack a systematic approach to integrating structure, processes and actions into strategic planning in order to meet their deficiencies into becoming fully sustainable (Clarissa de Oliveira et al. 2013). Agreements on a shared vision and understanding of the planning process, the fairness of decision making mechanisms of a collaborative nature, as well as the economic decisions are often the problems faced by most eco-villages (Christian 2003). A review of the relative literature, based on secondary data of published case studies and scientific articles was conducted in order to identify the most common sustainability themes and strategies employed by eco-villages, amongst the three sustainability dimensions.

2.1 Environment

Environmental or ecological sustainability in eco-villages can be characterized by strategies aiming towards infrastructure with low environmental impact (Loezer 2011); efforts towards restoring natural ecosystems (Bissolotti et al. 2006); utilization of permaculture design strategies (Clarissa de Oliveira et al. 2013); the connectedness with an urban area (Loezer 2011). With these strategies in mind, various objectives are being pursed by eco-villages in order to be ecologically sustainable (Andreas and Wagner 2012; Bissolotti et al. 2006; Loezer 2011) which include developing a sense of place; producing and distributing organic food; recycling; reducing consumption; generating less garbage; protecting and conserving water sources; using biological systems for sewer treatment; using renewable energy systems (solar, wind, geothermal, wave) for covering potential energy needs; using permaculture; employing bio-construction techniques (ecological construction aiming at lower levels of environmental impact that also utilize local recourses); minimizing the transit cost.

2.2 Social/Culture

Sustainability goals under the social context in eco-villages typically envelope developing and implementing a shared vision; strengthening of community interconnectedness; successfully developing and implementing methods for conflict-resolution and local self-governance (Loezer 2011). The social aspect of community life can bring challenges: balancing between personal and community life, communicating and having methods to reach consensus on issues and difference in income (Kirby 2003). Eco-villages have developed and established various strategies in order to achieve these goals and resolve conflicts as stated in Global Ecovillage Network's (GEN) website: *"use of alternative governance models, employing methods and structures to enhance interpersonal processes, sharing common resources and providing mutual aid, emphasizing holistic and preventive health practices, providing meaningful work and sustenance to all members, integrating marginal groups, promoting unending education, encouraging unity through respect for differences, fostering cultural expression"*.

2.3 Economy

As experimental sites, eco-villages also tend to utilize different economic structures. More simplistic models can be found in smaller Eco-villages, with no internal use of money and a value exchange system, and larger ones which have complex bureaucratic structures with more complex money flows to help them navigate the national legal systems while retaining some degree of personal freedom and security. Liftin (2014) mentions that in an extensive study among eco-villages, none of them had a truly sustainable economy. Their approach towards money and loans is based on solidary principles, such as issuing interest-free loans, an exchange and gift system, and a small wage gap (measured according to social criteria) (Andreas and Wagner 2012). Commonly owning land and buildings has been popular in this type of communities and usually leads to environmental and economic benefits such as savings on site and residential space, energy, resources, and consumption of goods (Loezer 2011). Almost all of the communities are not capital intensive, but they seek out ways to becoming economically sustainable; large enterprises within them are scare and the collective income usually comes from tourism, residential courses, educational programs, various retail and service businesses and partnerships with local institutions and governments, which often results in access to additional resources and improvements (Boyer 2015; Kirby 2003; Loezer 2011; Andreas and Wagner 2012). Also, by providing affordable housing and rental units, most communities encourage diversity and attract new residents who stimulate the local economy; reduction of waste and energy efficiency, by means of recycling, use of renewable energy sources, low energy housing, permaculture design, water management, composting seems to be common amongst eco-villages in their efforts to reduce community expenditure (Loezer 2011).

3 Sustainability Assessment

Sala et al. (2015) has argued that "*sustainability assessment (SA) is one of the most complex types of appraisal methodologies; not only this does entail multidisciplinary aspects (environmental, economic and social), but also cultural and value-based elements*". Also sustainability is understood differently between different contexts as several perceptions of this concept exist, but they are seldomly formulated or assessed very explicitly (Banuri and Opschoor 2007). Eco-villages have differing approaches to sustainable development and just a few communities have a strategic approach towards sustainability (Kazhura et al. 2005). One of the current challenges is the need for more effective and clear communication (Sakellari and Skanavis 2013), as people's understanding of sustainability within communities widely varies, as well as finding ways to achieve it (Clarissa de Oliveira et al. 2013).

Community sustainability assessment is becoming increasingly widespread and gradually the approaches to measure community sustainability have become more sophisticated. However, the diversity of different approaches that are in constant development, increases the challenge of identifying a single method towards evaluating community sustainability (MacKendrick and Parkins 2004). Several studies are being currently conducted which aim to identifying sets of indicators that measure a community's performance based on community sustainability conditions (MacKendrick and Parkins 2004). While these models attempt to assess sustainability in its social and economic aspects along with the ecological ones, in general though, as pointed out by Jackson (1998), it is much easier to quantify and assess the ecological and economic dimension. Many methods can be applicable, like measuring the ecological footprint, emergy or exergy analysis, Life-Cycle Analysis (LCA), Sustainability Index (SI) or embodied energy (NRE index) that are often used to evaluate materials flows, but not services or information (Siracusa et al. 2007). Social sustainability assessment frameworks are usually based on qualitative methods, which usually entail the identification of indicators relative to human social sustainability themes/areas consistently identified from literature and case studies (Colantonio 2009). Recurring themes between eco-villages that characterise their sense of social sustainability include "*a sense of belonging and communion with life, the awareness of one's place in the whole system, sense of community and supportive association with other humans, strengthened family and social ties, bonding among different generations*" (Kirby 2003), comfortable, equitable, secure and interesting living (Andreas and Wagner 2012). However, there are many ways in which human needs are being met within the system which are subject to the different relationships that may exist (Clarissa de Oliveira et al. 2013), making the evaluation of sustainability of an eco-village a complex and multidimensional process due to the wide range of parameters and interconnected relations that must be taken into account.

The selection of sustainability assessment criteria for eco-villages, could be informed by already successful practises of eco-villages to date, as well as back-casting from a future vision of sustainability for these communities. Certain frameworks have been developed which aim to guide eco-villages into adopting strategic planning processes towards sustainability (Kazhura et al. 2005; Clarissa de Oliveira et al. 2013). Most eco-villages don't utilize thinking and organising tools for strategic, long-term planning (Christian 2003). They have a reactive approach to sustainability, driven by crisis managing of the most prominent issue than planning ahead proactively (Kuznetsov et al. 2015; Clarissa de Oliveira et al. 2013). Strategies and objectives derived from such studies, could be employed as criteria to assess sustainability levels. Communities should be involved actively in the generation of indicators of local relevance that give a local context on quality-of-life concerns (MacKendrick and Parkins 2004).

3.1 Available Tools

As mentioned, in order provide eco-villages with a more strategic approach towards sustainability several organizations and/or authors have developed sustainability guides, frameworks and assessment tools, that assist communities in better understanding and achieving their sustainability goals or monitor related progress. Best practices could be put together to form a template, which can provide a simple framework that can be replicated by adapting to the local conditions (Gesota 2008). The creation of a tool-box of codified guidelines, processes, check-lists, engagement tools and suggested measures etc., that a community could draw from based on an understanding of the local context (Kazhura et al. 2005) has been proposed. An analysis of the relative literature was conducted in order to identify and review such available tools. Their theoretical foundation and methodological approaches were studied, serving as guides into synthesizing a new approach to community sustainability assessment.

3.1.1 Self-audit for Eco-villages and Communities

In the book "Creating Harmony: Conflict Resolution in Community" published by GAIA trust, eco-villages are characterised as sustainable settlements when they successfully intergrade physical structure (ecology), infrastructure, social structure and culture. Based on these principles and with the goal to clarify the objectives of sustainable settlements, Jackson (1998) developed a simple model for the self-assessment of communities that aspire to that core vision of sustainability. Each aspect of an eco-village (social, ecological, economic and spiritual) is represented with four circles, each of which was defined through a series of four statements regarding sustainability. For each statement, the user (community) is asked to mark the level of perceived achievement and sustainability, by shading the relative quadrants; by measuring their score (sum of the shaded quadrants), the community can assess their level of sustainability and compare it to a commonly agreed level, representative of the sustainability goals. The author does not provide any reasoning or employment of specific framework that led to the choice of the statements representing sustainability in each dimension, rather than his own experiences in the eco-village movement (Jackson 1998). This approach is quite simplistic, and not able to capture the complexities of realised sustainability actions in eco-villages.

3.1.2 Three-Tier Sustainability Indicator Model

Understanding the complexities of a community, Innes and Booher devised a three-tier system of sustainability indicators, which is based on viewing communities as self-organizing complex adaptive systems. This system of indicators was developed with the intention of assisting a community to be a self-organized learning system, which can adapt and respond to change and opportunity and can effectively address problems or potential system break-down based on feedback that flows among the participants of the system (Gesota 2008). The System Performance Indicator, measures key indicators that reflect the central values of concern to a community and are indicative for the overall health system. The Policy and Program Indicators reflect the activities and outcomes of various elements of the system or subsystems and allow for assessing the adjustment of programs. The rapid feedback indicators, help the different actors and stakeholders in making the best choices for their own daily actions (Gesota 2008). This type of model could potentially better reflect the values and goals of individual communities, and with a proper identification of key indicators regarding eco villages sustainability in all three tiers, could be a useful tool towards the communicating, monitoring and achieving sustainability goals.

3.1.3 Community Sustainability Assessment Tool

A more elaborate and comprehensive tool, the Community Sustainability Assessment Tool developed by the Global Eco-village Network (GEN) provides *"measuring rods for eco-villages and communities to compare their current sustainability status with ideal goals for ecological, social, economical and spiritual sustainability"* (https://gen.ecovillage.org). This tool expands each of the four dimensions in Jackson's model into a question-answer-score format which eventually add to a final score giving an indication of weak to strong sustainability (Gesota 2008). As it was designed in for use by a wide variety of communities, *"it takes into account their individual characteristics on how they achieve their sustainability objectives and encourages participants to share other ways in which they achieve sustainability, for future revisions of the document"* (https://gen. ecovillage.org). There are currently two versions available, the more elaborate and extensive one mentioned above, and a shortened version in the form of a questionnaire which acts as a preliminary self-assessment guidance tool, which incorporates the most frequent sustainability measures eco-villages employ to achieve sustainability. It gives a sense for their current situation and identifies areas of potential, growth and improvement. A new online-tool for sustainability assessment, which aims to showcase eco-village contribution to carbon sequestration and the United Nations Sustainable Development Goals by mapping successful practises by employing a broad base of participants, is currently being developed by GEN (https://gen.ecovillage.org).

3.1.4 Eco-village Sustainability Self Evaluation Test

One of the main results of the project "Eco-villages for Sustainable Rural Development" financed by the EU's Baltic Sea Region Programme 2007–2013, which started in February 2010, was the development of an online self-audit tool, addressed to leaders of established or initiated eco-villages (http://www.balticecovillages.eu). The Eco-village sustainability self-evaluation test is a tool aiming to assist in: (a) forming a vision of the eco-village; (b) diagnosing the real eco-village situation; (c) assessing distance between vision and reality; (d) identifying and improving aspects of eco-village governance (Vidickiene and Melnikiene 2012). There are two versions of the test, comprised by 12 and 36 questions respectively, which measure an eco-village's real and desirable situation on six main dimensions by using a 5 point scale. These dimensions include common property; diversity of values and interests of residents; agreements; collaboration of residents; changes in eco-village; and external relations (Vidickiene and Melnikiene 2012), which seem to be indicators of socio-economic sustainability within eco-villages. The test plots the results of both vision and reality of the eco-village on a Spider-gaph (Vidickiene and Melnikiene 2012), which makes the gap between them visually comprehensive and help identifying potential areas for improvement.

3.1.5 Wheel of Sustainability

The Wheel of Sustainability (WOS) was developed *"with the intention to properly illustrate the elements of a culture of sustainability, as well as the dynamics between them, and thereby encourage reflection between the relationships amongst them"* (Andreas and Wagner 2012). It is a tool for guiding the organizing process of social change for the culture of sustainability, which includes reflection, negotiation, implementation and evaluation, as well as developing new methods and approaches that aim in solving specific problems whilst keeping in sight their individual vision of sustainability (Clarissa de Oliveira et al. 2013). The challenge posed for a culture of sustainability, as defined in the publication of Brundland Report, Our common future (Oxford University Press) in 1987, is *"the alignment of human needs and lifestyles with the system requirements of sustainable development"*. The Wheel of Sustainability is represented as three concentric circular levels: the innermost level represents the human needs and living conditions necessary for a quality of life; the outermost level illustrates the societal sustainable development requirements and objectives; lastly, the middle level corresponds to the design process towards a culture of sustainability (Clarissa de Oliveira et al. 2013; Andreas and Wagner 2012). In order for the three levels to properly describe the dynamics of a sustainability culture, their contents are considered flexible to each other, allowing creation of new combinations among them (Andreas and Wagner 2012). This research currently aims to analyse and identify the relevant elements and dynamics of such social systems to strategically plan for a sustainability culture (Andreas and Wagner 2012). Their research has not uncovered any study regarding the actual use of the WOS by eco-villages. The WOS is a conceptual model, that provides a vision for the organizational process toward a sustainable culture; it could also be utilised as a guide for estimating the level of a culture of sustainability being implemented in different communities and how each community creates an individual vision of success by using the WOS's structure.

3.1.6 Direction Indicator for Sustainable Communities (DISC)

An assessment of the WOS using a strategic framework known as Framework for
Strategic Sustainable Development (FSSD), revealed some of its deficiencies
(Clarissa de Oliveira et al. 2013) in successfully leading communities to sustainability; collaborating with leaders, practitioners and experts in the eco-village field,
through interviews and a final validation survey (Clarissa de Oliveira et al. 2013),
the Direct Indicator for Sustainable Communities (DISC) was developed. It is
represented by three concentric rings with community vision and values at the core,
surrounded by the nine fundamental needs defining societal sustainability and the
seven permaculture principles accounting for ecological sustainability, with the
outer ring representing the biosphere (Guillen-Royo 2015). It is accompanied by a
set of guiding questions that are organised in four categories to be applied at each
level of the wheel, based on the ABCD process: awareness, evaluation, co-creation
and realization by backcasting from a sustainability vision (http://www.
thenaturalstep.org). It is designed to showcase relationships between human
needs, their possible satisfiers and the system boundaries (societal and ecological)
while being generic enough to fit within, and enhance, current planning processes
(Clarissa de Oliveira et al. 2013). The goal of the DISC was developing "*a communication and strategic planning tool which embodies a whole-systems perspective and strategic thinking that can provide a shared mental model for eco-
villagesfor improved communication and orientation in community planning and
decision-making processes to assist in their progress towards sustainability*"
(Clarissa de Oliveira et al. 2013) (Fig. 1).

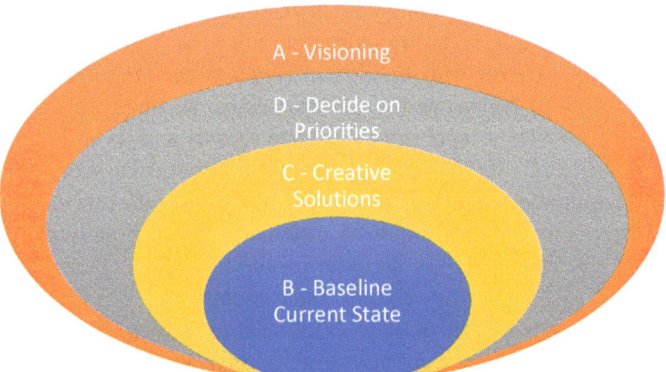

Fig. 1 The ABCD strategic planning process (http://www.thenaturalstep.org)

4 Methodology

4.1 Analysis of Current Tools

From the available self-assessment tools for eco-villages, we can derive that the most comprehensive one was GEN's Community Sustainability Assessment tool, which was modelled after the Self-Audit tool for eco-villages developed by Jackson (1998). Based on reviews and research done with many eco-villages across their network, it comprises of a long list of questions addressing various objectives that have been recognized to increase sustainability levels of eco-villages. All dimensions of sustainable development are addressed and a five-point scale is employed to as a way of to turn qualitative data into a scoring mechanism to easily assess sustainability levels. This is achieved by comparing them to defined limits that correspond to certain sustainability levels. The Eco-village Sustainability Self Evaluation Test covers significantly less sustainability objectives common among eco-villages and emphasizes on some attributes indicating sustainability practices. It also uses sets of questions and answers that correspond to a five-point scale, but with a significance difference: it has two sets of questions, one aimed to assess the sustainability goals of the eco-village and one to assess their current situation making it highly context specific. Visualizing the result using a Spider-graph makes it very easy to assess which areas an eco-village would need to focus in order to achieve their developmental goals and decide on priorities. This allows eco-villages to keep their overall vision in mind and be able to plan ahead in areas where sustainability is low, in a context specific manner. Its important to clarify that both systems are subjective, meaning that the evaluation of the degree of sustainability is subject to the user's experience and not based on objective sets of indicators that are able to measure socio-demographic, economic or ecological data.

Guidance tools that we reviewed, mainly the WOS and DISC, could be helpful complementary tools in sustainability assessment, as they help eco-villages to define clearly their vision of sustainability and guide them towards strategically planning for it. They suggest that a vision of sustainability can be achieved through the ethics of permaculture for sustainable design, which many eco-villages employ with different emphasis on its dimensions. MacKendrick's study for resource-based sustainable communities demonstrates the relevance of relating outcomes to community conditions and objectives; he identifies ecological integrity, economic vitality, civic vitality, physical and mental health, and recreational opportunities as outcomes of a sustainable community, which closely resemble those of permaculture design ethics, and sets of indicators for both processes and objectives. Clarissa de Oliveira's et al. (2013) study also recognizes the need for context-specific tools with an extensive library of strategies and relevant objectives that eco-villages could draw upon on, according to their needs. This pathway would enable them enhance awareness of their vision, assess their current state of achieving it and work on solutions to sustainability matters that arise.

4.2   Synthesis Approach for a New Tool

On the basis of a short reviewing of literature regarding community sustainability assessment, exploring the strategies and objectives of eco-villages towards sustainability and our overview of the relative assessment and guidance tools, we propose a synthesis approach to developing a new tool for self-assessment that employs the advantages of presented approaches and methods. Eco-village leaders have generally consented that DISC can help them provide a clearer vision for sustainability in their communities (Clarissa de Oliveira et al. 2013). This concept has been the basis for the model presented below. In order to assess if a community meets certain conditions that would realise their vision of sustainability, we employ Loezer's (2011) identified strategies, which then correlate to specific objectives communities usually apply (see Table 1).

Inspired by the way Vidickiene and Melnikiene (2012) tool can be context specific by determining the sustainability vision of each community, we assigned a weight value to the objectives which corresponds to a 1–5 scale. A weight function is a mathematical device used when performing a sum, integral, or average to give some elements more "weight" or excert influence on the result instead of having other elements in the same set doing it. By employing this tool, the assessment table aims to enable the user assess these objectives on the basis of their relative value in representing and addressing the important issues of their community. Then each strategies' relative sustainability score is derived as the weighted value average of sum of points corresponding to the objectives assessment. Also, communities are invited to add new objectives to the strategies, or new strategies altogether that best represent the practises of their community. The final results on individual strategies averaged a makeup value for the score in each "sustainability condition"; these results are then plotted in a spider-graph (see Fig. 2), which can then be used as an indication of sustainability levels.

5   Limitations

The evaluation of sustainability of an eco-village is not an easy process due to the multi dimensionality of parameters that must be taken into account. This paper presents an attempt at a preliminary synthesis approach in order to create a sustainability assessment tool, based on an overview of the most common practices and strategies applied by some eco-villages, with the employment of secondary data from published articles, scientific journals and websites. Many aspects of synthesizing a proper assessment framework, like further analysis of indicators or identifying the relative importance between possible trade-offs (for example, between increased economic activity and poorer environmental quality) have been overlooked in this paper for reasons of scope, length and time. Kusel (2001) has argued that when a set of indicators is applied at a community, it should use objective and

Table 1 Proposed tool

Environmental, social, economic and spiritual sustainability	Conditions for sustainability	Strategies	Vision assessment Weight value 0–5 scale	Various objectives (as specified)	Reality assessment Implementation level 0–5 scale
	Land and nature stewardship	Restore natural ecosystems			
		Provide infrastructure with net zero environmental impact			
	Technology	Use of renewable energy sources			
		Use of intelligent design			
	Education and culture	Strengthening community by connecting place and people			
		Increase social capacity			
	Land tenure and community governance	Provide for local self governance			
		Develop and implement a vision			
	Finance and economics	Increase efficiency and reduce waste			
		Attribute value to natural resources			
	Buildings	Provide infrastructure with net zero environmental impact			
		Civic vitality			
	Health and spiritual well being	Strengthening community by connecting place and people			
		Recreational opportunities			

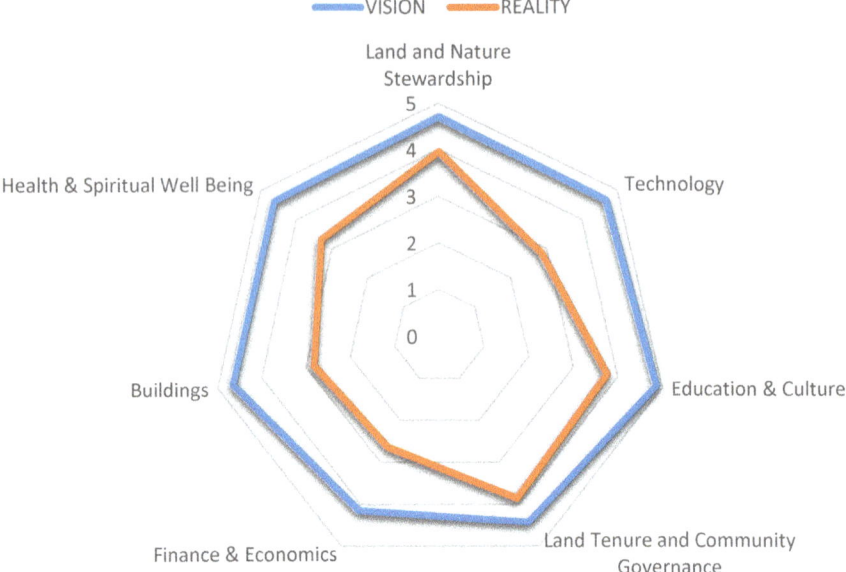

Fig. 2 Indicative spider-graph plot of relative vision and current situation sustainability assessment

subjective data together, such that the basic community conditions can be described and these conditions should be explained according to local social relationships and processes (MacKendrick and Parkins 2004). Further study for eco-village sustainability might concentrate on specific aspects of these communities in order to acquire more critical knowledge and generate sets of indicators that can accurately assess sustainability in eco-villages.

6   Conclusion

Many Eco-villages, through their vision of moving society to a culture of sustainability, have been acting as examples of demonstrating alternative systems and methods, with the potential of them transitioning and replicating at higher scales in society. With the intention of supporting their role as sustainability models, this paper reviewed some of their most common sustainability practises, as well as guidance and assessment tools aimed for eco-villages' support. Sustainability assessment is a complex and multi-dimensional issue, that also demands monitoring and resources. This review revealed the need for assessment tools that are have a whole-system approach while being context specific at the same time. The tools will serve as a simple way for eco-villages' assessment of their current sustainability status. We employed an approach, where we combined the strong qualities of currently existing models with data regarding sustainability objectives common in eco-villages. Based on the principle that certain conditions enable sustainability, we utilised the DISC's conceptual model and assigned to them strategies that are defined by certain sustainability objectives. To make it context specific, a weight value is assigned at the objectives that answers the question "how important is this objective for the community?"; the overall strategy then received a score that corresponds to the weighted value average. Two sets of data are then used to create a radar chart of the sustainability dimensions; the values expressing the 'vision' were plotted against the scores of the 'current situation' assessment. In that way it could be visually comprehensive and communicative which areas are not meeting their goals, and which goals should maybe re-evaluate in order to achieve sustainability. The list of available strategies and objectives could be expanded upon by future research to include more sustainability practises, which eco-villages could employ.

Appendix

I. List of sustainable measures practised by four North-American Eco-villages (Ithaca, Los Angeles, Village Homes, Enright Ridge)

See Fig. 3.

Sustainable Measures

Dimension	Strategy	Objectives
Social	Develop and implement a shared vision	decade of establishment
		visioning sessions
		design charrettes
		bought property
		received property as donation
		restricted to members
		faith based
	Provide for local self-governance	autonomy over local resources
		collaboration with/influence over local government
		community meetings for decision making
		use of indicators to measure trends or success
	Strengthen the community by connecting place with people	common areas and facilities
		physical environment encourages social interaction
		encourage outdoor and ecological recreation
		community events
		common meals
		public art
		online activity
		diversity of age
		diversity of income
		diversity of races
Environmental	Have an urban character and be connected to an urban area	integration with surroundings
		higher density
		discouragement of automobile use
		mixed-use
		connection to public transit
		car share program
		carpool program
		incentives to walking
		incentives to biking
	Restore natural ecosystems	ecosystem conservation and restoration programs
		permaculture
		support diversity of species
	Provide infrastructure with net zero environmental impact	water reuse systems
		water efficient appliances
		drought resistant landscaping
		local water treatment systems
		green roofs

Dimension	Strategy	Objectives
		raingardens
		bioswales
		stormwater pond
		solar energy
		wind energy
		geothermal
		solar water heating
		local energy production contributes for the grid
		energy efficient furnaces
		heat recovery systems
		natural ventilation
		high insulation
		agricultural land
Economic	Consume or provide local food	Community Supported Agriculture
		community gardens
		edible landscape
		reduce consumerism
	Attribute adequate value to natural resources	support policies for resource conservation
		measures to control pollution
		use of non-toxic materials
	Support the local economy	office/work space available
		residents work in the community
		local retail
		affordability
		availability of rental units
		educational programs
		partnership with universities
		partnership with local government
	Increase efficiency and reduce waste	waste reduction initiatives
		composting initiatives
		reuse of waste materials
		use of local construction materials
		recycling programs
		leasing instead of owning

Fig. 3 List of sustainable measures *Source* Loezer (2011)

G. Antonopoulos et al.

II. List of sustainable practises by three European Eco-villages (Sieben Linden, Findhorn, Vauban)

See Fig. 4.

	Sieben Linden	Findhorn	Vauban
Year Started	1997	1962	Mid-1990's
Situation	Rural	Rural	Urban
Income Sharing	Partially	No	No
Ideological focus	Toggles between Idealism and Pragmatism	Spiritual	None
Views itself as an education center	Yes	Yes	No
Organic farming	Yes	Yes	No
Renewable energy used	Yes (solar, wood)	Yes (wind, solar, wood)	Yes (wind, solar)
Grey Water recycling	Yes	Yes	In some units
Black Water recycling	Yes	Yes	In some units
Organic waste composting	Yes	Yes	Linked to the city of Freiburg
Rainwater harvesting	No	No	Yes, partially
De-emphasize cars	Yes	Partially	Yes
Sharing Resources	Yes	Partially	No
Food consciousness	Vegetarian and vegan	Vegetarian and vegan	Anything
Food co-operative	Yes	No	No
Emphasis on alternate medicine	Yes	Yes	No

Fig. 4 List of sustainable practises. *Source* Gesota (2008)

References

Andreas M, Wagner F (2012) Realizing Utopia: eco-village endeavors and academic approaches, RCC perspectives. Accessed 21 Aug 2017, no 8. https://doi.org/10.5282/rcc/5598
Andriopoulos C, Avgerinos E, Skanavis C. (2016) Is an eco-village type of living arrangement a promising pathway to responsible environmental behavior? In: Conference paper, innovation ARABIA health and environment, Dubai, UAE
Baltic Eco-villages website: http://www.balticecovillages.eu. Accessed 24 Aug 2017
Banuri T, Opschoor H (2007) Climate change and sustainable development. Economic & social affairs, DESA Working paper no 56. Accessed 24 Aug 2017. www.un.org/esa/desa/papers/2007/wp56_2007.pdf

Bissolotti PMA, Gonçalves Santiago A, Oliveira R (2006) Sustainability evaluation in eco-villages. In: The 23rd conference on passive and low energy architecture, Geneva, Switzerland. Accessed 24 Aug 2017. https://www.researchgate.net/publication/242509217_ Sustainability_Evaluation_in_Eco-villages

Boyer R (2015) Grassroots innovation for urban sustainability: comparing the diffusion pathways of three eco-village projects. Environ Plann A 47(2):320–337

Brundtland G (1987) Our common future. The world commission on environment and development. Oxford University Press, Oxford

Christian DL (2003, 2012) Creating a life together: practical tools to grow eco-villages and intentional communities. New Society Publishers. Accessed 22 Aug 2017. http://library. uniteddiversity.coop/Ecovillages_and_Low_Impact_Development/Creating_a_Life_Together-Practical_Tools_to_Grow_Ecovillages_and_Intentional_Communities.pdf

Clarissa de Oliveira A, Gallagher J, Orell P (2013) Reinventing the wheel to guide eco-villages towards sustainability. Master's thesis, School of Engineering, Blekinge Institute of Technology, Karlskrona, Sweden. Accessed 22 Aug 2017. https://www.divaportal.org/ smash/get/diva2:831042/FULLTEXT01.pdf

Colantonio A (2009) Social sustainability: a review and critique of traditional versus emerging themes and assessment methods. In: Sue-mot conference 2009: second international conference on whole life urban sustainability and its assessment. Loughborough University, Loughborough, pp 865–885. Accessed 10 Sept 2017. http://www.sue-mot.org/conference-files/2009/restricted/papers/papers/Colantonio.pdf

Dancing Rabbit Ecovillage: http://www.dancingrabbit.org/. Accessed 30 Sept 2017

Dawson J (2006) Eco-villages: new frontiers for sustainability. Schumacher briefings, UIT Press. Accessed 24 Aug 2017. http://tekobooks.com/downloads/eco-villages_new_frontiers_for_ sustainability_schumacher_briefings/

Ecovillage at Ithaca: http://ecovillageithaca.org/. Accessed 30 Sept 2017

Findhorn Foundation: https://www.findhorn.org. Accessed 30 Aug 2017

Gesota B (2008) Eco-villages as models for sustainable development: a case study approach. Master's thesis, Universitat Freiburg. Accessed 4 Sept 2017. https://portal.uni-freiburg.de/ globalstudies/research/gesota-2008-eco-villages.pdf

Gilman R (1991) The eco-village challenge: the challenge of developing a community living in balanced harmony—with itself as well as nature—is tough, but attainable. Context 29, 10–14. Accessed 24 Aug 2017. http://www.context.org/iclib/ic29/gilman1/

Global Eco-Village Network website: https://ecovillage.org. Accessed 26 Aug 2017

Guillen-Royo M (2015) Review of sustainability and wellbeing: human-scale development in practice Routledge. Ecol Econ 135:304–305

Jackson H (1998) What is an ecovillage? Paper presented at the gaia trust education seminar. Accessed 11 Sept 2017. http://www.gaia.org/gaia/ecovillage/whatis/

Kazhura Y, Bento Maffei de Souza B, Worosz H (2005) Sustainable community development in the Baltic sea region. Master's thesis, Blekinge Institute of Technology Karlskrona, Sweden. Accessed 28 Aug 2017. http://www.diva-portal.org/smash/get/diva2:829517/FULLTEXT01. pdf

Kirby A (2003) Redefining social and environmental relations at the eco-village at Ithaca: a case study. J Environ Psychol 23:323–332

Kusel J (2001) Assessing well-being in forest dependent communities—understanding community-based forest ecosystem management. CRC Press

Kuznetsov A, Dinwoodie J, Gibbs D, Sansom M, Knowles H (2015) Towards a sustainability management system for smaller ports. Mar Policy 54:59–68

Litfin K (2014) Eco-villages—lessons for sustainable community. Polity Press, Cambridge

Loezer L (2011) Enhancing sustainability at the community level: lessons from American eco-villages. Master's thesis, University of Cincinnati. Accessed 4 Sept 2017. http:// empowering-sustainability.weebly.com/uploads/3/0/2/6/30267909/leila-loezer.pdf

MacKendrick NA, Parkins JR (2004) Frameworks for assessing community sustainability: a synthesis of current research in British Columbia. CFS, Northern Forestry Centre. Accessed 27 Aug 2017. http://www.cfs.nrcan.gc.ca/bookstore_pdfs/24198.pdf

Sakellari M, Skanavis C (2013) Sustainable tourism development: environmental education as a tool to fill the gap between theory and practice. Int J Environ Sustain Dev 12(4):313–323

Sala S, Ciuffo B, Nijkamp P (2015) A systemic framework for sustainability assessment. Ecol Econ 119:314–325

Sevier L (2008) Eco-villages: a model life? Ecologist. Accessed 8 Sept 2017. http://gen.eco-village.org/iservices/publications/articles/ec_08_may_eco-villages.pdf

Seyfang G, Smith A (2007) Grassroots innovations for sustainable development: towards a new research and policy Agenda. Environ Politics 16(4):584–603

Siracusa G, La Rosa AD, Emiliano La Mola PP (2007) New frontiers for sustainability: emergy evaluation of an eco-village. Environ Dev Sustain 10(6):845–855

The German Advisory Council on Global Change website: https://www.eea.europa.eu/themes/climate/links/physical-science-on-climate/german-advisory-council-on-climate-change-wbgu. Accessed 24 Aug 2017

The International Plant Protection Convention website: https://www.ippc.int/en/. Accessed 29 Aug 2017

The Natural Step website: http://www.thenaturalstep.org/our-approach/. Accessed 26 Aug 2017

Toros T (2011) Ecological and sustainable urban design: eco-villages, eco-districts, and eco-cities, AIA, NCARB. Accessed 24 Aug 2017. https://www.academia.edu/8742172/Ecological_and_Sustainable_Urban_Design_Eco-Villages_Eco-Districts_and_Eco-Cities_2011_

Vidickiene D, Melnikiene R (2012) Eco-village sustainability self-evaluation test. Lithuanian Institute of Agrarian Economics, Baltic Sea region programme. Accessed 9 Sept 2017. http://www.balticeco-villages.eu/eco-village-sustainability-self-evaluation-test

Georgios Antonopoulos has a Master degree in Naval Engineering granted to him by the National Technical University of Athens (NTUA) in 2015. He is currently an M.Sc. candidate at the Department of Environment, University of Aegean, (Mytilene, Greece), and a researcher in the Research Center of Environmental Education and Communication, of the same department. His main interest lies with promoting communication tools, aiming to form positive attitudes towards sustainability.

Emmanouil Avgerinos received his Master degree in Civil Engineering from the National Technical University of Athens (NTUA) in 1999. Also he received a Master degree in Protection of Monuments and conservation and restoration of Historic buildings and sites from the NTUA in 2006. Currently he is Ph.D. candidate at the Department of Environment, University of Aegean, (Mytilene, Greece), focusing on Environmental Communication and Education, and sustainable development. Since 2000, he is a registered Professional Engineer in Technical Chamber of Greece and has been working in the area of conservation, restoration and Seismic Evaluation and Strengthening of Existing Buildings.

Kyriakos Antonopoulos is the President of the Board of Skyros Port Fund. He has created a multi awarded small marina at Linaria Port of Skyros Island. This Port has been identified as the Blue Port with a Shade of Green by United Nations. He is the author of several papers and has given a lot of presentations on the environmental excellence expected in the ports and how to reach such a state.

Dr. Constantina Skanavis is a Professor in Environmental Communication and Education at the Department of Environment, University of the Aegean (Mytilene, Greece). She is also the Head of the Research Centre of Environmental Education and Communication. She joined the University of the Aegean 15 years ago. Before that she was a Professor at California State University, Los Angeles. She has developed several courses on issues of environmental health and education. She currently teaches environmental education, environmental communication and environmental interpretation courses in undergraduate and postgraduate levels. Professor Skanavis has numerous publications on a international basis and has given presentations all over the world.

Climate Change Education Through DST in the Age Group "10–13" in Greece

Paraskevi Theodorou, Konstantina Christina Vratsanou,
Ilias Nastoulas, Effrosyni Sarantini Kalogirou
and Constantina Skanavis

Abstract This study attempted to demonstrate the extent at which the combination of a lecture given to the students, in order to educate them in difficult climate change concepts and a digital storytelling (DST) intervention tool named Pixton, were effective in teaching climate change science. The sample of the research consisted of 459 students in the 4th, 5th, 6th and 7th grades of school in Athens, during the course of Computer Science. The study assessed the rate of knowledge change, attitude and willingness to change behavior too, driven by pre-post questionnaires, which were given both at the start and at the end of the implementation. The questionnaires are differentiated only in the four questions concerning willingness to change behavior. Initially, the related work on the use of DST, the contribution of learner-generated comics and the use of the specific tool of Pixton, were discussed. The key implication of the findings is that information from climate change lectures is obtained with the aid of DST. The latter is a great tool for

P. Theodorou (✉) · K. C. Vratsanou
Research Center of Environmental Communication and Education,
Mytilene, Greece
e-mail: ptheod@env.aegean.gr

K. C. Vratsanou
e-mail: env12011@env.aegean.gr

I. Nastoulas
Environment and Sustainable Development, London's Global University,
London, UK
e-mail: env09062@env.aegean.gr

E. S. Kalogirou
Department of Environmental Communication and Education,
University of the Aegean, Mytilene, Greece
e-mail: sanykalogirou@gmail.com

C. Skanavis
Research Center of Environmental Communication and Education,
University of the Aegean, Mytilene, Greece
e-mail: cskanav@aegean.gr

© Springer Nature Switzerland AG 2019
W. Leal Filho et al. (eds.), *Addressing the Challenges in Communicating Climate Change Across Various Audiences*, Climate Change Management,
https://doi.org/10.1007/978-3-319-98294-6_20

teaching climate change issues and influences to some extent even the willingness to change in the future. Concerning the results, it is assumed that students cooperate more when learning is administered in a pleasant and interactive way. Success seems to take part due to the fact that students are given the possibility to be part of a learning experience creating their own content. Finally, the paper concludes with future guidelines in the field of other environmental issues such as recycling-reusing and energy.

1 Introduction

Today, climate change is one of the greatest menaces humanity is confronted with, due to its multiple implications for human survival. Human beings need a decent environment for effective and fruitful life (Offorma 2014) although it becomes readily apparent that human activities are those that influence climate. In order to achieve the latter, there is a need of innovative pedagogical approaches and tools that will allow us to design learning activities in which learners will be empowered to develop innovative, alternative interpretations of the concepts of sustainability and climate change in personally and collectively meaningful ways (Daskolia et al. 2015).

According to Tomhave (2010):

> education which is about "learning" how to learn, provides important opportunities for students to become engaged in real world issues that transcend classroom walls. They can see the relevance of their classroom studies to the complex environmental issues confronting our planet and they can acquire the skills they will need to be creative problem solvers. Additionally, it offers an antidote to the plugged-in lives of today's generation, which is the first to grow up indoors. Children who experience school grounds or play areas with diverse natural settings were more physically active, more aware of good nutrition, more creative and more civil to one another.

While education is proposed as the most efficient mechanism for changing behavior and improving climate literacy, it is unclear how to best deliver it (Mochizuki and Bryan 2015). Fletcher and Cambre (2009 as cited in Robin 2016) have found that DST can be a powerful classroom practice when used "as a pedagogical tool that brings the creator/student and the viewer together in a dialogue around the nature of representation, meaning, and authority embedded in imagery and narrative". DST has become a modern incarnation of the traditional art of oral storytelling. What is more, it allows almost anyone to use off-the-shelf hardware and software to weave personal stories with the help of still/moving images, music, and sound, combined with the author's creativity and innovation (Smeda et al. 2014).

There are many definitions of DST, but in general terms, DST is defined as all types of application that use digital media storytelling, to support the implementations (Schafer 2004). This characterization implies that comic authoring tools were classified as DST applications; since the output of these tools was digital stories in the form of comics (Azman et al. 2015).

Many educational subjects have been successfully adapted into DST. History is one of educational topics that is already adapted into this innovational teaching strategy (Wijaya and Sanjaya 2012) and mathematics another one (Febriani and Chandrawati 2012) Other implementations concern innovative courses in art education and visual culture which are taught through powerful ways of DST (Chung 2007). Additionally, the effectiveness of DST has been demonstrated for developing listening comprehension skills in elementary school English as second language learners (Tsou et al. 2006; Verdugo and Belmonte 2007).

DST activities in Environmental Education could not only lead to students becoming skilled in digital media, but also provide a cultural and environmental focus for the sharing of knowledge and practices between generations (Wyeld et al. 2007), which could support students in understanding the natural world and acquiring environmental awareness (Heo 2004).

There are studies that aim to contribute in the effort to provide an answer, by investigating if children can have the will and ability to act, not through the transfer of scientific knowledge, but through the expression and the communication of their own ideas (Tsevreni 2011). In the present study it is examined how this idea can fit to climate change education. Taking into account all the above, in this research it was attempted to understand better the impact on students learning when they take advantage of DST for their learning in the context of environmental education and awareness, in the field of climate change. Specifically, it concerns the effects of a lecture in climate change science of knowledge and the use of the web 2.0 tools in order to educate students on this environmental issue that is of great importance and it assesses knowledge, attitudes and willingness to change behaviors. In general, this activity operates as a supplementary method for students to absorb a difficult academic topic, polish their research skills in this field and raise awareness of principles in relation to climate change and the risks it evolves.

In the following section of this paper, background issues of this study are briefly presented including: Climate Change communication, Storytelling, DST, Learner-Generated Comic and Pixton tool. There follows a description of a sequence of online collaborative Digital Story learning activities using Pixton with special reference to the aforementioned issue of climate change. Finally, the design of this sequence is discussed, and conclusions and future research plans are drawn. The results showed that after the implementation students were more informed about climate change science and this significantly affected their opinions about climate change and subsequently their attitudes and willingness to change behavior towards it.

2 Background—Literature Review

Since the 1980s, a different path for environmental education has emerged aiming at children's participation and emancipation, not only through the transfer of knowledge but also through ownership and empowerment variables (e.g., in the US, Hungerford et al. 1980, 1983; Hungerford and Volk 1990). This approach has evolved and shaped the critical environmental education paradigm in the UK, Australia and Denmark (e.g., Robottom 1987; Fien 1993; Huckle 1993; Jensen 1997; Hart 2003, 2006; Breiting et al. 2009). In the UK, for example, Huckle (1991) mentions that environmental education should embrace the characteristics of critical pedagogy that focuses on active and experiential learning and critical reflection.

In the frame of critical pedagogy, pupils should develop their own power to shape their lives, comprehend the sources of beliefs and values and the interests they support, and reflect on the forces that restrict their lives and on democratic alternatives (Bangay and Blum 2010). If the role of education is to help learners of all ages to build up the knowledge, skills and capacities, which enable them to think critically, to deal with problems, and to address uncertainty, then the focus of climate change interventions should be placed on more holistic ways of confronting climate change reality (Bangay and Blum 2010).

The study focused on exploring the combination of collaborative creativity and critical thinking of DST as a learning process lecture in the field of climate change. This approach thus puts forth a constructionist perspective of collaborative creativity as an activity occurring within a learning context that involves and results in the generation of new understandings and tangible artefacts on the sustainability concept (new, at least for the students themselves), along with their collaborative construction of the digital stories (Daskolia et al. 2015).

2.1 Climate Change—Communication

Climate change is no longer the exclusive domain of scientific experts. It calls for action from all citizens (Skanavis et al. 2017) and this is achieved through the expansion of the audience of climate change literacy. If climate change education is to move beyond a mere analytical exercise and become a force for change, educational policies shouldn't focus solely on the causes and impacts of climate change. Curricular objectives should shift from vague "critical thinking" outcomes to a solutions-focused "action competence" model. Students should actively engage with the solutions they generate rather than simply demonstrate that they have the knowledge to generate them (Vaughter 2016).

Considerably, active citizens operate with responsible environmental behavior through their awareness, which is developed from the knowledge that they are provided, and this gives them the opportunity to practice responsible environmental actions and develop new environmental behaviors (Dressner et al. 1994; Hudson 2001).

According to Skanavis (2004)

> By promoting responsible environmental behavior, we can develop the bases for the active participation of citizens of each local society in environmental issues. By the term active participation, we refer to the participation of citizens in the process of decision-making concerning such issues. In order for citizens to be driven into active decision-making

participation, they need to have an informed opinion on the environmental issue of concern, climate change in this research, something that can only be achieved with proper environmental education/briefing.

2.2 Storytelling

Throughout the history of mankind, storytelling has been used as a tool for the transmission and sharing of knowledge and experience, because it is a natural and yet powerful technique of communicating and exchanging knowledge (Smeda et al. 2014). Also in relation to the use of storytelling in the classroom Behmer stated, "Storytelling is a process where students personalize what they learn and construct their own meaning and knowledge from the stories they hear and tell" (Behmer 2005).

As regards Environmental Education, storytelling has been proposed as a key teaching strategy for the achievement of the goals of Education for a sustainable future (UNESCO 2007). Properly-planned environmental storytelling activities could serve as a basis for the development of speculation on ecological, environmental, traditional or contemporary issues at local or global levels and contribute to the construction of conceptions, ethics, values and attitudes that favor a more balanced human-human/human-nature/environment relationship, approaching both ecological and cultural sustainability (UNESCO 2008; Agelidou and Tsilimeni 2009).

2.3 DST

DST enhances students' motivation and provides learners with an educational environment conducive for story construction through collaboration, reflection and interpersonal communication (Smeda et al. 2014). In other words, DST tools assist students develop enhanced communication skills by learning how to organize their ideas, to ask questions and to express their opinions. The use of DST exploits on the creative talents of students as they begin to research and tell stories of their own, learn to use the web to search rich, deep content while analyzing and synthesizing a wide range of information and opinions (Robin 2016).

2.4 Pixton (Comic Authoring Tool)

As most comic authoring tools, Pixton is an application that allows the user to generate digital comic from scratch. The user can select predefined objects, characters and backgrounds, which have a default setting but can still be altered,

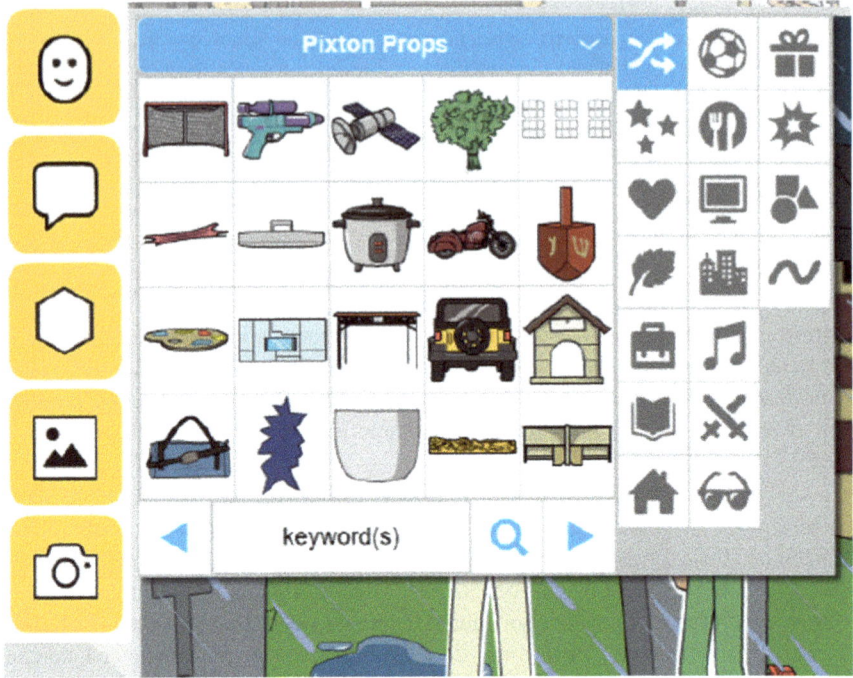

Fig. 1 Pixton drawing area

to make digital stories (Azman et al. 2015). It should be mentioned that all the following figures of this research are authentic creations of the researchers of the paper and the students of the sample (Fig. 1).

Pixton offers unlimited ways for users to creatively compose a variety of comic strip layouts with customized action and emotion. Users were able to animate the comic characters to a certain degree, switch colors, sizes, and distances, adding printed text and recording voices to fulfill the digital story. It should be mentioned that registered members in the community were symbolized as character avatars (Azman et al. 2015) (Fig. 2).

Pixton also supports constructivist learning theory in that it allows users to actively construct knowledge (Jonassen 1999). Users were able to articulate their previous knowledge, explore, speculate, manipulate the environment in order to construct and test their theories, reflect on what they did, and think about what they have learned from the activities. Students can use this web based tool to showcase their previous knowledge while acquiring new knowledge as they create their own comic and interact with other students.

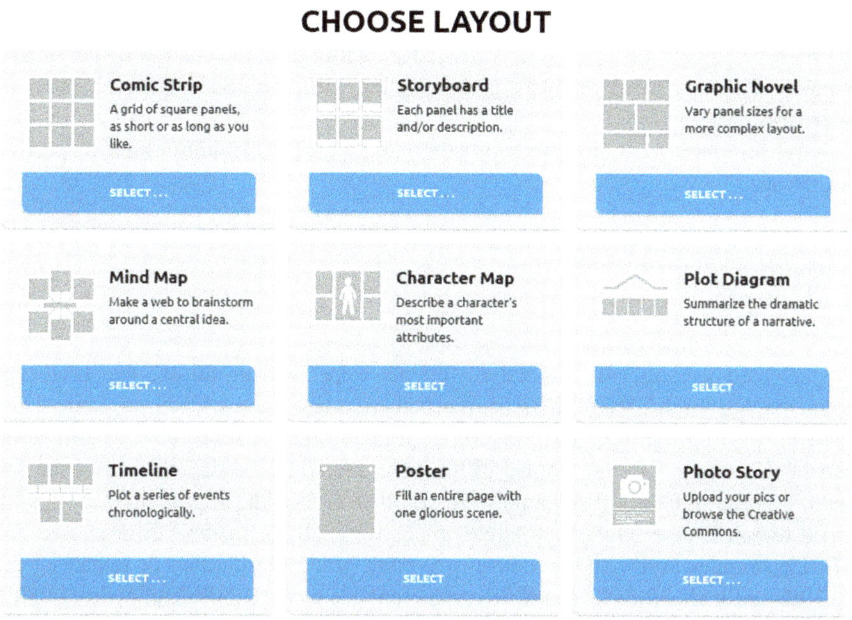

CHOOSE LAYOUT

Comic Strip
A grid of square panels, as short or as long as you like.

SELECT...

Storyboard
Each panel has a title and/or description.

SELECT...

Graphic Novel
Vary panel sizes for a more complex layout.

SELECT...

Mind Map
Make a web to brainstorm around a central idea.

SELECT...

Character Map
Describe a character's most important attributes.

SELECT

Plot Diagram
Summarize the dramatic structure of a narrative.

SELECT

Timeline
Plot a series of events chronologically.

SELECT...

Poster
Fill an entire page with one glorious scene.

SELECT

Photo Story
Upload your pics or browse the Creative Commons.

SELECT...

Fig. 2 Pixton available layouts

3 Methodology

3.1 Approach—Methods—Research Performed

The present study aims to discover the effectiveness of climate change lectures combined with instructional comics with the means of DST.

The particular implementation lasted two or three teaching hours depending on the level of the class and/or the perceived attention that students showed during the elaboration of the implementation. The role of the teacher was directive and accommodative.

Specifically, a pre test questionnaire was given to the students to examine their former knowledge, willingness to change behavior and attitude of climate change. Afterwards, a short lecture was given in order to transcend the concept of climate change. The lecture was based on a PowerPoint 2016 presentation with 55 slides covering:—basic climate change science clearly is differentiating between natural and anthropogenic climate change and best-studied impacts of climate change. The PowerPoint was a visual means with simple animations and diagrams demonstrating how climate change operates in the atmosphere. Each lecture was allotted 45 min for narration, allowing approximately 30–40 s for each slide to be shown. Students, who were following given guidelines, developed their own comics, storyboards and mind maps with the use of Pixton tool.

Afterwards, the same questionnaire was given for the second time. The measurements showed the change of knowledge, willingness to change behavior and attitude. The statistical analysis is conducted with the tool of Excel 2016 of Microsoft Office.

After the implementation students continued their course within their respective disciplines.

4 Sample

The selected schools and respective students were all from an average socioeconomic background, 75% of them from urban areas and the rest from rural ones. The students were in the age group of 10–13 years of age. In total, 459 students took part in this study. On the whole, 152th were attending the 4th class, 138 the 5th class, 129 the 6th and 40 students were enrolled in the 7th. In particular, in the 4th grade females were 39.47%, whereas males were 60.53. In the 5th grade we had 57.97% females and 42.03% males, in the 6th grade the number of females were 50.39% and male 49.61%. In the 7th grade females were 52.50% and males 47.50% (Figs. 3, 4, 5, and 6).

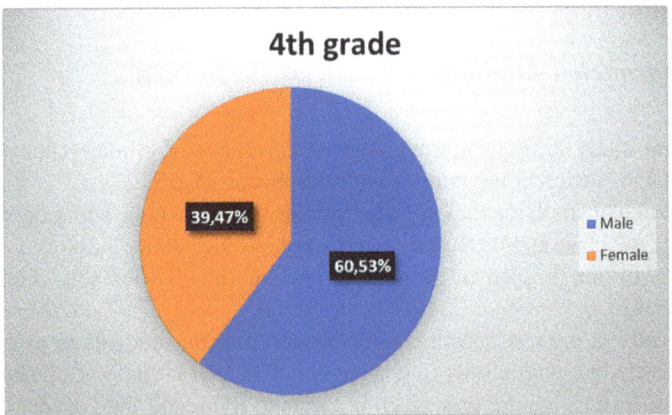

Fig. 3 Gender of 4th grade

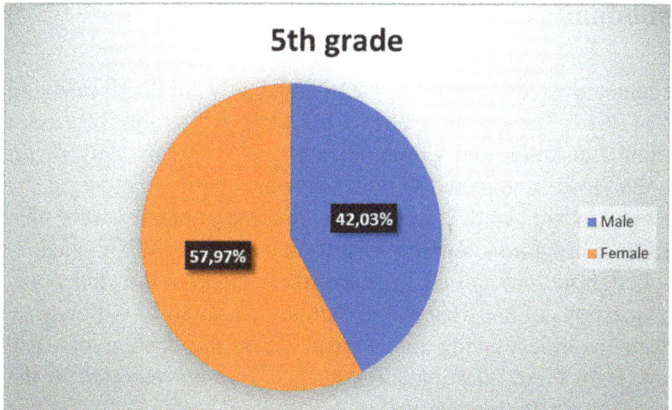

Fig. 4 Gender of 5th grade

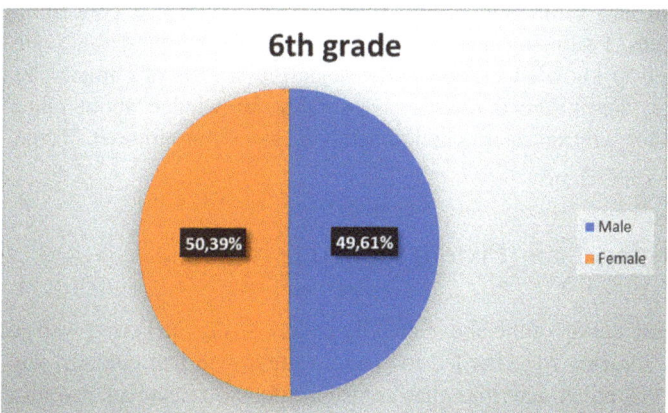

Fig. 5 Gender of 6th grade

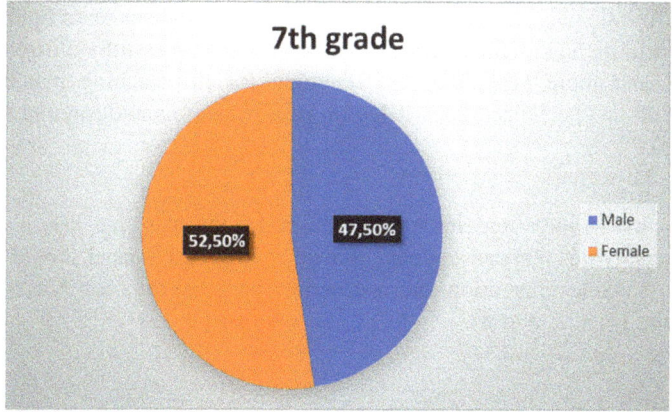

Fig. 6 Gender of 7th grade

5 Questionnaires

The questionnaires were composed of 22 items and were distributed and collected during fall of 2017. The questionnaires were approved by the Research Center for Environmental Education and Communication, of the University of the Aegean. The survey had to be completed within a 45-min school-time period. It should be mentioned that students were informed that the questionnaires were not a test, but part of a research program in order to assess their background knowledge of climate change. Although the pre and post questionnaire given to the students is the same concerning questions of knowledge and attitude, questions of willingness to change behavior were modified taking into consideration the content of each question. Furthermore, from these questionnaires results were not reported to the students, unless they asked for their assessment.

Students were asked to put in an anonymous tracking code, gender, age and class. Specifically, in the questions 1–7, students dealt with demographic, personal information and estimations about their former experience concerning environmental issues. Following this, the next items were multiple-choice questions and were related to knowledge about socio-scientific aspects of climate change. More specifically, there were 9 questions assessing knowledge about climate change science, 4 for willingness to change behavior and 2 for attitudes (Table. 1).

6 Learning Tool (Pixton) and Activities

The implementation of the above activity is done using the Pixton web software. At this stage, students were invited to create a mind map that contains various direct and indirect human activities that contribute to the release of greenhouse gases. Once the mind map has been completed, students that were using the same pixton tool were asked to create a storyline table illustrating the imminent effects of climate change on a global scale (e.g. sea level, ice melting, desertification, various forms of life, etc.). Additionally, in the context of these efforts, students learn how to search for useful information on the internet. The tool can be successfully utilized to create alternative and more appealing representations of the teaching materials from teachers with no specialized knowledge on painting or comic designing (Lazarinis et al. 2015).

Below, the scenario is described:

A. In the first activity students have to build a mind map that lists a variety of human activities that cause global climate change. Each panel should include a title, a suitable illustration and a description of how it contributes to global warming (Figs. 7 and 8).

Table 1 Questionnaire

Questions	Type	Topic
Q8	Knowledge	Do you know what the phenomenon of climate change is?
Q9	Knowledge	Is your school generally participating in environmental programs?
Q10	Knowledge	Which of the following phenomena were related to climate change?
Q11	Willingness to change behavior	A way to reduce the greenhouse effect is recycling. Are you willingness to recycle?
Q12	Knowledge	Are the greenhouse effect and the ozone hole the same thing?
Q13	Knowledge	Our planet is a large greenhouse, as plants grow on its surface. Were we used to saying that the "greenhouse effect" is presented on earth?
Q14	Knowledge	Do atmospheric pollutants form a glass surface around the earth, which constantly increases its temperature?
Q15	Knowledge	What is the definition of climate?
Q16	Willingness to change behavior	Which of the following habits do you intent to adopt in your everyday life?
Q17	Knowledge	Does the rise in earth's temperature raise the sea level?
Q18	Willingness to change behavior	What can you do to help limit the influences of climate change?
Q19	Knowledge	Does changing the Earth's climate cause floods and droughts in some parts of the world?
Q20	Attitude	If you were told that every day it would be obligatory for two or three hours not to have power supply and water, what would your opinion about that be?
Q21	Attitude	To stop the change in the earth's climate, all of us should help with our actions
Q22	Willingness to change behavior	Do you intent to persuade others to do things to protect the environment?

Note Although the pre and post questionnaire given to the students is the same concerning questions of knowledge and attitude, questions of willingness to change behavior were modified taking into consideration the content of each question

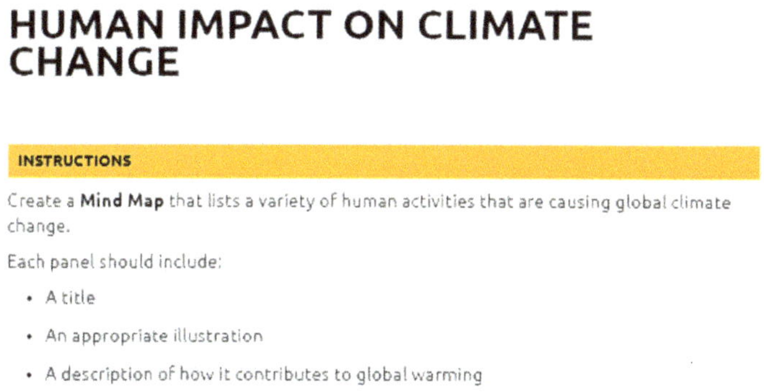

HUMAN IMPACT ON CLIMATE CHANGE

INSTRUCTIONS

Create a **Mind Map** that lists a variety of human activities that are causing global climate change.

Each panel should include:

• A title

• An appropriate illustration

• A description of how it contributes to global warming

Fig. 7 First activity instructions

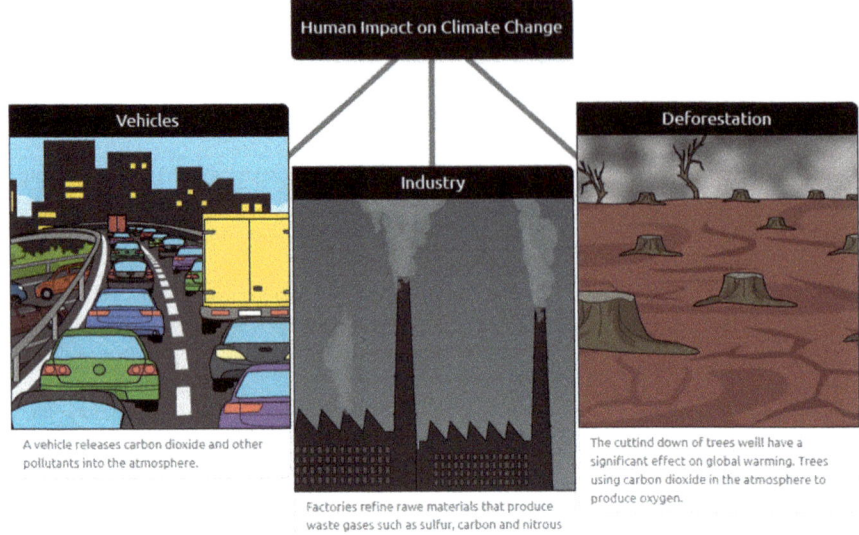

Fig. 8 First activity (mind map)

B. In the second activity students have to build a storyboard that illustrates the various impacts that global climate change is creating, identifying the activity, including appropriate description or dialogue and an appropriate picture for each table (Figs. 9 and 10).

C. In the third activity students build a poster that depicts all ways of reducing carbon emissions (Fig. 11).

7 Results

The participants' perceptions regarding climate change concept were measured in three axes that included the knowledge, attitude and willingness to change behavior.

The subjects were asked to answer questions related to their general knowledge on climate change related problems. The tables and graphs were listed below and concern the part of the questionnaire that included the total knowledge score. In total, it is estimated that from 459 students, 40.9% changed their knowledge, 44.4% the willingness to change behavior and 27.4% their attitude.

For all the questions concerning knowledge, it is recorded that the change for the 4th grade was 46.8%, the change for 5th grade were 45.6%, for the 6th grade the change were 42.9%, and the change for the 7th grade were 28.3%.

The next factor that is examined is the change in the willingness to change behavior of the participants on climate change related problems. For all the

Fig. 9 Second activity (1nd storyboard)

Fig. 10 Second activity (2nd storyboard)

Fig. 11 Third activity (posters)

questions concerning willingness to change behavior, it is recorded that the change were 44.2% for the 4th grade, 43.3% for 5th grade, 45.1% for the 6th grade, and 45% for the 7th grade. The change of attitude category was 13.5% for the 4th grade, 30.05% for 5th grade, 52.7% for the 6th grade and 13.9% for the 7th grade (Figs. 12 and 13).

8 Discussion

The results indicate that the production of digital comic strips can improve students' climate change knowledge, willingness to change behavior and attitude. While there are numerous technical policies and solutions for mitigating and reacting to climate change, changing the behavior of individuals will be the most critical component of the process (Vaughter 2016). An engaging, multimedia-rich Digital Story that serves as a bridge between existing knowledge and new material (Ausubel 1978), operates as an anticipatory set or hooks to capture the attention of students (Robin 2006). Therefore, the following results are also aligned with what has been reported in the literature.

The production of the stories proved that students were able to assimilate the content of climate change. Essentially, they showed through personalization and communication, their experiences about climate change. To clarify, they used unconventional content, varied resource and well-chosen points of view,

Total	Before (%	After (%)	Change (%)
KNOWLEDGE	48,6	89,5	40,9
WILLINGNESS TO CHANGE BEHAVIOR	47,0	91,4	44,4
ATTITUDE	69,2	96,6	27,4

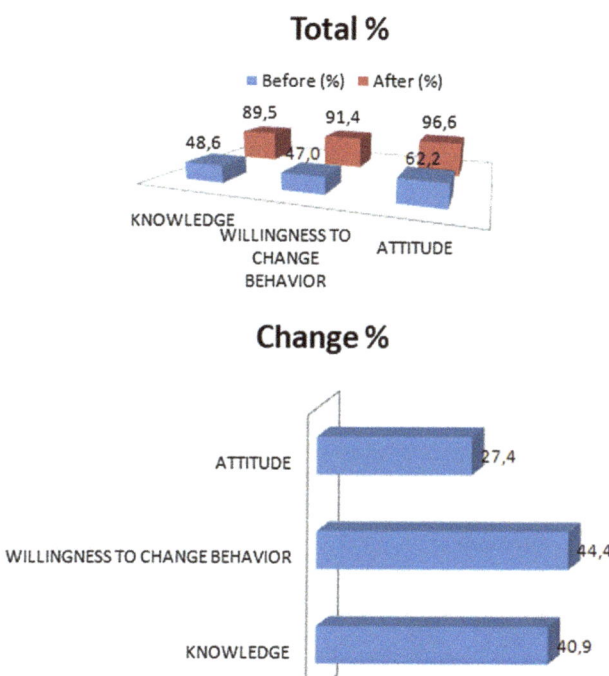

Fig. 12 Total changes

which reflected on their own thoughts and engagement with climate change subject. On the other hand, there were a few distracted digital effects and images that showed inadequate preparation and understanding. But this is assumed that happened due to the adequate technological equipment of some laboratories, or lack of a powerful internet connection. However, in total students really enjoyed the collaboration with each other and produced excellent work.

In particular, the total results concerning knowledge showed that there was a big difference before and after the implementation. The way to reach this state of excellence is through environmental education from an early age. Young people, by all available means, must learn to care for the planet, be familiar with nature and be part of an environmentally active community (Plaka and Skanavis 2016).

Comparatively, it is observed that the total knowledge of the fourth grade of all schools was increased in a level of 46.8%, respectively in fifth, the increase percentage was 45.6%, in the sixth grade was 42.9% and in the seventh 28.3%.

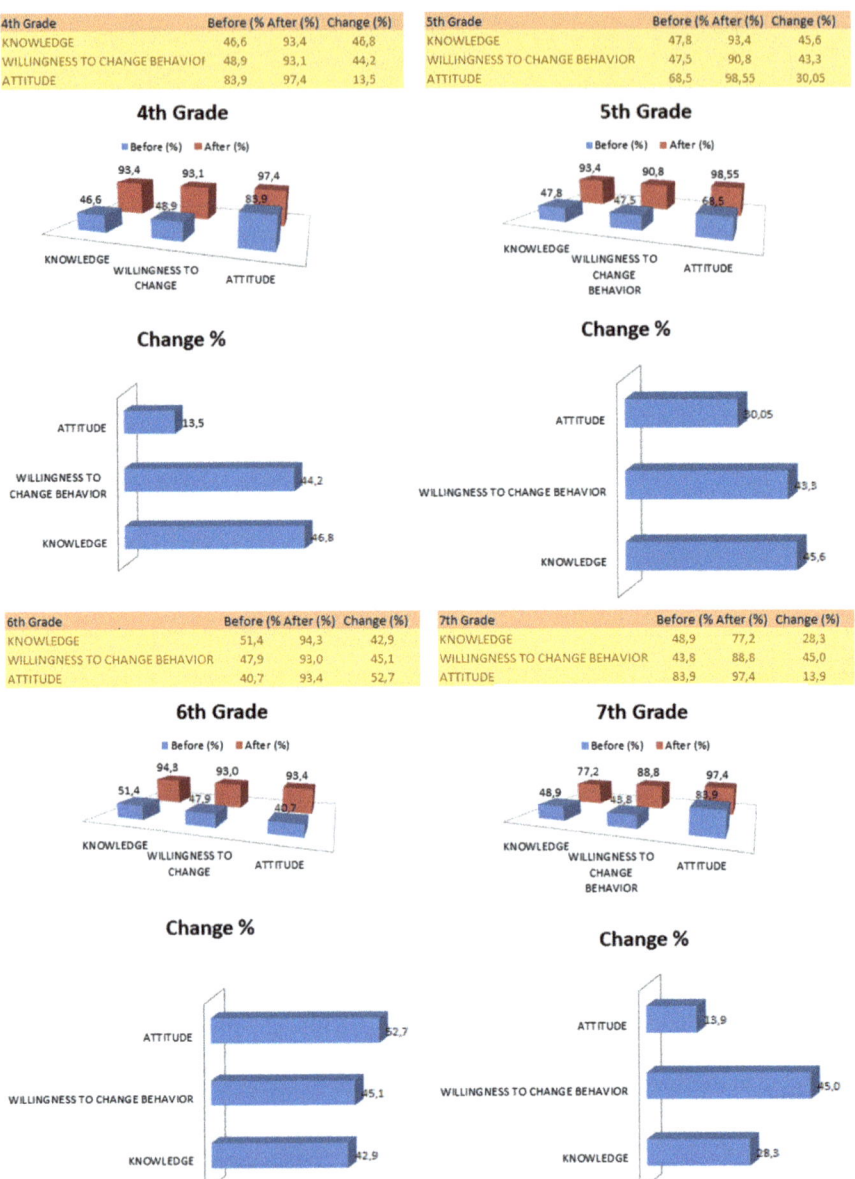

Fig. 13 Total changes in primary school and middle school

These results were based on the fact that students enjoyed the procedure and became real creators of their own stories. Students also majorly demonstrated that educational, digital comics could assist them to understand difficult technical and scientific content. These findings align with prior claims that comics were able to assist students' comprehension (Mallia 2007; Recine 2013; Yıldırım 2013).

Concerning the axe of willing to change behavior, it is observed that there is a stable total change in all the grades from 43.3 to 45.1%. That means that our study findings are in agreement with previous studies, that suggested that the digital stories were more effective in influencing behavioral intentions, but had unstable influence on two factors known to be strong determinants of behavior—knowledge about a topic and attitude toward it.

It was remarkable that in the sixth grade of primary school, there was the highest success rate of 52.7% concerning questions of attitude. Students had a strong belief that climate change couldn't be reversed. Of course, there is an exception at the seventh grade whose environmental attitude was improved only by 13.9%. That is caused because they were attending programs concerning environmental education, "Zero Waste".

The 7th grade showed stability with a percentage of 75.5% on the results concerning the question if greenhouse effect and the ozone hole were the same thing. Students responded this way, because that specific time they were being taught about the climate change in the school subject of geography. It is noticeable that in Greece according to the curriculum, the subject of environmental education is being taught only up to the fourth grade of primary school. After this grade, scattered topics of climate change were concluded in geography, biology and chemistry subjects.

All in all, students-produced stories appeared to be successful, because they met many of the pedagogical and technical attributes of digital stories. One of the main findings is that despite difficulties, DST projects could increase students' comprehension of climate change content. It also changed the attitudes of these students for the best as well as it created them the willing to change behavior with the intention of participating actively in climate change campaign (Sadik 2008).

According to Dogan and Robin almost half of the teachers indicated that the highest ranked perceived barrier to using digital stories in the classroom would be "Time issues". In fact, DST proves to be very time consuming and some educators believe that it is an inconvenient effort. That is, because it may take students several attempts at creating digital stories before they demonstrate technological proficiency and an understanding of their selected topic (Dogan and Robin 2017). As with all new instructional methods, students will need time to learn what is expected of them as they begin using DST (Dogan and Robin 2017). This result was expected as time is commonly cited as a barrier to incorporating technology in the classroom (Dexter et al. 2002; Ringstaff and Kelley 2002; White et al. 2002) and specifically with DST (Dogan et al.; Dogan and Robin 2017). This barrier existed in this particular research alongside with the slow internet connection that caused problems to the speed and the quality of the process.

Students' Critical thinking, which scholars since Dewey (1910) have emphasized as a major goal for education is developed through DST. According to Freire (1972), by reflecting and acting, children activate their critical consciousness. As children face the problems regarding their relationships both within and towards the world, they may feel responsible for providing solutions. Critical thinking makes students engaged to their goals and causes active participation in climate change campaign and communications, which is the scope of this conference.

9 Conclusion

Given the revolutionary and rapid growth of digital technologies and their impact on all dimensions of our lives, it is not surprising to find DS entering the whole education mainstream; however, its place in the classroom is still unclear (Lowenthal 2009). There is a lack of current research on effective use of DST education that address the use of DST as an effective instructional tool in the classroom, its effects on student learning, and potential problems that may arise in the implementation process and even more in the field of environmental education (Dogan 2012). Specifically, using DST in environmental education on a Greek scale of primary school it has not been investigated at all. That is why the results of this study are important for researchers in order to understand how DST is used for educational purposes in primary school, particularly in the field of climate change education and communication.

Another issue is that students weren't taught critical thinking in their formal learning environment and were not encouraged to experience or consider scientific concepts in novel and creative ways which could enhance their understanding and opinion cultivation (Robinson 2001, 2011; Simons and Hicks 2006). This study will trigger upcoming, future research to provide greater insights and understanding in how DST can engage, inform and enlighten new generations of students and educators in order to use critical thinking and positive in finding ways to integrate these stories in their classroom activities.

Conclusively, DST has proven to be a springboard progress in the field of teaching climate change communication. The results of this study will be investigated in further detail through additional implementations in the field of other environmental concepts such as recycling, reusing and renewable sources of energy.

Acknowledgements We express our gratitude to the school communities we visited (teachers, headmasters, students) and the parents of the students that gave us the consensus in order to fulfill our research.

References

Agelidou E, Tsilimeni T (2009) Storytelling as a learning tool in environmental education—activities and suggestions for pre-school and primary education. Kastaniotis Editions, Athens, Greece
Azman FN, Zaibon SB, Shiratuddin N (2015) Digital storytelling tool for education: an analysis of comic authoring environments. Advances in Visual Informatics, pp 347–355. https://doi.org/10.1007/978-3-319-25939-0_31

Ausubel DP (1978) In defense of advance organizers: a reply to the critics. Rev Educ Res 48: 251–257

Bangay C, Blum N (2010) Education responses to climate change and quality: two parts of the same agenda? Int J Educ Dev 30:359–368

Behmer S (2005) Literature review DST: examining the process with middle school students

Ben Tomhave (2010) Education training and awareness—there's a difference. http://www.secureconsulting.net/2010/05/education-training-and-awareness.html

Breiting S, K Hedegaard, F Mogensen, K Nielsen, K Schnack (2009) Action competence, conflicting interests and environmental education—the Muniv programme Copenhagen. Danish School of Education

Chung SK (2007) Art education technology: DST. Art Educ 60(2):17–22

Daskolia M, Kynigos C, Makri K (2015) Learning about urban sustainability with digital stories promoting collaborative creativity from a constructionist perspective. Constr Found 10(3)

Dewey J (1910) How we think. Heath & Co, Boston

Dexter S, Anderson RE, Ronnkvist A (2002) Quality technology support: what is it? who has it? and What difference does it make? J Educ Comput Res 26(3):287–307. Retrieved 6 Feb 2011 from http://sdexter.net/Vitae/Qual_final_with-ForGates.pdf

Dogan B (2012). Educational uses of DST in K-12: research results of DST contest (DISTCO), pp 1353–1362

Dogan B, Robin RB (2017) Climate change education from critical thinking to critical action. Institute for the advanced study of sustainability. Accessed 1 Dec 2017. https://ias.unu.edu/en/news/climate-change-education-from-critical-thinking-to-critical-action.html#files

Dressner M, Gill M (1994) Environmental education at summer nature camp. J Environ Educ 25(3):35–41

Febriani VW, Chandrawati TB (2012) Shooting game can be an education games for children, pp 23–24

Fien J (1993) Education for the environment. Critical curriculum theorizing and environmental education. Deakin University Press, Victoria

Fletcher C, Cambre C (2009) DST and implicated scholarship in the classroom. J Can Stud 43(1): 109–130

Freire P (1972) Pedagogy of the oppressed. Penguin, London (Revised in 1993), pp 75–78

Hart P (2003) Teacher's thinking in environmental education consciousness and responsibility. Peter Lang, New York

Heo H (2004) Storytelling and retelling as narrative inquiry in cyber learning environments. In: Atkinson R, McBeath C, Jonas-Dwyer D, Phillips R (Eds) Beyond the comfort zone: proceedings of the 21st ASCILITE Conference, pp 374–378

Huckle J (1991) Education for sustainability: assessing pathways to the future. Aust J Environ Educ 7:43–62

Huckle J (1993) Environmental education and sustainability: a view from critical theory. In: Fien J (ed) Environmental education: a pathway to sustainability. Deakin University Press, Victoria, pp 43–68

Hudson S (2001) Challenges for environmental education: issues and ideas for the 21st century. Bioscience 51(4):283–288

Hungerford HR, Volk TL (1990) Changing learning behaviour through environmental education. J Environ Educ 21(3):8–21

Hungerford HR, Peyton RB, Wilke RJ (1980) Goals for curriculum development in environmental education. J Environ Educ 11(3):42–47

Hungerford HR, Peyton RB, Wilke RJ (1983) Yes EE does have definition and structure. J Environ Educ 14(3):1–2

Jensen B (1997) A case of two paradigms within health education. Health Educ Res 12(4): 419–428

Jensen BB, Schnack K (2006) The action competence approach in environmental education. Environ Educ Res 12(3/4):471–486

Jonassen D (1999) Designing constructivist learning environments. In: Reigeluth C (ed) Instructional design theories and models, vol II. Lawrence Erlbaum, Mahwah

Lazarinis F, Mazaraki A, Verykios VS, Panagiotakopoulos C (2015) E-comics in teaching: evaluating and using comic strip creator tools for educational purposes. In: 2015 10th international conference on computer science and education (ICCSE) https://doi.org/10.1109/iccse.2015.7250261

Lowenthal PR (2009) Digital storytelling: an emerging institutional technology? In J Hartley, K McWilliam (Eds) Story circle: digital storytelling around the world, pp 252–259. Wiley-Blackwell, Oxford

Mallia G (2007) Learning from the sequence: the use of comics in instruction. Imagetext Interdiscip Comics Stud 3(3):1–10

Mochizuki Y, Bryan A (2015) Climate change education in the context of education for sustainable development: rationale and principles. J Educ Sustain Dev 9(1):4–26

Offorma GC (2014) Climate change and the need for new curriculum development in Nigerian Universities. Nigeria Clim Change Agric High Edu Multidiscip Issues Perspect 2(1):158–170. Available at: https://www.researchgate.net/publication/280554166. Accessed 9 Apr 2016

Plaka V, Skanavis C (2016) The feasibility of school gardens as an educational approach in Greece: a survey of Greek schools. Int J Innov Sustain Dev 10(2):141. https://doi.org/10.1504/ijisd.2016.075546

Recine D (2013) Comics aren't just for fun anymore: the practical use of comics by TESOL professionals. University of Wisconsin, River Falls

Ringstaff C, Kelley L (2002) The learning return on our educational technology investment. West Ed RTEC, San Francisco. Retrieved 6 Feb 2011 from http://www.wested.org/online_pubs/learning_return.pdf

Robin B (2006) The educational uses of digital storytelling. In Crawford C et al. (Eds) Proceedings of society for information technology & teacher education international conference 2006, pp 709–716. Chesapeake, VA: AACE

Robin BR (2016) The power of DST to support teaching and learning, digital education. University of Houston, USA

Robinson K (2001) Out of our minds: learning to be creative. Capstone Publishing Ltd., Chichester

Robinson K (2011) Out of our minds: learning to be creative. Capstone Publishing Ltd., Chichester

Robottom I (1987) Environmental education: practice and possibilities. Deakin University Press, Victoria

Sadik A (2008) DST: a meaningful technology-integrated approach for engaged student learning. Educ Technol Res Dev 56(4):487–506. https://doi.org/10.1007/s11423-008-9091-8

Schafer L (2004) Models for DST and interactive narratives. In: 4th international conference on computational semiotics for games and new media, Split, pp 148–155

Simons H, Hicks J (2006) Opening doors: using the creative arts in learning and teaching. Arts Humanit High Educ 5:77–90

Skanavis C, Kounani A, Ntountounakis I (2017) Greek universities addressing the issue of climate change. In: Leal Filho W (ed) Climate change research at universities. Springer, Cham

Smeda N, Dakich E, Sharda N (2014) The effectiveness of DST in the classrooms: a comprehensive study. Smart Learn Environ 1(1). https://doi.org/10.1186/s40561-014-0006-3

Tsevreni I (2011) Towards an environmental education without scientific knowledge: an attempt to create an action model based on children's experiences, emotions and perceptions about their environment. Environ Educ Res 17(1):53–67. https://doi.org/10.1080/13504621003637029i

Tsou W, Wang W, Tzeng Y (2006) Applying a multimedia storytelling website in foreign language learning. Comput Educ 47:17–28

UNESCO (2007) Teaching and learning for a sustainable future, storytelling, Module 19, in http://www.unesco.org/education/tlsf/, May 2009

UNESCO (2008). Final recommendations. In: 4th International conference on environmental education, 24–28 Nov 2007, Ahmedabad, India. Ihttp://www.tbilisiplus30.org/. Accessed 20 June 2010

Vassilikopoulou M, Retalis S, Nezi M, Boloudakis M (2011) Pilot use of digital educational comics in language teaching. Educ Media Int 48(2):115–126

Vaughter P (2016) Climate change education from critical thinking to critical action institute for the advanced study of sustainability. Accessed 1 Dec 2017. https://ias.unu.edu/en/news/climate-change-education-from-critical-thinking-to-critical-action.html#files

Verdugo DR, Belmonte IA (2007) Using digital stories to improve listening comprehension with Spanish young learners of English. Lang Learn Technol 11(1):87–101

White N, Ringstaff C, Kelley L (2002) Getting the most from technology in schools. WestEd, San Francisco

Wijaya NA, Sanjaya R (2012) History lesson using game as the tool. Int J Comput Internet Manag 19(9):6

Wyeld GT, Carrol J, Ledwich B, Leavy B, Gibbons C, Hills J (2007) The ethics of indigenous storytelling: using the torque game engine to support Australian Aboriginal cultural heritage. Proceedings of DiGRA Conference, pp 261–267

Yıldırım AH (2013) Using graphic novels in the classroom. J Lang Lit Educ 8:118–131

P. Theodorou is a Ph.Dc. student in the field of Environmental Science at the University of Aegean. She has an M.A. in Special Education and a M.Sc. in Digital Systems and Telecommunications. Her research and writing work includes publications in prestigious scientific journals, chapters in books and international and national conferences. Her research interests and activities focus mainly on climate change, sustainable development, immigration issue, environmental education and communication regarding the use of ICTs in order to promote responsible environmental behavior.

K. C. Vratsanou is Junior Researcher of Research Center of Environmental Communication and Education.

I. Nastoulas is Junior Researcher of Research Center of Environmental Communication and Education.

E. S. Kalogirou is Undergraduate student of Aegean University, Department of Environment.

Constantina Skanavis is a Professor in Environmental Communication and Education at the Department of Environment, University of the Aegean (Mytilene, Greece). She is also the Head of the Research Centre of Environmental Education and Communication. She joined the University of the Aegean 15 years ago. Before that she was a Professor at California State University, Los Angeles. She has developed several courses on issues of environmental health and education. She currently teaches environmental education, environmental communication and environmental interpretation courses in undergraduate and postgraduate levels. Professor Skanavis has numerous publications on a international basis and has given presentations all over the world.

Klima|Anlage—Performing Climate Data

Katharina Groß-Vogt, Thomas Hermann, Martin W. Jury,
Andrea K. Steiner and Sukandar Kartadinata

Abstract The urgent need to inform the general public about climate change is evident. Typically, this is done with the aid of visual and textual interpretations of findings of climate research. Other modes of perception might attract more attention. Sonification is a relatively new means of perceptualizing data by translating it into sound. This paper describes the Klima|Anlage, a walk-in sound installation "performing climate data". The climate data for this purpose were obtained from a global climate modeling experiment providing climate projections for the latest assessment report of the Intergovernmental Panel on Climate Change. Climate data from 1950 to 2100 can be chosen interactively by the listener for twelve selected regions of the world. The installation is based on four sound generators: a drip device with controlled drip rate for precipitation data, a record player with marble disks for wind data, a tetrachord instrument that is excited by radiation data, and three thunder sheets that play air temperature data. In addition, purely electronic sounds convey data of the global greenhouse gas concentrations. The Klima|Anlage has been exhibited at several locations since 2015, and excerpts of the sound recordings have been broadcast on Deutschlandradio, a German radio station. Sound and video examples may be accessed at http://klima-anlage.org/ and as

K. Groß-Vogt (✉)
Institute of Electronic Music and Acoustics, University of Music
and Performing Arts Graz, Inffeldg. 10/3, 8010 Graz, Austria
e-mail: vogt@iem.at

T. Hermann
Ambient Intelligence Group, CITEC, Bielefeld University, Bielefeld, Germany
e-mail: thermann@techfak.uni-bielefeld.de

M. W. Jury · A. K. Steiner
Wegener Center for Climate and Global Change, University of Graz, Graz, Austria
e-mail: martin.jury@uni-graz.at

A. K. Steiner
e-mail: andi.steiner@uni-graz.at

S. Kartadinata
Free Music Instrument Builder, Glui, Berlin, Germany
e-mail: sk@glui.de

© Springer Nature Switzerland AG 2019   339
W. Leal Filho et al. (eds.), *Addressing the Challenges in Communicating Climate
Change Across Various Audiences*, Climate Change Management,
https://doi.org/10.1007/978-3-319-98294-6_21

supplementary material to this paper (http://doi.org/10.4119/unibi/2914786). This paper contributes to a greater understanding of how to communicate complex scientific data to the public, using innovative communication channels. Conclusions on the design of the Klima|Anlage can be generalized to other sound installations at the border of science and media arts.

1 Introduction

Current climate change demands for an accurate understanding of posed challenges, which is commonly provided by climate research. The results of scientific research continue to be disseminated within academic circles, with each discipline using its own "pidgin" (Galison 1997) meaning scientific language that is not easily comprehensible to the general public. Nowadays, additional forms of dissemination are being sought by universities as well as public and private funding agencies—the public should be informed in a generally understandable way; however, this is a daunting task, as the authors know from their personal experience in trans-disciplinary research. Furthermore, the public is already saturated with classical means of information, i.e., textual and visual interpretations of research results, and yet remains inadequately informed or unconvinced. New perceptual modes of communication are called for climate research.

Sonification is the translation of information to the auditory perception; an acoustic analogue to visualization. More concisely, it has been defined as "the data-dependent generation of sound, if the transformation is systematic, objective and reproducible, so that it can be used as scientific method" (Hermann 2008). Research on sonification has been conducted for over 25 years within the International Community on Auditory Display, ICAD.[1] Various research showed that sounds have an affective value that facilitates deeper personal access to the implications of the data than standard information channels. Additional benefits of sonification include the ability to listen to multi-dimensional, and multi-rhythmical data sets as climate data often are. These benefits should be exploited in order to communicate climate change more directly to the public and lead to a greater understanding of the topic.

The Klima|Anlage uses sound to display climate data and it is designed in a way that is directly accessible and interesting to a general public. The installation does not intend to display aggregated research findings, but rather to give perceptual access to an entire data set stemming from a global climate model; in this way climate change is experienced aurally and possible skepticism is met by a neutral point of view. Previews of Klima|Anlage have been accompanied by scientific input lectures, thus the questions raised by the installation can be addressed to scientists.

The Klima|Anlage is the cooperative undertaking of sonification experts, climate scientists, media artists, and broadcasters (see acknowledgements). The cooperation

---

[1]www.icad.org (last accessed September 2017).

was established in two predecessor projects: The first was the award-winning[2] online installation #tweetscapes which connected sonification research to media arts (Hermann et al. 2012). Second, in the project "SysSon—A systematic procedure to develop sonifications" the open source software SysSon[3] has been developed for the purpose of turning climate data into sound (Rutz et al. 2015). The main authors of this paper have participated in one or the other of the projects mentioned above. The trans-disciplinary character of the team highlighted different aspects of climate change communication: climate scientists needed to partly rethink and reformulate the description of their research and findings on a more basic level. By the process of translating information to a different domain (here, the sonic domain) different pragmatic, artistic, and technically informed constraints had to be followed, as described in this paper. The resulting consensus is one possible way of performing climate data in a sound installation. Still, general considerations on this translation process can be drawn.

2 Auditory Display of Climate Phenomena

2.1 Displaying Data via Sound

Visual display is regarded as the predominant and most appropriate means of data representation in our society (Dombois 2002). Accordingly, scientists in different domains often focus on the visual presentation when analyzing and interpreting their data. This is also and especially true for climate scientists, as was shown in a survey by Nocke et al. (2008). Climate scientists depend heavily on visualization for exploration tasks and communication of their results. To do so, they mainly use standard visualization tools, e.g., time charts, bar charts, 2D maps, or scatterplots. These approaches do not meet the requirements of large, multivariate climate data and their heterogeneous user groups. More advanced examples have been presented by visualization researchers for developing their tools further. Exploration of climate data using interactive visualization has been carried out, e.g., by Ladstädter et al. (2009, 2010). However, new, state-of-the-art visualizations are not widely accepted within the community. The authors of the above mentioned survey suggest interactivity as an effective way to display climate data in different contexts of research and dissemination. We find that multimedia is another.

Sonification, or more generally auditory display, is a rather new means of displaying scientific data. The term sonification was coined in the 1990s by Kramer et al. (1999). Compared to visualization, sonification is still in its infancy. Basic methods and design principles are still being developed and tested in various fields.

---

[2]In 2012, #tweetscapes received the Award of Distinction in the Digital Musics and Sound Art category of Prix Ars Electronica.

[3]https://sysson.iem.at/ (last accessed September 2017).

A subjective overview of scientific sonifications comprises sounds of electroencephalography (EEG) brain waves (Hermann and Baier 2013), computational physics (Vogt 2010), and human movement [e.g., in sports (Cesarini et al. 2015) and rehabilitation (Vogt et al. 2010)]. One of the earliest attempts to sonify data in a scientific context used seismic data (Speeth 1961) and this field is still being explored by auditory means.[4]

Even if it is still uncommon to use sound to display data, there are good reasons to do so. First of all, human hearing is especially well developed to decipher the spatial (multi-dimensional) and temporal (multi-rhythmical) data that climate models often produce. One prominent example is the cocktail party effect—even in a noisy environment, we are able to follow one voice easily; a task which is still fundamentally challenging for speech recognition devices. The spectral and temporal range of human hearing exceeds that of human vision—we hear over a range of 10 octaves, 20 Hz–20 kHz, but we see only roughly one "octave", 380–780 nm. In monitoring situations, where the eyes are occupied, sound can be used as additional input. And while our eyes have a limited range of vision, our ears perceive the surrounding soundscape, with different degrees of spatial resolution in the azimuthal and elevation plane. We can hear alarms and sounds in the background without conscious effort. Finally, to arouse attention, sound is an ideal medium due to its affective nature.

Besides these advantages, sonification must be treated very differently from visualization for some reasons. First of all, a sonic "overview" in the sense of a static picture (with its accompanying legend) is not possible. Absolute values, e.g., in different audio parameters such as pitch or amplitude, are (individual) exceptions. Even worse, different parameters influence each other psychoacoustically, such that, for instance, the perceived loudness of a sound event depends on its pitch. Although this phenomenon is well known, many other psychoacoustic dependencies are obscure: while 2- or 3-dimensional graphs are orthogonal, sound parameters are not. Furthermore, listening habits are ecologically influenced and have strong cultural biases (e.g., references to music). For all those reasons, acoustic displays still challenge sonification designers and their listeners. A review of the field of sonification research and these challenges from a sociological perspective has been presented by Supper (2011).

2.2   Climate Related Media Projects

Few scientific sonifications of climate research exist. Individual metaphoric sounds, i.e., auditory icons (Gaver 1986), have been studied, e.g., in connection with rain

---

[4]A straightforward listening example of the Tohoku Earthquake in Japan in 2011 can be found at http://www.sonifyer.org/sound/erdbewegung/ (last accessed September 2017).

prediction (Halim et al. 2006) and a non-speech weather report for a local radio broadcaster (Hermann et al. 2003).

The first major effort to use sonification in climate science was the research project "SysSon", in which some of the authors of this paper explored a contextual design for developing a software that translates climate data (in NetCDF format) into sound (Vogt et al. 2012; Goudarzi 2015). With SysSon software, several sonifications of complex climate data were programmed. In the example of Rutz and Höldrich (Rutz and Höldrich 2017), data from satellite-based radio occultation observations (Steiner et al. 2011) have been used to sonify atmospheric variability, especially the Quasi Biennal Oscillation (QBO) (Wilhelmsen et al. 2018).[5]

On the other hand, many projects related to sound and climate at the border of science and the arts have been conducted. A. Polli, a US-based artist and researcher, uses sonification in her work. She sonified, e.g., storm data from weather models (Polli 2005). US artist/researcher M. Quinn focuses on scientific outreach by using sonification using musical sounds. His projects include a one-hour "Climate Symphony" (Quinn 2001) and a sonification of the polar ice caps on Mars for educational purposes (Keller et al. 2003). In 2016, "KlimArs", a contest for climate-related art work, was held in Graz, Austria, where several examples of sonification were presented.[6]

3 The Klima|Anlage

Most existing sonifications use computer-generated sound to represent data. The Klima|Anlage is a walk-in sound installation based on four physical sound generators. The setup is shown in Fig. 3. The sounds are electro-mechanically excited and amplified according to carefully chosen parameters of the climate data, and pure electronic sounds are added to the soundscape.

3.1 Climate Model Data

Triggered by the increased public interest in the impact of human activities on the Earth's climate, intensified research efforts have led to a multitude of data on the topic. Statements about the future climate are made possible by global climate modeling studies. In order to test the respective model's ability to reproduce the climate changes that have already been observed, these studies are driven by historical emission data. Additionally, as collective human behavior, for example

---

[5]Listening examples can be found at https://soundcloud.com/syssonproject/sets/ (last accessed Sept 2017).
[6]http://klimars2016.iem.at/ (last accessed September 2017).

future emissions of greenhouse gases (GHGs), is hardly predictable, possible futures are projected by forcing global climate models (GCMs) with different scenarios ranging from low to high GHG emissions. The current emission scenarios have been used to drive GCMs within the Coupled Model Intercomparison Project phase 5 (CMIP5) (Taylor et al. 2012), which subsequently provided the data base for statements made about the changing climate in the latest Intergovernmental Panel on Climate Change (IPCC) assessment report (Stocker et al. 2013). The Paris climate accord (UNFCCC 2015) which was signed by virtually all countries of the world aims at holding global average temperatures below 2 °C in comparison to pre-industrial levels, which approximately corresponds to climate projections forced with the moderate emission scenario RCP4.5.[7]

Against this background, data used for the Klima|Anlage stems from two datasets, one representing past and future GHG concentrations (forcing of climate models) and one providing information on past and future climate properties including temperature, precipitation, wind, and radiation (climate model output). The GHG concentrations are accounted for in carbon dioxide equivalents, and stem from different sources of observations and the mentioned RCP4.5 emission scenario (Wise et al. 2009). Climate change and global warming are reflected in a multitude of parameters. Here, we used the parameters easily accessible to human perception: air temperature near the surface (TAS) and precipitation rate (PR) accounting for changes in the atmosphere's sensible and latent heat content, respectively. The additionally trapped energy in the climate system leads to changes in planetary circulation systems, as reflected for instance in horizontal wind components. In particular, we account for these alterations by including eastward wind in the upper troposphere and lower stratosphere (at the levels of 200 and 50 hPa). Further, the emission of GHGs in the atmosphere induces changes in the global energy balance that are reflected in the Earth's radiation budget. To account for these changes we included the net downward radiation flux at the top of the atmosphere in the Klima| Anlage. In addition, we included the rather constant incoming solar radiation as represented in the parameter top of the atmosphere incident shortwave radiation. This data set stems from one representative GCM participating in the CMIP5, the MPI-ESM-LR (Giorgetta et al. 2013) of the Max-Planck-Institute for Meteorology in Hamburg, Germany,[8] which has been forced with historical emission data (from 1950 to 2005) and with GHG data of the emission scenario RCP4.5 (2006 until 2100). GCM data have been obtained as monthly mean values and have been bilinearly interpolated to a 2.5° × 2.5° regular Gaussian grid in the case of temperature and precipitation and to 10° zonal means in the case of eastward wind, in order to facilitate the transferability of the Klima|Anlage to other climate models. In addition, radiation data have been used as global annual means.

[7]RCP denotes representative concentration pathway, the number informes about the radiative forcing at the end of the century in W/m².
[8]www.mpimet.mpg.de (last accessed September 2017).

3.2   Selecting the Data Ranges for the Klima\Anlage

Even for an auditory display, that may handle more parallel data streams than classical visualizations approaches, the amount of data to be performed in Klima| Anlage is large: Fig. 1 shows the 72 latitudes times 144 longitudes, i.e. 10,368 cells, that were used as geographical span, for two aggregated time slices as only two dimensions of the data can be plotted in a meaningful way within the scope of a paper. Figure 2 gives an idea of the timely resolution—a period of 150 years given in 1800 monthly means, where the monthly data shows much more statistical noise than aggregated years. For each cell, we chose two climate parameters (precipitation and temperature near the surface). Additionally, the eastward wind (i.e., westwind) is given at two different pressure levels at 18 latitudinal means. Finally, we use global data for the GHG emissions.

For acoustical transparency, we decided to further aggregate the regions to regional subsets. In the installation, twelve regions can be chosen interactively via a graphical user interface (GUI) on a touchpad interface (Fig. 3a). The selected

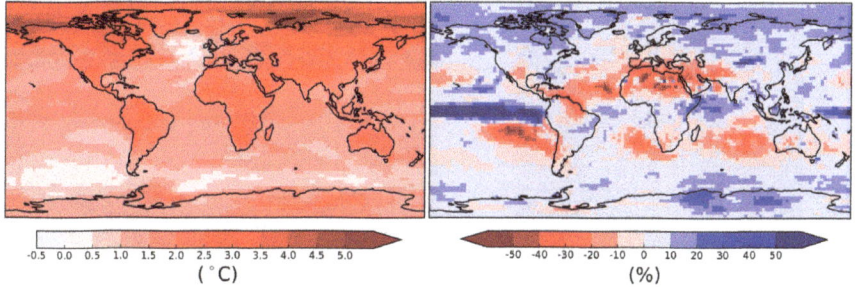

Fig. 1 Projected climate change for temperature near the surface (in °C, left) and precipitation rate (in %, right) until 2071–2100 relative to the climate state of 1976–2005

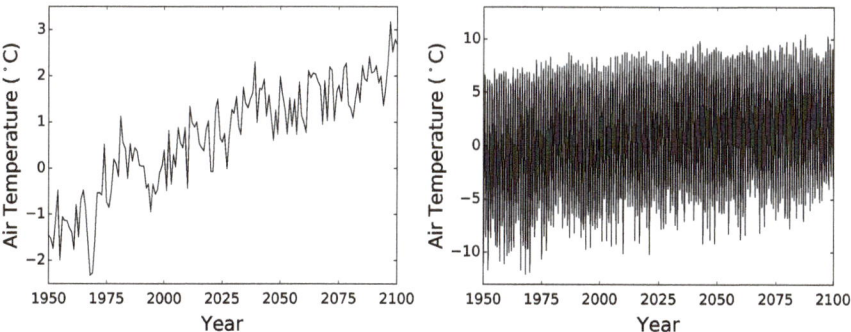

Fig. 2 Temperature change over time (in °C) relative to the state of the climate from 1976–2005 for the Northern Europe region (see Table 1). Annual aggregations (left) show less variability than monthly aggregations (right), the latter are utilized in the Klima|Anlage

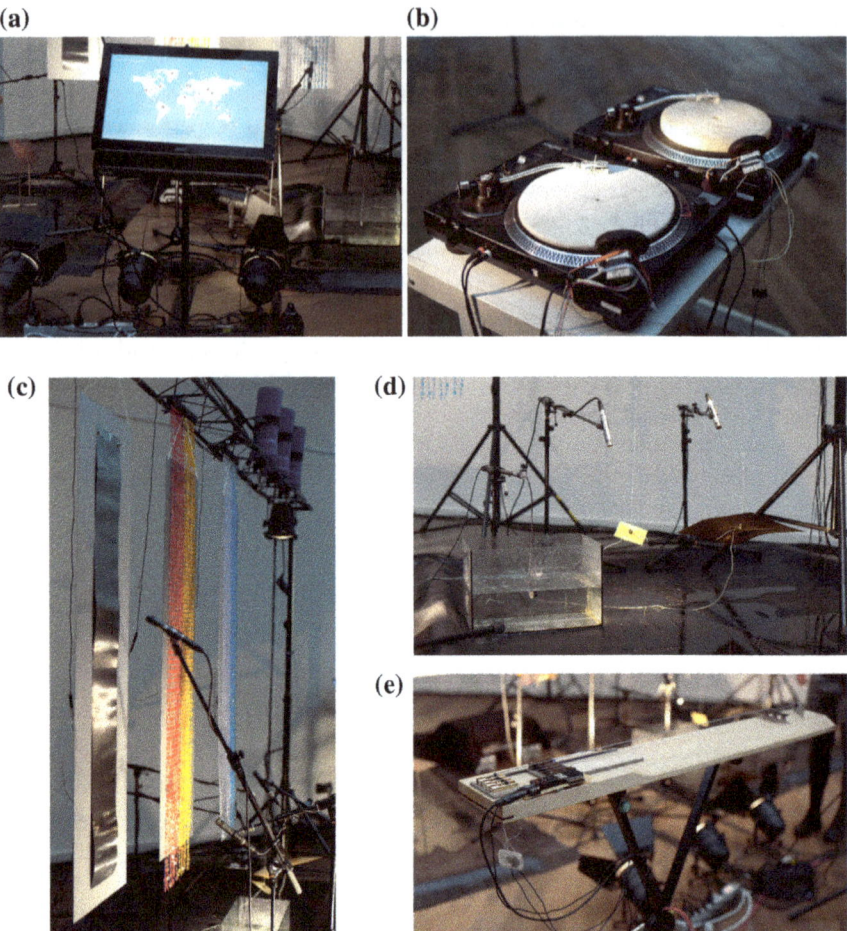

Fig. 3 The Klima|Anlage as presented at the preview in September 2015 in Berlin. **a** GUI—touch interface, **b** record player with two marble disks, **c** three thunder sheets, **d** drip device with different resonators (wooden and metal plates, and water), and **e** the tetrachord. © Deutschlandradio

regions are based on the Fourth Assessment Report of the IPCC (IPCC 2007). Each region stands for a different climate zone and thus provide the opportunity to compare subsets. The selected regions are shown in Table 1.

After the prototype of the Klima|Anlage was built, it turned out that it was difficult to hear subtle changes. For this reason, we decided to cut the data ranges below their 2.5 percentile and above their 97.5 percentile. The omitted data constitute extreme statistical outliers, as the data is timely and spatially averaged (in months and cells of $2.5° \times 2.5°$, respectively). Despite their rarity, these outliers would dominate the sound and shrink the dynamic range of the data. The remaining

Table 1 Selected regions from the fourth IPCC report (IPCC 2007)

Region	Latitude	Longitude	Selected area-center	Label in GUI
Sahara	18N–30N	060W–065E	Agadez	"Sahara"
Southern Europe Mediterranean	30N–48N	010W–040E	Graz	"Middle Europe"
Northern Europe	48N–75N	010W–040E	Berlin	"Northern Europe"
Southern Africa	35S–12S	010E–052E	Johannesburg	"South Africa"
Tibetan Plateau	30N–50N	075E–100E	Lhasa	"Tibet"
Southeast Asia	11S–20N	095E–155E	Jakarta	"Indonesia"
Southern Australia	45S–30S	110E–155E	Canberra	"Australia"
Central Asia	30N–50N	040E–075E	Almaty	"Central Asia"
East Canada, Greenland and Iceland	50N–85N	020–060W	Nuuk	"Greenland"
East Canada, Greenland and Iceland	50N–85N	103W–010W	Montreal	"East Canada"
Western North America	30N–60N	130W–103W	San Francisco	"Western North America"
Tropics	15S–15N	180W–180E	Manaus	"Amazon"

The regions vary in size; in order to ensure comparability, we chose one city within each region and defined a smaller area of 3 × 3 cells around the city's coordinates. These areas were played in the installation. The last column gives the region label as used in the GUI

data could now be mapped to the sound parameters with greater variational detail, as their resolution is also limited due to technical and psychoacoustical factors. Detailed ranges for the data are given in Table 2.

Time was mapped to sonification time so that one year corresponds to 6 s in the Klima|Anlage. With this choice, on the one hand one month is played at half a second. This scale represents an order of magnitude that can be well resolved psychoacoustically. On the other hand, the 15-min playing time of the whole data set is reasonable in a public exhibition setting.

3.3 Generating the Sound

The sonification and sound design for the Klima|Anlage were developed in four workshops by the core project team, comprising some part of the authors and workshop participants. The following requirements were taken into account:

- the installation should give *direct insight* into the chosen climate model data, with little or no data pre-processing, except for required technical compatibility between climate data and sound synthesis/ interaction software systems;

Table 2 Overview of climate parameters: the given and restricted data ranges (which are due to the omission of statistical outliers), and the respective sound parameters with ranges used

Climate data parameter	Given data range	Restricted data range	Sound parameter	Range of sound parameter
Precipitation	[0, 6.66 × 10^{-4}] kg/m²s	[0, 1.2 × 10^{-4}] kg/m²s	Drop period	[50, 0.08] s
Eastward wind at 200 hPa	[−16.2, 53.5] m/s	[−5.84, 40.2] m/s	Spatialization: velocity	Stereo: left–right Min–max[a]
Eastward wind at 50 hPa	[−24.5, 59.1] m/s	[−17.09, 39.06] m/s	Spatialization: velocity	Stereo: left–right Min–max[a]
Temperature near the surface	[−75, 46] °C	[−48.5, 30.1] °C	Excitation amplitude of three thunder sheets for three regimes of temperature (low, medium, and high third of the data set)	See Fig. 4
Solar radiation (top of atmosphere incident shortwave radiation)	[340.19, 340.58] W/m²	[340.22, 340.56] W/m²	Excitation of strings of tetrachord: A2 D4 E4 F#4	Triggered successively if exceeding 25% 40% 50% 60% of the data range
Radiation balance (net downward flux at top of model)	[−1.41, 2.43] W/m²	[0.043, 2.035] W/m²	Detunement of 4th string of the tetrachord	[−150, 90] Hz
Global GHG concentration	[282.35, 585.25] ppm	–	Rate of electronic beeps	[3, 0.5] s

[a]The eastward wind is also mapped to the velocity of the record player, which is an internal parameter of the device itself and only given as maximum and minimum

- the chosen climate parameters should have a real-world correspondence that is *familiar* to the public; air temperature, wind, precipitation, solar radiance, and GHG concentrations were selected to be displayed;
- the sound synthesis should mainly be done with *physical/ acoustical instruments*. Compared to pure electronic sounds played over loudspeakers this choice gives a holistic impression, as the instruments can not only be heard but also watched and walked around;
- the used materials, their visual appearance and excitation mode of each instrument should *refer to the data* it performs. These references were designed

taking cultural and ecological connotations into account (e.g., the rotating of the Earth is represented by a record player turning a marble disk);

- each instrument should have a *distinct sound* in order to reduce masking effects; thus the overall soundscape consists of different sounds that complement each other and change over time.

Four physical sound generators were developed: a drip device, a record player with marble disks, thunder sheets with additional rattle elements, and a tetrachord. In addition, we used purely electronical sound synthesis. The sound generation and the mapping of the data are described in the following. Details of the mapping are shown in Table 2.

(a) *Drip device (Precipitation data)*

The drip device (Fig. 3d) is based on biomedical drip devices that usually regulate the drip rate for infusions. In the Klima|Anlage, the drip rate is varied according to the precipitation data. The water drops fall on resonant plates made of wood and metal, that are suspended from a small stand each. Their sounds are recorded with Piezo microphones. Furthermore, one drip device is placed above an aquarium filled with water. These sounds are recorded with a hydrophone from within a water basin. All drip sounds are amplified and played over loudspeakers. Evidently, in this very direct mapping, small amounts of precipitation produce a few random drops that are spread over different devices. Heavy precipitation, in contrast, produces a dense grain cloud of many metal, wooden, and watery drop sounds.

(b) *Record player—varying velocity and spatialization of marble disks (Wind data)*

A record player (Fig. 3b) turns two marble disks—a thicker is used to represent a higher atmospheric level (50 hPa or approximately 20 km altitude), and a thinner one represents a lower level (200 hPa or approximately 12 km of altitude). Playing a marble disk with a slightly rough surface on a standard record player results in a natural, noisy sound. This raw sound signal is band-pass filtered and varies the position in the stereo panning, respectively according to wind speed and direction.

(c) *Thunder sheets—exciting sheets with mobile rattle elements (Temperature data)*

Three metal plates are suspended from a large stand (Fig. 3c), with two having mobile rattle elements placed immediately in front of them. The plates are excited by transducers at their resonant frequencies. If the sheets are excited a little, the rattle elements are not touched and the sound is a low grumbling, mapping low temperatures. If the excitation is high, the rattle elements add a higher pitched jitter that corresponds to high temperatures. Characteristic transfer functions have been pre-defined for each plate that define how temperature values translate into an actuation for each individual thunder sheet as depicted in Fig. 4.

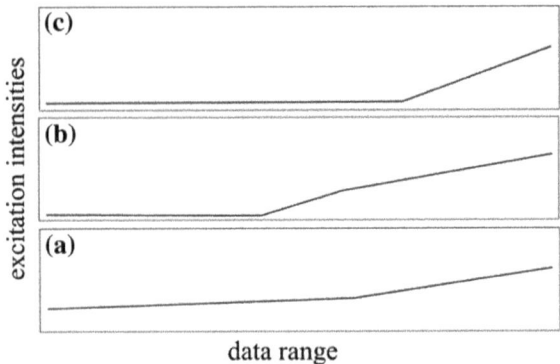

data range

Fig. 4 Excitation of thunder sheets. Characteristic lines of excitation of the three thunder sheets depending on three temperature regimes (**a** low, **b** middle, **c** high thirds of the data). The lines were designed by trial and error according to the mechanical response of the sheets and psychoacoustic considerations in order to achieve a smooth transition between the sheets

(d) *Tetrachord—Exciting strings (Radiation data)*

Envisioned in analogy to the ancient monochord, our tetrachord (Fig. 3e) is a wooden bar on which four strings are mounted. The strings are tuned to the musical notes of A2, D4, E4, and F#4. The strings can be excited individually by electromagnetic drivers. With rising solar radiation, the strings are excited successively from the lowest to the highest, and are triggered as soon as the radiation reading reaches the string-dependent thresholds given in Table 2. The resulting sounds are pitched sounds with a given harmonic structure that open up to a fuller chord with more radiation. The eleven-year solar cycle is clearly audible as a slow pulsation within one minute. Contrary to this harmonic basis, the radiation balance is used to detune the lowest string, with the result that the greater the imbalance is, the greater is the discomfort of the listening experience.

(e) *Electronic sounds (GHG concentration data)*

A synthesized "beep" sound is used to convey global GHG concentrations and provide orientation over time. The beeping rate increases with rising concentrations. The sound is deliberately simple and recognizable as an electronic sound.

Summarizing, the overall soundscape of the Klima|Anlage has distinct sounds that can be followed over time and compared for different regions (one at a time). The eleven-year solar cycle leads to a slowly pulsating harmonic spectrum on a tetrachord instrument, that is slowly detuned according to radiation balance data. Thunder sheets follow shorter, yearly cycles of the air temperature, and oscillate between a low grumbling and high intensity rattling. The noise of two marble plates placed on a record player mimic large wind intensities in the higher atmosphere, changing from left to right in the stereo panorama as the wind direction reverses. Finally, resonant water drops vary in their rate from random single drops to heavy rain sound.

4 Exhibitions

To date, the Klima|Anlage has been demonstrated live on three occasions: the preview in September 2015 was followed by a one-week exhibition in the main building of Deutschlandradio Kultur in Berlin. It was also shown in the course of a climate symposium in the Haus der Bundespressekonferenz, Berlin, in November 2015. A two-week exhibition took place in Gemeindehaus Ruhrort, Duisburg, in June 2016.

On Deutschlandradio, the Klima|Anlage was covered in September 2015 with a first broadcast performance, and continuously until today within the series "Sonarisationen". Short sound snipets of the installation are used in these series to fill time slots of varying length at the end of broadcast radio plays.

The project is well documented online at http://klima-anlage.org/. The website includes various audio examples, background information, and a making-of video. Selected media examples provided with this publication are made available as supplementary material at http://doi.org/10.4119/unibi/2914786.

5 Discussion and Outlook

This paper presented the background, design process, and final layout of the Klima| Anlage, a walkable sound installation "performing climate data". Four physical sound generators represent different parameters of the climate and their change over time: air temperature, wind, precipitation, and solar radiance. As an "artifical" element, human GHG emissions control a synthetic sound that becomes faster as the concentrations increase. The Klima|Anlage has been shown on three occasions. Future exhibits are envisaged. They shall be accompanied by a scientific framework, e.g., introductory talks as has been practiced in the past exhibitions. Furthermore we will use the opportunity of one of the next exhibits to conduct a more formal evaluation, e.g., in the form of semi-structured interviews.

Lessons we learned concern both this final layout of the Klima|Anlage and the trans-disciplinary design process. A successful sonification design is often characterized by a fruitful translation process between one domain, here climate science, to another, i.e. sound. We believe that the result of this process formulates the given scientific problem in a way that makes it generally better understandable. In short, trans-disciplinary projects should be encouraged to better communicate climate change, despite their larger organizational effort. Another experience that can be generalized is the basic design decision of using perceptible entities based on an ecological approach: all listeners may experience as they know—either from a basic empirical background, (e.g., "rain" on the drip device), their general knowledge (e.g., the rotation of the Earth on marble disks), or even from a profound musical acquaintance (e.g., using and detuning the harmonic structure in the tetrachord). In short, we encourage to translate climate change in an ecologically valid way.

The limitations we see in this work are discussed in the following. It was our deliberate intention to play climate data as directly as possible in a sonification, and to avoid any exaggeration in the data-to-sound mapping that would dramatize climate change. This direct approach has advantages and shortcomings. The sound of the Klima|Anlage is a rich, atmospheric soundscape. However, a general public may not easily perceive the many overlapping periodicities. We do not think that the installation in an exhibition visit will enable the public to draw its own conclusions on climate change. But a main outcome of the encounter with the Klima| Anlage is the playful engagement with the workings of climate and the provenience of climate data. *How are climate parameters—sun, rain, wind, temperature, gas emissions— related to each other? What are the differences between the various regions? How do climate models work? What is climate variability and what is climate change?*

During the development of the Klima|Anlage, the trans-disciplinary cooperation provided a good basis for feedback and design. The installation was well received at all exhibition events. However, a structured evaluation would be interesting to the authors, as it might contribute to improvements in the installation. Therefore, in a future display of the installation an evaluation is envisaged.

The presented installation is one way of making the public more aware of results of climate research. General conclusions can be drawn from the sonification of climate phenomena, e.g., the use of "tangible" data parameters that are mapped as directly as possible to meet the realm of experience of the listeners. Multimedia applications often attract more attention than classical communication channels, especially if sound is involved. It has to be noted, however, that new media also require new interfaces in order to work more effectively. The Klima|Anlage is a large, multifaceted installation. It is laborious to transport and set up, and greater effort is required for its maintenance (e.g., due to water management), compared to simpler media installations. This effort requires larger financial support which is difficult to obtain even for a highly relevant topic as climate change. For this reason, the Klima|Anlage has only been shown a few times until now. On the other hand, the richness of this multisensory installation provides a unique opportunity for the public to gain direct and affective insight into climate data.

Acknowledgements Marcus Gammel initiated the project and led it with great care and involvement. Werner Cee was responsible for the main sound layout and for creating the instruments. Esther Schelander accompanied the project workshops and created the feature for Deutschlandradio. Valdis Wish of the Bureau for Digital Good is responsible for the website. Thanks to Hanns Holger Rutz for help with the preparation of data. We acknowledge the World Climate Research Programme's Working Group on Coupled Modelling for the provision of CMIP5 model output, as well as the climate modeling group of the Max-Planck-Institute for Meteorology in Hamburg, Germany, for producing and making available their model output. This research was partly supported by the Cluster of Excellence Cognitive Interaction Technology 'CITEC' (EXC 277) at Bielefeld University, which is funded by the German Research Foundation (DFG). This work was funded by the Austrian Science Fund (FWF) under research grant P24159 (SysSon). Many thanks to Hank Fullenwider for proofreading and giving feedback for this manuscript.

References

Cesarini D, Ungerechts B, Hermann T (2015) Swimmers in the loop: sensing moving water masses for an auditive biofeedback system. In: 2015 IEEE sensors applications symposium (SAS) proceedings, IEEE, https://doi.org/10.4119/unibi/2718039

Dombois F (2002) In: Schürmann A, Weiss B (eds) Wann hören? Vom Forschen mit den Ohren

Galison P (1997) Image and logic. A material culture of microphysics. The University of Chicago Press, Chicago. ISBN 9780226279176

Gaver WW (1986) Auditory icons: using sound in computer interfaces. Hum Comput Interact 2:167–177. https://doi.org/10.1207/s15327051hci0202_3

Giorgetta MA, Jungclaus J, Reick CH, Legutke S, Bader J, Böttinger M, Brovkin V, Crueger T, Esch M, Fieg K, Glushak K, Gayler V, Haak H, Hollweg H-D, Ilyina T, Kinne S, Kornblueh L, Matei D, Mauritsen T, Mikolajewicz U, Mueller W, Notz D, Pithan F, Raddatz T, Rast S, Redler R, Roeckner E, Schmidt H, Schnur R, Segschneider J, Six KD, Stockhause M, Timmreck C, Wegner J, Widmann H, Wieners K-H, Claussen M, Marotzke J, Stevens B (2013) Climate and carbon cycle changes from 1850 to 2100 in MPI-ESM simulations for the coupled model intercomparison project phase 5. J Adv Modeling Earth Syst 5:572–597. https://doi.org/10.1002/jame.20038

Goudarzi V (2015) Designing an interactive audio interface for climate science. IEEE Multimed 22:41–47. https://doi.org/10.1109/MMUL.2015.4

Halim Z, Baig R, Bashir S (2006) Sonification: a novel approach towards data mining. In: 2006 International Conference on Emerging Technologies, Peshawar, IEEE, 548–553. https://doi.org/10.1109/ICET.2006.336029

Hermann T (2008) Taxonomy and definitions for sonification and auditory display. In: Susini P, Warusfel O (eds) Proceedings of the international conference on auditory display (ICAD 2008), IRCAM

Hermann T, Baier G (2013). Sonification of the human EEG. In: Franinović K, Serafin S (eds) Sonic interaction design. MIT Press, Cambridge, pp 285–297

Hermann T, Drees JM, Ritter H (2003) Broadcasting auditory weather reports—a pilot project. In: Brazil E, Shinn-Cunningham B (eds) Proceedings of the international conference on auditory display, Boston University Publications, Production Department

Hermann T, Nehls AV, Eitel F, Barri T, Gammel M (2012) Tweetscapes—real-time sonification of twitter data streams for radio broadcasting. In: Nees MA, Walker BN, Freeman J (eds) Proceedings of the international conference on auditory display (ICAD 2012)

Keller JM, Prather E, Boynton VW, Enos LH, Jones VL, Pompea S, Slater T, Quinn M (2003) Educational testing of an auditory display regarding seasonal variation of martian polar ice caps. In: Brazil E, Shinn-Cunningham B (eds) Proceedings of the international conference on auditory display (ICAD 2003)

Kramer G, Walker B, Bonebright T, Cook P, Flowers J, Miner N, Neuhoff J, Bargar R, Barrass S, Berger J, Evreinov G, Fitch W, Gröhn M, Handel S, Kaper H, Levkowitz H, Lodha S, Shinn-Cunningham B, Simoni M, Tipei S (1999) The sonification report: status of the field and research agenda. Report prepared for the national science foundation by members of the international community for auditory display. ICAD, Santa Fe

Ladstädter F, Steiner AK, Lackner BC, Kirchengast G, Muigg P, Kehrer J, Doleisch H (2009) SimVis: an interactive visual field exploration tool applied to climate research. In: Steiner AK, Pirscher B, Foelsche U, Kirchengast G (eds) New horizons in occultation research: studies in atmosphere and climate. Springer, Berlin

Ladstädter F, Steiner AK, Lackner BC, Pirscher B, Kirchengast G, Kehrer J, Hauser H, Muigg P, Doleisch H (2010) Exploration of climate data using interactive visualization. J Atmos Ocean Technol, 667–679. https://doi.org/10.1175/2009jtecha1374.1

Nocke T, Sterzel T, Böttinger M, Wrobel M (2008) Visualization of climate and climate change data: an overview. Digital earth summit on geoinformatics 2008: tools for global change research (ISDE'08). Wichmann, Heidelberg, pp 226–232

Polli A (2005) Atmospherics/weather works: a spatialized meteorological data sonification project. Leonardo 38:31–36. https://doi.org/10.1162/leon.2005.38.1.31

Quinn M (2001) Research set to music: the climate symphony and other sonifications of ice core, radar, DNA, seismic and solar wind data. In: Proceedings of the international conference on auditory display (ICAD 2001)

Rutz HH, Höldrich R (2017) A sonification interface unifying real-time and offline processing. In: Proceedings of the 14th sound and music computing conference (SMC), Espoo, SMC

Rutz HH, Vogt K, Höldrich R (2015) The SysSon platform: a computer music perspective of sonification. In: Vogt K, Andreopoulou A, Goudarzi V (eds) Proceedings of the 21st international conference on auditory display (ICAD 2015)

Speeth S (1961) Seismometer sounds. J Acous Soc Am 33:909–916

Steiner AK, Lackner BC, Ladstädter F, Scherllin-Pirscher B, Foelsche U, Kirchengast G (2011) GPS radio occultation for climate monitoring and change detection. RadioSci 46. https://doi.org/10.1029/2010rs004614

Stocker T, Qin D, Plattner G-K, Tignor M, Allen S, Boschung J, Nauels A, Xia Y, Bex V, Midgley P (eds) 2013. Climate change 2013: the physical science basis. Contribution of working group I to the fifth assessment report of the Intergovernmental Panel on Climate Change. Cambridge University Press, Cambridge. https://doi.org/10.1017/cbo9781107415324

Supper A (2011) The search for the "killer application": drawing the boundaries around the sonification of scientific data. In: Pinch T, Bijsterveld K (eds) The Oxford handbook of sound studies. Oxford University Press, Oxford. https://doi.org/10.1093/oxfordhb/9780195388947.013.0064

Taylor KE, Stouffer RJ, Meehl GA (2012) An overview of CMIP5 and the experiment design. Bull Am Meteor Soc 93:485–498. https://doi.org/10.1175/BAMS-D-11-00094.1

UNFCCC (2015) Adoption of the Paris agreement. http://unfccc.int/resource/docs/2015/cop21/eng/l09r01.pdf

Vogt K (2010) Sonification of simulations in computational physics. University of Music and Performing Arts Graz

Vogt K, Goudarzi V, Höldrich R (2012) SysSon—a systematic procedure to develop sonifications. In: Proceedings of the international conference on auditory display (ICAD 2012)

Vogt K, Pirrò D, Kobenz I, Höldrich R, Eckel G (2010) PhysioSonic—evaluated movement sonification as auditory feedback in physiotherapy. In: Ystad S, Aramaki M, Kronland-Martinet R, Jensen K (eds) Auditory display. Lecture notes in computer science, vol 5954. Springer, Berlin. https://doi.org/10.1007/978-3-642-12439-6_6

Wilhelmsen H, Ladstädter F, Scherllin-Pirscher B, Steiner AK (2018) Atmospheric QBO and ENSO indices with high vertical resolution from GNSS radio occultation temperature measurements. Atmos Meas Tech 11:1333–1346. https://doi.org/10.5194/amt-11-1333-2018

Wise M, Calvin K, Thomson A, Clarke L, Bond-Lamberty B, Sands R, Smith SJ, Janetos A, Edmonds J (2009) Implications of limiting CO_2 concentrations for land use and energy. Science 324:1183–1186. https://doi.org/10.1126/science.1168475

Dr. Katharina Vogt is Senior Scientist at the Institute of Electronic Music and Acoustics, University of Music and Performing Arts, Graz, Austria. She has been working on sonification of various data, ranging from physiotherapy to climate modeling and computational physics (for which she received her Ph.D. in 2010). She was chair of the International Conference on Auditory Display (ICAD) 2015 and served on the Board of Directors of ICAD (2012–2016).

Dr. Thomas Hermann is director of the Ambient Intelligence Group within the Center of Excellence in Cognitive Interaction Technology (CITEC) at Bielefeld University, Germany. His research interests include sonification, ambient intelligence, human-computer interaction, and cognitive interaction technology. After his physics degree he received a PhD in computer science from Bielefeld University. He served on the Board of Directors of ICAD (2004–2013), as vice-chair of the EU COST Action Sonic Interaction Design (2007–2011) and coordinated the working group on Sonification therein. He cofounded together with Andy Hunt the triennial Interactive Sonification Workshop series and coedited The Sonification Handbook (Logos Verlag, 2011) together with Andy Hunt and John Neuhoff.

Dr. Martin W. Jury is PostDoc scientist in the Regional Climate research group at the Wegener Center for Climate and Global Change, University of Graz, Austria. He holds a degree in environmental system sciences and is currently working in the field of statistical and dynamical downscaling of global climate model data.

Prof. Dr. Andrea K. Steiner holds a Ph.D. degree in meteorology and geophysics from the University of Graz. She studied in Biosphere 2, AZ, USA, and was visiting scientist at the Danish Meteorological Institute (DMI), Copenhagen, Denmark, the University Corporation for Atmospheric Research (UCAR) and the National Center for Atmospheric Research (NCAR), Boulder, CO, USA. Lecturing since 2003 at the University of Graz, she obtained in 2013 the venia docendi in geophysics and environmental system sciences and was appointed tenured professor in 2017. She is Vice Director of the Wegener Center for Climate and Global Change, University of Graz, Austria, and Vice Head of its Atmospheric Remote Sensing and Climate System (ARSCliSys) Research Group. Her research in atmospheric and environmental physics focuses on atmospheric remote sensing and use for climate research. She is an expert on radio occultation and its application for atmosphere and climate.

Sukandar Kartadinata is a musical instrument builder with a focus on electronic and software designs, although traditional wood and metal working can play a role too. He has helped many musicians and artists implement their projects that required custom technology, often involving sensors and spatialization. He has worked at ZKM (Karlsruhe), STEIM (Amsterdam), and CNMAT (Berkeley), but has been self-employed in Berlin for many years. Kartadinata studied computer science at FH Karlsruhe. He is also active as a musician.

Media Based Education and Motivation Through Phrasing: Can They Affect Climate Change Willingness?

Konstantinos Tsamopoulos, Kalliopi Marini
and Constantina Skanavis

Abstract In the field of environmental communication, short and memorable phrases are often used to convey a message, either in the form of assertive commands or in suggestions. In this article, we are discussing the effects that a video can cause on the willingness of the subjects to adopt proenvironmental behaviors, as well as the effects of assertive and non-assertive slogans on motivation in this area. The sample of the research consisted of 103 students studying in the School of Environment at the University of the Aegean. The results were calculated by measuring how each student perceives the importance of his/her actions related to environmental protection, by answering a questionnaire about their daily habbits; how they contribute to climate change and whether they are willing to change their behavior, before and after viewing an educational video about climate change. The video used was the winner of the Film4Climate competition of 2016. After the video, the subjects were asked to answer again knowledge and attitude questions to measure the changes that occured. Then, they were asked to choose 1 out of 3 possible options regarding the type of a slogan for the video. The first slogan was assertive, the second was non assertive and the third option stated that the language form is irrelevant/the video alone would suffice. The key implication of the findings is that the video use is a great tool for environmental communication since it affects not only the knowledge of the subjects, but also their possible intentions to change their behavior too. The video has an effect on both knowledge, but also a large effect in their intentions to partake in environmental decision making. It was also observed that women were influenced more than men. The present research con-

K. Tsamopoulos (✉) · K. Marini
Department of Environment, University of the Aegean, Mytilene, Greece
e-mail: mar12074@marine.aegean.gr

K. Marini
e-mail: kmarini1@gmail.com

C. Skanavis
Department of Environment, Research Center of Environmental
Communication and Education, University of the Aegean, Mytilene, Greece
e-mail: cskanav@aegean.gr

© Springer Nature Switzerland AG 2019
W. Leal Filho et al. (eds.), *Addressing the Challenges in Communicating Climate
Change Across Various Audiences*, Climate Change Management,
https://doi.org/10.1007/978-3-319-98294-6_22

357

firms the existing research and emphasizes the use of media on environmental education and communication.

1 Introduction

Climate change is one of the most serious challenges, humanity is facing today (qtd. in Fernandez et al. 2014). As stated in UNDP (qtd. in Fernandez et al. 2014), climate change is an enormous threat to human development and is already halting efforts to reduce extreme poverty. It also has hampered attempts to achieve the "Millennium Development Goals". As indicated in Fernandez et al. (2014), climate change presents a global challenge, which we have not encountered previously and it's impacts have implications on socioeconomic and political stability and security from local to international levels.

As IPCC states, climate change is a problem with implications registered on local to global scales, a problem which seems not specific enough to many people (Dulic et al. 2016).

Our ultimate goal is to help communities move towards more sustainable lifestyles, using methods to communicate climate change action interactively and in culturally relevant ways (Dulic et al. 2016).

1.1 Climate Change Communication

With the anthropogenic climate change first emerging on the public agenda in the 1980s, public communication of climate change and how to effectively communicate it, has witnessed a steep rise (Moser 2010). Climate change communication, developing out of science communication, aims to bring knowledge obtained through experimentation into the public arena, in order that the findings are visible and relevant to everyday life (Dulic et al. 2016).

Today, there is consensus within the scientific community about the anthropogenic causes of climate change, its severe consequences and the need for mitigation and adaptation actions. Hulme (qtd. In Schroth et al. 2014) takes a closer look at the question of why we disagree about climate change and comes to the conclusion that we receive multiple and conflicting messages about climate change which are also interpreted in different ways. Hulme presents alternative communication models, specifically rejecting the traditional "deficit model" where scientists inform the public through seemingly neutral media in a one-way communication process (Schroth et al. 2014).

Climate change communication—after years of practice without a solid foundation of research—is now of keen interest to those interested in increasing public engagement, and has emerged as a field of research in its own right.

Great effort has gone into developing and implementing campaigns to persuade people to adopt environmental behaviors (Chang 2012; Hartmann and Apaolaza-Ibanez 2009; White and Simpson 2013). These campaigns often use assertive language, which is known to convey a single meaning that can open the advocated view to critical scrutiny (O'Keefe 1997) and leave no doubt as to its intentions (Miller et al. 2007). In particular, assertive forms messages are frequently used to promote environmental messages (Baek et al. 2015). It is very often though, that campaigns use videos in order to promote a message since it is a way proven (Pearson et al. 2011) to motivate and educate.

1.2   Media-Based Education and Behavior Change

The media provides a crucial avenue for environmental education, which is often limited beyond formal schooling (Ballantyne and Packer 2005). Huckle (1995) points out that 'it is by watching television that many of the world's people acquire an awareness and understanding of environment and environmental issues near and far. Images and sounds from television are increasingly significant in shaping their beliefs, attitudes and identities ...'. This is supported empirically, with many studies reporting relationships between types of television viewing, news media coverage, and documentary watching with higher levels of environmental knowledge, concern, and/or action (Barbas et al. 1991, 2009; Eagles and Demare 1999; Holbert et al. 2003). The power of media-based communication is the fact that it can reach very large audiences. For this reason, television in particular, has been used widely to communicate tailored mass-media messages on behalf of governments and health organisations designed to influence important health behaviors (Pearson et al. 2011). Successful outcomes have included decreasing potentially harmful behaviors such as speeding or smoking, as well as increasing desired behaviour such as seat-belt use and exercise (Bauman et al. 2006; Delaney et al. 2004; Schar and Gutierrez 2001). However, the capacity for mass-media campaigns to create similar learning and behavioural change in the area of biodiversity conservation, and to communicate messages on behalf of conservation organisations, remains relatively unknown (Pearson et al. 2011). Similarly, there is a paucity of research which considers the outcomes of internet based information provision. This is increasingly important as 3.8 billion of the world's population are connected to the internet (more than half of the global population) (Internet World Stats 2017), representing an enormous audience that can be targeted with conservation messages online. The importance of targeting internet users was posed by Pearson et al. (2011). Health related campaigns have proven to be successful (Pearson et al. 2011), so in this case, we are following a similar approach for environmental problems.

1.3 Compliance with Assertive Language

The protection of the environment and its resources is a matter of utmost importance for consumers, businesses, governments and generally, the society (Banerjee et al. 2003; Grinstein and Udi 2009; Menon and Menon 1997; Peattie and Peattie 2009). However, in several cases, it is not perceived like that, from many individuals or groups (Lord 1994). This creates the need of ways to persuade them that a proenviromnental behavior is needed.

A great amount of effort has gone into developing campaigns to create proenvironmental behaviors. The majority of them use assertive slogans. Some of these slogans are: the Ad Council's "Only YOU can prevent forest fires"; Greenpeace's "Stop the catastrophe"; Plant for the Planet's "Stop talking. Start planting"; and Denver Water's campaign "Use only what you need."(Kronrod et al. 2012). An assertive request is one that uses the imperative form, such as "do," "go," and so forth, or one that leaves no option for refusal, such as "you must go" (Brown and Levinson 1987; Vanderveken 1990). Existing research in the fields of communications, consumer behavior and psycholinguistics (Dillard and Shen 2005; Dillard et al. 1997; Edwards et al. 2002; Gibbs 1986; Holtgraves 1991; Quick and Considine 2008; Wilson and Kunkel 2000) suggests that, assertively phrased requests decrease compliance with the message in comparison to non-assertive phrases.

Assertive language that uses imperatives rather than propositions or indirect suggestions explicitly pressures individuals to conform to commands or orders (Miller et al. 2007).

Examples of assertively phrased ads and slogans include Nike's 'Just Do It,' Sprite's 'Obey your thirst,' and Wendy's 'Do what tastes right' (Kronrod et al. 2012). Highly assertive, directive messages could be viewed as controlling and might contribute to a sense of helpless dependence, rather than confident independence (Lanceley 1985).

Many pieces of research (Bensley and Wu 1991; Dillard and Shen 2005; Quick and Considine 2008; Quick and Stephenson 2007; Wilson and Kunkel 2000) have highlighted the effects of message assertiveness on compliance seeking requests. Compelling evidence shows that when assertive messages are used frequently, consumers are less likely to comply. Since assertive messages clearly and directly tell individuals what to do, the messages tend to be perceived as a threat to freedom and to trigger reactance, (qtd. Kronrod et al. 2012).

Reactance produces a "boomerang effect" and actually causes recipients to behave in opposite ways from the behaviors endorsed by the persuasive communication (O'Keefe 1997; Yoon et al. 2011). To illustrate, Dillard and Shen (2005) investigated how a perceived threat to freedom, prompted by assertive language, influenced the effectiveness of health messages advocating flossing and responsible drinking. Their study demonstrated that reactance, comprising anger and negative cognitions, mediates the interactive effects of the perceived threat to freedom and reactance traits on attitudes toward the message and behavioral intentions. Similarly, controlling language has been shown to be perceived as a greater threat to freedom, generating greater levels of anger and more negative evaluations of message fairness (Miller et al. 2007).

However, according to Kronrod et al. (2012), when the environmental problem is perceived as important, the messages supporting this issue must be in assertive language, since they result in higher compliance. Fazio (qtd. Kronrod et al. 2012) argues that assertive language is more likely to be used in cases in which it is in line with already-formed attitudes. In contrast, weak and polite requests in this context might be considered irritating or "too polite" (qtd. Kronrod et al. 2012) and as a result, this may reduce compliance because nonassertive language is not in tune with the issue's perceived importance.

In this article, we are discussing the effects that a video can cause on the willingness of the subjects to adopt proenvironmental behaviors as well as the effects of assertive and non-assertive slogans following a video.

2 Methodology

The sample of the research consisted of 103 students from the Environmental Department of the University of Aegean. The ages of participants were between 18 and 30 and they were undergraduate students, postgraduate students and doctoral candidates. The study was conducted during September, 2017 and lasted 10 days. The first questionnaires were handed out on the 3rd of September and the final questionnaire was answered on the 13th of September.

Participants had to fill an online questionnaire that was posted on student groups from the University of the Aegean on Facebook. In order to maintain the questionnaire's confidentiality, it was posted anonymously. In the beginning of the questionnaire, there was a disclaimer stating that in order to complete it, you must be under 30 and understand English. It was also noted that the answers were anonymous. The questionnaire began with demographic questions such as Gender, Age, Education level, place of birth and place of growing up. We then proceeded to ask about the subjects' general knowledge about environmental problems using Yes/No questions, divided into two categories. The first category included questions that helped establish a baseline on the subjects' knowledge. The second category included the questions that were to be asked after the video again, in order to measure a change on the knowledge. Following, the subjects had to answer some questions that measured their environmental attitude. For these questions, we used a 1–7 Likert scale, (1-Not at all, 7-Very much). In addition to these questions, the subjects had to answer some questions based on their daily habits which reflected their behavior towards the environment. The answers were also on a Likert scale

(1-Not at all, 7-Very much). Despite no specific guidelines exist as to the range of Likert scale when measuring environmental attitude or behavior, we used the 7-scale in accordance with Ajzen's (2013) theory of planned behavior question-naires. Additionally, they were asked questions about their involvement in volun-teer actions, non-profit organizations. Before viewing the video, they also had rate their involvement in environmental decision-making, on a 1–7 Likert scale.

Before proceeding to the next part, the subjects had to watch a video about climate change. The video used was the winner of the Film4Climate 2016 com-petition (https://www.youtube.com/watch?v=VrzbRZn5Ed4). The jury of the competition was comprised of famous directors, journalists, environmental activists and environmental journalists, ensuring not only the quality of the video as a medium, but the quality of the content from an educational perspective. After the completion of the video, the subjects had to answer again the previous knowledge answers again. Likewise, they had to answer questions about their intentions to change their behavior and involvement on environmental matters. Finally, they were given the decision to chose a slogan to accompany the video considering which language form they felt that it moved them more. There were 3 possible options. The assertive slogan was "The planet needs you. You must help", followed by the non assertive "The planet needs you. Could you too help?". The final option stated that the language form of the slogan was irrelevant or that the video would suffice, so there was no need for a slogan. The language forms were chosen according to Brown and Levinson and Vanderveken (Brown and Levinson 1987; Vanderveken 1990).

Men and women comprised the 97% (100) of the pool, 48.5% (50) each, while the remaining 3% (3) defined their sex as "Other". The subjects had the option to decline to choose between male and female, if they didn't identify, by choosing other. Concerning the age groups, 9.7% were between the ages of 18–20, 31,1% between the ages 24–26, 6.8% between the ages of 27–29 while the majority of 52.4% was between the ages of 21–23. The Undergraduate students represented the 88.3% of the sample pool, 10.7% were Postgraduate students and only 1 subject was a doctoral candidate. Finally, the 88.3% of the pool was born and raised in an urban area, as opposed to the rest that were born and raised in rural areas. 93% were currently residing in an urban area while the rest in a rural area (Figs. 1, 2, 3 and 4).

Fig. 1 Gender (percentage)

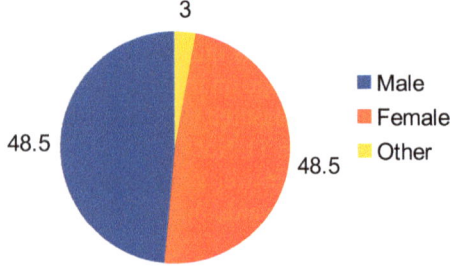

Fig. 2 Age (percentage)

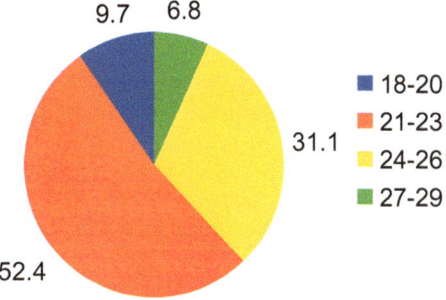

Fig. 3 Academic level
(percentages)

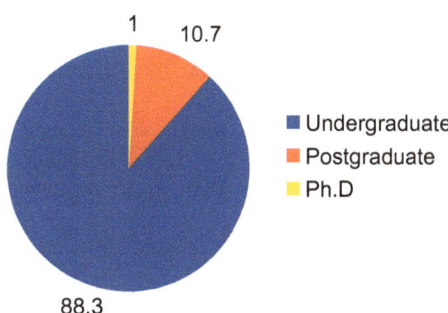

Fig. 4 Place of birth/growing
up and residence (percentage)

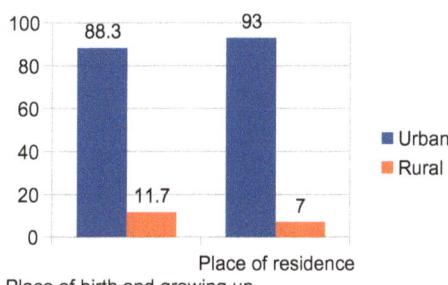

3 Results

The subjects were asked to answer questions related to their general knowledge on climate change related problems. Table 1 depicts the right and wrong answers, as well as the percentages.

From the total of 1133 answers given by the subjects, the 79.87% were correct. The question with the most right answers concerned the decrease of biodiversity (96.1%), followed by the question about the pollution of air by industries (94.1%). The questions with the lowest scores were about the decomposition rate of

Table 1 Knowledge answers

	Right	Wrong	Right (%)	Wrong (%)
Plastic bags do not decompose in landfills	85	18	82.5	17.5
Aluminium decomposes faster than iron or steel	52	51	50.5	49.5
Fish contain mercury in greater volumes than allowed	88	15	85.4	14.6
Industry is responsible for a great percentage of air pollution	97	6	94.1	5.9
Electric devices on sleep mode, do not consume electricity	82	21	79.6	20.4
Diesel pollutes less than gasoline	65	38	63.1	36.9
Climate change causes an increase in the number and power of storms	73	30	70.9	29.1
Climate change causes an increase in the number and power of hurricanes	77	26	74.8	25.2
Climate change causes an increase in the number of droughts	96	7	93.2	6.8
Climate change causes an increase in the number of wildfires	91	12	88.3	11.7
Climate change causes an decrease of biodiversity	99	4	96.1	3.9

aluminium in comparison to iron and steel (50.5%) and the pollution attributed to diesel in comparison to gasoline (63.1%) (Fig. 5).

In order to measure the knowledge change that the video caused, we calculated the difference in the right and wrong answers before and after the video (Table 2). The statistical significance of the differences as well as the effect size of the video intervention is shown detailedly below in the section Paired-samples t-test

There has been an increase in the number of correct answers by 36 points (8.72%) after watching the video. There has been an increase on all correct answers except the one concerning the decrease of biodiversity, in which the correct answers were fewer by one (Fig. 6; Table 3).

The subjects thought that environmental laws are crucial to the protection of the environment (M = 6.81, Std. = 0.44), while they advocated the existence of national parks (M = 6.67, Std. = 0.6). The increase of quality of life through protecting the environment also appeared to be a common belief (M = 6.53,

Fig. 5 Percentage of right and wrong scores of general knowledge answers

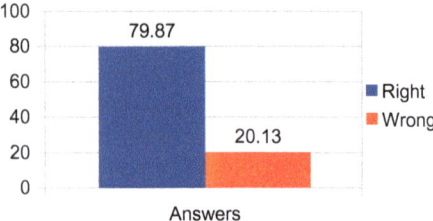

Table 2 Knowledge answers before and after video

Climate change causes	Increase in number and power of storms	Increase in power and number of hurricanes	Increase of droughts	Increase of wildfires	Decrease of biodiversity
Right before video	73	77	96	91	99
Wrong before video	30	26	7	12	4
Right after video	88	88	100	100	98
Wrong after video	15	15	3	3	5

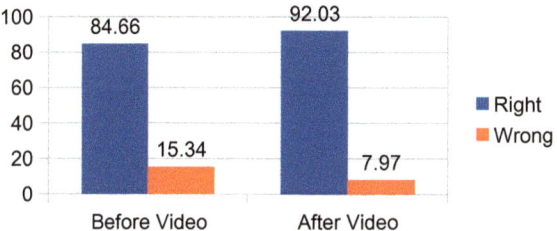

Fig. 6 General knowledge answers before and after the video

Table 3 Proenvironmental attitude

	Min	Max	M	Std.
Laws that protect the environment must exist	5	7	6.81	0.44
There must be national parks	4	7	6.67	0.6
Effectively protecting the environment would increase the quality of life	2	7	6.53	0.78
Small human interference in nature may have great effects	1	7	5.81	1.27
Natural energy resources must be protected in order to maintain high living standards	2	7	5.73	1.15
I am concerned by trash management	1	7	5.64	1.42
Most of the technological advances so far, have contributed to the increase of climate related problems	1	7	5.02	1.22
Reusing contributes more to the protection of the environment than recycling	1	7	5	1.67
There must be technological advancements even at the expense of the environment	1	7	3.53	1.43
The utilization of coastal areas to achieve the growth of tourism is pretty important, so the laws concerning the matter must be less strict	1	7	3.11	2.12

Table 4 Means and standard deviations of proenvironmental behavior before and after the video

	Before	Standard deviation	After	Standard deviation
Use means of mass transportation	4.92	2.04	5.59	1.7
Usage of unbleached paper	3.54	1.24	5.6	1.61
Meat consumption	5.34*	1.66	4.29	2.1
Turning off water tap	5.69	1.69	6.36	1.24
Usage of energy saving lightbulbs	5.74	1.39	6.25	1.31
Recycling	5.19	1.74	6.27	1.21
Usage of plastic bags	4.72*	1.71	6.09	1.35
Usage of palm oil products	3.2*	1.43	5.7	1.53

*The higher mean before reflects a negative proenvironmental lifestyle

Std. = 0.78). The subjects rated with the lowest score the utilization of coastal areas over the flexibility of laws on environmental protection (M = 3.11, Std. = 2.12) and the advancement of technology at the expense of the environment (M = 3.53, Std. = 1.43).

From the answers given in Table 4, in the questions 3,7 and 8, a higher mean before the video reflects a negative proenvironmental lifestyle and vice versa. In Meat consumption, the mean after watching the video is lower, indicating that the subjects expressed a decrease in their intention to continue consuming meat (Fig. 7).

We calculated a total score for proenvironmental behavior before video, proenvironmental behavior after video and the total proenvironmental attitude (Table 5).

Fig. 7 Means of proenvironmental behavior before and after the video

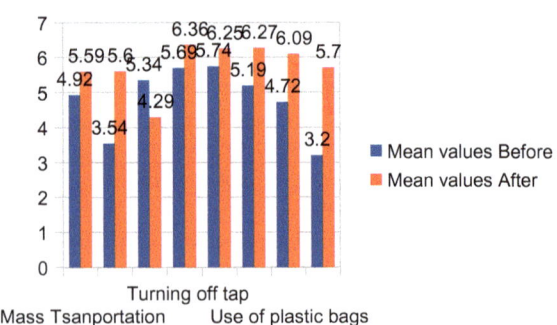

Table 5 TPEB/TPVPEB/Tenvatt

	N	Min	Max	M	Std. deviation
TPVPEB	103	8	56	46.26	9.51
TPEB	103	21	46	35.82	5.58
Tenvatt	103	34	68	56.56	5.24

The mean of total proenvironmental behavior (TPEB) was M = 35.82 (SD = 5.58). The mean of total environmental attitude (Tenvatt) was M = 56.56 (Std. = 5.24). The mean of total post video proenvironmental behavior was M = 46.26 (Std. = 9.51).

These totals were calculated in order to investigate any possible relations between them as well as to run the statistical controls (before and after video) needed for the research questions.

3.1 Correlation Between TPEB/TPVPEB/Tenvatt

The relationship between total proenvironmental attitude (Tenvatt), total proenvironmental behavior (TPEB) and total post video proenvironmental behavior (TPVPEB) was investigated using Pearson product-moment correlation coefficient. There was a medium positive correlation between total environmental attitude and total proenvironmental behavior [r = 0.41, N = 103, Sig. = 0.000] with high levels of total environmental attitude associated with high levels of total proenvironmental behavior. as well as between total environmental attitude and total post video proenvironmental behavior [r = 0.328**, N = 103, Sig. = 0.001]. There was also a high positive correlation between total proenvironmental behavior and total post video proenvironmental behavior [r = 0.503**, N = 103, Sig. = 0.000] (Table 6).

3.2 Paired Samples t-Test

The following tests show us the statistical significance of the difference of proenvironmental behavior before and after the video.

A paired-samples t-test was conducted to evaluate the impact of the video intervention on subjects' scores on this intention for proenvironmental behaviour (proenvironmental behavior before video (TPEB) versus intention for proenvironmental behavior after video (TPVPEB). There was a statistically significant increase

Table 6 Correlation between TPEB/TPVPEB/Tenvatt

		Tenvatt	TPEB	TPVPEB
Tenvatt	Pearson correlation	1		
	Sig.(2-tailed)			
TPEB	Pearson correlation	0.412**	1	
	Sig.(2-tailed)	0.000		
TPVPEB	Pearson correlation	0.328**	0.503**	1
	Sig.(2-tailed)	0.001	0.000	

**Correlation is significant at the 0.01 level (2-tailed)

Table 7 Volunteer actions, organization memberships and active participation before the video, and the intention to change it, after watching the video

	Yes	No
Do you partake in environment-related volunteer actions?	57	46
Were you inspired by the video to join volunteer proenvironmental actions?	77	26
Are you a member of an environmental organization?	30	73
Were you inspired by the video to join an environmental organization?	55	48

in TPVPEB scores before video (M = 35.82, SD = 0.58) and after video (M = 46.26, SD = 9.50), t = −12.82, p < 0.0005. The eta squared statistic (0.62) indicated a large effect size.

Table 7 depicts the participation of the subjects in environment-related volunteer actions and their involvement with proenvironmental organizations as well as the effect that the video had on these factors.

There has been an increase of 40% to the intention of partaking in volunteer actions. We also noticed an increase of 66.36% in the intention of actively participating in environmental decisions (Fig. 8).

The subjects also had to rate on a Likert scale of 1–7 (1-Not at all, 7-Very much), if they actively participate in environmental decision making before the video and if they intend to partake in environmental decision making in the future.

The mean before watching the video was 3.36 with a standard deviation of 1.7, whereas after the video, the mean of the intention to participate, rose to 5.09 with a standard deviation of 1.63 (Table 8).

Fig. 8 Volunteer actions, organization memberships and active participation before the video, and the intention to change it, after watching the video

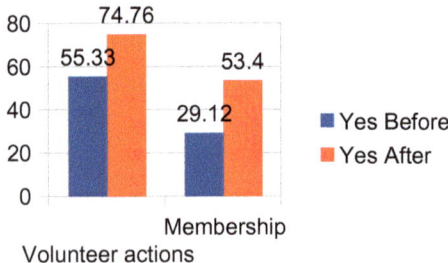

Table 8 Active participation and intention to actively participate

	M	Std.	Min	Max
Active participation in environmental decision making	3.3689	1.70346	1	7
Were you inspired by the video to partake in environmental decision making?	5.0971	1.63008	1	7

3.3 Paired-Sample t-Test

A paired-sample t-test was conducted to evaluate the impact of the video intervention on subjects' scores on their intention for taking part in environmental decision-making before the video versus their intention after the video. There was a statistically significant increase in the scores before the video (M = 3.36, SD = 1.70) and after the video (M = 5.09, SD = 1.63), t = −10.15, p < 0.0005. The eta squared statistic (0.50) indicated a large effect size.

3.4 Independent-Sample t-Test Between Males and Females Concerning Their Willingness for Active Participation in Environmental Decision Making

Independent samples t-tests were conducted to compare the active participation in environmental decision making for males and females before video as well as the intention for active participation in environmental active decision making in males and females after the intervention of the video. There was no significant difference in scores in active participation in environmental decision making for males (M = 3.18, SD = 1.58 for) and females [M = 3.64, SD = 1.78; T = −1.36, p = 0.17] before video. However, there was a significant difference in scores for males (M = 4.74, SD = 1.94) and females [M = 5.44, SD = 1.19; T = −2.16, p = 0.03]. The magnitude of the differences in the means was very small (Eta squared = 0.045).

3.5 Independent-Sample t-Test Between People that Grew up in Rural Versus Urban Areas

Independent samples t-tests were conducted to compare the active participation in environmental decision making for those grew up in rural areas versus those who grew up in urban areas. There was no statistically significant difference in scores (Table 9).

For the slogan, phrases with similar vocabulary had to be used so the subject can choose based only on the language form that deemed more suitable. The majority of

Table 9 Slogan influence

	The planet needs you. You must help	The planet needs you. Could you too help?	I do not believe that the slogan's form is of significance/it wouldn't influence me more than the video alone
Frequency	54	10	39

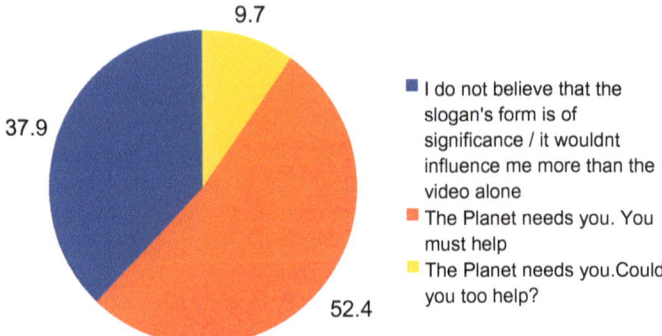

9.7

37.9

52.4

■ I do not believe that the
slogan's form is of
significance / it wouldnt
influence me more than the
video alone
■ The Planet needs you. You
must help
■ The Planet needs you.Could
you too help?

Fig. 9 Slogan influence

the subjects leaned towards the assertive phrasing (52.4%), while the non assertive
scored only 9.7%.

3.6   Gender and Choice

Chi-square criterion was used to investigate possible differences between slogan
influence and gender. There were no statistically significant differences between the
two variables but because of the fact that the lowest expected frequency in the cells
was violated, the result should be interpreted as uncertain (Fig. 9).

4   Discussion

The results indicate that educational videos are a great tool in the arsenal of
environmental educators. We can derive that not only can they be used to increase
knowledge and awareness, but they seem to influence the subjects' intentions for
behavioral change and even more their intention for active participation in environmental decision making, the ultimate goal of environmental communication. In
addition to the findings, we were able to confirm Kronrod's et al. (2012) research on
language forms and compliance.

It is apparent that there was an increase in knowledge of 8.72%, proving that
videos can be used in education. However, it must be examined whether the
knowledge they acquired will be remembered for long periods of time after only
one viewing, creating the need for repetition.

The findings also indicate that environmental attitude is linked to proenvironmental behavior, although an already preexistent proenvironmental behavior is the
foundation that we must build upon, in order to achieve higher involvement in
environmental decision making. The subjects' place of origin and growth did not

seem to have a statistically significant impact on the extent that it can be considered a factor relevant enough to change his behavior.

Genderwise, there were no significant differences in the scores before the video, although still visible, indicating that gender is not such an important factor when measuring the baseline participation in environmental decision making. However, it is apparent that women are more susceptible to the influence of the video concerning the aforementioned factor.

To get a better understanding of the reason of this phenomenon, we have to follow a sociopsychological approach. Women are associated with nature, the material, the emotional, and the particular, while men have been associated with culture, the rational, and the abstract (Sakellari and Skanavis 2013). The notion that women are "closer to nature," caring for land, water, forests, and other aspects of the environment, has held powerful sway in development theories since the 1980s, grounded in varieties of ecofeminist analysis (Sakellari and Skanavis 2013) from Shiva's "feminine principle" (Sakellari and Skanavis 2013) to ecospiritualists orientations (Sakellari and Skanavis 2013). Women report stronger environmental attitudes and behaviors than men. It was found that females were consistently more likely than males to be empathetic toward a conceptualized other and reported a significantly stronger ethic of care to take responsibility for ameliorating environmental problems (Sakellari and Skanavis 2013). Their findings suggest that women may be more concerned about environmental issues, but their limited biographical availability, for example, personal constraints, as entailed in the demands of the "double day" of paid and domestic work, present barriers to their activism. However, because many environmentally friendly behaviors can be undertaken in the context of domestic labor and everyday routines, biographical availability does not constrain their responsible environmental behavior. (Sakellari and Skanavis 2013)

Based on the aforementioned facts and the findings of this paper, we can safely assume that women are slightly more inclined to have a proenvironmental attitude than their male counterparts.

In addition to that, when exposed to a "trigger", a video in this particular study, women seem to be affected more. However, it is not clear if the difference of the responses to the stimuli between men and women is an effect of women being predisposed to adopt a proenvironmental behavior or they are more susceptible to such stimuli, and to what extent, thus it should be examined further.

As proven by Kronrod et al. (2012), the presence of a video can influence the form of language used to promote an environmental problem. The subjects in that study preferred an assertive slogan over a non assertive, when the slogan followed a video. In addition to Kronrod's findings, we discovered that when promoting an environmental matter with a perceived importance, the subjects preferred the absence of a slogan over one given with a non assertive form, because the language was not in tune with how crucial the matter was.

It is also important to note that the effect of a video, despite it's intensity, usually has a short-term effect on the subject. With that in mind, environmental educators must be able to motivate and inspire the subjects continuously in order to create a long-term environmental behavior.

5   Conclusion

The findings indicate that the most effective way to convey an environmental message about an important problem through media, is to accompany it with an assertive slogan, since the viewers can perceive the problem's importance, due to the serious nature of the assertive form. The target audience should primarily be women, although not focused only on them, since they seem to be more invested to environment and it's protection as well as they tend to be influenced more by media based proenvironmental campaigns. There must be a repetition of the stimuli in order to cultivate and maintain a long lasting proenvironmental attitude and behavior.

This research can be used as a foundation to build towards understanding the effects of videos in environmental communication and education, as well as distinguishing the differences between genders on environmental matters. It's importance lies in the fact that while being able to ascertain previous researches on these matters, it emphasizes on the success of videos in climate change education and communication and the importance of the correct use of these means, while providing a new approach on the gender differences, when approaching environmental matters.

6   Limitations

The tests conducted on the subjects can only measure the short term effects of the video. Because of that, we were not able to measure the long term effects on their behavior but their willingness to change their behavior.

References

Ajzen I (2013) Theory of planned behaviour questionnaire. Measurement instrument database for the social science. http://psycnet.apa.org/doiLanding?doi=10.1037%2Ft15482-000
Baek TH, Yoon S, Kim S (2015) When environmental messages should be assertive: examining the moderating role of effort investment. Int J Adv Rev Mark Commun
Ballantyne RR, Packer JM (2005) Promoting environmentally sustainable behaviour through free —choice learning experiences: what's the state of the game? Environ Educ Res 11:281–295
Banerjee S, Iyer E, Kashyap R (2003) Corporate environmentalism: antecedents and influence of industry type. J Market 67:106–122
Barbas TA, Paraskevopoulos SS, Stamou AG (2009) The effect of nature documentaries on students' environmental sensitivity: a case study. Learn Media Technol 34:61–69
Bauman A, Smith BJ, Maibach EW, Reger-Nash B (2006) Evaluation of mass-media campaigns for physical activity. Eval Program Plann 29:312–322
Bensley LS, Wu R (1991) The role of psychological reactance in drinking following alcohol prevention messages. J Appl Soc Psychol 21(13):1111–1124
Brown P, Levinson S (1987) Politeness: some universals in language use. Cambridge University Press, Cambridge

Chang CT (2012) Are guilt appeals a panacea in green advertising? the right formula of issue proximity and environmental consciousness. Int J Adv 31(4):741–771

Delaney A, Lough B, Whelan M, Cameron M (2004) A review of mass media campaigns in road safety. Monash University Accident Research Centre Report. http://www.monash.edu.au/muarc/reports/muarc220.pdf

Dillard JP, Shen L (2005) On the nature of reactance and its role in persuasive health communication. Commun Monogr 72(2):144–168

Dillard JP, Wilson SR, Tusing JK, Kinney TA (1997) Politeness judgments in personal relationships. J Lang Soc Psychol 16(3):297–325

Dulic A, Angel J, Sheppard S (2016) Designing futures: inquiry in climate change communication. Futures 81:54–67

Eagles PFJ, Demare R (1999) Factors influencing children's environmental attitudes. J Env Edu 30:33–37

Edwards SM, Li H, Lee JH (2002) Forced exposure and psychological reactance: antecedents and consequences of the perceived intrusiveness of pop-up ads. J Adv 31(Fall):83–95

Fernandez G, Thi M, Thi T, Shaw R (2014) Climate change education: recent trends and future prospects. Part of the disaster risk reduction book series (DRR), Education for Sustainable Development and Disaster Risk Reduction pp 53–74 (Chapter 4)

Gibbs RW (1986) What makes some indirect speech acts conventional? J Mem Lang 25 (April):181–196

Grinstein A, Udi N (2009) Demarketing, minorities and marketing attachment. J Mark 73:105–122

Hartmann P, Apaolaza-Ibanez V (2009) Green advertising revisited: conditioning virtual nature experiences. Int J Adv 28(4):715–739

Holbert RL, Kwak N, Shah DV (2003) Environmental concern, patterns of television viewing, and pro-environmental behaviors: integrating models of media consumption and effects. J Broadcast Electron Media 47:177–196

Holtgraves T (1991) Interpreting questions and replies: effects of face-threat, question form, and gender. Soc Psychol Quart 54:15–24

Huckle J (1995) Using television critically in environmental education. Environ Educ Res 1:291–304

Internet World Stats (2017) https://www.internetworldstats.com/stats.htm

Kronrod A, Grinstein A, Wathieu L (2012) Go green! should environmental messages be so assertive? J Mark 76(1):95–102

Lord KR (1994) Motivating recycling behavior: a quasiexperimental investigation of message and source strategies. Psychol Mark 11(9):341–359

Lanceley A (1985) Use of controlling language in the rehabilitation of the elderly. J Adv Nurs 10 (9):125–135

Menon A, Menon A (1997) Enviropreneurial marketing strategy: the emergence of corporate environmentalism as market strategy. J Mark 61:51–67

Miller CH, Lane LT, Deatrick LM, Young AM, Potts KA (2007) Psychological reactance and promotional health messages: the effects of controlling language, lexical concreteness, and the restoration of freedom. Hum Commun Res 33(2):219–240

Moser SC (2010) Communicating climate change: history, challenges, process and future directions, vol 1. Wiley, pp 31–53

O'Keefe DJ (1997) Standpoint explicitness and persuasive effect: a meta-analytic review of the effects of varying conclusion articulation in persuasive messages. Arg Adv 34(1):1–12

Pearson E, Dorrian J, Litchfield J (2011) Harnessing visual media in environmental education: increasing knowledge of orangutan conservation issues and facilitating sustainable behaviour through video presentations. Environ Edu Res 17(6):751–767

Peattie KA, Peattie S (2009) Social marketing: a pathway to consumption reduction? J Bus Res 62 (2):260–268

Quick BL, Considine JR (2008) Examining the use of forceful language when designing exercise persuasive messages for adults: a test of conceptualizing reactance arousal as a twostep process. Health Commun 23(5):483–491

Quick BL, Stephenson MT (2007) Further evidence that psychological reactance can be modeled as a combination of anger and negative cognitions. Commun Res 34(3):255–276

Sakellari M, Skanavis C (2013) Environmental behavior and gender: an emerging area of concern for environmental education research. Appl Environ Edu Commun 12(2):77–87

Schar EH, Gutierrez KK (2001) Smoking cessation media campaigns from around the world: recommendations from lessons learned. World Health Organization Report. http://www.euro. who.int/document/e74523.pdf

Schroth O, Angel J, Sheppard S, Dulic A (2014) Visual climate change communication: from iconography to locally framed 3D visualization. Environ Commun 8(4):413–432

Vanderveken D (1990) Meaning and speech acts, vol 2. Cambridge University Press, New York

White K, Simpson B (2013) When do (and don't) normative appeals influence sustainable consumer behaviors? J Mark 77(2):78–95

Wilson SR, Kunkel AW (2000) Identity implications of influence goals: similarities in perceived face threats and facework across sex and close relationships. J Lang Soc Psychol 19(2):195–221

Yoon S, Choi Y, Song S (2011) When intrusive can be likable: product placement effects on multitasking consumers. J Adv 40(2):63–75

Konstantinos Tsamopoulos is an undergraduate student in the Department of Marine Sciences of the University of Aegean and a researcher in the Department of Environmental Communication and Education. His involvement with fine arts and technology lead him to research their application into environmental education and awareness. He is a person with broadened research interests, fact that can be seen on the various researches that he is involved.

Kalliopi Marini Psychologist-Psychotherapist Experienced in Psychotherapy and in multiple applications of Applied Psychology including mostly Psychological Testing, Professional Training and Social Research.

HR Consultant in the fields of Recruitment, Training and Development of employees and executives.

Since 2003 she' s been working as a free lancer, she's taken part in the Olympic Organizational Committee as a Venue Staffing Manager and is one of the basic co-operators of ISON Psychometrica Ltd. During the last 10 years she's been working privately as a Cognitive—Behaviorial Psychotherapist.

Holds a B.Sc. in Psychology, an M.Sc. in Social/Organizational Psychology and is a certified Cognitive—Behavioral Psychotherapist. She is currently a Ph.D. candidate in the department of Environment in the Aegean University.

Constantina Skanavis is a Professor in Environmental Communication and Education at the Department of Environment, University of the Aegean (Mytilene, Greece). She is also the Head of the Research Centre of Environmental Education and Communication. She joined the University of the Aegean 15 years ago. Before that she was a Professor at California State University, Los Angeles. She has developed several courses on issues of environmental health and education. She currently teaches environmental education, environmental communication and environmental interpretation courses in undergraduate and postgraduate levels. Professor Skanavis has numerous publications on a international basis and has given presentations all over the world.

The "Paris Lifestyle"—Bridging the Gap Between Science and Communication by Analysing and Quantifying the Role of Target Groups for Climate Change Mitigation and Adaptation: An Interdisciplinary Approach

Stephan Schwarzinger, David Neil Bird and Markus Hadler

Abstract World society and decision makers are running out of time to implement measures on climate change mitigation and adaptation. Incomplete knowledge and vast challenges in communicating climate change are crucial factors in this problem. In order to increase people's awareness of their role in climate change, highly specific communication strategies are necessary. Besides insufficient information on group-specific realities of life, existing strategies are often limited by the absence of quantitative data that could give decision makers the opportunity to estimate the potential and evaluate the success of communication measures. In order to meet these requirements, energy use and corresponding emissions must be analysed in relation to behavioural patterns and technology choices of relevant social groups. This perspective leads to a more detailed understanding of how energy use and the responsibility for greenhouse gas emissions are distributed within society. This paper presents an interdisciplinary approach for providing the required knowledge within a single research process and describes its most relevant features as compared to previous methods. We describe the empirical development of an impact based "Energy Lifestyle" typology for the Austrian society and describe the six identified groups in detail with special focus on the challenges that might evolve in

S. Schwarzinger (✉) · D. N. Bird
Centre for Climate, Energy and Society, Joanneum Research
Forschungsgesellschaft mbH, Waagner-Biro Straße 100,
8020 Graz, Austria
e-mail: stephan.schwarzinger@joanneum.at

M. Hadler
Department of Sociology, University of Graz, Universitätsstraße 15/G4,
8010 Graz, Austria
e-mail: markus.hadler@uni-graz.at

© Springer Nature Switzerland AG 2019
W. Leal Filho et al. (eds.), *Addressing the Challenges in Communicating Climate Change Across Various Audiences*, Climate Change Management,
https://doi.org/10.1007/978-3-319-98294-6_23

group specific communication. Thereafter, we set the six Energy Lifestyles in context with the name-giving concept *"Paris Lifestyle"* and discuss its role for evaluating the succession towards the goals set out in the Paris Agreement.

1 Introduction

In order to achieve the aim of limiting global warming to a maximum of 2 °C compared to the preindustrial stage, as set out in the Paris Agreement of 2015,[1] parties to the agreement have announced ambitious milestones over the upcoming decades. The European Union, for example, intends a transformation to a "low-carbon economy" by 2050, which is, amongst others, defined by a reduction of domestic emissions to 80% below the level of 1990. By 2030 and 2040 emission cuts by 40% and 60% respectively are planned.[2] Significant change in core areas of human action will be necessary to meet such targets. Emissions are in a large part determined by the way people satisfy their personal needs in all areas of life. Efficient communication measures addressing this behaviour are always connected to the need of extensive knowledge, which gives rise to the essential question of lifestyle research in the area of climate and energy: Who consumes how much of what and why?

It is obvious that these questions cannot be assessed by per-capita numbers on a national level due to a lack of detail. Even if one intends to generate a more detailed picture by taking sociodemographic parameters like age, sex and socioeconomic status into account, the central questions cannot be answered satisfactorily. An observable difference in energy demand and greenhouse-gas emissions between groups defined by sociodemographics usually exists, but the lives of individuals within such groups are, to the major part, too diversified to be understood as collectives with a homogeneous role for climate and energy. Of course, the problem mentioned is not limited to questions linked to energy consumption. The need for the identification of target groups less contradictory than those provided by sociodemographic parameters has always been a driving factor for lifestyle-research, and many methods have found their way to target group management, marketing and advertisement. The number of lifestyle-related academic literature is especially large in publications from Central Europe, where most of the constituent contributions were developed and elaborately discussed (Hartmann 1999; Hermann 2004; Hradil 1987; Meyer 2001; Otte 2005; Schulze 2005; Spellerberg 2001). Many authors refer to Pierre Bourdieu's work "Distinction: A Social Critique of the Judgement of Taste" (Bourdieu 1987), which was published in 1979. His work is closely connected to the idea of certain societal groups having specific habits, patterns of behaviour, norms and values. The common use of the

---

[1]UNFCCC (2017).
[2]European Commission (2017).

term "lifestyle" and the effort to identify distinctive and sociologically relevant lifestyle-groups are two interlinked arenas that partially evolved from the discussion of Bourdieu and his predecessors. Until now several authors have developed and tested many different empirical procedures to identify such lifestyle-groups. Some of them radically incorporate the assumption of lifestyles being widely independent from sociodemographic factors like social class or age and being mainly determined by individual preferences, values and attitudes. However, as Tomlinson set out, this perspective clearly disagrees with Bourdieu's understanding of lifestyle as class specific (Tomlinson 2003). Also, in regard to energy relevant behaviour, it seems doubtful that sociodemographics, including income, should play only a very subordinate role. Since both, the use of simple sociodemographics as well as lifestyle identification based on attitudinal parameters, have clear limitations when one is interested in the corresponding behaviour and target group oriented communication, we suggest the procedure presented in the following paper:

In a first step we put our focus directly on the behavioural impact itself and identify groups based on their energy demand profiles. Contextual and possible factors used to explain the profiles are identified in the second step and used to create an empirically sound understanding of "Energy Lifestyles". This strategy combines three advantages: Firstly, it discovers a more plausible approximation of the ways different social groups conduct their daily life than sociodemographic segmentation and attitude based approaches do. Secondly, with lifestyles being constituted by climate relevant parameters, groups with a maximal differentiation in their energy demand and climate impact are identified, which makes sure that the major part of the empirically existent energy lifestyles spectrum is revealed. Thirdly, the identified groups can in addition be analysed with regard to their motivations and parameters characteristic of their individual situation, which leads to the coverage of all three components of lifestyles, "*Performance*", "*Situation*" and "*Mentality*" (Lüdtke 1996, Reusswig 2002). The resulting in-depth knowledge about the climate impact and driving factors of relevant lifestyle groups allows (1) the design of highly specific communication instruments, (2) a more valid estimation of effects to be expected by mitigation strategies and (3) the selective design of policy measures like consumer carbon-taxes for the funding of adaptation measures following the polluter pays principle.

In a more general way, besides tangible starting points for communication and policy strategies, the results may provide a more differentiated understanding of the human role in climate change: By highlighting the fact, that individual behaviour and technology choices can sum up to massive differences in lifestyle specific climate impacts, the approach identifies the responsibility of every single person. It demonstrates that the national context and the associated economic and political frameworks do not rigidly determine the climate impact of citizens, but obviously leave plenty of space for decisions to move towards a sustainable lifestyle.

The paper is organised as follows: After this introduction, the development of our impact based lifestyle approach is described. Thereafter six Austrian lifestyle groups, identified by this approach, are characterised in terms of behavioural patterns and sociodemographics. On this basis, implications for communication and policy making are discussed.

2 Development of Impact-Based Lifestyle Identification Approach

In the following, the reasons for conducting an impact-based approach are described, starting from the main reasons why conventional models based on attitudinal parameters are not our first choice.

2.1 Limitations of Attitude-Based Models

As mentioned initially, general lifestyle approaches based on attitudinal variables tend to not satisfactorily identify a broad spectrum of empirically existent climate and energy relevant lifestyles in terms of energy relevant behaviour and their respective impact.[3] Hierzinger and colleagues developed an energy specific typology based on attitudinal variables and identified five "Energy Styles" (Hierzinger et al. 2011). However in many areas these five groups show questionable characteristics. For example the group "Eco-Responsible" ("*Ökologie-Verantwortlicher*") has a below average share of respondents who use the bicycle as their main mode of transport, while the group "Disoriented Polluter" ("*Orientierungsloser Umweltsünder*") has an above average share of people who cycle in everyday life. In terms of heating, some findings are similarly controversial: While the group "Eco-Responsible" has a below average number of people using heat pumps, "Unworried Spenders" ("*Sorgloser Verschwender*") have an above average share of respondents who heat their homes using heat pumps. Bohunovsky and colleagues used a cultural sociological model (Schulze 2005) and identified four lifestyle groups who all had similar patterns of energy demand and a total direct energy demand between 11,000 and 12,000 kWh per capita and year (Bohunovsky et al. 2011).

Obviously, attitude-based approaches have substantial limitations:

Firstly, they are of limited value for the design of communication strategies and policy measures. Secondly, they do not provide a suitable view on how different lifestyles contribute to the overall national emission records and mitigation goals. Thirdly, they do not provide a detailed basis for the development of targeted carbon tax models.

---

[3]One factor that obviously limits the benefit from such models is the attitude-action gap, which is a common problem in research on environmental behaviour.

The absence of an integrated approach that covers these special requirements within a single research process led to the development of the procedure presented here. In contrast to assessments on a national level, where average per-capita numbers are calculated in a top-down manner based on national inventories, we calculated the lifestyle group specific impacts bottom-up, based on the respondents' individual data.

2.2 Methodology and Added Value of an Impact-Based Model

The behavioural impact can, depending on the research interest, be assessed by different measures, such as energy demand. In the course of the impact calculation, the multitude of parameters that determine the respective impact is reduced to one output-variable for each analysed area of life. In terms of dimension reduction and information loss this step can be compared to factor analysis, with the impact-variable being the output-factor and underlying parameters being a variety of items determining the value of the factor-variable.

We used the direct annual per capita energy demand as impact variable. It was calculated for seven areas that are especially relevant for direct energy demand: Motoring, Air Travel, Heating, Appliances, Cooking, Hot Water, and Lighting. The sizes of and relations between these seven impact variables are the determinant for the identification of groups with similar patterns of energy use. The resulting "raw"-model is only descriptive and depicts only the demand profiles of the identified respondent-groups. Variables that explain group membership or group specific factors with an influence on energy demand are introduced and analysed in a second step.

The advantage of identifying target groups that significantly differ in their energy relevant behaviour and behavioural impact, goes hand in hand with a method-ological benefit: The impact-based assessment is independent from the cultural context, which means that the comparison of "raw"-models from different national or cultural entities is widely undistorted. The regularly encountered challenge of low measuring equivalence in internationally comparative research was extensively discussed by Johnson (1998). The need for a straight basis becomes even clearer, when the partially heterogeneous findings on the linkage between attitudinal parameters and behaviour are taken into account.

For example, Poortinga and colleagues found home and transport energy use to be "more strongly related to sociodemographic variables" than to attitudinal vari-ables (Poortinga et al. 2004). Likewise, Abrahamse and Steg found energy use to be mainly "determined by socio-demographic variables", and psychological factors to be more relevant for changes in energy use (Abrahamse and Steg 2009). Holden found "green" households to have a lower per capita ecological footprint, but not as a result of their ecological awareness, but merely because of more people living in such households (Holden 2004). Tabi found no significant difference in energy demand and CO_2 emissions between environmentally aware and environmentally unaware consumers and motivation-driven activities to be offset by structural factors (Tabi 2013).

2.3 Energy Demand Versus Greenhouse Gas Emissions

As mentioned initially, one essential component of the Paris Agreement is a massive cut in greenhouse gas emissions. Why are we then discussing "Energy Lifestyles" instead of "Emission Lifestyles"? This question is as legitimate as it is quickly answered: Energy demand as such is in many cases a better indicator for the individual lifestyle than emissions are. A certain type and size of dwelling for example requires a more or less constant amount of energy to be heated; the associated emissions instead strongly depend on the technology in use. On the one hand the need for a warm home can be satisfied using a heating system fully powered by renewables, like a heat pump operated on electricity from photovoltaics. On the other hand, an oil-fired heating system supplied with electricity from a fossil fuel power station can come to use.

Thus, under constant technological progress towards energy efficiency, "Energy Lifestyles" are by far more stable social categories than "Emission Lifestyles" would be. However, the ratio between energy demand and associated emissions [emission intensity (kg CO_2/kWh)] is an important indicator for the spread of sustainable technologies and the emission level is the main target size for communication strategies and policy measures. Thus the three values *direct energy use per capita*, *direct emissions per capita* and *emission intensity* are considered in the characterisation of "Energy Lifestyles".

In the following, our impact-based approach is described in more detail and the resulting groups are characterised in terms of socio-demographics.

3 Research Methods, Data, and Analysis Strategy

3.1 Dataset Used

In order to develop an impact-based model, an empirical dataset meeting special requirements is needed. Besides common quality criteria such as representativeness, it must contain data on energy-relevant behaviour and contextual factors in such detail that the direct energy-demand of the respondents' whole life can be estimated.[4] Databases open to public access usually do not provide information of that

---

[4]Indirect energy demand or emissions based on a Lifecycle-Assessment of food, products and services is not covered by this Analysis.

kind. Collecting data in such detail tends to be problematic in terms of response-rate and data-quality when it is done via an online-survey by self-recruited respondents. Since personal interviews are resource-intensive, they are normally not conducted to capture data for methodological experiments. Our colleagues from Sustainable Europe Research Institute (SERI) kindly provided us the dataset that was used in a past study (Bohunovsky et al. 2011). Therefore the results from the past study and our approach can be directly compared.

The dataset contains the answers of 1014 household-interviews that were conducted in Austria between May and June 2009. All households were selected by a random-route procedure. Each case represents one respondent who answered *in deputy* of the respective household. In order to gain data of highest possible validity, it was the aim to interview that person with the most knowledge concerning energy use and energy demand. The interviews were conducted as computer assisted personal interviews (CAPI) and usually took between 30 and 40 min.

3.1.1 Sociodemographics and Attitudinal Variables

In addition to variables focussing on the "Performance" component (behaviour and equipment), the dataset contains variables on sociodemographics and attitudes towards energy saving. While the group identification is fully based on the former, sociodemographics ("Situation") and attitudes ("Mentality") play a crucial role for the development of an-depth understanding of different lifestyles.

While "Performance" and "Situation" are covered satisfactorily, the "Mentality" component is only covered by the respondents' subjective relevance of energy-saving in different areas (In general, Appliances, Electricity, Hot Water, Heating, Cooking, Lighting, and Mobility). The variables on subjective relevance of energy-saving proved to be one-dimensional, which is why a mean-score was calculated.

3.2 Model Development

As mentioned above, energy demand is determined by a multitude of parameters in behaviour and equipment. Instead of creating a model based on all relevant parameters in seven areas of life, as a starting point the respondents' individual energy demand in the seven areas was calculated. The resulting variables representing primary energy demand caused by direct energy use were added to the dataset. Information on public transport was provided by the dataset, but not included because of a high level of uncertainty. Therefore the raw model is based on demand variables (kWh per capita and year) for the following areas of life:

(1) Motoring (e.g. fuel-type, fuel consumption, driving-distance)
(2) Air Travel (e.g. frequency, distance flown)
(3) Heating (e.g. type and size of building, room temperature, heating system)
(4) Appliances (e.g. type and number of appliances, efficiency-class, mode of use)
(5) Cooking (e.g. frequency, stove type, other appliances)
(6) Hot Water (e.g. type and energy source of boiler, usage patterns)
(7) Lighting (e.g. number and types of lamps, usage patterns).

3.2.1 Choice of Clustering Method and Model

In order to maintain a high resolution of data and because energy demand in each area is quasi-independent from other areas, we decided against the use a further dimension reduction in the form of a factor-analysis. To ensure straight and transparent modelling we chose Latent Class Analysis, which keeps the number of parameters that may affect results at a minimum (Bacher et al. 2010). Because the algorithm needs variables to be categorical or ordinal positive integers, the metric demand variables had to be recalculated into 11 intervals between "0" and the outlier-cleaned maximum. Using the package "poLCA" for R Statistics, a spectrum of solutions from 2 to 9 classes was calculated from the categorised demand-data.

Based on model statistics (Kuha 2004) (Table 1) and contextual considerations, the 6-class model was chosen and will be the basis for our analysis on group level (Fig. 1).

The exact group average values of energy demand for the seven considered fields are shown in Table 2.

4   Characterisation of 6 Lifestyle Groups

In the following, the six groups identified above are characterised by their energy profiles (Fig. 2; Table 2), by sociodemographics and by data with regard to mobility and residence (Table 3) in order to gain a basic understanding for their realities of life. The groups will from now on be termed "Lifestyles".

Beforehand, one essential finding should be mentioned: There is no significant correlation between the respondents' subjective relevance of energy saving and the respondents' actual calculated energy demand or CO_2 emissions. This matches with earlier findings discussed above.

Table 1 Model statistics for calculated solutions

	2 classes	3 classes	4 classes	5 classes	6 classes	7 classes	8 classes	9 classes
Cases total	1014	1014	1014	1014	1014	1014	1014	1014
Cases completed	740	740	740	740	740	740	740	740
Parameters estimated	131	197	263	329	395	461	527	593
df	883	817	751	685	619	553	487	421
Maximum-LL	−12,667.6	−12,559.7	−12,476.9	−12,406.6	−12,347.3	−12,289.4	−12,244.1	−12,194.8
AIC	25,597.2	25,513.5	25,479.7	25,471.2	25,484.5	25,500.7	25,542.2	25,575.7
BIC	26,242.0	26,483.0	26,774.1	27,090.5	27,428.6	27,769.6	28,135.9	28,494.2
Repetitions	100	100	100	100	100	100	100	100

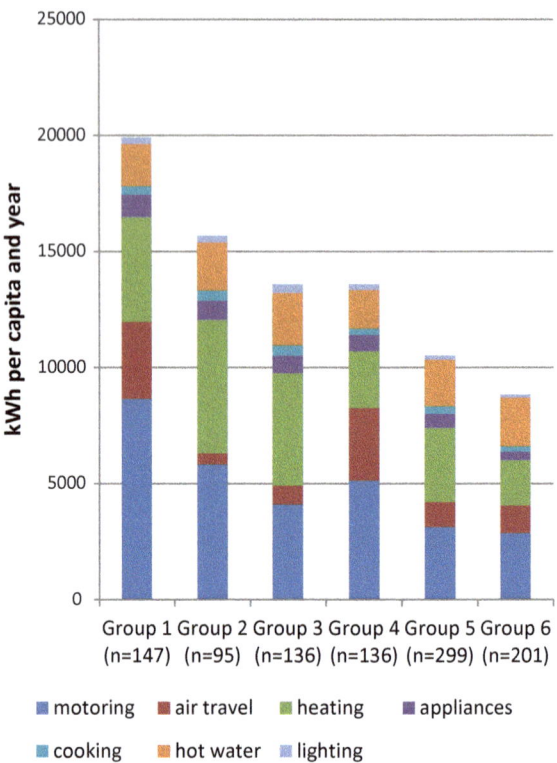

Fig. 1 Cumulative energy demand of 6 latent classes

Table 2 Group average primary energy demand in kWh per capita and year

	Group 1	Group 2	Group 3	Group 4	Group 5	Group 6	Full sample average
Motoring	8676.2	5842.4	4123.5	5145.3	3154.1	2892.8	4518.7
Air travel	3299.3	482.3	791.4	3127.3	1064.6	1168.8	1595.0
Heating	4519.8	5744.4	4856.7	2440.6	3217.0	1960.2	3506.7
Appliances	956.2	830.4	748.4	688.1	588.5	378.1	702.8
Cooking	364.4	433.6	439.3	268.9	325.4	234.2	331.0
Hot water	1836.6	2075.0	2284.6	1702.1	2003.4	2090.5	2000.2
Lighting	262.4	275.5	355.8	225.6	170.7	106.6	212.2
Total	19,914.9	15,683.7	13,599.7	13,597.9	10,523.6	8831.1	12,866.6

Fig. 2 Group relative profiles

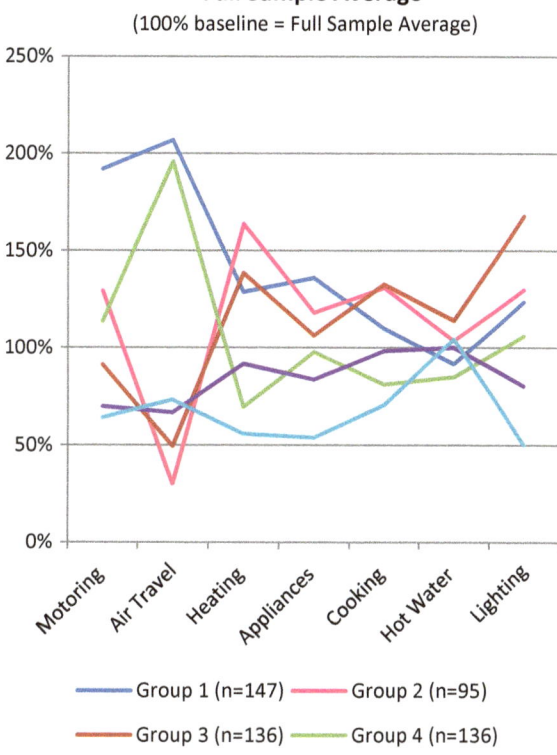

6 Group Profiles relative to
Full Sample Average
(100% baseline = Full Sample Average)

——— Group 1 (n=147) ——— Group 2 (n=95)

——— Group 3 (n=136) ——— Group 4 (n=136)

——— Group 5 (n=299) ——— Group 6 (n=201)

4.1 Lifestyle 1: "Mobile Performers" (14.5%)

The group with the highest average direct energy consumption and CO_2 emissions is named "Mobile Performers" for two main reasons: Firstly, the name reflects the remarkably high amount of energy that is needed for mobility and the fact that the number of cars per capita is higher than in any other group (Tables 2, 3). Secondly respondents assigned to this group have the second highest OECD-modified equivalence income and the second highest share of higher education (Table 3). Having the highest energy demand for Motoring, Air Travel and Appliances, and above average energy demand in nearly all areas allows no alternative to having the highest overall energy demand (Tables 2, 3, Fig. 3).

In addition to these name-giving characteristics, the group contains the largest share of people living in cities with more than 50,000 inhabitants and the smallest share of people living in small villages of 5000 or below. "Mobile Performers" live in significantly smaller households than the sample average. They also have a noticeably small number of children under 14 living in their households. The

Table 3 Lifestyles in comparison

	Lifestyle 1	Lifestyle 2	Lifestyle 3	Lifestyle 4	Lifestyle 5	Lifestyle 6	Full sample
Primary energy demand from direct energy use per capita/year (kWh)	19,915[c] [18,754–21,076]	15,684 [14,135–17,232]	13,600 [12,642–14,558]	13,598 [12,886–14,310]	10,524 [9992–11,055]	8831[d] [8311–9351]	12,867 [12,458–13,252]
Lifecycle CO_2 emissions from direct energy use per capita/year (kg)	5482[c] [5082–5882]	3985 [3424–4547]	3178 [2870–3485]	3641 [3413–3868]	2533 [2366–2700]	2158[d] [1997–2319]	3254 [3126–3382]
Lifestyle emission factor ($kgCO_2$/kWh)	0.272[c] [0.263–0.281]	0.236 [0.217–0.254]	0.231[d] [0.218–0.245]	0.266 [0.259–0.274]	0.238 [0.229–0.248]	0.238 [0.229–0.246]	0.246 [0.241–0.250]
Age	46.8 [44.2–49.3]	50.4 [46.5–54.3]	53.2[c] [50.4–56.0]	41.6 [39.6–43.6]	47.9 [45.9–49.8]	41.5[d] [39.6–43.4]	46.6 [45.6–47.6]
Male respondents (%)	54.4	49.5	45.6	49.2	46.1	46.3	48.0
Net. equivalence income (€)[a]	1656 [1571–1741]	1405 [1309–1502]	1421 [1339–1503]	1664[c] [1585–1742]	1360 [1310–1409]	1267[d] [1208–1327]	1438 [1408–1468]
Higher education (%)	30.8	19.8	17.9[d]	32.2[c]	20.1	18.3	23.6
Household size	1.68 [1.56–1.80]	1.21[d] [1.12–1.30]	1.63 [1.52–1.75]	2.87 [2.71–3.03]	2.28 [2.20–2.42]	3.57[c] [3.42–3.72]	2.35 [2.28–2.42]
Number of children under 14	0.04 [0–0.08]	0.01[d] [0–0.03]	0.03 [0–0.06]	0.47 [0.35–0.59]	0.25 [0.19–0.31]	0.95[c] [0.81–1.08]	0.34 [0.29–0.38]
Living area per capita (m²)	61.3 [57.8–64.8]	72.5[c] [65.9–79.0]	62.5 [58.6–66.4]	37.3 [35.6–38.9]	42.8 [41.1–44.4]	29.8[d] [28.1–31.5]	47.6 [46.1–49.0]
Share single family homes (%)	49.6	35.8[d]	50.8[c]	50	44.1	44.8	46
Private cars per HH-member over 14 years age[b]	0.87[c] [0.82–0.92]	0.55 [0.45–0.65]	0.60 [0.52–0.67]	0.69 [0.64–0.74]	0.48[d] [0.44–0.52]	0.53 [0.49–0.57]	0.59 [0.57–0.62]

(continued)

Table 3 (continued)

	Lifestyle 1	Lifestyle 2	Lifestyle 3	Lifestyle 4	Lifestyle 5	Lifestyle 6	Full sample
Hometown <5000 (%)	52.8[d]	58.4	61.5	70.0[c]	64.0	67.6	62.8
Hometown <20,000 (%)	76.0	81.8	87.7[c]	83.0	76.9	75.0[d]	79.0
Hometown < 50,000 (%)	83.2[d]	93.5	94.5[c]	91.0	83.4	89.2	87.7
Renewable main heating energy source (%)	12.2	6.3	13.2	4.4[d]	14.7[c]	12.4	11.5
Mean-score "energy saving attitude" [1–10]	6.3	6.3	6.2	6.7	6.8	6.6	6.6

[a]OECD-modified scale (Weights: 1st adult = 1, each further adult = 0.5, each child: 0.3): http://www.oecd.org/eco/growth/OECD-Note-EquivalenceScales.pdf (last retrieved: September 14, 2017)
[b]Dataset only contains the "No. of children under 14"—thus, the number of adults is not fully accurate
[c]Highest value of all groups
[d]Lowest value of all groups

Fig. 3 Energy demand of
"mobile performers" versus
average sample (*Highest
value of all groups)

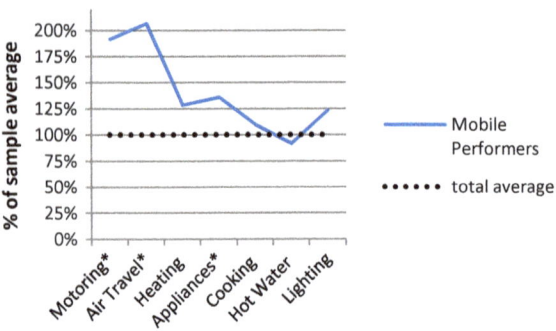

Table 4 Climate impact of
"mobile performers"

Direct energy use per capita/year	19,915 [18,754–21,076]
Direct CO_2 emissions per capita/year	5482 [5082–5882]
Lifestyle emission factor $kgCO_2/kWh$	0.272 [0.263–0.281]

average living area per capita is the third largest of all six identified groups. The share of single family homes is above average. Interestingly the share of male respondents is remarkably high amongst "Mobile Performers" (Table 3).

"Mobile Performers" have a significantly higher energy demand than any other group and a Lifestyle Emission Factor that is significantly higher than in the full sample. The group has significantly higher direct CO_2 emissions per capita and year than any other group (Table 4).

4.2   Lifestyle 2: "Uncommitted" (9.4%)

This group has the second highest average direct energy consumption and is named "Uncommitted" because its respondents live in households with the smallest number of people and the lowest number of children under 14 (Table 3). The group has the largest living area per capita, which likely contributes to also having the highest per capita energy demand for heating (Table 2, Fig. 4). Additionally, the small household size minimises potential scale effects for driving and in-home activities (Table 3).

The group has the smallest share of single family homes and an OECD-modified equivalence income close to the full sample average. The average age of "Uncommitted" respondents is slightly higher than in the full sample. The share of respondents with higher education is smaller than in the full sample. The group has

Fig. 4 Energy demand of "uncommitted" versus average sample (*Highest of all groups; **Lowest of all groups)

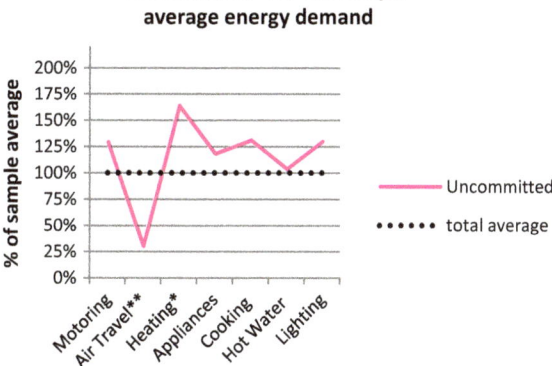

Table 5 Climate impact of "uncommitted"

Direct energy use per capita/year	15,684 [14,135–17,232]
Direct CO_2 Emissions per capita/year	3985 [3424–4547]
Lifestyle emission factor $kgCO_2/kWh$	0.236 [0.217–0.254]

a below average share of people living in villages smaller than 5000 inhabitants but an above average share of respondents that live in cities up to 20,000 and up to 50,000 (Table 3).

"Uncommitted" have above average direct energy use and direct CO_2 emissions significantly higher than the full sample. The groups Lifestyle Emission Factor is close to the average (Table 5).

4.3 Lifestyle 3: "Settled" (13.4%)

The group is called "Settled" for three reasons: Firstly, it has the highest average age of all six groups (Table 3). Secondly, the energy demand for housing and in-house activities is higher than average. Thirdly, respondents assigned to this group use a below average amount of energy for Flying and Motoring (Table 2, Fig. 5). In addition to that, it has the highest share of single family homes. The group has about the same OECD-modified equivalence income as the full sample and below average household size. "Settled" have the smallest share of respondents with higher education. The number of children under 14 is very low, for which age of respondents is most probably a contributing factor. "Settled" have the second largest living area per capita and more cars than the average respondent. The group has the lowest share of people who live in a hometown with more than 20,000 or more than 50,000 inhabitants (Table 3).

Fig. 5 Energy demand of "settled" versus average sample (*Highest of all groups)

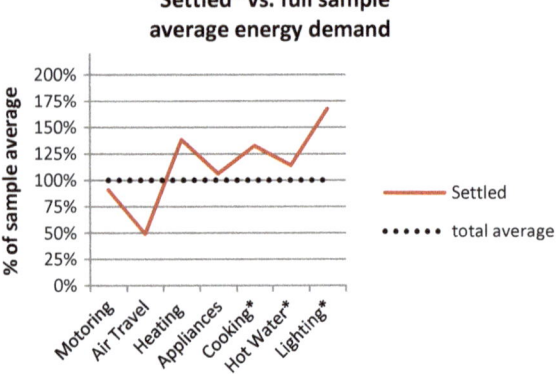

Table 6 Climate impact of "settled"

Direct energy use per capita/year	13,600 [12,642–14,558]
Direct CO_2 emissions per capita/year	3178 [2870–3485]
Lifestyle emission factor kgCO₂/kWh	0.231 [0.218–0.245]

"Settled" have a total direct energy demand and direct emissions similar to the full sample average. Although the emission factor has the smallest average value of all groups, it is not significantly lower than the full sample average (Table 6).

4.4 Lifestyle 4: "Educated Cosmopolitans" (13.4%)

The name "Educated Cosmopolitans" was chosen because the group has the highest share of higher education, the highest OECD-modified equivalence income (Table 3) and the second highest energy demand for Air Travel. The below average per capita energy demand at home might support the interpretation of being a consequence of extensive travelling activities (Table 2, Fig. 6). However this hypothesis could not be proven and the low energy demand at home could be also caused by scale effects connected to the over average household size, number of children and living area per capita (Table 3).

The share of single family homes and the number of cars per capita are higher than in total average. Regarding urban or rural living, the group has below average shares of respondents who live in towns with less than 2000 and more than 50,000 inhabitants (Table 3).

"Educated Cosmopolitans" have an overall energy demand close to the full sample average, but significantly higher direct CO_2 emissions, which is caused by a higher emission factor (Table 7).

Fig. 6 Energy demand of "educated cosmopolitans" versus average sample (**Lowest of all groups)

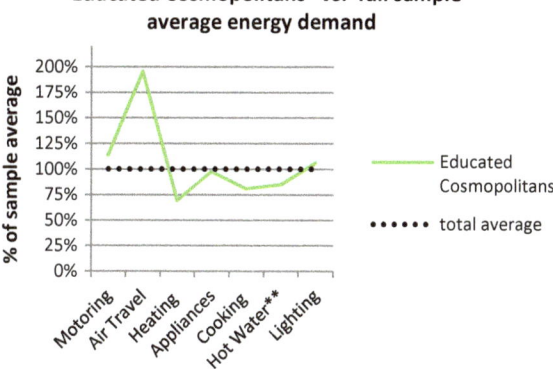

Table 7 Climate impact of "educated cosmopolitans"

Direct energy use per capita/year	13,598 [12,886–14,310]
Direct CO_2 emissions per capita/year	3641 [3413–3868]
Lifestyle emission factor $kgCO_2$/kWh	0.266 [0.259–0.274]

4.5 Lifestyle 5: "Economically Restricted" (29.5%)

The group was named "Economically Restricted" because the OECD-modified equivalence income is lower than the sample average (Table 3). In terms of energy demand the group is below or quasi equivalent with the sample average in all seven areas (Table 2, Fig. 7). Age and household size are close to the sample average, while education, the number of children, living area per capita, the share of single family homes and the number of cars per capita are below average (Table 3).

Regarding rural versus urban living, "Economically Restricted" have the highest share of people living in a hometown with more than 50,000 inhabitants (Table 3).

"Economically Restricted" have an overall energy demand and direct emissions significantly lower than the average. Their Lifestyle Emission Factor is close to average (Table 8).

4.6 Lifestyle 6: "Underprivileged" (19.8%)

At a first sight, the label "Underprivileged" might sound discriminatory, especially in consideration of the fact that it has the lowest per capita energy demand and CO_2 emission of all identified groups (Table 3, Fig. 8). Isn't that the way we are

Fig. 7 Energy demand of "economically restricted" versus average sample (**Lowest of all groups)

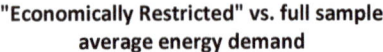

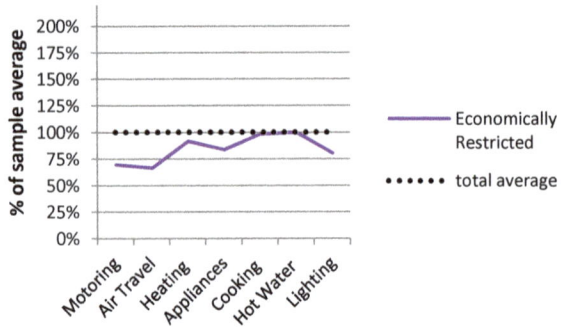

Table 8 Climate impact of "economically restricted"

Direct energy use per capita/year	10,524 [9992–11,055]
Direct CO$_2$ emissions per capita/year	2533 [2366–2700]
Lifestyle emission factor kgCO$_2$/kWh	0.238 [0.229–0.248]

supposed to go? From an impact perspective it definitely is, but seen in a societal context, the group is facing some special challenges.

The group has the lowest OECD-modified equivalence income and the second smallest share of higher education. By average respondents assigned to this group are younger than the full sample. The shares of "Underprivileged" respondents who live in a hometown with less than 2000 or more than 50,000 inhabitants are smaller than of the full sample (Table 3).

Respondents assigned to this group live in households with the smallest living area per capita. This might partly reflect the fact that the group has the highest number of children under 14 and the largest average household size (Table 3). In lack of an applicable equivalence scale, the comparison of living space under different household settings is problematic. If the weights from the "OECD-modified scale" (1st adult = 1; each further adult = 0.5; each child under 14 = 0.3) are used as a workaround for estimating an equivalence size, the group "Underprivileged" has still a significantly lower living area than any other group.

This group has the lowest direct energy demand and emissions, both significantly smaller than the average. The Lifestyle Emission Factor is about average (Table 9).

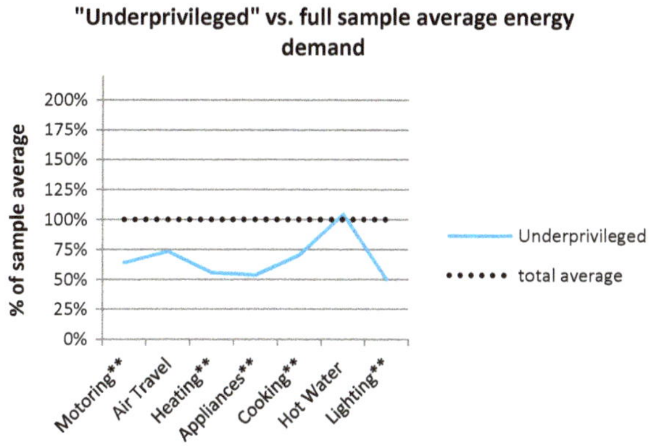

Fig. 8 Energy demand of “underprivileged” versus average sample (**Lowest of all groups)

Table 9 Climate impact of “underprivileged”

Direct energy use per capita/year	8831 [8311–9351]
Direct CO_2 emissions per capita/year	2158 [1997–2319]
Lifestyle emission factor $kgCO_2/kWh$	0.238 [0.229–0.246]

5 The Way to a “Paris Lifestyle”

As set out in the introduction, meeting or missing the goals of the Paris Agreement is not mainly a question of technological progress, but is in large part determined by the way every single one of us satisfies his or her needs. What became obvious in the discussion of the six identified Austrian “Energy Lifestyles” is that “average Austrians” as a relevant target group for communication do virtually not exist. This is, of course, only moderately surprising. However, in lack of a better alternative, climate change communication is still often based on national averages and designed without having an empirically profound picture of how different societal groups impact climate and energy systems throughout their whole life and what the life realities of these groups are. It seems natural that such a limitation of information complicates the design of effective and efficient communication strategies. In nearly any aspect of climate change, information is highly complex, which is one more challenging factor for communication. In order to reduce this complexity and enable individuals as parts of groups to becoming active in terms of climate protection, we suggest expressing the individual role for climate change mitigation quantitatively. The individual emission footprint—differentiated into main activity areas—can then be set into relation to the “Paris Lifestyle”, which is the benchmark of a lifestyle compatible with the goals of the Paris Agreement. A target-actual comparison of this kind has three potential benefits:

(1) Individuals supplied with both domain specific and overall quantitative feedback on the emission impact of their "whole life" could be less likely to develop an overcompensation of savings as e.g. discussed by Buchanan and colleagues (Buchanan et al. 2015). Also the problem of "compensatory green beliefs" (CGB) (Kaklamanou et al. 2015), could potentially be limited by the provision of an easily comprehensible breakdown.

(2) It is likely that a significant share of people accepts the challenge of comparing their progress with the achievements made by other members of the reference group (e.g. Lifestyle group) or towards goals like the "Paris Lifestyle". A review of 26 studies that analysed the effects of "Gamification" in the domain of domestic energy consumption found that about 88% of the identified effects were positive (Johnson et al. 2017).

(3) An empirically sound information basis about, if and how different social groups succeed with limiting their greenhouse gas emissions is the only way to quantitatively evaluate the impact of communication and policy measures in detail. This kind of feedback after certain periods of time (exemplarily shown in Fig. 9) allows the realignment of strategies based on solid criteria.

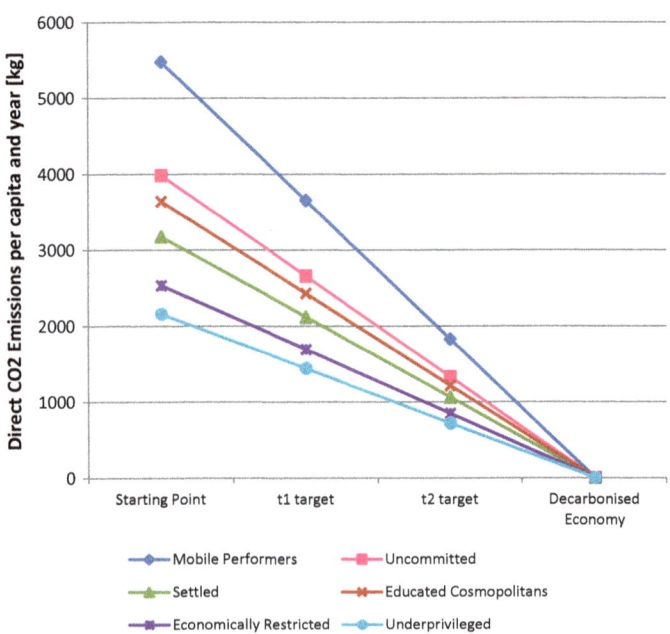

Fig. 9 Hypothetical reduction paths of 6 "energy lifestyles"

5.1 "Paris Lifestyle" Is not (yet) a Zero-Carbon Lifestyle

Having virtually no CO_2 emissions already today is perceived as unrealistic by most people, especially if the current socio-technical setting is taken into account. Fortunately "Zero-Carbon" Lifestyles do not need to be implemented overnight in order to comply with the Paris Agreement. Instead all social groups need to aim for the common goal of having a "Zero-Carbon" lifestyle as part of a decarbonised economy. Individual reduction paths are the logical consequence of different starting points. This means that a lifestyle compatible with the aims of the Paris Agreement is not universally defined, but mainly a question of how well the respective group or individual stays on track towards having zero net emissions.

6   Discussion

We identified six different types of "Energy Lifestyles" based on their profiles of direct energy use from seven areas of life. The findings show a broad spectrum of living situations and behavioural patterns and a correspondingly wide range of resulting energy demand and climate impact. We showed that the groups do not only differ in their behavioural patterns and impacts, but also in their characteristics regarding sociodemographics. Similar to the results from earlier studies, no significant correlation between attitudes towards energy saving and actual energy demand or CO_2 emissions could be found.

The added value for effective communication strategies provided by the approach became obvious by identifying groups like "Settled" and "Educated Cosmopolitans", who have practically the same overall energy demand, but entirely different behavioural patterns. In addition, they significantly differ in age, income and education, which is why the latter are most likely more receptive to communication strategies via digital channels like social media. It was also shown that the impact based approach is capable of identifying empirically existent lifestyles with a large spectrum in direct energy demand and emissions. "Mobile Performers" have 2.26 times the direct energy demand and 2.71 times the direct emissions of "Underprivileged".

7   Conclusion

By identifying lifestyles with distinct patterns of energy relevant behaviour and a large spectrum of energy demand and emissions, the approach presented in this paper turned out to be a reasonable alternative to general lifestyle models.

Future research with such a focus on intra-group effects promises a more detailed understanding of different social group reactions caused by changes in external parameters. It will also allow a modelling of different decarbonisation pathways with a high input-resolution. The combination of (1) individual level, (2) group level and (3) policy level will then facilitate the design of even more specific communication strategies and policy measures.

Yet, our analysis of individual energy demand and emissions also pointed to severe shortcoming in measuring the per capita emissions in different household settings. Families with children probably benefit in terms of having lower per-capita energy demand and emissions due to the smaller quantitative impact of children's needs on the emissions. Thus, an equivalence scale similar to scales on household expenditures that provide different weight factors for household members of different ages needs to be developed and is currently in progress.

Acknowledgements The work presented in this paper is part of the dissertation of Stephan Schwarzinger, who is the leading author regarding the identification of impact based lifestyle groups.

Finally, we would like to thank our colleagues at SERI (Sustainable Europe Research Institute) who provided the dataset used in our analysis.

References

Abrahamse W, Steg L (2009) How do socio-demographic and psychological factors relate to households' direct and indirect energy use and savings? J Econ Psychol 30(5):711–720. https://doi.org/10.1016/j.joep.2009.05.006

Bacher J, Pöge A, Wenzig K (2010) Clusteranalyse. Gruyter, de Oldenbourg

Bohunovsky L, Grünberger S, Frühmann J, Hinterberger F (2011) Energieverbrauchsstile Datenbank zum Energieverbrauch österreichischer Haushalte: Erstellung und empirische Überprüfung. Endbericht

Bourdieu P (1987) Distinction: a social critique of the judgement of taste. Harvard University Press

Buchanan K, Russo R, Anderson B (2015) The question of energy reduction: the problem(s) with feedback. Energy Policy 77:89–96. https://doi.org/10.1016/j.enpol.2014.12.008

Hartmann P (1999) Lebensstilforschung. Darstellung, Kritik und Weiterentwicklung. Leske & Budrich

Hermann D (2004) Bilanz der empirischen Lebensstilforschung. KZfSS Kölner Z Soz Sozpsychol 56(1):153–179. https://doi.org/10.1007/s11577-004-0007-2

Hierzinger R, Herry M, Seisser O, Steinacher I, Wolf-Eberl S (2011) Energy styles. Klimagerechtes Leben der Zukunft—Energy Styles als Ansatzpunkt für effiziente Policy Interventions. Endbericht zum Projekt Energy Styles. Klima- und Energiefonds

Holden E (2004) Towards sustainable consumption: do green households have smaller ecological footprints? Int J Sustain Dev 7(1):44–58. https://doi.org/10.1504/IJSD.2004.004983

Hradil S (1987) Sozialstrukturanalyse in einer fortgeschrittenen Gesellschaft. Sozialstrukturanalyse in einer fortgeschrittenen Gesellschaft. Leverkusen: Leske + Budrich. https://doi.org/10.1007/978-3-322-97175-3

Johnson D, Horton E, Mulcahy R, Foth M (2017) Gamification and serious games within the domain of domestic energy consumption: a systematic review. Renew Sustain Energy Rev 73 (Suppl C):249–264. https://doi.org/10.1016/j.rser.2017.01.134

Johnson TP (1998) Approaches to equivalence in cross-cultural and cross-national survey research. In: Harkness J (ed) Cross-cultural survey equivalence, vol 3. Harkness, Janet, pp 1–40. Retrieved from http://nbn-resolving.de/urn:nbn:de:0168-ssoar-49730-6

Kaklamanou D, Jones CR, Webb TL, Walker SR (2015) Using public transport can make up for flying abroad on holiday. Environ Behav 47(2):184–204. https://doi.org/10.1177/0013916513488784

Kuha J (2004) AIC and BIC. Soc Methods Res 33(2):188–229. https://doi.org/10.1177/0049124103262065

Lüdtke (1996) Methodenprobleme der Lebensstilforschung. Probleme des Vergleichs empirischer Lebensstiltypologien und der Identifikation von Stilpionieren. https://doi.org/10.1007/978-3-322-99689-3_7

Meyer T (2001) Das Konzept der Lebensstile in der Sozialstrukturforschung—eine kritische Bilanz. Soz Welt 52(3):255–271. Retrieved from http://www.jstor.org/stable/40878354

Otte G (2005) Construction and test of an integrative lifestyle-typology for Germany. Z Soz 6:442–467

Poortinga W, Steg L, Vlek C (2004) Values, environmental concern, and environmental behavior. Environ Behav 36(1):70–93. https://doi.org/10.1177/0013916503251466

Reusswig F (2002) Lebensstile und Naturorientierungen. Gesellschaftliche Naturbilder und Einstellungen zum Naturschutz. In: Rink D (ed) Soziologie und Ökologie, vol Lebensstile und Nachhaltigkeit. Konzepte, Befunde und Potentiale. VS Verlag für Sozialwissenschaften, Wiesbaden, pp 156–180

Schulze G (2005) Die Erlebnisgesellschaft. Kultursoziologie der Gegenwart. Campus Verlag, Frankfurt am Main

Spellerberg A (2001) Peter H. Hartmann: Lebensstilforschung. Darstellung, Kritik und Weiterentwicklung. KZfSS Kölner Z Soz Sozpsychol 53(1):170–171. https://doi.org/10.1007/s11577-001-0011-8

Tabi A (2013) Does pro-environmental behaviour affect carbon emissions? Energy Policy 63:972–981. https://doi.org/10.1016/j.enpol.2013.08.049

Tomlinson M (2003) Lifestyle and social class. Eur Sociol Rev 19(1):97–111. https://doi.org/10.1093/esr/19.1.97

Online Sources

European Commission, 2050 low-carbon economy, viewed 28 Sept 2017, https://ec.europa.eu/clima/policies/strategies/2050_en

OECD-modified Income Scale, viewed 14 Sept 2017, http://www.oecd.org/eco/growth/OECD-Note-EquivalenceScales.pdf

UNFCCC (2017) The Paris agreement, viewed 28 Sept 2017. http://unfccc.int/paris_agreement/items/9485.php

Mainstreaming Climate Change Adaptation in Infrastructure Planning— Lessons Learned from Knowledge Transfer and Communication

Alexandra Jiricka-Pürrer, Markus Leitner, Herbert Formayer, Thomas F. Wachter and Andrea Prutsch

Abstract Incorporating climate change (CC) impacts and adaptation into planning and development of large-scale infrastructure projects is facilitated by a wide range of international and national guidance material, but is still a challenge, as several studies point out. A number of guidance documents and studies show that Environmental Impact Assessment (EIA) could provide a good entry point for incorporating considerations of CC impacts and adaptation into project planning. This paper presents results of a research project aimed at developing recommendations for mainstreaming CC into EIA. Multiple levels of "knowledge brokerage" were undertaken in an actors-based participatory approach. The different formats, which all ensured diverse levels of interaction (knowledge brokerage), proved to be beneficial both regarding the exchange of know-how as well as raising awareness for the consideration of CC in EIA. This paper discusses benefits of knowledge brokerage processes as well as limitations to the mainstreaming of CC in project planning.

A. Jiricka-Pürrer (✉) · H. Formayer
University of Natural Resources and Life Sciences Vienna/Universität
Für Bodenkultur, BOKU, Vienna, Peter-Jordanstraße 92, 1190 Vienna, Austria
e-mail: alexandra.jiricka@boku.ac.at

H. Formayer
e-mail: herbert.formayer@boku.ac.at

M. Leitner · A. Prutsch
Environment Agency Austria (EAA)/Umweltbundesamt GmbH, Vienna, Austria
e-mail: markus.leitner@umweltbundesamt.at

A. Prutsch
e-mail: andrea.prutsch@umweltbundesamt.at

T. F. Wachter
Dr. Wachter Büro für Umweltplanung, Hamburg, Germany
e-mail: wachter@wachter-bfu.de

© Springer Nature Switzerland AG 2019   399
W. Leal Filho et al. (eds.), *Addressing the Challenges in Communicating Climate Change Across Various Audiences*, Climate Change Management,
https://doi.org/10.1007/978-3-319-98294-6_24

1 Introduction

Climate change (CC) impacts pose a serious challenge to the precautionary planning of large scale projects and to the assessment of impacts on the project environment. A wide range of international and national guidance material (e.g. Department of Transport 2014; Bles et al. 2015; Dallhammer et al. 2015) provides support on how to incorporate CC impacts and adaptation into planning and development of large-scale infrastructure projects, but the matter is still challenging (Boyle et al. 2013; EC 2013; Larsen 2014; Wachter et al. 2017a).

Several studies and guidelines (e.g. Walker et al. 2013; Byer et al. 2012; Agrawala et al. 2011; Runge et al. 2010) show that Environmental Impact Assessment (EIA) is able to address CC issues and could provide a good entry point for incorporating considerations of CC impacts and adaptation within the already existing modalities of project design, approval and implementation. The European level has responded to this possibility with a revised EIA Directive (2014/52/EU) also aiming to foster mainstreaming of CC. Since May 2017, both CC mitigation and adaptation should be considered in EIA in EU Member States. In practice, the "voluntary" consideration of CC adaptation in EIA has, thus far, been very limited (Fischer and Sykes 2009; Larsen 2014; Hands and Hudson 2016; Enríquez-de-Salamanca et al. 2016).

Examining studies on the consideration of CC in EIA, certain challenges quickly become apparent. In mainstreaming CC into EIA, Birkmann and Fleischhauer (2009) foresee new efforts regarding complexity and changes without linear predictability, which could complicate the adherence to threshold values and legal requirements commonly practiced in EIAs, or could even require such practice to be reconsidered entirely. Applied knowledge about the best ways to consider CC impacts and risks in planning practice is still limited (Rannow et al. 2010). While the direct and indirect CC impacts on projects and the project environment are recognised at a very abstract level, the transfer of this knowledge (e.g. in the form of concrete impact models/vulnerability assessments) to the regional and even local planning context still remains a challenge for the different stakeholders involved.

Against this background, enhanced transdisciplinary knowledge transfer and communication appear necessary to provide support for project developers, environmental consultants and competent authorities when considering CC impacts in practice (e.g. Wachter et al. 2017a; Jiricka et al. 2014; Balla et al. 2017; Fischer et al. 2011). Transdisciplinary knowledge transfer/brokerage is a key factor, as it challenges the actors concerned with EIA to co-produce new knowledge on mainstreaming CC into EIA together with scientific experts in the field. The challenge is to bring these different actors together and to create a space which fosters new knowledge that is easily comprehensible (Lemos et al. 2014) and of use to the target groups (Akerlof et al. 2012; Roeser 2012; Moser 2014; Wirth et al. 2014).

Considering these research gaps and needs, this study was conducted to help overcome the "science-policy-practice-divide" (Swart et al. 2014; Moser 2016, 2017) in mainstreaming CC into EIA. In detail, the study aimed to gain a better understanding of entry points for CC into the EIA planning process in Austria. It highlighted the relevance of CC impacts with regard to the specific regional/local conditions and project-related environmental issues. Furthermore, the objective was to provide support for effective decision-making despite given uncertainties through the joint development of guidance tools. The following key question of the study is discussed in this paper: how to best design and conduct the transdisciplinary transfer of knowledge between science and the different groups of actors in EIA (project developers, planning authorities, environmental authorities, consultants, EIA assessors/practitioners) in order to strengthen the capacity among key actors to mainstream CC and adaptation into EIA?

The two-and-a-half-year study (funded under the Austrian Climate Research Program ACRP) was designed to be transdisciplinary and participatory, bringing together key actors from (i) private project developers, (ii) planning offices, and (iii) public EIA authorities (including experts on CC adaptation in the federal authorities), as well as experts on CC (Zentralanstalt für Meteorologie und Geodynamik (ZAMG)). Scientists from climatology as well as adaptation planning were present in all phases of the process, supported by international scientists specialised in CC adaptation and impact assessment.

Various forms of knowledge brokerage from information to engagement (as defined e.g. by Michaels (2009) and Adelle (2015) were implemented, depending on the key steps of CC integration into EIA (e.g. identification of entry points for CC, or methods to assess CC impacts on a local project level). Furthermore, the influence of German EIA actors in the process—particularly during the stakeholder workshops—as well as further international expertise provided by the project advisory board (experts from UK and Scandinavia), will be discussed, as all EU Member States are currently implementing the new EIA Directive. However, concrete implementation is strongly influenced by the mainstreaming of CC adaptation into national policies as well as by the awareness raised by national and regional adaptation strategies in the specific countries.

2 Background

Considering CC impacts and adaptation in large-scale infrastructure projects requiring EIA is an increasingly discussed topic in international studies and reports, due to the large investment volumes and long time frames of such projects (Walker et al. 2013; EC 2013; Byer et al. 2012). This applies to the projects themselves (keyword "climate proofing"), but also to the impact on the environmental issues surrounding the project. McCallum et al. (2013) point out possible cumulative impacts which can evolve when the "baseline environment" is influenced by CC. In an approach [extended from Runge and Wachter (2010) and Runge et al. (2010) Birkmann et al. (2012)], Jiricka et al. (2014) indicate CC-related consequences for the EIA-procedure in different directions:

- direct impacts of CC on projects (e.g. increasing number of extreme heat days may cause damage to road or railway infrastructure, storm events can damage the infrastructure)
- impacts on projects which are indirectly associated with the project environment (e.g. drought and periods of extreme heat days might increase the likelihood for forest fires which can lead to an indirect impact on projects, storm events can cause windfall of nearby trees and consequently damage the infrastructure)
- direct impacts on the environmental issues surrounding the project (drought can impact wetland habitats or decrease the efficiency of mitigation measures for wetland habitats; e.g. barrier effects could have a stronger impact on amphibians moving between different wetland habitats).

To date, only few attempts have been made to assess concrete spatial and temporal impacts of CC on projects and their associated environments (Runge et al. 2010; EBA 2014; Kunze et al. 2014; Balla et al. 2017). A major challenge is connected to the question of which data on CC can actually be used in EIA. With regard to EIA, because of its role as commissioning process, the Austrian and German EIA system is strongly oriented along legal requirements (Margelik and MCCallum 2014; Jiricka et al. 2016a). A lack of standards and data regarding CC projections/possible impacts—commissioned/recommended by federal or regional authorities (e.g. vulnerability analyses, impact models)—was often addressed as a major reason for inaction with regard to CC adaptation in EIA in the past (e.g. Jiricka et al. 2016b; Wachter et al. 2017a).

In order to progress with "evidence-based" decision-making despite given uncertainties, knowledge brokerage between science, policy and practice has recently been recognised as a promising path forward (Mimura et al. 2014), which could also be applied to foster the mainstreaming of CC into EIA under consideration of the challenges acknowledged above. Research on how to broker CC adaptation knowledge to policymakers has emerged in the last few years, mainly focusing on conceptual aspects (Hering 2016). Nevertheless, empirical cases as well as cases involving the private sector are still widely missing (Prutsch 2017).

Research has shown that different knowledge brokerage strategies, comprising linear and interactive approaches, are needed, as decision makers are themselves dealing with a range of issues requiring different kinds of knowledge (Michaels 2009; Hegger and Dieperink 2015; Prutsch 2017). As the notion of knowledge brokerage is still somewhat indistinct, we apply a definition provided by Prutsch (2017), describing knowledge brokerage as an intermediary between science and policy/practice (comprising the private sector) that aims to facilitate decision-making processes with various brokering techniques from informing to capacity-building (based on Litfin 1994 in Michaels 2009; Dilling and Lemos 2011). Linear and interactive knowledge brokerage strategies each have their advantages, but one must consider that they lead to divergent results (cf. Fig. 1).

These aspects are discussed in chapter four and five against the results of applied knowledge brokerage strategies in order to increase integration of CC and adaptation into EIA planning processes.

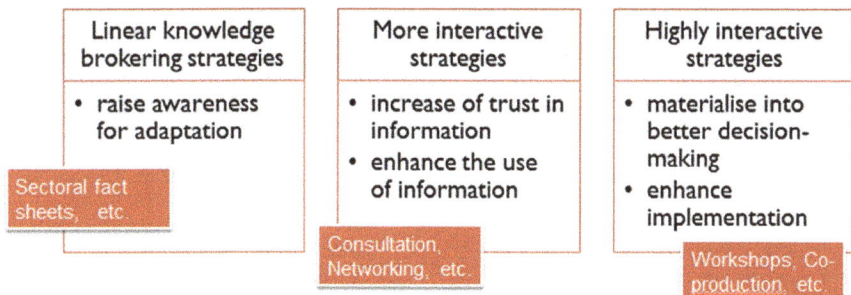

Linear knowledge brokering strategies	More interactive strategies	Highly interactive strategies
• raise awareness for adaptation Sectoral fact sheets, etc.	• increase of trust in information • enhance the use of information Consultation, Networking, etc.	• materialise into better decision-making • enhance implementation Workshops, Co-production, etc.

Fig. 1 Three knowledge-brokering clusters to differentiate the benefits accordingly (Prutsch 2017)

3 Methodological Approach

Various knowledge brokerage strategies were applied in this study, focusing on the applicability of information about CC impacts and adaptation measures at the specific project level (road, rail and power grid infrastructure), and focusing on identifying special data requirements and considering the usability of CC models as input into EIA. Relevant actors concerned with EIA (project developers, planners/ EIA consultants and authorities) and climate scientists worked together in a transdisciplinary manner.

3.1 Research Design

In an actor-based, multi-level approach (see Table 1), thematic entry points for the consideration of CC adaptation were first identified through comprehensive analysis of recent EIA and project planning documents from Austrian and German EIA practice. The aim was to identify present and potential future meteorological phenomena and impacts related to CC, as well as to illustrate the relevance of considering them in EIAs of large-scale infrastructure projects. The results of this content analysis (based on search terms identified by international literature and checked by experts on CC impacts and adaptation in the fields) were consolidated in a handout which was sent to the actors as preparatory information for the interviews and stakeholder workshops. It first included general information on CC scenarios for Austria as well as potential impacts relevant to EIA, and, secondly, information on concrete entry points related to the examination of current EIA practice. Existing data was analysed in parallel, according to its applicability in EIA. The outcomes of the two main research programs on CC in Austria (Austrian Climate Research Programme (ACRP) and StartClim) were examined. Particular

Table 1 Application of different levels of knowledge brokerage in the research process

Actors involved	Objective	Format	Brokering strategy
EIA consultants, EIA authorities/environmental authorities, project developers, researchers (planning and impact assessment, climate change adaptation and communication, climatology)	• Findings on thematic entry points for the consideration of CC impacts	Handout on CC impacts and entry points identified by the analysis of EIA practice Development of an online toolkit, which is suited to the targets and needs of the different groups of actors	Inform
Environmental authorities, Researchers from previous research projects who were consulted regarding their research outcome (e.g. impact models)	• Identification of data (impact models and spatially referenced research outcomes) suitable for the consideration of CC impacts in EIA • Link with the entry points identified in the first step, to consider future changes due to climate change	Analysis of research data —consultation via mail and phone	Consult
EIA consultants, EIA authorities/environmental authorities, project developers, researchers (planning and impact assessment, climate change adaptation and communication, climatology)	• Feedback to the content analysis • Identification of barriers and entry points	Interviews	Consult
EIA consultants, EIA authorities/environmental authorities, project developers, researchers (planning and impact assessment, climate change adaptation and communication, climatology)	• Discussion about entry pointes, barriers and possible solutions in scenario technique approach • Identification of specific entry points into framing conditions (standards, guidance, vulnerability assessments/open data sources)	Stakeholder workshops Co-development of an online toolkit, which is suited to the targets and needs of the different groups of actors	Engage

focus was set on impact models, maps and decision support systems applicable for the whole state territory of Austria, and on regional data and models which could be transferred to other regions with feasible effort (e.g. information on vulnerable species). Furthermore, relevant authorities of the federal states (both Austria and

Germany have a federal governance system) were consulted with regard to data or on-going research likely to produce information relevant for application in EIA.

In a second step, together with the actors (project developers, planners and authorities), specific thematic as well as procedural entry points were discussed, first individually, in twenty-one expert-interviews, and then jointly, in two stakeholder workshops with around thirty participants each. At the stakeholder workshops, the project team first presented the thematic entry points identified from current EIA practice, upon which the participants discussed these entry points as well as further thematic aspects with relevance to an altered sensitivity of the environmental issues (project environment). Further, the applicability of data (e.g. impact models, maps, decision support systems) with relevance to the environmental issues was discussed, based on examples. During this process, the overview of existing data evolved from the afore-mentioned initial analysis of all research projects funded under the two Austrian CC research programs and the consultation of federal authorities in the field.

In a third and final step, the outcome of the project, an online "climate-fit toolkit", was developed together with the actors from planning practice. It includes a decision support system (DSS) to help evaluate the relevance of potential CC impacts, and supports users in identifying specific important topics regarding the topographic and climatic conditions in relation to the sensitivity of the environmental issues and the project type concerned. The actors involved in the process evaluated the first concept of the online toolkit at the second workshop, and were engaged in the continued development process through feedback loops, even after the workshop.

The acting knowledge brokers in this process were the Environment Agency Austria, the Ministry of the Environment as well as universities specialised in planning and impact assessment as well as CC adaptation research (such as BOKU Vienna). Additionally, some of the actors involved in EIA could also be considered as knowledge brokers in the process—such as the environmental consultants (EIA consultants in planning offices) as well as the specific environmental authorities involved in the process of scoping and issuing environmental statements, which communicate the relevance of topics to be considered in EIA to the project developer. The methodological approaches for each step are described in detail in the following.

3.2 Finding Thematic Entry Points—Content Analysis of EIA Practice

The aim of the content analysis of EIA documents was to ascertain the level of consideration of CC impacts in EIA to date, as well as potential entry points for doing so in the future, and to make the significance of CC-relevant aspects more tangible for the actors. Between June 2016 and March 2017, a total of 23 EIA procedures in Austria and 28 procedures in Germany (concerning the areas of rail, road, high-voltage/extra-high-voltage power lines) underwent an "ex-post evaluation" (content analysis) in order to identify consideration of potential CC impacts to date (EIA practice from 2005 to 2015) as well as possible approaches for the future. For the Austrian procedures, documents from all phases of the EIA process (EIS and expert reports, authority opinions, technical reports, administrative rulings, etc.) were examined. For German procedures, the focus was on the application documents (due to the limited availability of documents from other EIA phases).

The content analysis consisted of a two-stage process; first, documents were subjected to a search for direct references to CC, in order to determine relevance. The search terms (in German), used in this first phase were "climate change", "climatic change", "climate change impacts", "climate change adaptation", "vulnerability", "scenario" (connected with climate change).

Next, a content analysis of certain meteorological phenomena (e.g. "strong storms" or "heavy precipitation") and associated impacts (e.g. "windthrow" or "slope slumping"), including combinations thereof, was conducted. A comprehensive catalogue of search terms was compiled in preparation for the ex-post evaluation. The compilation of possible meteorological phenomena is based on international (European Commission 2013) and national adaptation strategies (BMLFUW 2017; BMU 2009, 2011) as well as on results of previous research projects in the two countries, combined with an analysis of the relevant literature. Furthermore, experts from the Environment Agency Austria as well as the BOKU University Vienna were consulted. The meteorological phenomena correspond with those discussed in other studies on CC adaptation, for example by the European Environment Agency (2017).

3.3   Identifying Barriers and Entry Points—Interviews with EIA Actors

Twenty expert interviews were conducted with EIA seven authorities, four project applicants and eight planning offices/technical report authors in Austria and Germany between March and May 2017 in preparation for the stakeholder workshops. Since there were multiple people present at many of the interviews, the total number of interviewees was 34.

Interviews were structured, the interview guideline comprising three thematic blocks:

– Personal and institutional area of responsibility
– Experience with CC adaptation
– Evaluation of future development.

Wherever possible, the interviews were conducted in person, or by telephone if necessitated by distance. All interviews were transcribed and documented and submitted to the interviewees for verification. Upon receipt of potential corrections, the interviews were coded and evaluated according to answer categories.

3.4   Engagement in the Development of Support Tools— Stakeholder Workshops

On 13 June 2017, representatives from authorities, planning offices and project applicants met for the first project workshop at the Environment Agency Austria. In-depth discourse on the issue of CC adaptation in EIA was promoted by an interactive mix of methods (KETSO, Mindmaps, different evaluation approaches, …). Entry points for consideration of CC adaptation in EIA on various levels were discussed—from the adoption of information from higher-level planning projects, over consideration during the examination of environmental issues, to the concrete option of CC adaptation during the planning and monitoring of measures. In a second step, priority aspects for the implementation of CC adaptation in EIA were identified and evaluated by the three actor groups.

During the second workshop on 13 November 2017, the focus was on the concrete implementation of CC adaptation in EIA, with regard to the individual procedural steps and with regard to the potentially affected environmental issues. The focus here was on reviewing the presentation of information and data on CC adaptation for EIA, with a particular focus on the online "climate-fit toolkit"—the joint product of the project. Once more, the actors' involvement in the content of the toolkit continued even after the second workshop, through feedback loops.

3.5   Limitations

The stakeholder workshops were evaluated with regard to the perceived benefits in terms of knowledge brokerage, as well as to the expectations and limitations of consideration of CC adaptation in EIA (noted by the participants). However, the overall knowledge brokerage process was not evaluated empirically.

Mainstreaming CC adaptation into the relevant policy documents at a national level was still an ongoing process during the study lifetime. Austrian general elections took place halfway through the project lifetime. Thus, the implementation of the EIA directive into national law was protracted and political support for the process as a whole was discontinuous (as the EIA system in general became subject to discussion).

4 Results

In the following, the results of the various knowledge brokerage strategies applied are presented with regard to the thematic and procedural entry points of CC into EIA jointly identified with the involved actors. Furthermore, barriers and opportunities to overcome those barriers identified in the process are outlined.

4.1 Inform

The central format to inform relevant stakeholders concerned with integrating CC into EIA was a handout (see Chap. 3). Its aim was to inform firstly about CC projections for Austria. Furthermore, it included information on the consideration of meteorological phenomena and future aspects relevant to CC derived from the analysis of the status-quo.

The analysis of the status-quo revealed that direct references to CC adaptation are still limited in EIA (Wachter et al. 2017b). It also showed that extreme weather events already receive frequent consideration. Until now, however, future changes and especially uncertainties and unforeseen events are barely integrated in the consideration of environmental issues. When examining the environmental issues addressed together with the search terms used in the initial analysis of EIA documents, strong differences between environmental issues were revealed regarding the consideration they receive for potential future CC relevance. A detailed analysis of results for all search terms, and particularly for the most relevant meteorological phenomena ("heavy rainfall", "storms" and "aridity") plus related impacts ("erosion", "flooding", "wind throw", "drought", "slope slumping" and "wind erosion") mentioned in the EIA documents was presented to the actors with exemplary quotes from the qualitative analysis.

Connecting general information on CC with information about its past consideration in EIA helped to establish links to the practice in order to raise awareness for future CC impacts. According to the feedback of the actors, the link to possible thematic entry points highlighted the relevance of CC adaptation in the future. Furthermore, the handout served as a common basis for all actor groups (for the engagement phase).

4.2 Consult

In the consultation phase, expert interviews were conducted with all three groups of actors. In parallel, climate researchers were consulted regarding their research results (via e-mail, telephone or in person) to identify actual links for the integration of their data in EIA.

The interviews confirmed the results of the analysis of status-quo regarding EIA practice. As a matter of fact, CC adaptation is scarcely considered in EIA until now. Furthermore the consultation revealed that several aspects which were discussed in international (European Commission 2013) and national adaptation strategies (BMLFUW 2017; BMU 2009, 2011) have rarely been mentioned in EIA thus far (e.g. "forest and slope fires", "temperature fluctuation", "heavy snowfall with wet snow" or "freezing/thawing weather" with an increased potential for "slope instability and erosion"). Those "thematic gaps" were prepared for discussion in the engagement phase (stakeholder workshops) using concrete examples with regard to the environmental issues.

Regarding the environmental issues, the interviews confirmed that additional aspects pertaining to plants/animals/habitats and humans/health could become relevant in the future, which have scarcely been considered thus far. Conversely, the consultation of researchers and institutions revealed that—precisely regarding the area of flora/fauna/habitats—data on climate-sensitive species and habitat changes already exist, which could either be integrated in EIA directly or could be transferred to other regions (e.g. information from Germany; Streitberger et al. 2016). Here, the consultations of both practitioners and researchers complemented each other ideally to consolidate both perspectives and sets of information into the third step, engagement. The information from the practitioner interviews regarding the missing aspects mentioned above, together with impact models and further data revealed by the consultation of external researchers, were presented to all actors in the engagement phase to reach a common understanding about the necessity for further capacity-building in this field.

The first two steps of the survey already identified methodological entry points. The analysis of past EIA documents showed that several common mitigation and compensation measures could be re-dimensioned or adapted (e.g. by the selection of more heat and drought resistant tree species) whereas, for some potential CC impacts, new strategies would need to be adopted and/or different approaches to monitoring the efficiency of the existing measures be elaborated (e.g. with regards to neobiota, wetland habitats or endangered alpine species).

Barriers and opportunities

The interviewed project applicants and proponents, in particular, did already see the relevance of considering CC in their technical planning and in ongoing operations. They saw no relevance, however, of gaining information about these topics from EIA, or of addressing them in EIA. Furthermore, some Austrian authorities considered climate proofing to be within the project applicants' and proponents' own responsibility, and disconnected from EIA. German authorities, on the other hand, emphasized the potential to be gained from an interaction between the examination of environmental issues in EIA and the technical project planning, regarding climate proofing. This in-depth examination of the different estimations of relevance during the consultation phase formed an excellent basis on which to test whether these estimations changed during the third phase.

The fast-paced dynamics and potential need for greater flexibility were often pointed out in the consultation phase as concrete challenge, as was the handling of given uncertainties in general. In particular, the actors addressed the differentiation between assuming a worst-case probability of potential incidents and catastrophes, and considering other potential CC impacts, with the possibility of adapting at a later date. Above all, however, spatial datasets, such as causation models, are desirable. The availability of "open data" (free data that can be used by actors within the framework of EIA) was noted as being very desirable in this respect. All of these aspects mentioned were incorporated in phase three, especially in the discussion for the co-production of the online toolkit.

4.3 Engage

Both workshops, which were attended largely by the same participants, built upon each other structurally. During the first workshop, the participants' different levels of experience and knowledge regarding CC adaptation were more than apparent. While the coordinators in consulting offices and authorities had often had few points of professional contact with the area of CC adaptation (except those specifically active in that very field), participants from the technical divisions of those offices and authorities had often already accumulated more knowledge through their professional practice, and could introduce it into the discussion. In this case, the exchange between the different groups of actors was highly advantageous. They complemented and enriched one another regarding their fields of knowledge. Project applicants already had in-depth knowledge regarding starting points for climate proofing from previous (often internal) research projects. Potential impacts on the project environment, however, were a new topic for them. On the other hand, participants from consulting offices and authorities found the relation between the altered sensitivity of a project environment and the potential impacts on the project to be a new aspect, particularly during the first workshop. At the end of the first workshop, a short presentation by a German authority, which has already integrated CC adaptation into the mandatory guideline in their field, offered a valuable example of how concrete mainstreaming of CC adaptation in EIA is possible.

During the research process, the actor groups were directly involved in the development of a guidance tool to assist in considering CC adaptation in EIA. In a first step, during the first workshop, participants compiled a list of environmental issues particularly affected as well as methodological approaches and EIA steps particularly suited for an integration of CC adaptation, in their view. In the second workshop, the project team presented the draft of an "online toolkit", and partici-pants evaluated (individually using forms, as well as collective evaluation in actor groups) its individual components regarding their suitability in practice, discussed and amended them. After the workshop, further conceptual and technical feedback regarding the components of the online toolkit was given in written feedback loops. These multiple levels of feedback provided the researchers with valuable information for the further development of the guidance tool, and also ensured that the actors were already familiarized with the documents.

Barriers and opportunities

The results of the workshops confirm those of the expert interviews regarding barriers and potential solutions. Participants still expressed reservations as to the actual feasibility, but they also acknowledged that solution approaches were conceivable in good time and in cooperation with other actors.

Across all actor groups, the stakeholder workshops also showed that long-term monitoring accompanying the dynamic processes (while implementing mitigation and adaptation measures) should be a high priority. Regarding the question of how to deal with the uncertainties that are typical of causation models based on climate scenarios, transparency in the presentation of potential risks and transparency regarding the limits to considering potential CC impacts in EIA evaluation and planning were shown to be of particular importance in the view of the workshop participants.

Across actor groups, participants determined that environmental issue-specific "standards", above all, are required in order to properly and uniformly consider the current state of knowledge on CC impacts and to consider CC adaptation in EIA. In terms of knowledge brokering, this challenge could only be met in the future on the level of a common agreement and through "capacity building", with a supportive exchange between the technical employees and coordinators of consulting offices and authorities.

5   Discussion and Outlook

Through the interactive process applied in this transdisciplinary study, several brokerage levels were applied in order to mainstream CC adaptation into EIA and project planning. One aim of the underlying research project was the provision of specific knowledge on CC impacts and adaptation potentials relevant for EIA and project planning (linear transmission of information as described by Ballantyne 2016). Additionally, the project aimed to illustrate the relevance for consideration of CC adaptation in EIA. The different formats—which ensured diverse levels of participation (knowledge brokerage)—proved to be beneficial for both the exchange of know-how as well as the raising of awareness for consideration of CC impacts and adaptation in EIA. Communication between different groups of actors, in particular, was enabled—a beneficial aspect of communicating CC adaptation highlighted e.g. by Pearce et al. (2015).

The first two phases (information and consultation) laid excellent groundwork for the third phase (engagement), in this respect. The three phases (see Chap. 2, Fig. 1) were tightly interlocked (Michaels 2009; Adelle 2015) and complemented each other ideally to identify both chances as well as barriers together with the actors, but also (and especially) to raise awareness for the potential to consider CC adaptation in EIA. The handout, which was tailored to the practice partners, allowed all actors to enter into the participation process on the same level of knowledge (Wirth et al. 2014). At the same time, the significance of CC adaptation was addressed from the outset through an early discussion of concrete entry points for EIA and planning practice.

All participants were able to find specific examples from their respective fields (e.g. partial analyses for different project types, partial and qualitative analyses of potential impacts on project environment/environmental issues, and analyses of application during different steps of the EIA process). Thanks to the conversations during the interviews, the participants already felt individually understood in their respective viewpoints—particularly since even critical quotes were included in the presentations at the beginning of the workshops, as a basis for discussion during the workgroup phases. Introducing links from previous EIA practice has proven to be a good starting point for the work in small workgroups.

The effect of the exchange within and between actor groups during the workshops was that even those participants, who had seen little or no relevance for the issue of CC adaptation in EIA during the interviews, began to see the question as more significant and began to make their own suggestions for implementation in practice. The joint "co-design" of the "climate-fit toolkit" provides the chance of increasing acceptance for future consideration of CC impacts on environmental issues and projects in EIA. Ideally, the actors will be able to identify more strongly with the support tool because they themselves were involved in its development. Since many individual technical divisions within the actor groups are affected in EIA (e.g. technical divisions of the regional authorities), one of the aims was that the workshop participants themselves become valuable knowledge brokers within their own actor groups (e.g. EIA coordinators towards technical authorities). This aspect could be important to ensure communication for the "long haul" and a long-term engagement (Moser 2016, 2017). Here, one of the project's limitations also became apparent: the aim had been to include the technical divisions as well, but most of the participating institutions "only" sent their coordinators to the workshops, for reasons of economy. Furthermore, the amended EIA Directive has not yet been implemented into national law in Austria—thus there is, as yet, no legal foundation for the topic and for future consideration of CC adaptation.

Since barriers were already identified earlier on (from scientific literature and during the consultation phase) suggested approaches towards solutions could be introduced by the project team as a basis for discussion during the course of the process (e.g. pointing out datasets on specific CC impacts that are already available for consideration in EIA). The respective actors' evaluation of implementation relevance of the strategies, as well as additional requests for guidance material, could be discussed directly with the decision-makers from the authorities. In parallel, the project team was involved in the development of an environmental impact statement (EIS) guideline for Austria. Representatives of the responsible federal ministry also participated in the project workshops. The aim was to integrate recommendations for consideration of CC impacts and possibilities for CC adaptation directly into the national "binding" EIS guideline.

Of course, the need for a consideration of CC aspects in the technical standards (e.g. red lists) relevant for EIA can only be communicated as a recommendation within the framework of the research project. Likewise, the proposal to anchor technical and methodological references to CC adaptation in the technical sections of the EIA (e.g. water/flood protection/hazardous area planning/vulnerability analyses) remains a future step that exceeds the brokerage process within the framework of the project. Furthermore, it was shown that the results of research projects on CC impacts and CC adaptation are currently only of very limited usability in EIA practice. In response, the project also attempts to provide clues as to the transferability of research approaches and results (e.g. which spatial CC impact models/vulnerability analyses would be of particular relevance for the consideration of CC impacts in EIA).

The inclusion of international actors introduced implementation examples into the process (e.g. anchoring of CC adaptation in guidelines, nationwide vulnerability analyses) that were enriching for the Austrian actors. The fact that these approaches were presented directly by representatives of German authorities and planning offices (and not by the research team) increased their impact and was most advantageous for the overall process.

6   Conclusion

The chosen approach of mainstreaming CC adaptation into EIA, as carried out in this study through an actors-based participatory approach, revealed strong benefits in terms of raising awareness for the topics, and of communicating potential thematic and methodological entry points for different groups of actors in practice. Different formats of knowledge brokerage, which together ensured diverse levels of interaction, helped to identify challenges and barriers perceived by the actors at an early stage and to discuss opportunities provided by a scenario-based approach.

However, limitations in the communication process could be observed. One of the main limitations to the process was the small number of actors from technical divisions who could be involved directly in all stages of the knowledge brokerage process. The co-design of an online tool-kit on CC adaptation in EIA by the actors allowed a strong involvement and engagement in the communication of know-how. Mainstreaming CC adaptation into the relevant policy documents at the national level was still an ongoing process during the study lifetime. The involvement in the knowledge-brokerage process of the persons in charge of the legal implementation and development of guidance documents was beneficial for both the consideration of relevance and the communication of barriers and opportunities. One unfortunate limitation—also observed in previous studies—was the political impact at the highest level, which could yet influence the national implementation significantly.

References

Adelle C (2015) Contexualising the tool development process through a knowledge brokering approach: the case of climate change adaptation and agriculture. Environ Sci Policy 51:316–324

Agrawala S, Matus Kramer A, Prudent-Richard G, Sainsbury M (2011) Incorporating climate change impacts and adaptation in environmental impact assessments, opportunities and challenges. OECD. https://doi.org/10.1787/5km959r3jcmw-en

Akerlof KE, Maibach W, Fitzgerald D, Cedeno AY, Neuman A (2012) Do people "personally experience" global warming, and if so how, and does it matter? Glob Environ Change 23 (1):81–91

Balla S, Peters H-J, Schönthaler K, Wachter T (2017) Überblick zum Stand der fachlich-methodischen Berücksichtigung des Klimawandels in der UVP. Analyse, Bewertung und Politikempfehlungen zur Anpassung nationaler rechtlicher, planerischer und informatorischer Politikinstrumente an den Klimawandel—2. Teilbericht zu Arbeitspaket 4 des FE-Vorhabens FKZ 3713 48 105 im Auftrag des Umweltbundesamts, Dessau-Roßlau i.V (Climate Change 2017)

Ballantyne AG (2016) Climate change communication: what can we learn from communication theory? vol 7. Climate Change, Wiley Periodicals. https://doi.org/10.1002/wcc.392

Birkmann J, Fleischhauer M (2009) Anpassungsstrategien der Raumentwicklung an den Klimawandel: „Climate Proofing"—Konturen eines neuen Instruments. Raumforsch Raumordn 67(2):114–127

Birkmann J, Schanze J, Müller P, Stock M (2012) Anpassung an den Klimawandel durch räumliche Planung. Grundlagen, Strategien, Instrumente, E-Paper der ARL Nr. 13, Hannover

Bles T, Bessembinder J, Chevreuil M, Danielsson P, Falemo S, Venmans A (2015) Roads for today, adapted for tomorrow guidelines. The CliPDaR Consortium. URL: http://www.transport-research.info/sites/default/files/project/documents/ROADAPT_integrating_main_guidelines.pdf

BMLFUW (Eds.) (2017) Die österreichische Strategie zur Anpassung an den Klimawandel. Teil 1 —Kontext, Wien

BMU—Bundesministerium für Umwelt, Naturschutz und Reaktorsicherheit, (2009) Dem Klimawandel begegnen. Die deutsche Anpassungsstrategie, Berlin

BMU—Bundesministerium für Umwelt, Naturschutz und Reaktorsicherheit (2011) Aktionsplan Anpassung der Deutschen Anpassungsstrategie an den Klimawandel vom Bundeskabinett am 31. Aug 2011 beschlossen, https://www.umweltbundesamt.de/themen/klima-energie/klimafolgen-anpassung/anpassung-auf-bundesebene/aktionsplan-anpassung

Boyle J, Cunningham M, Dekens J (2013) Climate change adaptation and Canadian infrastructure. A review of the literature. International Institute for Sustainable Development, Winnipeg

Byer P, Cestti R, Croal P, Fisher W, Hazell S, Kolhoff A, Kørnøv L (2012) IAIA statement on climate change and impact assessment. In: IAIA—climate change in impact assessment: international best practice principles. Special publication series no. 8

Dallhammer E, Formayer H, Jiricka A, Keringer F, Leitner M, McCallum S, Schmied J, Stanzer G, Völler S (2015) ENVironmental impact assessment satisfying adaption goals evolving from climate change, publish-ready final report. Financed by: ACRP, Klima- und Energiefonds, p 31

Department for Transport (2014) Transport resilience review. A review of the resilience of the transport network to extreme weather events. Secretary of State for Transport, UK

Dilling L, Lemos MC (2011) Creating usable science: opportunities and constraints for climate knowledge use and their implications for science policy. Glob Environ Change 21:680–689

EBA—Eisenbahnbundesamt (2014) Umwelt-Leitfaden zur eisenbahnrechtlichen Planfeststellung und Plangenehmigung sowie für Magnetschwebebahnen. Teil III Umweltverträ glichkeitsprüfung—Naturschutzrechtliche Eingriffsregelung

EC—European Commission (2013) Guidance on integrating climate change and biodiversity into environmental impact assessment

Enríquez-de-Salamanca A, Martín-Aranda RM, Díaz-Sierra R (2016) Consideration of climate change on environmental impact assessment in Spain. Environ Impact Assess Rev 57:31–39

European Environment Agency (2017) Climate change adaptation and disaster risk reduction in Europe. Enhancing coherence of the knowledge base, policies and practices.https://www.eea.europa.eu/publications/climate-change-adaptation-and-disaster/at_download/file

Fischer TB, Sykes O (2009) The new EU territorial Agenda—indicating progress for climate change mitigation and adaptation? In Davoudi S et al (eds) Planning for climate change, Earthscan, pp 111–124

Fischer TB, Potter K, Donaldson S, Scott T (2011) Municipal waste management strategies, strategic environmental assessment and the consideration of climate change in England. J Environ Assess Policy Manage 13(4):541–565

Hands S, Hudson MD (2016) Incorporating climate change mitigation and adaptation into environmental impact assessment: a review of current practice within transport projects in England. Impact Assess Proj Appr 34(4):330–345

Hering JG (2016) Do we need "more research" or better implementation through knowledge brokering? Sustain Sci 11:363–369

Hegger D, Dieperink C (2015) Joint knowledge production for climate change adaptation: what is in it for science? Ecol Soc 20(4):1

Jiricka A, Völler S, Leitner M, Formayer H, Fischer TB, Wachter TF (2014) Herausforderungen bei der Integration von Klimawandelfolgen und -anpassung in Umweltverträglichkeitsprüfungen— ein Blick auf die Planungspraxis in Österreich und Deutschland. UVP-Rep 28 (3, 4):179–185

Jiricka A, Bösch M, Völler S (2016a) Learning from the past and upcoming challenges—the implementation of the amendment of the EIA directive in Austria

Jiricka A, Völler S, Fischer TB, Leitner M, Formayer H, Schmidt A, Wachter TF (2016b) Consideration of climate change impacts and adaptation in EIA practice—perspectives of actors in Austria and Germany. Environ Impact Assess Rev 57:78–88. UVP-Rep 3:143–151

Kunze K, v Haaren C, Reich M, Weiß C (2014) Kompensationsflächenmanagement im Klimawandel—Anpassungsmaßnahmen im Bremer Feuchtgrünland zur Erhaltung von Ökosystemleistungen und Empfehlungen für die Eingriffsregelung. In: Korn H, Bockmühl K, Schliep R (eds) Biodiversität und Klima—Vernetzung der Akteure in Deutschland X— Ergebnisse und Dokumentation des 10. Workshops. BfN-Skripten 357

Larsen SV (2014) Is environmental impact assessment fulfilling its potential? The case of climate change in renewable energy projects. Impact Assess Proj Appr 32(3):234–240

Lemos MC, Kirchhoff CJ, Kalafatis SE, Scavia D, Rood RB (2014) Moving climate information off the shelf: boundary chains and the role of RISAs as adaptive organizations. Weather Clim Soc 6(2):273–285. https://doi.org/10.1175/WCAS-D-13-00044.1

Margelik E, McCallum S (2014) Umweltverträglichkeitsprüfung in Österreich—Einblicke in ein umfassendes Dokumentationssystem. UVP-Rep 28(3, 4):128–132

McCallum S, Dworak T, Prutsch A, Kent N, Mysiak J, Bosello F, Klostermann J, Dlugolecki A, Williams E, König M, Leitner M, Miller K, Harley M, Smithers R, Berglund M, Glas N, Romanovska L, van de Sandt K, Bachschmidt R, Völler S, Horrocks L (2013) Support to the development of the EU strategy for adaptation to climate change: background report to the impact assessment, part I—problem definition, policy context and assessment of policy options. Environment Agency, Vienna, Austria

Michaels S (2009) Matching knowledge brokering strategies to environmental policy problems and settings. Environ Sci Policy 12(7):994–1011

Mimura N, Pulwarty RS, Duc DM, Elshinnawy I, Redsteer MH, Huang HQ, Nkem JN, Rodriguez SRA (2014) Adaptation planning and implementation. In: Field CB, Barros VR, Dokken DJ, Mach KJ, Mastrandrea MD, Bilir TE, Chatterjee M, Ebi KL, Estrada YO, Genova RC, Girma B, Kissel ES, Levy AN, MacCracken S, Mastrandrea PR, White LL (eds) Climate change 2014: impacts, adaptation, and vulnerability. Part A: global and sectoral aspects. Contribution of working group II to the fifth assessment report of the intergovernmental panel on climate change. Cambridge University Press, Cambridge, UK, New York, NY, USA, pp 869–898

Moser SC (2014) Communicating adaptation to climate change: the art and science of public engagement when climate change comes home. WIREs Clim Change 5:337–358

Moser SC (2016) Reflections on climate change communication research and practice in the second decade of the 21st century: what more is there to say? Wiley Interdisc Rev Clim Change 7(3):345–369

Moser SC (2017) Communicating climate change adaptation and resilience. Oxford Research Encyclopedia of Climate Science

Pearce W, Brown B, Nerlich B, Koteyko N (2015) Communicating climate change: conduits, content, and consensus. Wiley Interdisc Rev Clim Change 6(6):613–626

Prutsch A (2017) Knowledge brokerage and the communication connected with adaptation to climate change. Thesis in fulfilment of the requirements for the doctoral degree at the University of Natural Resources and Life Sciences, Vienna (BOKU), Vienna

Rannow S, Loibl W, Greiving S, Meyer BC, Gruehn D (2010) Potential impacts of climate change in Germany. Identifying regional priorities for adaptation activities in spatial planning. Landscape Urban Plann 98(34):160–171

Roeser S (2012) Risk communication, public engagement, and climate change: a role for emotions. Risk Anal 32:1033–1040

Runge K, Wachter T (2010) Umweltfolgenprüfung von Klimaanpassungsmaßnahmen. Nat Landsch 42(5):141–147

Runge K, Wachter T, Rottgart E (2010) Klimaanpassung, climate proofing und Umweltfolgenprüfung. UVP-Rep 27(4):165–169

Streitberger M, Ackermann W, Fartmann F, Kriegel G, Ruff A, Balzer S, Nehring S (2016) Artenschutz unter Klimawandel: Perspektiven für ein zukunftsfähiges Handlungskonzept, Hrsg. Bundesamt für Naturschutz, Naturschutz und Biologischen Vielfalt, vol 147. Bonn-Bad Godesberg

Swart R, Biesbrook R, Capela Laurencu T (2014) Science of adaptation to climate change and science for adaptation. Front Environ Sci, 02 July 2014. https://doi.org/10.3389/fenvs.2014. 00029

Wachter T, Balla S, Schönthaler K (2017a) Methodische Empfehlungen zur Berücksichtigung des Klimawandels in der Umweltverträglichkeitsprüfung. UVP-Rep 31(3):213–223

Wachter T, Leitner M, Kleinbauer I, Czachs C, Jiricka-Pürrer A (2017b) WP2 Bericht—Literaturanalyse SPECIFIC UVP und Klimawandelanpassung aktuell

Walker WE, Haasnoot M, Kwakkel H (2013) Adapt or perish: a review of planning approaches for adaptation under deep uncertainty. Sustain J 5(3):955–979

Wirth V, Prutsch A, Grothmann T (2014) Communicating climate change adaptation—state of the art and lessons learned from ten OECD countries. GAIA 23(1):30–39

A Mobile-Guided Smart-Safari on an Extracurricular Location

Sascha M. Henninger and Tanja Kaiser

Abstract Associated with climate change, it is necessary to enable young people to develop adaption strategies and assess impacts on ecosystems and agriculture. This requires an education based on skills rather than pure knowledge. Particularly motivational aspects can be an impulse for future active and responsible decision-making. Thus, primary experiences and problem-oriented learning in extracurricular activities could enforce commitment and readiness for action. Working with living plants and discovering their adaption strategies towards harsh climatic conditions strives for a fundamental understanding of the evolutionary processes that have led to these strategies. This deeper understanding implies also to think about the long period of time needed for the plants to develop genetically fixed adaption methods. This paper describes the development of an action-oriented and problem-based interdisciplinary learning environment for senior grade students in a botanical garden using digital media. Along 16 interactive stations, students can discover adaption strategies of various plants towards climatic conditions with regard to their morphology or physiology. Multimedia learning materials explaining climatic factors in the polar, temperate, subtropical and tropical climatic zone are provided by a web-application, which has been designed for this particular setting. Guided by tablet PCs, the students discover limiting factors like frost, high temperature and insolation, water shortage and snow cover in the various glasshouses autonomously in partner work. Using scientific methods like observing and describing morphological characteristics, they generate hypotheses at first. The web-app offers climate diagrams, maps and information as text or video to come to a solution. Hands-on stations with experiments will help to explain physiological adaption strategies of the plants.

S. M. Henninger (✉) · T. Kaiser
Department of Spatial and Environmental Planning, Technische Universität
Kaiserslautern, Pfaffenbergstraße 95, 67663 Kaiserslautern, Germany
e-mail: sascha.henninger@ru.uni-kl.de

T. Kaiser
e-mail: tanja.kaiser@ru.uni-kl.de

© Springer Nature Switzerland AG 2019
W. Leal Filho et al. (eds.), *Addressing the Challenges in Communicating Climate Change Across Various Audiences*, Climate Change Management,
https://doi.org/10.1007/978-3-319-98294-6_25

1 Introduction

The strongly subject-based structure of secondary education in Germany (consisting of chemistry, physics, geography and biology) make a transdisciplinary approach to global environmental issues like climate change rather difficult (Storksdieck 2011). In the federal state of Rhineland-Palatinate, the subject geography in grade 11 focuses on climatological elements and factors, the climate zones and classifications, as well as on the impact on human systems (Ministerium für Bildung, Wissenschaft und Weiterbildung Rheinland-Pfalz 1998b). The subject biology in grade 12 explains ecosystems and focuses more on their biotic factors (Ministerium für Bildung, Wissenschaft und Weiterbildung Rheinland-Pfalz 1998a). Extracurricular activities such as field trips are a popular method for introducing students to experiences that cannot be provided in a classroom environment. This is particularly important for transdisciplinary areas of teaching, such as environmental education.

Botanical gardens show an overview of the world's phytodiversity. Especially botanical gardens of universities serve for science and education rather than recreation. They provide background information as text via brochures or signs for every visitor or offer guided tours for special groups. Some botanical gardens even have „green classrooms" offering learning programs for schools at different levels and topics.

The idea of the tablet-guided learning program "In 80 Minuten um die Welt" (*In 80 minutes around the World*) in the botanical garden of the Technische Universität Kaiserslautern combines providing multimedial bits of information with the idea of „green laboratories"as it also offers small experiments in the plant beds next to the respective plant directly. Thus, the learner is not only recipient of given information, but also actively making own discoveries.

1.1 Geobotany and Scientific Discovery as a Link Between the School Subjects Geography and Biology

The concept of the learning program uses the field of geobotany to connect students' knowledge from both school subjects (see Fig. 1) by not only examining the morphology or physiology of a plant, but also its abiotic location factors, particularly climatic conditions (Wittig 2012).

As a transdisciplinary scientific field, geobotany explores the relations between the distribution of plants, their vegetation units and the correlation with biotic and abiotic factors (Pott 2005). During evolution, each plant species has adapted to these environmental conditions with regard to morphology and/or physiology (Cox and Moore 1987). Thus, the morphogenesis of a plant is not only an endogeneous process—these control the development of specific characteristics of a species—but modifications are affected by environmental conditions. A variety of these characteristics allows plants to cope with environmental stress achieving a state of being

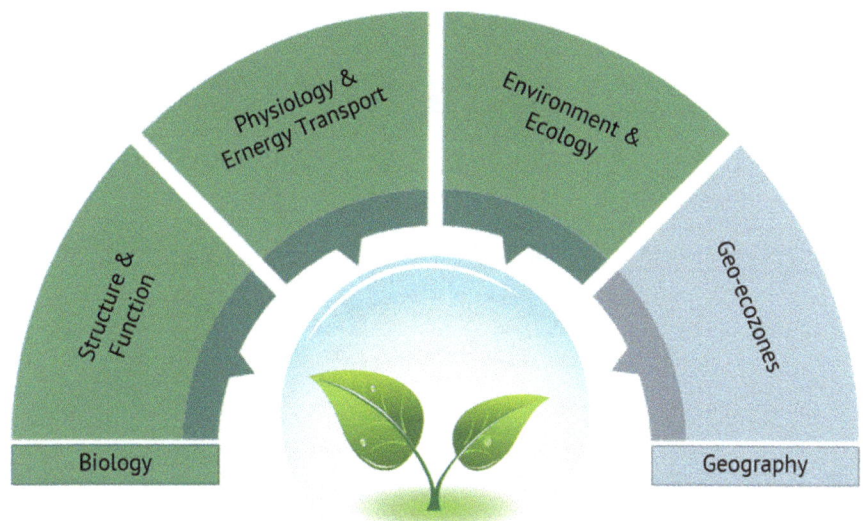

Fig. 1 Geobotany as a linkage between the curricula in biology and geography in grade 11 & 12 in German high schools in the state of Rhineland-Palatinate

adapted (Strasburger et al. 2014). The learning program focuses only on evolutive characteristics, which are irreversible and embedded in the genetic code of the plant, e.g. hygromorphoses (adoptions to high or low humidity), trophomorphoses (nutrition), and thermomorphoses.

Along 16 stations, the students discover those adaption strategies in the glass-houses of the botanical garden tempered according to the conditions of the tropics, subtropics, moderate and polar climate zones.

The tropics have the most impressive stations, since the wet heat in the glass-house gives a vivid impression on the meteorological conditions of this climate zone. Here, the focus is mainly on the limiting factor of sunlight in the tropical forests and adaption strategies of various plants to life in the canopy. Another station raises the question why sugar is produced by two plants in different zones: sugar cane in the tropics and sugar beet in the temperate zone. Examining the differences between C3 and C4 photosynthesis reveals the evolutionary newer C4 photosynthesis as an adaption strategy to high temperature conditions (Ehleringer and Cerling 2002).

Coping with arid conditions is the stations' main theme of the subtropical zone. Examining the function of the waxy layers on leaves or a photo safari documenting different types of succulent plants are on focus. Minimizing the plants´ surface is another strategy to reduce transpiration and save water. However, the cross section of some cacti resembles a star shape – not a circle. In an experiment, the students find out the advantages of the star shape (see Fig. 2).

Fig. 2 Students discovering the advantage of star shaped cacti during an experiment (Photo by the author)

In the moderate zone, one focus is on soil conditions in the habitat of carnivorous plants. At first students have to search for different trap mechanisms to describe how these catch insects. During the following experiment, they examine the soil with regard to the nutrition supply and conclude why carnivorous plants have to hunt on insects. Another learning station deals with the impact of climate change towards viticulture. Analysing climatological data, the learners describe the prospected increase in temperature and the expected development of the vegetation period in general. Comparing maps of the current and prospected areas where viticulture will be possible with regard to temperature development, the students get an idea of anticipatory planning in agriculture.

The stations of the polar zone deal with the harsh temperature conditions and adaption strategies of plants as well as the impact of rising temperatures on permafrost soil structure and vegetation. An introductory video shows some craters in Siberia. The students have to develop a hypothesis on the formation process of these craters. In an experiment, they heat frozen soil containing essential oils, so they can smell the escaping scent representing the methane gas hidden in the permafrost. Besides the impact of the explosions of methane plumes in the permafrost, the learners also deal with the effects of the thawing soil on buildings, vegetation, and infrastructure.

Beside the thematic linkage between the subjects of geography and biology, both also use a similar pool of scientific methods, e.g. observations, comparisons, descriptions, measurements, and experiments (see Fig. 2). Thus, the program uses the *Scientific Discovery as Dual Search-Model* by Klahr (2000) describing the research for hypotheses as a first step in scientific research (Hammann 2007). The students make observations and are guided to generate their own hypothesis or question. In a second step they can test their presumptions by running experiments. Subsequently, they analyse and interpret their results and draw their conclusions.

Students face discrepancies between theory and practice to arise higher involvement. Thus, they get an idea of the limits of theoretical models. According to Raunkiaer's life forms, for example, the phanerophytes and therophytes are the forms with the highest frost resistance (Larcher 2001). The students have to sort some examples of life forms with regard to their limiting temperature. Afterwards, they analyse a table revealing the actual distribution of those life forms in the subpolar region, revealing that the "Top 2" of the frost-resistance ranking rarely exists in this region. During following experiments, students discover the reasons for this discrepancy.

2 Guided Discovery with Digital Media

The learning program has been implemented as a web-application to allow an independent discovery and working in the learner's own pace with the original plants without a formal guide and frontal instructions. After a short introduction, the learners discover the stations in teams of two or three students guided by a tablet PC.

The introductory page shows the site map of the stations with regard to the climatic zones. Although it looks like a game board, there is no competitive perspective in the learning program (see Fig. 3). The renunciation of competition provides the opportunity to have time for own discoveries and questions.

At first, the display shows a simplified ground plan of the botanical garden as well as pictures, which will help the user to find the right station (see Fig. 3) embedded between the plants in the glasshouses. Further, it provides media to arise interest and contextualise the station followed by instructions for their scientific work. Analysing study materials like maps, climatic graphs and infographics help to solve the given questions.

Fig. 3 (Left site) Navigation within the app: Home page of the learning application according to the four climatic zones. (Right site) Helpful for orientation: a location map and picture sections (own illustration)

2.1 Designing the Learning Program: Challenges, Methods and Principles

Designing the program to fit in limited time and finding a balance between the subjects is rather challenging. A further aim is to reduce the cognitive load rised by unknown surroundings and working with tablet PCs without personal guidance.

Primarily, one problem of the development of this tablet-guided learning program is the heterogeneous level of prior knowledge and interest. Although the program addresses only students between grade 11 and 12 (age between 16 and 17)—not every visitors of the botanical garden—there are strong differences regarding the basic knowledge depending on their study level in biology and geography as well. Therefore, it is a challenge to find the appropriate balance within the instructional design to rise curiosity, to allow self-determined discovery and reduce cognitive load.

3 Evaluation Methods

The design-based research (DBR) is a useful framework for designing educational interventions. Similar to engineering processes (see Fig. 4) educational DBR aims at solving applied problems related to the design of learning environments in authentic educational settings (The Design-Based Research Collective 2003; Reinmann 2005; Jen et al. 2015). It helps to explain their effectiveness or ineffectiveness using the results of reflective inquiry to refine the innovative learning environment where possible.

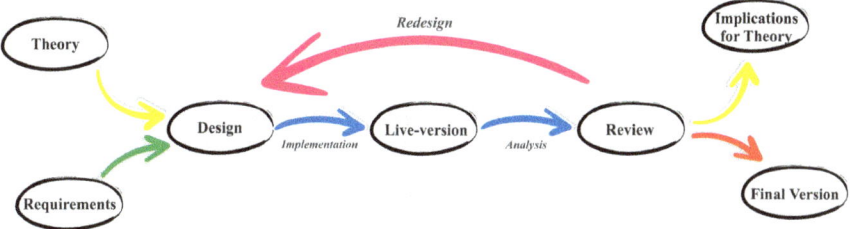

Fig. 4 Schema of the design-based research framework (compiled by the author): the central part of design, implementation, analysis and redesign is an iterative process aiming at improving the learning program as well as generating implications for theory

Key features of the DBR applied in the learning program:

- design is driven by following methodical principles like real world encounter, students' own activity and transdisciplinarity (Rinschede 2009)
- design is conducted in the real world setting of a botanical garden as an extracurricular activity
- multiple types of dependent variables (e.g. students' variability like knowledge level, media affinity, interest, system variables like air temperature, attractiveness of the plants in summer and winter respectively)
- processes are iterative cycle of analysis, design, implementation, and redesign
- use of mixed methods.

The primary concept of the learning program was strongly linked to the content of the respective curricular topics. The availability of plants, a time limit of five to ten minutes work per station as well as limitations of connections to electricity and Wi-Fi had to be considered. Thus, the initial version was bound to content and conditions.

The primary revision focused on theoretical guidelines of the cognitive load theory (CLT, Sweller et al. 2011) and the „Cognitive Theory of Multimedia-Learning" (CTML, Mayer 2012) in order to stage the original plants in the garden by improving media like videos, maps and graphics. Information given as text was also strongly revised according to the principles of comprehensibility (Kerres 2000; Langer et al. 2011; Scheiter et al. 2014; Horz and Ulrich 2015). Unfortunately, giving feedback on students' text answers is not possible, but proper solutions are often necessary to continue the tasks. Thus, additional information to solve previous questions as well as contextualise upcoming steps were added to the program.

Whereas these primary steps were close to content, technical and organizational conditions, the further development of the instructional design focuses not only on effective tasks, but is strongly linked to students' interests and learning emotions.

4 Mixed-Methods Design

Since it is not possible to revise all of the sixteen stations at once, the students' feedback is highly welcome. After finishing the learning program, the students evaluate the stations with regard to interest and difficulty. The summary of feedback results in a priority ranking; e.g. stations considered difficult and/or not interesting have to be refined first.

The next step is a qualitative analysis of the students' performance. The quality of the answers saved in the database as well as the analysis of video recordings of their dialogues and behaviour reveal deficiencies of the instructional design.

Besides this **qualitative usability test** of existing stations, there is also the need to develop additional stations as a result of the difficulty, that some plants are not available all over the year and some experiments are bound to cool surrounding temperatures. Regarding the interests of the students while (re-)designing the program, they complete a questionnaire in a pre-post-test design.

5 Results

Between January 2017 and February 2018, twenty courses with 363 students took part in the program. During the first six month, there was a priority on implementing the video study. Thus, the pre-post-questionnaires started in June with a total of 183 students.

Based on the first observations, it is obvious that in particular students with limited knowledge in biology face difficulties. Whereas geographical competencies of how to read a map or analyse a climate graph are rather easy, physiological processes like photosynthesis seem to be difficult with little prior biological knowledge. Figure 5 shows a first impression on the heterogeneity: 59 students (33%) were studying biology only at a basic level at school or even had not taken the subject at all. The qualitative part of the study will reveal their performance on the tasks.

		biology			Total
		advanced level	basic level	subject not chosen at all	(valid)
geography	advanced level	77	22	17	116
	basic level	32	8	5	45
	subject not chosen at all	10	7	0	17
Total (valid)		119	37	22	178

Fig. 5 Crosstabulation: students' study level in geography and biology

The pre-test focuses on **students' interests in methods** used in geography as well as in biology. For the field of geography, the items were selected according to a study on interests in geographical topics and methods run in 1995 and 2005 by Hemmer (2010). From 17 items in the study, six fit to the study.

The program with its basic principle of "hands-on" seems to match with the students' interests (see Fig. 6), both in geography and biology. Action-oriented

Fig. 6 Students' interests in selected geographical and biological methods using a five-point Likert-scale with a range from 1 = "not interested" to 5 = "very interested"

methods of gaining information on processes like experiments, computer simulations as well as using instruments are more popular than traditional methods and media. For further improvement, the use of computer simulations embedded in explanatory videos, working with more measuring instruments as well as the microscope will be considered.

6 Utility: Challenges of the Botanical Garden

Technical challenges like the legibility of the instructions while sunlight is shining through the glass roof, Wi-Fi connectivity as well as battery power of the tablet PCs require some compromises. Operating equipment like ventilators and pumps have to be shut off during the visit of the students. Minimizing these background noises is not only important for the students to listen to the videos, but also for the qualitative video study.

In contrast, cutting back plants or even moving them to another place is easy to assess—uploading new photos or an update of the location map is quite comfortable.

7 Future Prospects and Conclusion

The methods and concept of the learning program seems to meet students' interest and the fact that schools try to attend the program when it fits to their curriculum and not as a "day-off" or in the period before holiday, is a strong motivation to develop and improve the program. As mentioned above, the focus will be on working with measurement instruments and microscope, if possible as well as illustrating difficult processes by videos. Foresight, on some of the 16 stations it is required to produce "silent" videos, because of the surrounding noise. Further, the interactivity has to be improved. Drag-and-drop implementations as well as concepts for digital tools to analyse climate graphs or maps are on the schedule. Due to renovation activities in the botanical garden, some stations have to be replaced.

The qualitative video study will certainly reveal weaknesses of the instructional design and also might give hints on how the students cope with cognitive conflicts, perform in experiments and tackle with the original plants.

Guided discovery causes learners to actively deal with a problem, to gain own experiences by performing appropriate experiments and finally come to new insights into complex issues that form the basis for connectable knowledge.

The deeper understanding of adaption strategies by primary experiences with the plants sensitizes for the effects of climate change on plants, vegetation and agriculture. Being aware that adaption strategies of these highly specialised plants are the result of long evolutionary processes, the impact of upcoming climate change with fast changes of temperature rates becomes more than obvious. The selection pressure will have a crucial impact on single species as well as the species composition in the regional vegetation cover and ecosystem structure. Some plants will stay in place and adapt to changing conditions whereas others will have to move or become extinct. This emphasises the explosive nature of the issue of climate change in general and sensitizes that only man can take countermeasures against possible loss of essential ecosystem services, but must also develop adaptation strategies to ensure sustainable livelihoods.

References

Cox CB, Moore PD (1987) Einführung in die Biogeographie: Uni-taschenbücher Biologie, Geographie, vol 1408. Fischer, Stuttgart

Ehleringer JR, Cerling TE (2002) C3 and C4 photosynthesis. In: Munn T (ed) Encyclopedia of global environmental change. Wiley, Chichester, pp 186–190

Hammann M (2007) Das Scientific discovery as dual search-modell. In: Krüger D, Vogt H (eds) Theorien in der biologiedidaktischen Forschung: Ein Handbuch für Lehramtsstudenten und Doktoranden. Springer, Berlin, Heidelberg, pp 187–196

Hemmer I (2010) Schülerinteresse an Themen, Regionen und Arbeitsweisen des Geographieunterrichts: Ergebnisse der empirischen Forschung und deren Konsequenzen für die Unterrichtspraxis. Geographiedidaktische Forschungen, vol 46. Selbstverl. des Hochschulverbandes für Geographie und ihre Didaktik, Weingarten

Horz H, Ulrich I (2015) Lernen mit Medien. In: Reinders H, Ditton H, Gräsel C, Gniewosz B (eds) Empirische Bildungsforschung: Gegenstandsbereiche. VS Verlag für Sozialwissenschaften, Wiesbaden, pp 25–39

Jen E, Moon S, Samarapungavan A (2015) Using design-based research in gifted education. Gifted Child Quarterly 59:190–200. https://doi.org/10.1177/0016986215583871

Kerres M (2000) Information und Kommunikation bei mediengestütztem Lernen: Entwicklungslinien und Perspektiven mediendidaktischer Forschung. ZfE 3:111–130. https://doi.org/10.1007/s11618-000-0008-5

Klahr D (2000) Exploring science: the cognition and development of discovery processes. MIT Press, Cambridge, MA

Langer I, Schulz von Thun F, Tausch R (2011) Sich verständlich ausdrücken. Reinhardt, München

Larcher W (2001) Ökophysiologie der Pflanzen: Leben, Leistung und Streßbewältigung der Pflanzen in ihrer Umwelt; UTB für Wissenschaft, vol 8074. Ulmer, Stuttgart

Mayer RE (ed) (2012) The Cambridge handbook of multimedia learning. Cambridge handbooks in psychology. Cambridge University Press, Cambridge

Ministerium für Bildung, Wissenschaft und Weiterbildung Rheinland-Pfalz (1998a) Lehrplan Biologie: Grund- und Leistungsfach Jahrgangsstufen 11 bis 13 der gymnasialen Oberstufe (Mainzer Studienstufe)

Ministerium für Bildung, Wissenschaft und Weiterbildung Rheinland-Pfalz (1998b) Lehrplan Gemeinschaftskunde: Grundfach und Leistungsfach mit dem Schwerpunkt Geschichte, mit dem Schwerpunkt Sozialkunde, mit dem Schwerpunkt Erdkunde in den Jahrgangsstufen 11 bis 13 der gymnasialen Oberstufe (Mainzer Studienstufe)

Pott R (2005) Allgemeine Geobotanik: Biogeosysteme und Biodiversität. Springer, Berlin

Reinmann G (2005) Innovation ohne Forschung? Ein Plädoyer für den design-based research-Ansatz in der Lehr-Lernforschung: innovation without research? Arguments for design-based research in educational research. Beltz Juventa, Weinheim

Rinschede G (2009) Geographiedidaktik, Grundriss allgemeine geographie, vol 2324. Schöningh, Paderborn

Scheiter K, Schüler A, Gerjets P, Stalbovs K, Schubert C, Huk T, Hesse FW (2014) Welche Rolle spielt neben Merkmalen des Instruktionsdesigns die fachspezifische und aufgabenspezifische Motivation beim Lernen mit multimedia im naturwissenschaftlichen Unterricht? Z Erziehungswiss 17:279–296. https://doi.org/10.1007/s11618-014-0516-3

Storksdieck M (2011) Field trips in environmental education. Schriftenreihe Umweltkommunikation—Band 3. BWV Berliner Wissenschafts-Verlag, Bremen

Strasburger E, Noll F, Schimper AFW, Kadereit JW, Körner C, Kost B, Sonnewald U (2014) Lehrbuch der Pflanzenwissenschaften. Springer Spektrum, Berlin

Sweller J, Ayres P, Kalyuga S (2011) Cognitive load theory, explorations in the learning sciences, instructional systems and performance technologies, vol 1. Springer Science+Business Media LLC, New York, NY

The Design-Based Research Collective (2003) Design-based research: an emerging paradigm for educational inquiry. Edu Res 32:5–8

Wittig R (2012) Geobotanik, vol 3753. Haupt, Stuttgart, Bern u.a, UTB GmbH

SPONSORED BY THE

Federal Ministry of Education and Research

"In 80 Minuten um die Welt"as a sub-project of "U.EDU: Unified Education – Medienbildung entlang der Lehrerbildungskette" is part of the "Qualitätsoffensive Lehrerbildung", a joint initiative of the Federal Government and the *Länder* which aims to improve the quality of teacher training. The program is funded by the Federal Ministry of Education and Research."

Creating Change in the United States' Museum Field: Using Summits, Standards, and Hashtags to Advance Environmental Sustainability and Climate Change Response

Sarah Sutton

Abstract There are 35,000 museums and historic sites, estimated, in the United States, contributing $50 billion in USD to the Gross Domestic Product (GDP), including $6 billion USD to trade, transportation, and utilities. Every year people make 850 million visits to museums. If the sector were to track its GHG emissions it could no longer ignore its direct impact on climate change. How can we determine, and can we reduce, the sector's impact on climate through GHG emissions from energy use? What will mobilize museums to use their valuable relationship with the public to foster climate awareness in ways that lead to broader individual action and support for policies engendering positive climate impacts? This paper examines the slow process of building environmentally-sustainable practice in the museum field in the United States, explains existing programs for monitoring GHG emissions, and identifies how the future of sustainable and resilience action lies with collaboration and cross-institutional movements. It explores the roles of supporting, cross-institutional approaches such as Keeping History Above Water, #NotAnAlternative, and #MuseumsforParis, and cross-sector approaches of #WeAreStillIn. It concludes that, based on field-wide Summits, the success of other standards, and the growth of hashtags as social evidence of a movement, the field can no longer avoid its responsibility to climate. The changes the human world needs most are all related to our changing climate. If humans address the causes and opportunities of that changing climate, we can build a more just, verdant, and peaceful world. Museums have limitless value in building that world.

S. Sutton (✉)
Sustainable Museums, 67-616 Kahui Street, Waialua, HI, USA
e-mail: sarah@sustainablemuseums.net

© Springer Nature Switzerland AG 2019
W. Leal Filho et al. (Eds.), *Addressing the Challenges in Communicating Climate Change Across Various Audiences*, Climate Change Management,
https://doi.org/10.1007/978-3-319-98294-6_26

429

1   Introduction

There are an estimated 35,000 museums and historic sites in the United States, contributing $50 billion in USD to the Gross Domestic Product (GDP), including $6 billion USD to trade, transportation, and utilities. If the sector were a state, it would rank just behind the city of Albany, the capital of New York State. If the sector were to track its GHG emissions it could no longer ignore its direct impact on climate change. That ability to suspend responsibility is coming to end.

Americans visit those 35,000 museums in person at a rate of 850 million times in a year, and another 524 million times online. That is more in-person visits than attendance at all major league sports and theme parks combined (AAM 2017). Americans trust those museums more than any media format, including National Public Radio (a private nonprofit media corporation which receives some federal funding), more than politicians and even academia (AAM 2013). That credibility gap between museums and the media is widening. That trust is where we find our public authority, and an opportunity to impact climate change.

The physical impact of US museums is significant enough to warrant changed behavior. The intellectual impact of US museums is as well. To leverage any of that capacity requires coordinated efforts to make environmentally sustainable behaviors and practices a priority among museums nationwide. To do this we must at least:

- Help professionals to understand the science behind climate change causes and effects, and behind environmentally sustainable decision-making. This will allow them to share it with the public, and to use it in making decisions in professional practice.
- Start and sustain important conversations that address professional biases and long-held institutional practices that are resistant to change; create recognition of shared values that create and galvanize communities of supporters; and create and sustain conversations with the public and with policy makers to foster change.
- Prepare our field to manage climate mitigation, and prepare for climate-related disasters, by working within the structures of the professional associations—nationally or globally—to create and supports standards that guide museums in aligning understanding of climate change, environmentally-sustainable behavior, and public engagement with institutional missions.

This paper examines the slow process of building environmentally-sustainable practice in the museum field in the United States (US), explains existing programs for monitoring GHG emissions, and identifies how the future of sustainable and resilience action lies with collaboration and cross-institutional movements. It explores the roles of supporting, cross-institutional approaches such as Keeping History Above Water, #NotAnAlternative, and #MuseumsforParis, and cross-sector approaches of #WeAreStillIn.

2 Energy Use and Greenhouse Gas Emissions

In the US, institutions can voluntarily participate in a number of programs that allow them to track GHG emissions and receive recognition. One of those is the free, online portal of the EnergyStar® Portfolio Manager Program® at the federal Environmental Protection Agency. Users can record energy and water use data, assess GHG emissions and reductions, and share that information publicly if they wish. When leadership sets institutional commitments, staff can design reduction goals and monitor performance by measuring use and emissions. After twelve months' worth of data, the program also provides an EnergyStar® rating based on that use compared to benchmarked buildings. This rating can be used to earn points toward certification in the widely-used Leadership in Energy & Environmental Design (LEED) program led by the United States Green Building Council's (USGBC), a private nonprofit and not a government agency.

Unfortunately, the museum field is slow to engage in this practice. One reason is the absence of a Museum building category in Portfolio Manager®. Many professionals feel that the expected high-intensity energy use of museums' conditioned spaces for objects and visitors, or living species, disadvantages museums in acquiring a true rating. Benchmarking data from cities requiring this documentation indicate that museums' Energy Use Intensity (EUI, energy use per square foot) "… is generally well above the EUI of other buildings…the high end of EUI is 6 or 7 times of the lowest consumers." The wide performance range of under 50 to over 300 highlights "multiple opportunities for efficiency improvement and design innovation." (Lee 2017) According to the Commercial Building Energy Consumption Survey (CBECs), conducted by the US Energy Information Administration, the median (not mean) EUI reports from a grouping including museums, movie theaters, convention centers, and performing arts institutions, puts the number at 45.3. Clearly the field needs more responsive data. The EnergyStar® program has been actively recruiting a cross-regional representation of museums to build a benchmarked category based on our own data, yet the response is slow. Fewer than 300 museums are participating, the minimum number for benchmarking. Why? Anecdotal evidence indicates that most decision-making about participating, with or without a reduction goal, rests with the time-pressed facilities managers who appear to be embracing HVAC system and lighting efficiencies offered to them, do not feel encouraged to focus on the broader environmental picture and contribute time to field-wide assessments.

There is currently no field-wide expectation that institutions should perform to any standard in energy use and GHG emissions conservation or measurement.

The only standard consistently referenced is that which leads to our highest consumption of energy: the vigorously-challenged climate control assumption that quality care for collections requires 70°F ± 2°, and 50% RH ± 5%. There is gathering evidence that not all materials require such tightly controlled conditions, or such narrow ranges of temperature and relative humidity, and that relaxing these requirement, and in some cases allowing seasonal fluctuation, would support conservation for both objects and energy. This discussion is known colloquially as the Plus Minus Dilemma. This, ahead of gallery lighting, is where much of the energy is consumed. Current, adaptive practice in this area is informed by the work of conservators, curators, and researchers committed to developing and sharing a more responsive set of parameters by collections type, and more effective active and passive environments for collections care and display (Hatchfield 2011). Those working on changing from an over-consumptive, non-science-based practice to a more informed and efficient approach include the members of the Association of Art Museum Directors working to develop and adopt more responsive loan agreements that do not specify the 70°T/50% RH parameters without regard to materials type; the researchers at Image Permanence Institute who test and verify materials' response to T/RH; and the National Endowment for the Humanities' Sustaining Cultural Heritage Collections grant program which supports the design and implementation of more sustainable collections care conditions in museums.

    If the museum profession had standards requiring energy use and impact assessments as an ethic or a practice, it would accelerate the efforts of facilities managers and the curators, conservators, and researchers; and improve the response of the field.

3   The Standards Discussion

Because there is currently no internal guide, the most visible sustainability advances have been adopted form outside our domain, namely building design, construction, and management. Professionals from outside the field have created and tested standards programs without input by museum professionals. In addition to EnergyStar® Portfolio Manager®, this includes LEED, Green Globes, and Living Building Challenge. As energy use regulations expand, these programs and requirements will be created without our field's input, and without our ability to inform their design and value. So, what is the resistance to standards?

    The American Alliance of Museums' professional network on environmental sustainability, PIC Green, has attempted to highlight metrics systems, and encourage adoption of energy efficiency and other standards to guide the field. In particular, the impressive reach of the USGBC's LEED program has spurred widespread discussion about its intent and format, and its applicability for the museum field. To build awareness and understanding, and gain input from museum professionals, PIC Green hosted three "Summits on Sustainability Standards in Museums" at the AAM annual conferences between 2013 and 2015. The first year included speakers representing the best-known metrics programs: United States Green Building Council, Sustainable Sites Initiative, Green Globes, Living Buildings, and ExhibitSEED presenting their metrics programs. In subsequent years there were presentations by professionals who had used those metrics systems at their sites, and convenings to discuss developing standards.

We discovered continued concerns from museum professionals. From both summits we heard concerns about creating standards (AAM and McGraw 2014):

- Where will the funds, time, and expertise to implement a standards program come from at both the national and the institutional levels?
- Will designating resources for sustainability pull funding from other functions?
- As a field, do we have the ability and the protocol in place to agree on what best practices would look like?
- Are leaders of our institutions and associations ready to carry sustainability standards projects forward?

Practitioners said they needed a shared language around sustainability, and that the field would benefit from a format for sharing information more widely. Participants wondered how sustainability standards and best practices—and related evaluation programs—will be used: "Will they be used to encourage museums to improve, regulate and evaluate an organization for recognition purposes?"

We left these sessions understanding that practitioners needed tools, leadership commitment, and financial support if we and they were to move ahead. Finally, it seemed that there was some interest in an approach that was responsive to museums' needs, provided it brought with it a shared language and necessary tools, and included some level of agency weight to it—whether professional endorsement or monetary support, preferably both. The white paper *Museums, Environment and our Future* noted:

> The concept of a neat, fixed set of standards and best practices appears black and white in concept, but is much messier in actual application. Museums are so diverse in terms of mission resources, location, facilities, collections care needs, lifecycle, etc., that creating a single set of voluntary standards and best practices is an ever-present challenge—as is maintaining a program that effectively evaluates a museum against them. New issues and new thinking are also continually impacting practices and standards, and we must keep up.

The paper concluded that "AAM has the experience—from almost 45 years of running accreditation and assessment programs—and expertise in PIC Green, to take a leadership role in shaping the future of defining and measuring excellence in sustainability."

When AAM asked for a broader mandate beyond attendees at the national conference, PIC Green volunteers agreed that, whether the field chose to adopt LEED or LEED-like practices, create its own standards, or choose none, an informed decision required more perspective to combine with broader participation. There was a need for a convening of peers from inside and outside the museum domain. To save carbon and cash, PIC Green and AAM sponsored *Tides of March*, a series of national web interviews and discussions in March 2016. There were presenters from major museums or zoos; association leaders with and without standards systems; and non-museum nonprofit leaders with experience in field-wide change. At the close of the series, AAM invited PIC Green to recommend Characteristics of Excellence for the field.

4 Creating Characteristics of Excellence

At the time there were, and still are, thirty-eight Characteristics of Excellence listed on the AAM website, considered the "core standards," adaptable to "museums of all types and sizes, with each museum fulfilling them in different ways depending on their unique circumstances". Topics include mission, resources, governance, collections. PIC Green created a set of characteristics, vetted by the speakers from the *Tides of March* programs, and submitted a draft of six characteristics (PIC Green 2016):

Draft Characteristics of Excellence

Environmental Responsibility in Museums May 13, 2016

1. *The museum states its role in stewardship of the environment, demonstrates how its choices align with that role, and actively addresses related environmental concerns within the community.*
2. *The museum measures and makes public its environmental impacts; sets goals for continuous improvement; and evaluates progress and effectiveness.*
3. *The museum demonstrates leadership by exceeding environmental codes, regulations, and professional standards as appropriate, e.g. higher efficiency, or more effective practices.*
4. *The museum has reviewed its investments and set a timeframe for investing in a socially responsible portfolio.*
5. *The museum has a plan and timeframe for becoming climate neutral, then climate positive.*
6. *The museum identifies risks resulting from climate change, and takes steps to anticipate and mitigate risks and damage for itself and, in collaboration, on behalf of the community.*

AAM applied to the Institute of Museum and Library Services for funding to support the formalization of a standards-development process using environmental sustainability as the test case project. The grant was denied in 2017, and the project has paused. Meanwhile, the American Association for State and Local History (AASLH) held a listening session on environmental sustainability at the 2017 annual conference and then called for a Task Force to make recommendations to its Council on how to support and guides the country's 15,000 historical societies, history museums, and historic sites in incorporating environmental sustainability to reflect standards of responsible stewardship. It may make progress where the broader museum field has not for at least three reasons: demonstrated support by a governance entity, AASLH; energy in new volunteers; and a more homogenous membership. This sector's awareness and experience of disaster-related risks due to storm events and sea level rise is heightened by members' locations on historic coastlines and along riverine transportation routes.

5 How Can We Create and Accelerate a Movement?

Clearly the discussion of standards continues to expand, and there are tools for measuring and reporting impacts, but still little progress. What can we do to accelerate the change? Solving such a wicked problem requires an approach that engages multiple perspectives and formats; and is patient enough to accept failure and support continuous learning. Likely, the answer is a well-orchestrated implementation of any and all available resources:

- unrelenting yet continuously-adapting messaging
- building shared belonging to the movement by social and official incentives or mandates as appropriate
- improved preparation for the increase in climate events threatening museum and community resources, comfort, and safety.

First, what do we know about how humans are able to ignore something many of us see as a real and present danger; then, how can we change their minds and then their behavior?

There are many possible explanations; two stand out: we are social creatures, and we are still on an evolutionary journey. Elsewhere during this conference and in these proceedings, you will read and hear from George Marshall, author of *Don't Even Think About It: why our brains are wired to ignore climate change*, so there is no need to explain his work to you. There is a concept, though, that I would like to highlight as it may describe the delay in the adoption of standards in the museum field: in the gap between the activists or pro-climate response folks, and the deniers or resisters, are the people who are unsure. "They…are deliberately making a calculation about how they wish to interpret these arguments, knowing very well that if their emotional brain becomes too involved, they are likely to feel anxious and worried…. they tend to adopt a position of wait and see. Their rational brain is sufficiently aware that they know there is a problem; but their emotional brain is sufficiently engaged that it is looking out for social cues about how they should respond. And both of their brains are sufficiently detached that they do not have to deal with the problem unless actively compelled to do so" (Marshall 2014).

Daniel Kahneman in describing similar human responses in *Thinking, Fast and Slow* also discusses the evolutionary aspect of humans which we can translate to climate denial (2011). Since humans respond to immediate and imminent threats, and to threats with greater relative cost and degree of surprise, the slow threat of a changing climate barely raises interest. He explains that psychologist Robert Zajonc argued "To survive in a frequently dangerous world, an organism should react cautiously to a novel stimulus, with withdrawal and fear. Survival prospects are poor for an animal that is not suspicious of novelty. However, it is also adaptive for the initial caution to fade if the stimulus is safe." So, when nothing bad happens when we hear or see something familiar, the "stimulus becomes a safety signal." By extension the warnings of climate change become casual conversation or not conversation at all; and high tide in the street and the basement of your historic house becomes the new normal.

The field requires tools to move beyond these barriers in our professional brains. Since those tools do not yet include a standards system or even an expectation of behavior, the field should consider a mantra, a simple shorthand guide for behavior as it progresses toward more substantial tools, such as standards, that articulate and encourage change.

6 A Mantra

In the US there is a healthy food movement, and a particularly popular writer, Michael Pollan, who simply and effectively describes his solution to the wicked problem of how Americans can change to healthier, more sustainable food systems (2008). Pollan is the author of many best-selling, books, including *In Defense of Food*. He writes that he can sum up all he has learned about sustainable food and eating healthy in seven words: "Eat Food, Mostly Plants, Not Too Much."

A similar phrase to help us address the human psyche and the climate challenge would be "Raise Awareness; Boost Contagion; Count on Disaster."

Raise Awareness: We must continue to make connections between museum work and environmental systems and human impacts. To be useful or effective, the tools, the standards, the incentives, and the mandates all require awareness first, but it alone is not enough to overcome the inertia of fear or risk avoidance among professional practitioners. Our Canadian peers in the Working Group for Museums & Sustainability in Communities reported in Dumouchel and Worts (2017), after seven years of study and advocacy, that "there is a great deal of inertia in the status quo of museum operations—and therefore not a great appetite or comfort for the notion of experimentation…" that would lead to sustainability engagement. That "inertia" is most evident among those who do not see a connection between environment and climate, with the mission of museums. To raise awareness within the sector requires the combined efforts of professional networks within the national associations, and from outside the sector. #MuseumsforParis (described later) is an attempt from within the sector to use the seventeen global goals, the United Nations Sustainable Development Goals (UNSDGs), to highlight areas of environmental and social need, and museum alignment.

Boost Contagion: What mechanism replicates the connection, that many of us already have, that keeps us pushing forward? The staying power and scale of individual actions is not enough, and so collective participation becomes critical. We must explore two types of "viral" approaches: one, the establishment of standards and practices facilitated either by incentive or mandate; and the other, movements attracting individuals to collective activities that seed change. The social media campaign #NotAnAlternative, that disparages museum endowment investment in fossil fuel, has had successes. Its demonstrations and media outreach have forced change by calling for divestment and for removal of trustees at natural history museums who are climate deniers, and likely triggered many more pre-emptive conversations and conversions at other museums nation-wide. Certainly, it has elevated discussion in the field, and we are seeing change spread. The Characteristics of Excellence, if adopted, can provide standards and practices to be pursued through strategic planning or Accreditation requirements. Whether considered an incentive or a mandate, GHG emissions measurement and reduction commitments, as components of Accreditation, could spread museum compliance.

Count on Disaster: Disasters are coming in the forms of new and costly mandates, and actual physical disasters. They will create change where awareness and expectations alone have not spread the movement significantly. Museums respond with climate adaptation and mitigation often only after a serious and damaging storm event, or the threat of rising daily tides raises awareness in their coastal communities. Urgency appears only after a wildfire, or an extended heat wave or cold snap, or a hurricane and flood.

There is great risk that if we do not self-regulate, then environmental practice may be mandated without our input. There will be costs associated with acquiring the equipment and creating the infrastructure to reduce energy use and manage water. Museums in the US will be forced into climate response, forced to comply to others' standards, to pay fees imposed for non-compliance, and to borrow for significant costs of compliance in a too-short timeframe.

7   Hashtags, in the Absence of Standards

While the US museum field waits for standards, there is work to be done internally to illuminate the alignment of missions and climate change response, and to create external pathways, encouragement, and recognitions for that response. The hashtag movement of #MuseumsforParis represents the internal work; #WeAreStillIn provides external motivation to create change.

#WeAreStillIn

The cross-sector, national forum of the #WeAreStillIn movement satisfies all three aspects of the mantra introduced above: it creates high-level *awareness* that may lead to standards; the national and international recognition through reporting and goal-setting create *contagion*; and its origin is in response to the *disaster* of the announcement that the US would withdraw from the Paris Agreement.

#WeAreStillIn is "an unprecedented network of networks" of institutions from five sectors: Businesses & Investors, Cities & Counties, States & Tribes, Higher Education, and Faith Communities, who have committed to support the Paris Agreement and its associated goals, primarily GHG reduction. "We Are Still In is the largest cross-section of local leaders in support of climate action in the United States." It is coordinated by World Wildlife Fund (WWF), the American Sustainable Business Council, B Team, Bloomberg Philanthropies, Center for American Progress, Ceres, CDP, Climate Mayors, Climate Nexus, C40, C2ES, Environmental

Defense Fund, Environmental Entrepreneurs, Georgetown Climate Center, ICLEI, National League of Cities, Rocky Mountain Institute, Second Nature, Sierra Club, The Climate Group, We Mean Business, and Sustainable Museums.

On March 24th, 2018, the day the World celebrated Earth Hour, #WeAreStillIn announced the inclusion of the cultural institutions sector in the fight against climate change. The #WeAreStillIn movement, its structure, and its media value, encourage wider measurement of GHG emissions and concrete goals for reduction. It provides a format for measuring and reporting, and worldwide recognition for participation. The Web portal is where museums, zoos, gardens, aquariums, and historic sites share success stories, and report on the initiatives we have set in place for individual institutional goals. Here we record our contribution to national goals to reduce climate change, and highlight tangible measurements of value to our communities. At #WeAreStillIn we raise the field's expectations of performance.

#MuseumsforParis and United Nations Sustainable Development Goals

For museum professionals who do not feel climate response is unrelated to their work, #MuseumsforParis provides examples of museums' alignment with the United Nations' seventeen Sustainable Development Goals (Sutton 2018). Only one goal emphasizes GHG emissions; together all seventeen highlight the need to mitigate climate change and its adverse effects on life on the planet, through the protection of land and sea, cultural heritage, universal human rights, peace and justice, health and welfare—and the collaborations that advance and scale them all. Museums support the UNSDGs when they:

- Share art, fostering cultural awareness and understanding, social and human empathy, and inspiration and creativity. Creativity, awareness, and empathy are necessary for exploring difficult and/or challenging discussions, and encouraging social justice and appreciation of cultural heritage.
- Share cultural history, fostering cultural awareness and understanding, exploration of past and present injustices, and sharing stories of courage and heroism by individuals and groups. The relevance of historic experiences to the present day prepares us to address inequality and promote peace.
- Share science and nature, fostering awareness of nature and of scientific practice and understanding. They can narrate stories of discoveries and solutions; and provide understanding of how humans interact with the physical world, and how the natural world came to be what it is today and may be in the future.
- Share natural history, exploring the past and present living worlds, and providing important clues to the development of the Anthropocene. Understanding the science of climate and ecosystems allows us to choose to change our damaging behavior while emphasizing restorative work.

Our Unique Value to Climate Change Response

Our sector encompasses "the diverse physical and intellectual resources, abilities, creativity, freedom, and authority to foster the changes the world needs most," enabling humans" to understand challenges and to be moved to overcome them in ways that expand well-being for everything around us" (Sutton et al. 2017). The sciences, arts, and humanities all have a role in addressing the many aspects of climate response. That US scientists should have to defend themselves and their work on climate change is outrageous. The science sector remains a powerful force, and if all manner of museums can support them in climate response, then we will increase the impact of our work and theirs. At the 2018 World Economic Forum's discussion on climate change, to leaders made statements that also highlight the value of our work. Anand Mahindra, leader of Mahindra companies (worldwide), thanked US Vice President Al Gore for the liberal arts approach in US education. He said, "The liberal arts will generate the poets who will call out" the naysayers, the deniers, and who will "make the protests" that bring about public response to climate change.

Washington state governor, Jay Inslee, said "What we treasure individually we can only save collectively." When we work together we learn, innovate, inspire, and achieve more. Only coalitions and collaborations can scale the work that we need. In addition to #WeAreStillIn and #MuseumsforParis, other collectives have been at work. They are gatherings of like-minded professionals committed to sharing understanding of and response to climate change.

The Canadian Coalition of Museums for Climate Justice is a marvelous source of energy, and information. The Coalition's role so far has been to *raise awareness,* and their example may move us all forward as its reach spreads in the US. Arguably the most successful program to change public awareness about environment in the US field to date is National Network of Climate Change Interpreters (NNOCCI), www. climateinterpreter.org. It is a network of professionals and institutions organizations in informal education, the social sciences, and climate sciences that use "evidenced-based communications methods" and provide each other "the social and emotional support needed to engage as climate communicators." Today there are 170 zoos and aquariums in 38 states who have become members. It is an alliance of the New England Aquarium, a conservation research nonprofit, New Knowledge, Inc., and a communication nonprofit, The Frameworks Institute. Collaboratively they developed and continue to test "the knowledge, techniques, community and confidence needed to empower our audiences." The goal is to "speak about climate change consistently across the country...to change public discourse to be positive, productive, and solutions-focused." The program is funded by two federal agencies, the National Oceanic and Atmospheric Administration (NOAA) and the National Science Foundation. NNOCCI's five-year report is an important resource for scaling the ability of educators in zoos and aquariums to *spread contagion* (Fraser et al. 2015).

Keeping History Above Water is a young conference program, that has twice gathered participants around sea level rise and stormwater events. Many of the attendees are not from museums but from government preservation agencies. The growing strength and resourcefulness of the Keeping History Above Water conferences' participants bring added expertise and urgency to the cultural sector's call for resilience measures. So many of them have experienced direct hits from disasters themselves that they are living proof that we can *count on disasters* to help us bring about change.

8 Conclusions

The likely physical impact of 35,000 US museums on the environment through GHG emissions is significant enough to warrant changed behavior. The intellectual impact of US museums on 850,000 annual visitors is as well. To date we have failed to leverage most of that capacity. To wait any longer is criminal. We must make coordinated, collaborative efforts to make environmentally sustainable behaviors and practices a priority among museums nationwide. Field-wide standards, self-designed or borrowed from other fields, offer opportunity, instruction, goal-setting, recognition, and expectations that can focus that effort toward lasting change. Based on field-wide Summits, the success of other standards, and the growth of hashtags as social evidence of a movement, the field can no longer avoid its responsibility.

The changes the human world needs most are all related to our changing climate. If humans address the causes and opportunities of that changing climate, we can build a more just, verdant, and peaceful world. Museums have limitless value in building that world.

Acknowledgements Colleagues at PIC Green, particularly Adrienne McGraw, Carter O'Brien, Stephanie Shapiro, Shengyin Xu, Beka Economopolous, Jim Richerson, Elizabeth Wylie, Don Meckley, and Roger Chang, across the last decade, have been important friends, peers, and advisors in this work we share. We continue to provide each other with encouragement, and an inspiring example just when we need it most. Along the way Ron Kagan, Gerry VanAcker, Patrick Kociolek, John Fraser, Bob Beatty, Karen Daly, Jerry Foust, Jeremy Linden, and Sarah Nunberg have added insights and encouragement that were invaluable. Creating a new path is a challenging adventure made easier with fellow travelers.

References

AAM (2013) Museum fast facts. American Alliance of Museums, Washington, DC, USA
AAM (2017) Museums as economic engine. American Alliance of Museums, Washington, DC, USA
AAM, McGraw A (2014) Museums, environmental sustainability, and our future, a call to action from the summit on environmental sustainability standards. American Alliance of Museums, Washington, DC, USA
Dumouchel C, Worts D (2017) Museums & sustainable communities—six things our working group learned, coalition of museums for climate justice. Web. Accessed 15 Jan 2018
Fraser J, Flinner K, Galvin L, Swim J (2015) NNOCCI's impacts after 5 years: community of practice and the strategic framing approach are helping educators activate public conversations about climate change solutions. New Knowledge Organization, New York, NY, USA
Hatchfield P (2011) Crack, warp, shrink, flake—a new look at conservation standards. Museum, January—February, American Alliance of Museums, Washington, DC, USA
Kahneman D (2011) Thinking fast and slow. Farrar, Straus and Giroux, New York, NY, USA
Lee J (2017) Energy star score for museums: you can manage what you measure. Green Building Information Gateway, Web. Accessed 15 Feb 2018

Marshall G (2014) Don't even think about it: why our brains are wired to ignore climate change. Bloomsbury USA, New York, NY, USA

Pollan M (2008) In defense of food: an eater's manifesto. Penguin Press, London, UK

Sutton S (2018) Museums and the Paris agreement. History News 72(4) (American Association for State and Local History, Nashville, TN, USA)

Sutton S, Wylie E, Economopoulos B, O'Brien C, Shapiro S, Xu S (2017) Museums and the future of a healthy world: 'just, verdant and peaceful'. Curator Mus J 60(2):151–174. https://doi.org/10.1111/cura.12200

Sarah Sutton, LEED-AP, Principal, Sustainable Museums was a founding member of AAM's PIC Green (recently renamed the Environment & Climate Network), a past co-chair, and has continued participating by leading the Standards effort. She is a founding co-chair of AASLH's Task Force on Environmental Sustainability. She trained originally in administration of cultural heritage sites at The Colonial Williamsburg Foundation and The College of William & Mary. Her time spent as fundraiser for history museums led her by chance to sustainability planning, and then to the study of how to translate that work to museums. Her consultancy, Sustainable Museums, is committed to world-wide efforts to align the good work of museums, zoos, gardens, historic sites, and other cultural and natural resource organizations in creating a just, verdant, and healthy world. She is the co-author, with Elizabeth Wylie, of two editions of *The Green Museum, a primer on environmental practice* (*2008* & *2013*), and author of *Environmental Sustainability at Historic Sites and Museums* (*2015*).

The Possible Museum: Anticipating Future Scenarios

Bridget McKenzie

Abstract In this paper, the author proposes that the climate emergency makes it imperative that museum teams should do more frequent and rigorous work of imagining the future. The context of climate change, and disinformation about it, are outlined in order to underline the urgency of this ethical and practical imperative. Envisaging future scenarios in both imaginative and evidenced ways, and then planning for adaptation according to what is imagined, could be called 'anticipatory work'. Museums should embrace this work because of their trusted role and their duty to preserve heritage into posterity. A museum committed to anticipatory work could be defined as a 'Possible Museum', or, one which serves the public by contributing to the possible continuation of biodiverse and civilised life. The future Possible Museum will take an ethical path, validating indigenous voices and proactively engaging communities to shift towards more regenerative lifeways. As the author is a consultant responding to needs of organisations through tendered contracts, she has been unable to carry out controlled academic studies. As such, this is a propositional paper drawing on 15 years of engaged reflection on culture and climate change, and on her promotion of ecologically-informed anticipatory work, in particular Scenario Planning, which she describes in detail. The author has researched to discover the current extent of anticipatory work in museums, internationally. While Scenario Planning is common across other sectors, the author has found little evidence of its use by museums. That said, she found that museums are beginning to anticipate futures in various other ways that are public-facing, rooted in places, imaginative and optimistic. McKenzie concludes that these examples, and initiatives such as the Happy Museums Project and growing awareness of Ecomuseums, increase the likelihood of more rigorous anticipatory work being done in future.

B. McKenzie (✉)
Flow Associates, 152 Waller Road, London SE14 5LU, UK
e-mail: bridget.mckenzie@flowassociates.com
URL: http://aboutbridgetmckenzie.wordpress.com

© Springer Nature Switzerland AG 2019
W. Leal Filho et al. (eds.), *Addressing the Challenges in Communicating Climate Change Across Various Audiences*, Climate Change Management,
https://doi.org/10.1007/978-3-319-98294-6_27

443

1 The Possible Museum Depends on the Possibility of a Liveable Planet

The author has proposed over the past 15 years that museums can become truly relevant to contemporary needs by being more future-facing in terms of mission, and that this requires contextualising collections in ways that advance systemic and ecological literacies. In addition, their relevance requires engaging audiences in ways that promote active stewardship—of cultural heritage, of biodiverse places, of peaceful society and a sustainable planet—informed by an ecological ethics. These propositions have been made in numerous forms and platforms, and to give one example, the author co-wrote the original manifesto of the Happy Museum Project, which called for museums to be stewards of the future as well as the past. Specifically, she has introduced the method of Scenario Planning—invoking multiple scenarios to anticipate the future—in several museum studies training courses and organisational change programmes.

As a consultant responding to needs of organisations through tendered contracts, she has been unable to carry out controlled academic studies. As such, this paper constitutes a proposition drawing on this engagement in the field of culture and climate change, and the promotion of ecological ways of thinking. The proposition is that museum teams should do more frequent and rigorous work of imagining the future, going beyond simply 'looking ahead'. Envisaging future scenarios in both imaginative and evidenced ways, and then planning for adaptation according to what is imagined, could be called 'anticipatory work'. The author has carried out desk research to discover the current extent and typical styles of anticipatory work in museums, internationally.

2 An Underlying Hypothesis

Underpinning the proposition is the following hypothesis:

- That the most **probable** future scenarios are those in which most, if not all, museums will be extremely challenged by ecological and social collapse.
- That the most **preferable** scenarios are those where museums are sustained as organisations by rapidly transitioning to become agents for systemic change in a regenerated planet.
- And that the **possible** scenarios lie between these two difficult extremes.

There is a wealth of evidence that the planet is in a state of emergency. Read, for example, the Warning to Humanity delivered by 16,000 scientists to coincide with the climate talks in Bonn, in November 2017. Clearly, it is preferable that museums play a role in mitigating this situation, for the sake of their own business continuity and reputation at least. Also, it is self-evident that 'the possible' lies between the probable and the preferable scenarios.

What needs more evidence is around the *extent* and the *qualities* of the scaffolding needed to push and pull museums to transition into identifying as 'The Possible Museum'. If museum staff are also members of public that, in many places, remain relatively unwilling to face the climate emergency, what will effectively guide them in this transition? The pull—or the intrinsic motivation—could be that essential and imaginative work is invigorating. The push—the extrinsic motivation—is that the possibility of museums existing in future depends on the planet's liveability.

Museums may be, in many ways, a grand fabrication, a utopian escape from the world. However, they are also real institutions, usually in real buildings, requiring real human staffing and, most importantly, looking after real material artefacts. Their core purpose is stewardship of cultural and natural heritage for posterity. If they are going to continue to exist and to enact their role as stewards, they need to consider quite how much they need to change in an era of radical disruption from long-established norms.

3   Museums Consist of Meshes of Imagination and Experience

Although it is hard for us to imagine the mental capacities of other animals, we can be sure that imagination is, if not a unique trait amongst species, a distinguishing one for humans. The downside of our imaginative nature is that the technologies arising from it have begun to threaten the entire global habitat for humans and other species, due to a combination of population growth and consumption based on extraction of natural resources, and then the unpredictable concatenations of us having breached the boundaries of the planet's operating system.

Our capacities for imagination have been extended by our practices of symbolisation—crafting objects by hand and ideas by tongue—fertilised by the crossings of cultures. In museums, the objects of our imagination and the material traces of our lived experience are interpreted using visual and verbal language, making museums crossing places of the imagination. Museums contextualise objects so that we can imagine past events or lost cultures more holistically, while also connecting these to our contemporary world. Museum workers have an ethical choice about the purpose of contextualising the past: Is it to let visitors escape the present through enjoyable immersive experiences, or to help them be more capable of creating a thrivable future? My experience as a museum educator has taught me that immersive entertainment on its own, when not intentionally designed for deep learning, is not usually effective enough to create impactful transformations.

One possible role for museums is to anticipate the future through public programmes to help visitors ponder some worst-case scenarios and to proactively generate more preferable ones, including re-generating positive scenarios from the past. Museums are best equipped to do this in ways that use the open-ended imagination of contemporary artists, or that draw on historical perspectives, particularly the lost or dwindling knowledge of indigenous peoples.

To give one of many examples of the importance of validating indigenous knowledge: Fateful Arctic explorer, John Franklin, refused to consult the Inuit because he viewed them as ignorant savages, leading to disaster for his crew, whereas the more successful Amundsen adopted the survival tactics of the Netsilik Inuit.[1] What can we learn from peoples who are masters of adaptation now that we are exploring the critical terrain of a climate-changed future? Museums can certainly experiment with ways of learning from these people and validate the voices of people continuing to express knowledge of their ancestors.

4 The Possible Museum Will Be Generated from What Is Latent Now

Traditionally, museums collect, research and interpret fragments of the world, usually extracted from civilisations or habitats, so that we can understand these times and places better. Increasingly museums collect what is surfaced in contemporary art, industry or demotic life, and species still being discovered. An emerging role is to provide a safe harbour for artefacts and knowledge at risk from disruption of lands and cultures caused by industrialisation, rapid regeneration, climatic damage and conflict. Museums can help by reconstructing missing context, making it visible and intelligible.

At best, they also recognise that these processes are therapeutic, providing expressive outlets of grief and hope for people in the midst of rapid change, assuaging for the extinction of species and languages, and recapitulating ancestral connections with nature and inherited myths. An example of this is 'Troubled Waters, Stormy Futures', the research project led by Dr. Sara Penrhyn Jones at Aberystwyth University, in collaboration with the Museum of Archaeology and Anthropology in Cambridge. This investigates heritage at a time of accelerated climate change, working with artists and communities in the Pacific island of Kiribati. Creative interventions include a film with the artist Natan Itonga who has dedicated his career to preserving 'maneaba', a leadership system for villages which ties into crafts, ritual and food production.

The author took part in Cambridge Museums' Climate Hack 2018, creating a participatory visitor experience to illuminate this Troubled Waters project. As part of this, visitors were invited to imagine future scenarios in which they, as residents of low-lying Fenlands, would have to leave their homes in a boat, taking only the most essential objects with them. This process invoked empathetic expressions and many pertinent questions from visitors. Stories generated can be seen on https:// ngarughostboat.wordpress.com/ the website created by the author during the Climate Hack.

---

[1]Brian Bethune interviewing Robert Boyd, on human adaptation to difficult environments (2017).

5   A Possible Pivot to Ecological Thinking About History

The study of History is vitally important at this critical juncture in time, but the author proposes that it must now be based on an ecological ethics, and be contextualised through democratic educational processes. History is essentially constituted from the stories we tell around the interactions of people with places, and museums focus our attention on the objects and documents related to those stories.

Within industrial civilisations, History is understood as events unfolding over time, and is usually rehearsed to assert (or counter) the dominion of certain peoples over places. This 'survival of the fittest' narrative of human progress underpins common notions of why we study History. This narrative has been more aggressively asserted, and then also more contested, the more that 'growth' has intensified destruction of human and other habitats. For example, when Michael Gove became England's Education minister in 2010, he pushed for the History National Curriculum to tell 'our island story' and to reclaim pride in the achievements of the British Empire. Histories told from indigenous perspectives tend to reinforce the continuity of community, and identity *with* place and with other non-human inhabitants. As elites have developed more sophisticated technologies for effective exploitation of nature and people, and for defence of their claimed territories, the ownership of histories has become more problematic.

Museum professionals are thinking much harder about these problems, as tensions are rising within communities stirred by ideological media. In the USA, museum educator Mike Murawski and La Tanya S. Autry have initiated the campaign 'Museums are Not Neutral'. They focus on cultural conflict but are aware of the broader context whereby people and nature have been exploited in tandem. Another response is 'The Past is Now: Birmingham and the British Empire', an exhibition at Birmingham Museum & Art Gallery that challenges the typical colonial narrative of Empire and examines the museum's own bias (Minott 2017). One of the themes explored by the invited curators is Environment, describing how one of the central features of the British Empire was a desire to control and exploit natural resources.

When museum narratives are informed by ecological understanding they can be more inclusionary of diversity, opening up discursive space to consider how peoples interact with each other and with other species. One tenet of ecological thinking is that interdependence results in symbiotic exchange and collaboration more than competition. Across species, habitats and eras, phenomena that evidence the 'sustainability of the fitting', a term explored by ecologist Alan Rayner, have been found to be much more common than 'survival of the fittest'. Seen from a contemporary perspective, one might imagine that human History has always been, everywhere, a story of conflict and colonisation (2011). An ecological approach to History can illuminate the establishment of cultures based on aggressive competition, as well as those cultures in which benign co-operation has held sway. Museum narratives often convey a hopeless sense that these benign lifeways are all but lost. However, it is possible that benevolent values can be resurgent and these lifeways re-established, and museums can play a role in this.

6 Climate Change, Predatory Delay and the Truth

Research, such as a 2013 study by BritainThinks for the Museums Association, has shown that there is high public trust in museums, seen as 'guardians of factual information'. Trust and truth are essential for a functioning democracy. However, national politics are increasingly corrupted by propaganda. Behind this corruption lies a cabal of 'oiligarchs' (or extractive capitalists) manipulating truths so that they can delay climate action and continue to profit from their ecocidal and iniquitous modus operandi. This is 'predatory delay', a term promoted by Alex Steffen, who writes, "The disruption of fossil fuels and unsustainable industries…[is now] by necessity the defining focus of any real climate strategy" (2017). Around this focus he proposes four main actions: Decoupling economic growth from consumption; Decarbonising energy and materials; Large-scale ecological restoration; Ruggedization of human systems to prepare for unprecedented conditions. However, these actions face a formidable co-ordinated resistance.

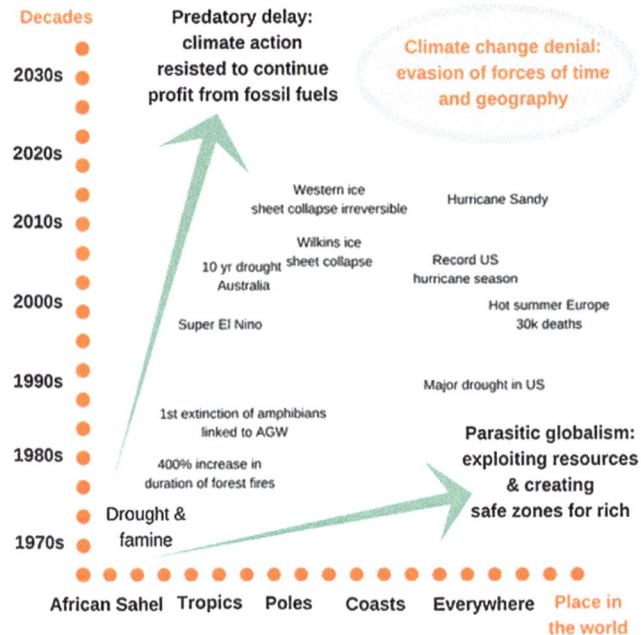

Severe effects of climate change began 50 years ago but a powerful group, led by fossil fuel interests, are delaying action in order to profit from continued extraction, aiming to accumulate enough resources to escape these impacts by investing in land in the safest places. *Source* Bridget McKenzie

Since 1860, with Tyndall's discoveries, scientists have known that CO_2 is a heat-trapping gas. Since the 1960s oil companies have been aware that fossil fuel emissions will have catastrophic results, and the first climate impacts were realised in the famines of the African Sahel in the 1970s. Denial orchestrated by the oil industry began at this point, and has reached its apogee in Trump's election, with his removal of the US from the Paris Accord and enabling of fossil fuel extraction in protected zones.

Our world is becoming increasingly 'VUCA'—Vulnerable, Uncertain, Complex, Ambiguous—due to the breaching of multiple boundaries of safety in the planet's operating system. It is important to understand the oil industry not just in relation to its CO_2 emissions but also the damaging impacts of its petrochemical derivatives such as plastics, fertilisers, pesticides and detergents. The uncertainty of these dynamically interacting impacts is being exploited to further confuse us, and online communications are being weaponised to these ends.

It is becoming apparent that this 'oiligarchy' and its political and media agents are engaged in a takeover of our algorithmic, social and broadcast media in order to manage our perceptions and destabilise our democratic and international institutions, using the passion-inducing power of nationalistic and religious tribalism as lever and cover. The term 'fake news' is more accurately described as 'weaponised media'.

Museums have a vital role to play as places of open enquiry, to enable the nuances of the never-perfect truth to be explored and to amplify this beyond their physical sites into online and broadcast media. To carry out this role they must avoid being vulnerable to weaponisation, for example, by resisting sponsorship by fossil fuel corporations as is encouraged by campaigners within the international Fossil Free Culture movement.

Another method of inoculation is for teams to think rigorously and frequently about the future in context of the climate emergency. Museums must do anticipatory work, both for themselves as a means for their organisations to adapt and survive, and with their communities so that they can contribute to regenerative action to resist the planetary crisis.

7 Thinking About the Radically Uncertain Future

The algorithm designed to measure Arctic warming in Alaska stopped recording because temperatures were never expected to go so high.[2] This reminds us that our technologies to predict and prepare for near-future events are based on the 'old normal'. California is experiencing both extreme drought (leading to forest fires) and extreme flooding. We are not well practised in imagining these dynamic

---

[2]http://www.independent.co.uk/environment/arctic-global-warming-rapid-computer-rejected-alaska-a8110941.html.

interactions of different weather phenomena that come with 'global weirding'. Similarly, we are not practiced in imagining potential interactions between climatic disasters, food and water resources, mass movements of people, conflicts and the manipulation of information.

The media tends to report on spectacular phenomena using linear narratives and agonistic debate, rather than covering systemic emergencies using multimodal enquiry techniques. This in turn hampers the public in developing habits of mind to cope with complexity and uncertainty. Museums, especially when enabled with partnerships with communicators and with technologies, are well placed to help us visualise these future possibilities and develop these mental habits. However, museum staff must rigorously question and reinvent their own interpretative practices so that these are fitting for challenges on the horizon. As a preparatory step to becoming Possible Museums by doing anticipatory work with audiences, museum teams should first speculate futures for their own working lives and their organisation.

8 Museums Imagining Their Futures: Breaking Out of Thinking Habits

The futurism often seen in media articles is not systematic enough as it typically describes phenomena based on trends—patterns emerging now—and then applies wishful thinking to project desirable futures. For example, we might read about how Virtual Reality heralds scenarios where we can deeply immerse ourselves in others' situations to better empathise with them. A more systematic approach might flip these idealistic visions to consider their shadow sides, to ensure preparedness for their risks. This would explore, for example, how Virtual Reality can be so escapist and compelling that we fail to serve those who most need help in reality. The problem is that, even with a negative flip, this still extrapolates from trends, and is more likely to draw on stock ideas of utopian and dystopian futures. Thinking about big picture change is, more often than not, informed by limited perspectives and received ideas, which can be misguided.

Cultural Leaders from the Oxford University Museums Partnership came together in February 2016 for a workshop on futures thinking. The blogged account of the day by Jessica Suess suggests that this was immensely helpful (2016). However, it offers an example of how important it is not to rest on assumptions. The facilitator "placed trends within the context of time—trends can vary significantly in their pace: there are slow trends like climate change; medium trends, for example changes in attitude towards marriage equality; and fast trends like self-monitoring (e.g. fitbit)." This suggests a simplification of the complexity, speed and scale of the impacts of Climate Change, which is now a rapidly unfolding emergency after two centuries of unchecked greenhouse gas emissions. The Arctic will very soon be free of sea ice in summers. Current anthropogenic Climate Change is a Gordian knot of slow, medium and fast trends, liable to explode in unpredictable ways. As such, it cannot be termed a trend but a 'critical uncertainty'. An aspect of the constructive denial of Climate Change is to return to default notions that its impacts will affect only future generations.

Museum staff should adopt and develop techniques for anticipatory work that are informed by the latest data and predictive models, combined with the imagination of multiple scenarios that take account of such critical uncertainties. They will also need to consider responses to these scenarios that are not simplified to one of either these stultifying extremes: 'It will be a catastrophe so there is no use in taking action' or 'Technology is leading us to paradise, so there is no need to take action'. The possible paths of action between these two will involve an enormous amount of tough, concerted and messy work.

9 Scenario Planning as Management Technique

Scenario Planning developed out of military strategy for 'VUCA' situations, and is now being adopted across many industries. Its purpose is to visualise multiple scenarios, in order to mitigate the threats and benefit from the opportunities they present. As Robert Janes describes it, in Museums in a Troubled World, "It is a technique to assist with the creation of new mental models that result in powerful stories about how the future might unfold" (2009). If used diligently and imaginatively, it scaffolds the process of future-gazing to avoid the easy projection of desired futures and/or extrapolation from existing trends. It is most helpful when strongly informed by a wide range of influence factors, and when participants are very clear-sighted about how critical these factors might be, and, crucially, how they dynamically interact with each other.

It will also be more effective if it is repeated regularly. Anticipatory work could be seen as Cognitive Behavioural Therapy for organisations, getting staff into habits of mental flexibility by entertaining multiple thoughts about what is and what could be.

A Scenario Planning process has been trialled by the author, as director of Flow Associates, as part of the Future Views project. This research and workshop tools can be found on www.futureviewstoolkit.com.

These are some steps museum teams might take if they want to try Scenario Planning:

- Ensure attendance in a workshop by a '**diagonal slice**' from an organisation.
- Establish a specific **field of action** (e.g. the museum, its locality, or the wider museum sector) so that the findings help answer a shared question.
- Participants should be invited to express their own ideals about the future, as well as their fears. The team can appreciate these, but then deliberately set them and associated emotions aside.

- Zoom out to take a broad view, accessing data from diverse but reliable sources, and together list some influential factors of change. Consider these factors of change by working through different lenses, so that you consider all possible dimensions.
- Separate these drivers of change into predictable **trends** that seem certain to continue (e.g. consumer demand, or use of some resources) and **critical uncertainties** that could affect change in unpredictable and drastic ways (e.g. social response to a disaster, or impacts of a referendum).
- Select the two drivers that you think will be most significant for your situation. Choose one that is a **trend** and one that is a **critical uncertainty**. Use these to plot the parameters for four different future scenarios, aiming for a pattern as in this diagram.

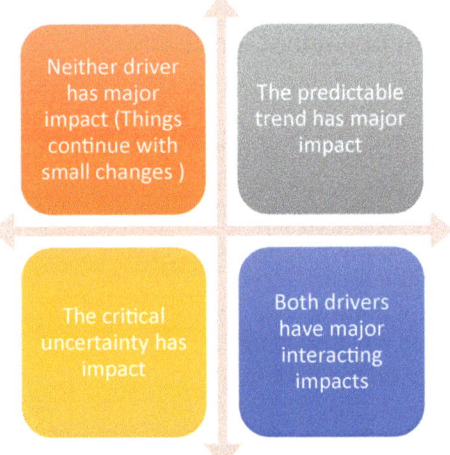

Matrix showing four scenarios, more opportunistic at the top, more threatening at the base.
Source Bridget McKenzie

- Next comes the creative part of the process. Give each scenario a name, and flesh it out, thinking about your 'field of action'. Pay less attention to the scenario in which drivers have less impact but keep it in mind as a reference. Use concrete and imaginative techniques: Consider what a typical working day will look like, or the contemporary artefacts museums might collect. Use drawing or collage, or invent characters. Bring other drivers for change from your longlist into play.
- Consider which scenario contains characteristics of being the most **probable**, and which the most **preferable**. Between these, what would the most **possible** scenario look like? The most possible scenario would be the most preferable, yet achievable with effort and luck. (You may want to adjust one of your developed scenarios to enhance its characteristics to be more possible.)

- If you have time, or if there are more groups of participants in the workshop, you could take two more from your initial set of drivers for change, and use these to sketch out four more scenarios.
- The final step is to discuss the actions needed to move on to your most possible scenarios. You could organise these into categories:

  - Grand challenges (actions beyond our capacity, requiring policy change or wholly different practices);
  - Strategies (actions we can collaborate on and plan for);
  - Tactics (actions we can take now, ourselves).

10   A Note on Climate Change as a Driver for Change

If Climate Change is chosen as a driver for change, it is important to take a very broad view of this complex factor. Typically, participants jump to consider proactive carbon-reducing impacts on organisations and audiences, without addressing the full range of impacts (e.g. increasing resource insecurity, conflict, refugees and infrastructure damage). One of the knock-on effects of these extremely negative Climate Change impacts could be a widespread failure to take positive action to tackle global warming emissions.

The author assumes that Climate Change would be selected as a top driver for change by most groups doing Scenario Planning but this assumption is based on the fact that she sees the world primarily through an environment lens. This is an unusual perspective. Our contextual culture is one in which the economic lens dominates, and the environmental lens is the least favoured. However, the environment should not strictly be defined as a lens, or as an issue. Instead, we need to continually hold in our consciousness the environment as our living world, and as the totality of systems we inhabit, interact with and co-create. The environment lens is relevant to all the others.

Working in a globally disrupted climate requires that we test our long-held assumptions that the future will necessarily involve the progress of civilisation, onwards and upwards. In fact, cultural evolution may emerge more through messy convolution of various adaptive actions than through straightforward revolution. These adaptive actions will need to be combinatory, interdisciplinary and poly-optical (or seeing from different perspectives).

11 Results: How Are Museums Doing Anticipatory Work Now?

The results are essentially the findings of the author's online research to discover how museums are currently carrying out anticipatory work. McKenzie has written and presented about how museums can use Scenario Planning in a context of climate change since 2010. When she wrote an essay about this subject in 2012, 'Seeing Museums in 2060', she could not find any real life examples of museums or museum professionals using this as a planning methodology. Five years later, researching for this paper, she still could not find any museums promoting their use of this methodology. Although it is recommended as a future-facing method by the Center for the Future of Museums, no examples are offered of museums trying it.

However, she found some instances of museums doing anticipatory work in other ways:

The Alliance of American Museums have used speculative fiction as a method to explore the future of museums. Contributors to the December 2017 issue of its journal, entitled Museum 2040, were asked to write as if they were in the year 2040 (Merritt 2014). Some of its features refer to climate change, suggesting that museums will both need to adapt to its impacts and play an active role in mitigating them.

The Carnegie Museum of Natural History is appointing a curator of the Anthropocene, and is showing a year-long exhibition entitled "We Are Nature: Living in the Anthropocene" (Thomas 2017). As part of this programme, visitors can vote for the celebrity extinct species of the future, casting their vote by making cash donations to the World Wildlife Fund.

Science Gallery Dublin has issued a call for artistic proposals towards an exhibition that explore extreme environments and encourage imaginative speculations. One question in their call asks: 'As our own planet grapples with more extreme and precarious environmental conditions, how could the constraints of extreme environments lead to creative new tools, methods, micro-organisms, and technologies?' See https://opencall.sciencegallery.com/life-edges for more.

The city of Rio de Janeiro, has built an entire museum of future scenarios. The Museum of Tomorrow offers a narrative about how we can live and shape the next 50 years. This is from its mission statement: 'Guided by the ethical values of Sustainability and Coexistence, essential for our civilization, the Museum also seeks to promote innovation, disseminate the advances of science and publish the vital signs of the planet. A Museum to enlarge our knowledge and transform our way of thinking and acting.' See https://museudoamanha.org.br/pt-br/sobre-o-museu for more.

These examples show that there is some environmentally conscious anticipatory work taking place, which is public-facing, imaginative and—in some ways—optimistic. There are some other initiatives which also help lay the ground work for museum staff to think in more holistic ways about the future role of museums, to give the following two examples:

The Happy Museum Project is a peer-learning network of museums in the UK actively exploring the role of museums for planetary and social wellbeing in the future. One of its six core principles is: 'Value the environment: Be a steward of the future as well as the past' which echoes the indigenous American 'Seventh Generation Principle'. This requires consideration, prior to any decision, of ancestors, other species and of seven generations into the future. There are plans for this principle to be explored in more depth in the next two years. For more case studies and resources about this, see www.happymuseum.org.

This idea of stewardship leads to the final example of Ecomuseums, a global network of small museums that aim to improve local resources by encouraging local people to take responsibility for their natural and cultural environment. They use ethical forms of enterprise to sustain themselves, for example, selling traditional artefacts that are made from local traditional materials. An article on Ecomuseums by Borrelli and Davies, 'How Culture Shapes Nature', offers numerous examples of ways that they contribute to shaping and sustaining places (2013). None of the examples refer to the use of any obvious mode of anticipatory work such as Scenario Planning. In itself, this is instructive because futures thinking is integrated invisibly into an ecological approach. Perhaps the label of 'futurism' can be extended to any cultural activity that imagines the possibility of continuing traditions of living harmoniously in one's habitat?

12 Conclusions: An Emerging Possibility of More Possible Museums

The author proposes that the Possible Museum is one that will continue to exist despite environmental challenges by doing anticipatory work in order to maintain its relevance. By imagining the future, museum teams will see that the path of relevance is an ethical one. They will proactively work with communities to shift towards more regenerative and circular economies. They will explore ethical and participatory forms of entrepreneurship in order to sustain themselves when or where public funding dries up. They will provide safe, inclusive spaces for envisaging possible futures, for learning from past and indigenous cultures and from the capacities of nature, and for helping communities take action for eco-social justice.

Internationally, there is still a lack of anticipatory work that uses structured methods such as Scenario Planning or which systematically faces the challenges of climate change. However, the rapidly changing context, and inspiring examples of emerging practice, suggest that it is likely museum teams will start to do more frequent and rigorous work of imagining the future, and planning for adaptation according to what they imagine. This work will be essential across all sectors, but successful museum teams will realise they should lead in it, due to their trusted role in communities and their duty to preserve heritage into posterity.

Happy Museums and Ecomuseums, in particular, offer routes that might be taken towards the Possible Museum. They provoke the consideration that the most important work for museums is to hold a vision of continuity of thriving places, while working with communities to adapt to change.

References

Bethune B (2017) 'Humans' capacity for culture is the key to our success, an anthropologist argues' Interview with Robert Boyd, on http://www.macleans.ca/author/brianbethune/, Canada

Borrelli N, Davies P (2013) How culture shapes nature: reflections on Ecomuseum practices. In: Nature + culture, vol 7. Helmholtz Centre for Environmental Research, UFZ, Germany, June 2013

BritainThinks (2013) Public perceptions of—and attitudes to—the purposes of museums in society. Museums Association Commissioned Report, United Kingdom

Janes R (2009) Museums in a troubled world: renewal, irrelevance or collapse? Routledge, United Kingdom

McKenzie B (2012) Seeing museums in 2060. The Learning Planet https://thelearningplanet.wordpress.com/2012/05/08/seeing-museums-in-2060/

Merritt E (2014) Questioning (some) assumptions. Center for the Future of Museums, Blogpost, Mar 2014. http://futureofmuseums.blogspot.co.uk/2014/03/throwback-thursday-questioning-more.html

Museum 2040 (2017) J Am Allian Mus, Dec 2017. Available online https://www.aam-us.org/docs/default-source/museum/museum2040.pdf?sfvrsn=2

Minott R (2017) The past is now: the exhibition is open. Birmingham Museums Blogpost, Dec 2017. http://www.birminghammuseums.org.uk/blog/posts/the-past-is-now-the-exhibition-is-open

Rayner A (2011) Sustainability of the fitting—bringing the philosophical principles of natural inclusion to the educational enrichment of our human neighbourhood. In: Keynote address at 8th world congress on action learning & action research, Sept 2010, Melbourne, Australia. www.alara.net.au/worldcongress/2010/objectives

Steffen A (2017) The real politics of the planetary crisis. The nearly now on medium, Dec 2017. https://thenearlynow.com/the-real-politics-of-the-planetary-crisis-216229324deb

Suess J (2016) The long view—futures thinking for museums. Account of workshop with Oxford Aspires Museums, Blogpost, Mar 2016. http://www.oxfordaspiremuseums.org/blog/long-view-futures-thinking-museums

Thomas M (2017) Carnegie museum of natural history initiates a discussion of the Anthropocene, Oct 2017. http://www.post-gazette.com/ae/art-architecture/2017/10/25/Carnegie-Museum-of-Natural-History-We-Are-Nature-Anthropocene-1/stories/201710240172

Bridget McKenzie is based in London, UK, and has 27 years of experience in researching, evaluating and managing museums and arts programmes. Past roles include Education Officer for Tate, co-ordinator of widening participation programmes for the University of the Arts London, and Head of Learning for the British Library. She founded Flow Associates in 2006, and has worked as an independent cultural consultant since that date. Current projects include evaluating the impacts of the Happy Museum and of the environmental art curation of Invisible Dust. She is a trustee for the ONCA Gallery in Brighton, an advisor for Culture Unstained, and is a Julie's Bicycle Creative Climate Leader. She has initiated a project to found a Climate Museum for the UK. She has a B.A. Hons and an M.A. in History of Art, and two post-graduate teaching qualifications.

The Views of Citizens on the Issue of Participation in Confronting Climate Change: The Case of Greece

Aikaterini Zerva, Georgios Tsantopoulos, Evangelos Manolas and Stilianos Tampakis

Abstract Climate change no longer constitutes a prediction for the future but it is already occurring. For this reason what is necessary is both the adaptation of citizens to new changes as well as action by the scientific community and by the bodies involved in the fight against climate change. Thus, the bodies will need to organize and undertake effective action which will encourage citizens to participate in such actions in order to adapt to future impacts. The aims of this research are, on the one hand, the investigation of the characteristics which influence participation in activities and, on the other hand, the discovery of specific characteristics with regard to citizen preferences. In order to achieve its aims this research used a structured questionnaire and 1536 questionnaires were collected from January 2014 to June 2015. The main results include that younger citizens greatly trust scientists, show great willingness for voluntary action and get their information on climate change through documentaries. Also, those with higher educational level trust the actions of non-governmental bodies while middle aged citizens and secondary education graduates trust the actions of governmental bodies while older citizens prefer activities for the reduction of pollutants.

A. Zerva · G. Tsantopoulos · E. Manolas (✉) · S. Tampakis
Department of Forestry and Management of the Environment
and Natural Resources, School of Agricultural and Forestry Sciences,
Democritus University of Thrace, 193 Pantazidou Street, 68200 Orestiada, Greece
e-mail: emanolas@fmenr.duth.gr

A. Zerva
e-mail: azerva@fmenr.duth.gr

G. Tsantopoulos
e-mail: tsantopo@fmenr.duth.gr

S. Tampakis
e-mail: stampaki@fmenr.duth.gr

© Springer Nature Switzerland AG 2019
W. Leal Filho et al. (eds.), *Addressing the Challenges in Communicating Climate Change Across Various Audiences*, Climate Change Management,
https://doi.org/10.1007/978-3-319-98294-6_28

457

1 Introduction

Climate change is probably the most important environmental problem facing humanity. Climate change is a result of both natural and anthropogenic causes which are responsible for the increase in Earth's temperature and the increase in greenhouse gas emissions (Stern and Kaufmann 2014). The creation of communication strategies for the environment constitutes one of the most crucial socio-political issues of our time (Brevini 2016). However, the designing of different policies and strategies and the different ideologies of leaders of countries (McCright et al. 2016) delay the adoption of measures for confronting climate change.

According to the predictions of the Intergovernmental Panel on Climate Change (IPCC) the upward trend of atmospheric temperature will continue during the 21st century in all areas of the planet accompanied by an increased trend of extreme weather conditions (IPCC 2013).

Climate change is not only an environmental issue but it is also linked with development at personal and socio-political level which means all citizens will need to make efforts in order to deal with the issue effectively (Lorenzoni and Pidgeon 2006). According to the report of the World Commission on Environment and Development (WCED 1987) climate change is closely linked with sustainable development and, therefore, the following goals should be pursued:

- a political system which will secure the effective participation of citizens in decision making processes
- an economic system which will be in position to provide advantages and strategies on an autonomous and firm basis
- a social system which will be effective in preventing tensions
- a system of production which will respect the preservation of ecological balance with respect to development
- a technological system aimed at a continuous search for new solutions
- an international system which will promote sustainable patterns with regard to trade and the economy
- an administrative system which will be flexible and capable of correcting problems automatically.

Therefore, the creation of these ideal conditions, being able to respond to the impacts of climate change, linking climate change with sustainable development as well as designing effective mitigation and adaptation policies are not at all easy issues to deal with. This is so because there are many factors which influence negatively the development of mitigation and adaptation policies and, which hamper the contribution of bodies, which are involved in taking action against climate. With regard to the above factors most important seem to be the factor public understanding concerning perception of danger, something which is of fundamental importance in relation to communicating the challenges of global climate change (Hagen 2016). Also, most important is the dynamic relationship between values and beliefs which incorporate moral concerns (Persson et al. 2015). These factors in combination with publicly available information, beliefs of society (Lorenzoni and Hulme 2009) which differ from geographical area to geographical area and with regard to the personal experience of citizens in local weather conditions (Egan and Mullin 2012) are factors which can influence the attitudes and views of citizens in relation to climate change.

Generally, at global level, most citizens worry more for other environmental problems, such as atmospheric pollution (77.0%) rather than climate change (69.0%) (Carrington 2011). In Greece, in particular, which is hit by economic crisis the last seven years, it was found that nine out of ten citizens (87.0%) regard the problem of climate change as a very important problem. This percentage is bigger than the average of all European countries. Compared to other important problems, Greek citizens think that the most important problem in Greece is poverty and climate change the fourth most important problem (48.0%) (European Commission 2015). This is probably a result of Greece's economic situation. Obani and Gupta (2016) think that citizens during economic downturns focus more on the recovery of the economy rather than environmental problems because they believe that giving priority to environmental problems is something which will delay the recovery of the economy (Zerva et al. 2018).

Thus, the aim of this research is, on the one hand, the investigation of the factors which influence the willingness of the citizens for participation in activities of various bodies involved with climate change and, on the other hand, to determine through these factors the specific features of citizens which, in order to confront the problem, are willing to engage in activities with regard to mitigation and adaptation issues and with regard to promoting sustainable development.

2 Review of Literature

Climate change policy addresses two different issues: (a) mitigation and (b) adaptation. However, most important for climate change are three elements: the cost of mitigation, the cost of adaptation and the cost of impacts which can neither be mitigated nor adapted (UNFCCC 2009). In particular, the concept of mitigation refers to anthropogenic intervention with regard to the reduction of sources or the strengthening of plans for reducing greenhouse gases. The concept of adaptation refers to the adaptation of natural or human systems in a new or changing environment. Adaptation may also include the planning of policy on making the public aware on the issue which according to Hulme (2010) does not refer to optimizing society to stand the test of dangers caused by extreme weather events but a compensation to unknown and some known future climate change events.

According to Gough (2008) pre-requisites to the successful application of climate change policies is public understanding and social acceptance of the problem by the public. However, progress with regard to the formulation and application of these policies is frequently blocked by various obstacles. These obstacles include the lack of citizen awareness on climate change, the lack of citizen trust towards the involved bodies, financial resources and political commitment (Clar et al. 2013). In particular, the last two decades European policy as well as the journalistic coverage of the issue by the means of mass communication (Ford and King 2015) has focused only on the creation and application of mitigation policy for the reduction of greenhouse gases while the concept of adaptation was a taboo.

Later and particularly in recent years, i.e. from mid 1990s, the issue of adaptation got attention by researchers, governments and the UNFCCC. All these bodies recognized the necessity of both mitigation and adaptation measures (Evans et al. 2014) simultaneously. Next, the European Union approved seven national strategies for facilitating the development of a national adaptation strategy which will link the scientific and technical support required for the development and application of such a strategy, the role of strategic information provision, communication and awareness with regard to the application of the suggested activities and strategies involved with the issue of incorporation and coordination with other sectors (Biesbroek et al. 2010).

However, a pre-requisite for the application of such strategies is the achievement of certain goals. According to Adger et al. (2005), the fundamental goals of public policy with regard to climate change adaptation, as referred to in most documents are mainly: (1) the protection of vulnerable populations through the reduction of danger, (2) the provision of danger with regard to programming and strengthening adaptation measures and (3) the protection of important public goods and heritage.

Also, for the evaluation of these aims and the application of adaptation policies on climate change certain reliable indicators of vulnerability have been developed which to a great extent differ both with regard to the types of indicators and the measures used as well as their approval by vulnerable countries (Eriksen and Kelly 2007). Therefore, adaptation to the impacts of climate change is a complex project not only for scientists but also for the bodies involved in decision making.

However, beyond the factors which influence the application and evaluation of climate policies, there are also other factors which will have a direct impact on consequences. These factors are usually the uncertainty of the problem and the impacts of climate change, the complexity of the issue which is apparent in multiple scales and which will need to include multiple levels of governance as well as the fact that adaptation problems do not appear in isolation but in the framework of social, demographic, political and economic changes (Hanger et al. 2013).

The contribution of bodies which are involved in the fight against climate change is particularly important for the participation and active involvement of citizens. This is so because citizen involvement in activities is directly influenced by the trust citizens have towards responsible bodies for activities they have undertaken for fighting climate change. In parallel, trust towards bodies and particularly governmental bodies is directly influenced by personal efficiency, environmental values (Lubell et al. 2007), ideological differences or by satisfactory arguments and provisions for action. In addition, another factor which influences the trust of citizens towards bodies are social conditions since it was found that the quality of the environment is not particularly good or if the government cannot secure the necessary economic security for citizens then citizens may trust the government less. However, in order to change and simultaneously in order to have reduction of unemployment in every country (Cin 2012) it will be necessary to have fair procedures and protect integrity (Terwel and Daamen 2012). For this reason a series of research projects were carried out which were related to the degree of trust citizens have towards governmental bodies in combination with the activities undertaken by these bodies in confronting climate change. The results of these projects with regard to governmental bodies were not satisfactory. In particular, in research carried in the USA, it was found that a big part of citizens does not trust the government in confronting climate change and despite the high levels of secondary education graduates among the public, citizens are not particularly familiar with climate change issues (Tjernström and Tietenberg 2008). In addition, in similar research in the Netherlands it was found that the trust of citizens towards bodies involved in confronting climate change was not particularly significant and may influence the attitudes of citizens with regard to carbon dioxide management (Midden and Huijts 2009). The same crack between government and public opinion was observed in Asia with regard to what the public understands on the issue of necessary climate prevention policies (Kim 2011).

Similar attitudes are displayed by Greek citizens in Iraklio, Crete, who do not trust the bodies involved in climate change and who think that the governmental bodies involved in the fight against climate change are not in a position to effectively manage future impacts of climate change (Jones et al. 2012). Thus, what Greek citizens understand with regard to the governmental bodies involved in the fight against climate change is something which worries scientists because such bodies play a key role in confronting problems which are caused by human interventions on the environment.

Therefore, in order for this to change and in order to improve the relationship between government and public opinion it will probably be necessary for the government to co-ordinate the various involved bodies which interact with one another in confronting climate change. It will be necessary to incorporate these bodies in the framework of regional and local action (House of Commons Environmental Audit Committee 2008). This way, these bodies will have the chance to improve their arguments or ideas in confronting climate change so that they can encourage the public to participate in activities (Pearce and Cooper 2011). At the same time, the opportunity will be created to increase the degree of citizen trust towards the government and other bodies with the ultimate aim being the minimization of climate change impact in the future. It has been found that the taking of timely actions is of fundamental importance for the stabilization of atmospheric concentrations even if the efforts of mitigation in the rest of the world are not realized on time (Jakob et al. 2012). On the contrary a delay with regard to action will be quite costly, if a mediocre policy for climate is approved in a short time in comparison with the taking of action (Bosetti et al. 2009).

In addition, the lack of action with regard to every country's adaptation and mitigation will bring serious economic cost as it is in the case of Greece. According to national policy regarding adaptation to climate change (YPEKA 2016) and the

Committee for the Study of Climate Change Impacts (EMEKA 2011), it has been estimated that the total damage from the impacts of climate change correspond to a huge economic cost for the country.

In particular, according to the EMEKA (2011) report regarding the estimation of future climate change impacts in Greece it is observed that agriculture, forests, fishing, transport, tourism, the built environment and water ecosystems will be directly influenced by the increase in temperature and drought. Coastal ecosystems, transportation and tourism will be also influenced by the increase in sea level, transport and extreme weather phenomena. In particular, from the above those which will have negative impacts for the economy will be agriculture, coastal ecosystems and tourism since they constitute the key pillar of the economy and the basic co-efficient for changes in Greece's Gross Domestic Product (GDP).

Of course, in order to successfully complete activities in confronting these impacts and in order to have an effective application of policy goals for climate for each government (Hanusch 2015) it will probably be necessary to separately analyze each key factor since each factor is linked to a different approach (Prins et al. 2010), different incentives and different possibilities for taking action but also different contribution to climate change. Also, it will probably be necessary to improve more action and measures at local level without taking into account if earth is heated or cooled (Jayawardena 2015). This will aim to the reduction of social and environmental vulnerabilities on the basis of cost-benefit analysis and dangers (Pielke 2005) with an emphasis in the investigation of different adaptive measures with regard to local extreme events (Flechsig et al. 2000).

However, the efficiency of local government depends on many factors. The two most basic factors are usually that local actions are mainly restricted by the reduction of gas emissions excluding the rest of the impacts and that they are influenced by the availability of resources and political support (Bedsworth and Hanak 2013). Also, the action of local bodies does not only depend on councils but also on partner bodies in cases where it is uncertain how much they can harmonize with agreements at local level and undertake responsibility for their application (Pearce and Cooper 2011). Also, in order to improve the degree of trust of citizens with regard to actions by national and local government it will be necessary that both the national and local government will make great efforts to co-ordinate action with the rest of the organizations which are active in the area.

3 Methodology

3.1 Research Area

The research area for this paper was the entire territory of Greece with the population, after the last census in 2011, being 10,815,197 from which 49.2% were male and 50.8% female. Greek economy mainly relies on the primary sector (agriculture, livestock farming, and forests) and tourism which contributes much to the Gross Domestic Product (GDP) and constitutes a source for foreign currency. With regard to climate change Greece is a special case since as a Mediterranean country is one of the most vulnerable areas of the Mediterranean both now and in the future. Every part of Greece is equally important for research on climate change since due to its geographic features (mainland, mountainous, coastal) Greece is predicted to suffer different impacts on every geographic section and, therefore, people who live in these areas have different perceptions of dangers due to climate change. This different perception of dangers by citizens in different geographical sections of the country might influence the views and attitudes of the citizens on climate change and their participation in activities. Therefore, as a result of such notable differences, the sample selected represented all geographic regions in the country. Also, the anxiety percentages of citizens which were found in the research may be influenced by the economic crisis which, in recent years, affects Greece.

3.2   The Questionnaire of the Research—Sample of the Research

This research is part of a wider research on the investigation of views, attitudes and knowledge of citizens on climate change. In particular, this research was carried out through the use of a wider closed questionnaire which had two sections. The first section refers to the satisfaction and trust of citizens with regard to the activities of the involved parties on climate change while the second part refers to demographic data of citizens (age, education, marital status, profession, population of areas of residence). The majority of the questions were expressed through the Liekert scale which is used for the study of behavior models or the investigation of principles regarding the behavior of a social group (Siardos 2005). They are expressed in simple words because they refer to citizens regardless of levels of education, age and geographical section. Therefore, the questions refer to all citizens and address their views, attitudes and knowledge on climate change. Before the beginning of the research a pilot research was carried out in order to investigate with accuracy the criteria regarding the satisfaction of citizens and the criteria which should be included in the questionnaire.

The sampling method applied was simple random sampling because of its simplicity and because, compared to other methods, it requires the least possible knowledge with regard to population (Damianou 1999; Kalamatianou 2000; Matis 2001). The population under "investigation" is the total of the country's households. The use of households constitutes the classic case of using as a sampling unit a group of individuals and not individuals. Indeed, the process of selecting a member (from the selected—random household) was in such a way organized so that to avoid selecting the same member all the time, i.e. always selecting the father or the wife etc. (Filias et al. 2000).

Although simple random sampling without replacement was used, the correction of finite population may be ignored because the size of the sample n is small in relation to the size of the population N (Pagano and Gauvreau 2000).

$$n = \frac{t^2 \cdot \bar{p} \cdot (1 - \bar{p})}{e^2} = \frac{1.96^2 \cdot 0.50 \cdot (1 - 0.50)}{0.025^2} \cong 1536$$

where t = the value of the distribution Student for probability $(1 - \alpha) = 95\%$ and n − 1 degrees of freedom. Since the size of pre-sampling is big (bigger than 50) the value t, regarding the desired probability, is taken from the normal distribution probability tables. In practice, for probability 95% the value is 1.96 (Matis 2001). p = estimated analogy.
e = the highest approved difference between the sampling mean and the unknown mean of the population. In order to calculate the size of the sample we had to conduct pre-sampling with the size of the sample being 50 people. So, for each variable the real population analogy was calculated. The use of the questionnaire is not restricted to the estimation of one only population variable but of more variables. Thus, we had to estimate the size of the sample for each one of the variables. The variable which produced the bigger size of the sample is "sex" since it was found from pre-sampling that women constitute 50% which means that p = 0.50. If the sizes of the samples estimated are similar and the size concerning all is within the financial possibilities of the sampling conducted, then as size of the sample the biggest is selected. This way, the variable with the highest variance is estimated with the desired accuracy, and the rest with a higher accuracy than originally determined (Matis 2001). This research was carried out in Greece from January 2014 until June 2015. Through the use of personal interviews 1536 anonymous structured questionnaires were collected.

However, the main problem in every sampling process is the creation of the sampling framework. This is so because of the need to create a voluminous catalogue of all citizens per prefecture in alphabetical order and give to every adult citizen a unique code number. As it becomes obvious the above is time-consuming, quite difficult and costly. Thus, we created a sampling framework in which each citizen has been given a number which determines the region, the prefecture, the municipality and the local community in which he/she is located. Knowing the name is necessary only in cases of selection of the particular number in our sample. We also accept that we must select individuals who are registered in the local municipality/community registers as well as reside in the particular municipality/ community. In the case we select an individual who was absent for more than two times but lives in the municipality or community in which sampling is carried out then if this person is not found then we select the person next in line (Tampakis 2000).

3.3   Statistical Evaluation of Data

As mentioned above this research is part of a wider research carried out in Greece. We will deal with the multi-theme variables "Trust of citizens towards the bodies involved with confronting climate change" (Q1), (Nine bodies were suggested for investigation which were assessed with 1 = Not at all satisfied and 5 = very much satisfied) "Activities which the citizens are prepared to engage in for confronting climate change" (Q2) (Ten bodies were suggested for investigation which were assessed with 1 = I do not wish to participate and 5 = I very much wish to participate) and "Means of communication which citizens use in order to obtain information on climate change" (ten means of communication were suggested for investigation which were assessed with 1 = not at all informed and 5 = very much informed).

For the evaluation of data the statistical program SPSS was used, the co-efficient α-Cronbach, descriptive statistics, Friedman's non-parametric criterion, factor analysis and cluster analysis.

Cronbach's α-coefficient is used to identify the internal consistency of a questionnaire, i.e. whether the data have the tendency to measure the same fact. It expresses the squared correlation between the score (observed) that a person is assigned on the given scale and the score that they would have obtained (true) if they had been asked about all issues (Siardos 1998). The α-Cronbach co-efficient for the multi-theme variables Q1, Q2 and Q3 was estimated.

Friedman's non-parametric test is used for comparing the values of three or more correlated groups of variables. The distribution of Friedman's test is the distribution X^2 with degrees of freedom (df) df = k − 1, where K is the number of groups or samples. This test classifies differently the values of variables for each criterion and calculates the average class of classification values for each variable (Freund and Wilson 2003; Ho 2006). The use of Friedman's non-parametric test produced the classification of values for the multi-theme variables Q1, Q2 and Q3.

Factor analysis is a statistical method which investigates the existence of common factors in a group of variables (Sharma 1996). More specifically, principal component analysis was used here, which is based on the spectral analysis of the variance (correlation) matrix (Murtagh and Heck 1987; Lupton 1993; Siardos 1998; Djoufras and Karlis 2001; Jolliffe 2002). The criterion used regarding the significance of the principal components is the one suggested by Guttman and Kaiser (Cattell 1978; Fragos 2004), according to which, the limit for the extraction of the suitable number of principal components is determined by the values of the characteristic roots which are equal or greater than one. Factor analysis was used in order to extract factors from the multi-theme variables Q1, Q2 and Q3.

The factor scores saved from previous factor analyses as variables were applied again in order to extract factors which would lead citizens to participate in activities against climate change. Thus, in the cluster analysis conducted we applied the final factor scores in order to investigate the possibility of existence of specific types of citizens from the factors extracted from the last factor analysis.

4 Results

From the socio-demographic characteristics of the sample we observe that most of the people asked, who constitute more than half of the sample, are adults of average age, not more than 40 years old who are at least graduates of secondary education. In particular, females are more than males and in the majority of the sample the respondents are mainly between 18–30 and 41–50 years old while those more than 51 years old are fewer. With regard to educational level, the majority of the respondents are graduates of tertiary education (45.9%) while those who are holders of graduate degrees or people who have completed technical and elementary education are significantly fewer. In addition, the majority of them are public servants (28.3%), employees of the private sector (16.4%), free lance professionals (13.2%) while those in the primary sector (farming and livestock farming) or those who only do household work are fewer. A particularly high percentage of citizens, about 15.5% of the respondents, remain unemployed. In addition, the majority of the citizens of the sample (29.2%) live in cities with more than 100,000 inhabitants. There is also a high percentage of citizens who live in provincial towns with less than 15,000 inhabitants.

4.1 Trust of Citizens Towards the Bodies Involved with Confronting Climate Change (Q1)

In order to find the most important body which citizens trust in order to participate in activities against climate change, Friedman's non-parametric test was used. The analysis showed that the majority of citizens, in order to participate in activities against climate change or their adaptation to new changes, respect more environmental groups (6.99). The reason for the citizens preferring environmental groups may be the bigger satisfaction they feel from the efforts such groups have made in confronting climate change. In addition, important bodies for the participation of citizens in activities were found to be scientists (6.85), groups of citizens (6.4) and non-governmental groups (6.04) since they were also found to have similar mean rank values. The rest of the bodies were found to have smaller mean rank values, i.e. between 5–2.5 (Fig. 1).

Next, in order to have high loadings in the factor analysis, which will create the factors, we found the value of the reliability coefficient which is 0.660 and is regarded satisfactory. The Keiser-Meyer-Olkin index has a value of 0.765. It is suggested that the KMO index should be greater than 0.80, however values over 0.60 are considered acceptable. In parallel, the Bartlett indicator was found to have a value of 4,776,919 and 36 degrees of freedom which rejects the null hypothesis. From these data two factors were extracted which explain about 55% of the total variance. Therefore, according to Table 1 which shows the loadings of the variables, it was found that the first variable (Q1_A) is comprised of the involved

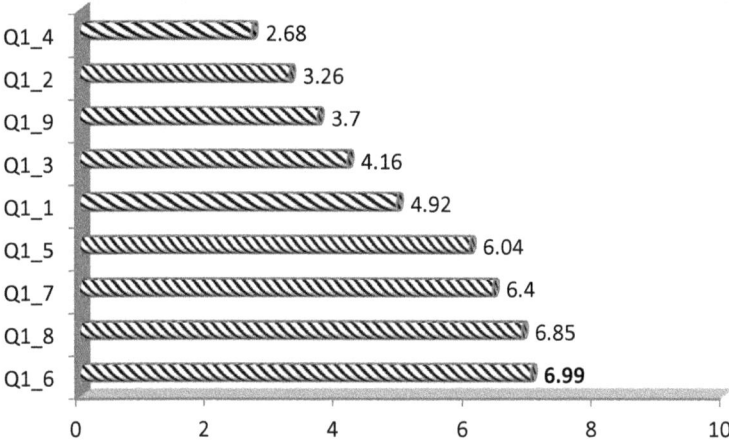

Fig. 1 Mean rank for bodies which citizens trust in order to participate in activities against climate change (Q1_1 = European Union, Q1_2 = government, Q1_3 = local government, Q1_4 = political parties, Q1_5 = non-governmental organizations, Q1_6 = environmental groups, Q1_7 = groups of citizens, Q1_8 = citizens, Q1_9 = means of mass communication; $N = 1536$ Chi-Square = 5508.89 df 8 Asymp. Sig. <0.001)

Table 1 Factor loadings before and after rotation with regard to bodies citizens trust in order to participate in activities against climate change

Variables	(Non-rotation)		(Rotated)	
	1	2	1	2
European Union	0.637	−0.225	**0.638**	0.221
Government	0.736	−0.495	**0.884**	0.072
Local government	0.704	−0.194	**0.671**	0.287
Political parties	0.618	−0.518	**0.806**	−0.018
Non-governmental organizations	0.613	0.405	0.226	**0.699**
Environmental groups	0.622	0.610	0.105	**0.866**
Groups of citizens	0.502	0.547	0.050	**0.741**
Scientists	0.553	0.409	0.176	**0.665**
Means of mass communication	0.614	−0.268	**0.648**	0.173

Bold signifies the factor that each variable belongs to

bodies such as the European Union, government, local government, political parties and means of mass communication. Therefore, these variables, since they have similar loadings they can be grouped and interpreted as "Trust towards government bodies and means of mass communication". On the contrary, in the second factor (Q1_B) high loadings also show the variables which are related to the involved bodies such as non-governmental organizations, environmental groups, groups of citizens and scientists which can be interpreted as "Trust towards non-governmental bodies".

4.2 Activities Citizens are Prepared to Engage in for Combating Climate Change (Q2)

In order to find the most important activities citizens are prepared to engage in for combating climate change, Friedman's non-parametric test was used. Thus, it was found that the most important activity the majority of citizens is prepared to engage in, with mean rank 6.69 is buying friendly products. In addition, citizens regarded as important recycling of household waste—composting (6.32) and installation of better insulation in houses (5.99). Also, similar mean rank values were found in activities such as replacing old devices with energy efficient new ones, the use of means of mass transportation, air-conditioning and the sole use of recyclable materials. The rest of the activities had even smaller mean rank values, i.e. between 5 and 3.5 (Fig. 2).

In order to apply factor analysis regarding the activities which citizens are prepared to engage in for combating climate change we initially found the value of the reliability co-efficient (0.824), the Kaiiser-Myer-Olkin (0.851 > 0.6) correlation indicator and Bartlett's indicator of sphericity with a value of 4199.331. Next, as a result of the factor analysis performed it was found that the values of the characteristic roots show that the total explained variance percentage is 50.4% which can be explained by two factorial axes.

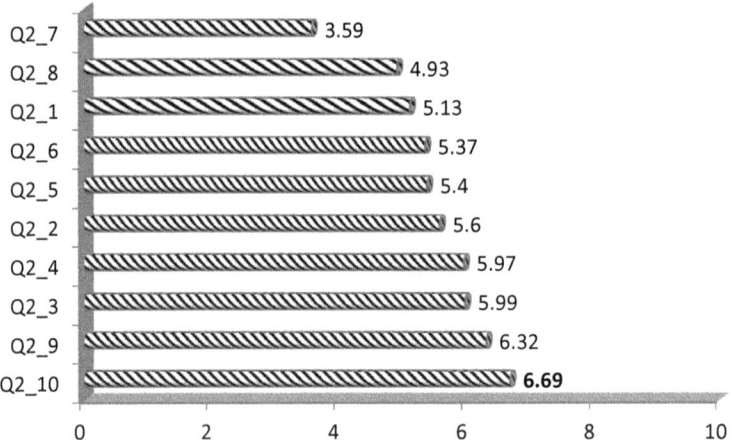

Fig. 2 Mean rank for each activity citizens are prepared to engage in for combating climate change (Q2_1 = select car which uses gas as a fuel, Q2_2 = Make better use of the means of mass transportation, Q2_3 = install better insulation at home Q2_4 = replace old devices with energy efficient new ones, Q2_5 = use less air-conditioning, Q2_6 = use only recyclable materials, Q2_7 = join and financially support environmental organizations, Q2_8 = actively support pro-environmental policies, Q2_9 = recycle household waste—composting, Q2_10 = buy environmentally friendly products; *N = 1536 Chi-Square = 1748.80 df 9 Asymp. Sig. <0.001*)

Table 2 Factor loadings before and after rotation regarding activities which citizens are prepared to engage in for combating the problem of climate change

Variables	Non-rotation		Rotated	
	1	2	1	2
Select car which uses gas as a fuel	0.491	0.258	**0.532**	0.158
Make better use of the means of mass transportation	0.555	0.185	**0.527**	0.255
Install better insulation at home	0.655	0.497	**0.816**	0.101
Replace old devices with energy efficient new ones	0.645	0.412	**0.750**	0.155
Use less air-conditioning	0.641	0.139	**0.556**	0.348
Use only recyclable materials	0.678	−0.171	0.367	**0.596**
Join and financially support environmental organizations	0.574	−0.530	0.042	**0.780**
Actively support pro-environmental policies	0.661	−0.462	0.151	**0.792**
Recycle household waste—composting	0.674	−0.165	0.368	**0.588**
Buy environmentally friendly products	0.675	−0.109	0.408	**0.549**

Bold signifies the factor that each variable belongs to

Thus, from the results of factor analysis and according to Table 2 which shows the loadings of the variables it was found that the first factor (Q2_A) is comprised of the high loadings of the first column which concern activities such installation of better insulation at home, replacement of old devices with energy efficiency new ones, using less air-conditioning, making more frequent use of the means of mass transportation and selecting a car which uses gas. Thus, these variables can be grouped and interpreted as "Activities for the reduction of pollutants and energy saving". With regard to the second factor (Q2_B), variables with bigger loadings are the ones which concern activities about active support of environmental policies, joining and financially supporting environmental organizations, using recyclable materials only, recycling household waste—composting and buying environmentally friendly products. These can be named as "Activities of voluntary service and recycling".

4.3 Means of Communication Which Citizens Use in Order to Obtain Information on Climate Change (Q3)

In order to find the means of mass communication citizens use in order to obtain information on climate change, Friedman'a non-parametric test was used. The analysis showed that citizens regard documentaries as important with mean rank 7.52. Also, regarded as important for purposes of obtaining information are specialized websites and books or magazines with mean rank values 6.13 and 5.69 respectively. Also, the rest of the means of mass communication, responsible bodies and individual members were found to have lower mean rank values, i.e. between 5.5 and 4.5 (Fig. 3).

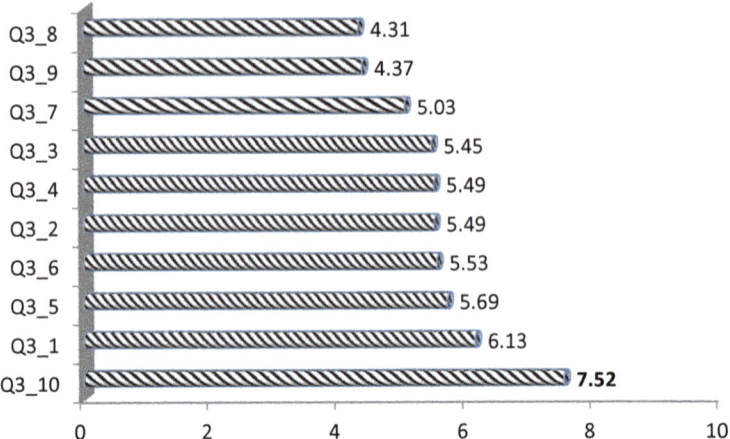

Fig. 3 Mean rank for each means of mass communication which citizens use in order to obtain information on climate change (Q3_1 = specialized websites, Q3_2 = general websites, Q3_3 = news websites, Q3_4 = television, Q3_5 = books or magazines, Q3_6 = environmental organizations, Q3_7 = family members or friends, Q3_8 = radio stations, Q3_9 = movies, Q3_10 = documentaries; *N = 1536 Chi-Square = 1702.08 df 9 Asymp. Sig. <0.001*)

In order to find the loadings of the means of mass communication which citizens use in order to obtain information on climate change initially Cronbach's alpha co-efficient was found which on the basis of the value of the characteristic root is 0.819 and is very high. Next, the rest of the necessary tests were conducted in order to apply factor analysis. In particular, the Kaiser-Myer-Olkin (KMO) correlation indicator was found, which is 0.816 > 0.80 which concerns how adequate factor analysis is and Bartlett indicator (X^2 = 4785.753 and df = 45). Next, the values of the characteristic roots found show that the total explained percentage of variance is about 63.7% and that they determine the number of dimensions (number is equal to 3). Therefore, according to the values of the loadings of the ten variables in the three main components showed in Table 3, the first main component which constitutes the first factor (Q3_A) is comprised of the loadings of the first column which concern the means of communication and personal contacts such as television, radio stations, movies and family members or friends. These variables have similar loadings and can be grouped as "Non-specialized means of mass communication and family members". In the second factor (Q3_B) bigger loadings also have the variables which concern environmental groups, specialized websites, books or magazines and documentaries which can be grouped as "Specialized means of mass communication and environmental organizations". In the third factor (Q3_C) the other two variables show bigger loadings which concern the general websites and news websites and can be grouped as "General means of mass communication".

Table 3 Factor loadings before and after rotation regarding the means of mass communication citizens use in order to obtain information on climate change

Variables	Non-rotation			Rotated		
	1	2	3	1	2	3
Specialized websites	0.545	−0.700	−0.020	−0.227	**0.708**	0.485
General websites	0.657	−0.277	0.545	0.149	0.226	**0.856**
News websites	0.692	−0.103	0.551	0.308	0.156	**0.821**
Television	0.500	0.541	0.282	**0.718**	−0.131	0.298
Books or magazines	0.688	−0.047	−0.340	0.394	**0.655**	0.085
Environmental organizations	0.656	−0.335	−0.390	0.149	**0.810**	0.124
Family members or friends	0.609	0.322	−0.242	**0.636**	0.359	0.003
Radio stations	0.654	0.395	−0.0127	**0.715**	0.282	0.094
Movies	0.578	0.393	−0.003	**0.662**	0.162	0.157
Documentaries	0.623	−0.105	−0.235	0.304	**0.581**	0.156

Bold signifies the factor that each variable belongs to

4.4 Factors Which Influence the Participation of Citizens in Activities on Climate Change

In order to investigate the views of the citizens regarding the factors which influence participation in activities on climate change, we applied again Principal Component Analysis (PCA) with varimax rotation of the factorial axes on the factors extracted from the factor analyses conducted on the above multi-theme variables. In particular, for the factors Q1A, Q1B, Q2A, Q2B, Q3A, Q3B and Q3C. Before applying factor analysis we conducted the necessary tests for the above multi-theme variable. With regard to motivating bodies the value of the Keiser-Meyer-Olkin is 0.534. In addition, Bartlett's test of sphericity ($X^2 = 643.561$, df = 21, $p < 0.001$) rejects the null hypothesis. The analysis highlighted three important factors or factorial axes which interpret 56.45% of the total variance (Table 4).

Therefore, with regard to strategies on the participation of citizens in activities which concern confronting climate change we have three factors. The first factor can be called "willingness for participation in activities of non-governmental organizations on climate change" (P1), the second factor can be called "willingness for participation in activities of governmental bodies" (P2) and the third "willingness for participation in activities for the reduction of pollutants and energy saving". In addition, we see that the citizens who show trust in activities undertaken by non-governmental bodies, use specialized means of communication in order to obtain information on climate change and prefer more participation in activities undertaken by environmental organizations. This is in contrast to citizens who show interest for governmental bodies and use non-specialized means of communication in order to be informed on climate change.

Table 4 Factor catalogue of second factor analysis which influence the willingness of citizens to participate in activities

Catalogue of principal factors—results of second factor analysis	Loadings	Eigenvalue	Variance (%)
Participation in activities of non-governmental bodies (P1)		1.506	21.516
Activities of voluntary service and recycling (Q2_B)	0.783		
Specialized means of mass communication and environmental organizations (Q3_B)	0.722		
Trust in non-governmental bodies (Q1_B)	0.579		
Willingness for participation in activities of governmental bodies (P2)		1.31	18.71
Trust in governmental bodies and means of mass communication (Q1_A)	0.779		
Non-specialized means of mass communication and family members (Q3_A)	0.665		
General means of mass communication (Q3_C)	0.467		
Participation in activities for the reduction of pollutants and energy saving (P3)		1.136	16.233
Activities for the reduction of pollutants and energy saving/mitigation policy (Q2_A)	0.863		
Total variance (%)			**56.458**
Keiser-Meyer-Olkin = 0.534. Bartlett X^2 = 643.561. df 21. $p < 0.01$			

4.5 Cluster of Citizens and Characteristics of Each Cluster

Next, and in order to investigate specific groups of citizens cluster analysis was applied from which three types of citizens were extracted: CL1, CL2 and CL3. In particular, as we see in Table 5, the citizens who belong to cluster CL1 are positively influenced by the first and second factor and negatively influenced by the third factor, the citizens who belong to the second cluster are positively influenced only by the first factor and negatively influenced by the other two factors and the citizens who belong to the third cluster CL3 are positively influenced only by the third factor.

Table 5 Average loadings in each cluster of citizens who are willing to participate in activities undertaken by governmental and non-governmental bodies against climate change

	CL_1 (27.7%)	CL_2 (41.1%)	CL_3 (31.2%)
P1	0.24689	0.19374	−0.47328
P2	1.17025	−0.58927	−0.26151
P3	−0.11339	−0.63273	0.93217

Table 6 Statistically significant differences among the three clusters of citizens who are willing to participate in activities undertaken by governmental and non-governmental bodies against climate change

Variable	Scale	CL_1 (27.7%)	CL_2 (41.1%)	CL_3 (31.2%)
Age	18–31	34.4	36.8	30.6
	31–40	27.3	27.7	32.9
	41–50	19.5	23.8	21.3
	51–60	14.6	9.4	9.8
	>60	4.2	2.4	5.4
Marital status	Unmarried	46.8	50.6	41.5
	Married	48.2	43.9	54.2
	Divorced (separated)	4.5	4.8	2.7
	Widowed	0.5	0.8	1.7
Education	Elementary school graduate	3.8	1.3	4.2
	Lower secondary school graduate	4.0	5.5	2.9
	Technical school graduate	9.9	8.7	9.8
	Upper secondary school graduate	19.5	31.1	25.2
	University graduate	35.1	27.4	30.2
	Technological Education Institute graduate	14.1	15.5	16.7
	Graduate degree	13.6	10.5	11
Profession	Public servant	30.4	25.7	29.8
	Private sector employee	15.5	19.3	14.8
	Free lance professional	12.9	12.5	14.4
	Farmer—livestock farmer	4.0	7.1	6.7
	Unemployed	15.8	14.4	16.3
	Household work	5.2	6.2	4.6
	Pensioner	8.2	4.4	8.5
	Student—pupil	8.0	10.3	5.0

In addition, in order to investigate the existence of statistical differences among the three clusters we applied the test of independence X^2 in relation to the variables which refer to age, marital status, education and profession (Table 6).

In particular, the citizens who belong to the first cluster CL1 (27.7%) which also constitutes the smallest cluster of citizens, are the youngest, married who are mainly university graduates and holders of graduate degrees. They are also public servants who are most willing to participate in activities undertaken by non-governmental bodies. The citizens who belong to the second cluster CL2 (41.1%) which is the biggest one, are usually citizens of middle age, unmarried and married, mainly graduates of upper secondary education, public servants and private sector employees who are against participating in activities against climate change and get their information from the means of mass communication. Finally, the citizens who belong to the third cluster of citizens CL3 (31.2%), are those who are from middle to older age, married and mainly university graduates who prefer, independently from the activities undertaken by governmental or non-governmental bodies in confronting climate change, activities which concern the reduction of pollutants and energy saving, i.e. mainly mitigation activities.

5 Discussion

Understanding the attitudes and views of the public regarding the trust of citizens towards the bodies involved in confronting climate change is quite significant since it constitutes the main guideline concerning the willingness of citizens to participate in activities in the future. Therefore, the aim of the research was to investigate the attitudes and views of citizens regarding trust in the activities of various bodies, trust in the means of mass communication and the activities which citizens are willing to follow in confronting climate change. Identifying the characteristics of the groups of citizens is something which will contribute positively to activities. The aim was achieved to a satisfactory degree since the three groups of citizens were extracted from the views and attitudes of a satisfactory sample of citizens of different sex, age, education, profession and area of residence. This way the restricting factor of different views and attitudes of citizens per area was dealt with satisfactorily since it created a holistic view on the activities citizens are willing to participate in for combating climate change. Previous research has shown that public information and beliefs of society (Lorenzoni and Hulme 2009) which differ from geographical area to geographical area in combination with local weather conditions (Egan and Mullin 2012) can influence the views and attitudes of the citizens. The same point is made by Howe et al. (2012), who support the view that the personal experience of citizens regarding the impact of climate change can play an important role in the formation of citizen behavior during the application of mitigation and adaptation policies which, this way, can influence international public opinion on climate change.

From the characteristics of the three groups found it seems that the majority of Greek citizens just like other research carried out in USA and Asia (Kim 2011; Tjernström and Tietenberg 2008), does not trust the government, political parties and local government in order to participate in activities against climate change. This shows that most of the citizens just like previous research carried out in Crete (Jones et al. 2012), think that there is no public body which is appropriate in confronting climate change.

These conclusions are probably produced as a result of the insecurity citizens feel with regard to the government and local government which may originate from the low quality of environmental measures undertaken by specific bodies in confronting climate change but also from financial difficulties the state faces (Cin 2012). The financial difficulties each state faces increase the insecurity of its citizens and at the same time decrease the degree of citizen participation in activities undertaken by the state. This insecurity is probably the cause which leads to the denial of citizens to participate in activities undertaken by governmental bodies.

The findings of the research show that there is a second group of citizens which are willing to participate in activities organized by governmental bodies and a third group of citizens which is linked by actions against climate change which are related to the management of pollutants such as management of carbon dioxide and energy (Midden and Huijts 2009) independently of the type of organization which organizes the activities (governmental and non-governmental). In addition, it seems that the majority of the citizens trusts more environmental groups, scientists and groups of citizens probably because they feel more secure in relation to bodies which are not directly influenced by governmental bodies and by individual or other interests.

Therefore, in order to improve the situation and increase the trust of citizens towards the activities of governmental bodies the state will initially need to organize the bodies involved in climate change (House of Commons Environmental Audit Committee 2008) and as a next step improve the activities or arguments ignoring all restricting factors. Therefore, in order to achieve this, the state could possibly organize strategic activities for which the majority of the citizens showed that they are willing to participate in them and which are mainly about buying environmentally friendly products, recycling household waste—composting and installing better insulation in houses.

However, except the above restricting factors, there is one more factor which influences the undertaking of activities by organizations and which concerns the means of mass communication (Shanahan 2007). In this research the respondents showed minimum confidence towards the activities undertaken by the means of mass communication in confronting climate change. This fact probably shows that the citizens think that the means of mass communication cannot produce objective information or undertake important activities against climate change. In particular, the means of mass communication such as television because they connect everyday reality with social experiences (O'Neill et al. 2013) can easily alter the content of issues and this way influencing the views of public opinion with wrong messages which do not correspond to reality. These messages can then be reproduced by third parties and this way strengthening the percentage of mistaken understanding by the public and thus creating a climate of misinformation. The means of mass communication themselves can transmit in two different ways the types of news, i.e. one focusing on politics and one focusing on science (Freudenburg and Muselli 2013). With regard to climate change, they have, in other words, the ability to falsify, misinterpret, distort and misinform the public in various degrees (Boykoff 2008).

Choosing the means of mass communication for informing citizens on climate change is something which influences directly the willingness of citizens to participate in activities against climate change which are undertaken by governmental bodies. The individuals who use specialized means of mass communication do not generally trust the means of mass communication in order to be informed on climate change and at the same time they are against the activities undertaken by governmental bodies. The means of mass communication run by professional publishers and journalists produce a type of new means of mass communication in a political, economic socio-economic and institutionalized landscape (Boykoff and Roberts 2007) which many times support particular political ideologies and produce directed messages altering, this way personal views on environmental issues. This was also found through the answers of citizens who mainly choose documentaries and specialized means of mass communication in order to be informed on climate change.

6 Conclusions

Climate change constitutes a serious environmental problem which has already obvious impacts, which will become worse in the future, and which requires direct action by all citizens and all bodies involved in confronting climate change. Regarding the undertaking of activities and the participation of citizens in them, more important for citizens seem to be the involved governmental bodies and particularly the state which has the power to undertake activities which will respond to international agreements on climate and will, at the same time, encourage citizens to participate in them aiming at their adaptation to the new changes. Despite all this, the insecurity felt by citizens for activities undertaken by governmental bodies in combination with the lack of trust towards the general means of communication regarding the information they obtain about climate change, these factors create a negativism and abstention from activities related to confronting the problem. Despite all this, it was found that the majority of citizens are willing, despite the economic crisis, to participate in actions which are related to confronting climate change mainly through non-governmental organizations. Therefore, the percentage of these citizens constitutes a satisfactory sample of citizens for confronting climate change and this during a period of economic crisis which can be increased in the future through environmental communication strategies.

In order to solve this problem and effectively deal with the vacuum between governmental bodies and the means of communication, it will be necessary that environmental bodies and the general means of communication undertake, organize or improve, ignoring every restricting factor, their activities with regard to recycling, the reduction of pollutants and energy saving. Finally, both the government and the means of mass communication will need to become more organized and strengthen their activities and the objectivity of the information they provide on climate change so that they increase the trust citizens feel towards them as well as encourage citizens to participate in activities on climate change. One way to achieve this would be by creating strategic activities related to voluntary service and recycling but also activities addressing the reduction of pollutants and energy saving.

References

Adger WN, Arnell NW, Tompkins EL (2005) Successful adaptation to climate change: perspectives across scales. Glob Environ Change 15(2):75–86

Bedsworth LW, Hanak E (2013) Climate policy at the local level: insights from California. Glob Environ Change 23(3):664–677

Biesbroek GR, Swart RJ, Carter TR, Cowan C, Henrichs T, Mela H, Morecroft MD, Rey D (2010) Europe adapts to climate change: comparing national adaptation strategies. Glob Environ Change 20(3):440–450

Bosetti V, Carraro C, Sgobbi A, Tavoni M (2009) Delayed action and uncertain stabilisation targets. How much will the delay cost? Clim Change 96(3):299–312

Boykoff MT (2008) Lost in translation? United States television news coverage of anthropogenic climate change, 1995–2004. Clim Change 86(1–2):1–11

Boykoff MT, Roberts JT (2007) Media coverage of climate change: current trends, strengths, weaknesses. Human Development Report 2007/2008. Available: http://hdr.undp.org/sites/default/files/boykoff_maxwell_and_roberts_j._timmons.pdf. Accessed 14 Sept 2017

Brevini B (2016) The value of environmental communication research. SAGE J 78(7):684–687

Carrington D (2011) Climate change concern tumbles in US and China. The Guardian

Cattell RB (1978) The scientific use of factor analysis in behavioral and life sciences. Plenum, New York

Cin CK (2012) Blaming the government for environmental problems: a multilevel and cross-national analysis of the relationship between trust in government and local and global environmental concerns. Environ Behav 45(8):971–992

Clar C, Prutsch A, Steurer R (2013) Barriers and guidelines for public policies on climate change adaptation: a missed opportunity of scientific knowledge-brokerage. Nat Resour Forum 37 (1):1–18

Damianou X (1999) Sampling methodology: techniques and applications, 3rd edn. Aithra, Athens

Djoufras I, Karlis D (2001) Elements of multivariate data analysis. Notes on course "Data Analysis I". Department of Business Administration University of the Aegean, Chios

Egan PJ, Mullin M (2012) Turning personal experience into political attitudes: the effect of local weather on Americans' perceptions about global warming. J Polit 74(3):796–809

EMEKA (2011) The environmental, economic and social impacts of climate change in Greece. Bank of Greece, Athens

Eriksen S, Kelly P (2007) Developing credible vulnerability indicators for climate adaptation policy assessment. Mitig Adapt Strat Glob Change 12(4):495–524

European Commission (2015). Citizen support for climate action, Special Eurobarometer 435/ Wave EB83.4. Commission: Brussels, Belgium. Available: https://ec.europa.eu/clima/sites/clima/files/support/docs/gr_climate_en.pdf

Evans L, Milfont TL, Lawrence J (2014) Considering local adaptation increases willingness to mitigate. Glob Environ Change 25:69–75

Filias V, Pappas P, Antonopoulou M, Zarnari O, Magganara I, Meimaris M, Nikolakopoulos I, Papachristou E, Perantzaki I, Sampson E, Psychogios E (2000) Introduction to social research methodology and techniques. Gutenberg Social Library, Athens

Flechsig M, Gerlinger K, Herrmann N, Klein RJT, Schneider M, Sterr H, Schellnhuber HJ (2000) Weather impacts on natural, social and economic systems (wise, ENV4-CT97-0448) German Report. PIK Rep

Ford JD, King D (2015) A framework for examining adaptation readiness. Mitig Adapt Strat Glob Change 20(4):505–526

Fragos CK (2004) Methodology of market research and data analysis with the use of the statistical package SPSS for windows. Interbooks Publications, Athens

Freudenburg WR, Muselli V (2013) Reexamining climate change debates: scientific disagreement or scientific certainty argumentation methods (SCAMs)? Am Behav Sci 57(6):777–795

Freund R, Wilson W (2003) Statistical methods. Elsevier, New York

Gough C (2008) State of the art in carbon dioxide capture and storage in the UK: an experts' review. Int J Greenhouse Gas Control 2:155–168

Hagen B (2016) Public perception of climate change: policy and communication. Routledge, London

Hanger S, Pfenninger S, Dreyfus M, Patt A (2013) Knowledge and information needs of adaptation policy-makers: a European study. Reg Environ Change 13(1):91–101

Hanusch F (2015) The role of norms in US foreign climate policy. In: Sommer B (ed) Cultural dynamics of climate change and the environment in Northern America. Brill, Leiden, pp 77–105

Ho R (2006) Handbook of univariate and multivariate data analysis and interpretation with SPSS. Chapman & Hall, Boca Raton

House of Commons Environmental Audit Committee (2008) Climate change and local, regional and devolved government, HC225. The Stationery Office, London

Howe PD, Markowitz EM, Lee TM, Ko C-Y, Leiserowitz A (2012) Global perceptions of local temperature change. Nat Clim Change 3:352–356

Hulme M (2010) Problems with making and governing global kinds of knowledge. Glob Environ Change 20(4):558–564

IPCC (2013) Summary for Policymakers. In: Stocker TF, Qin D, Plattner G-K, Tignor M, Allen SK, Boschung J, Nauels A, Xia Y, Bex V, Midgley PM (eds) Climate change 2013: the physical science basis. contribution of working group i to the fifth assessment report of the intergovernmental panel on climate change. Cambridge University Press, Cambridge, United Kingdom and New York, NY, USA

Jakob M, Luderer G, Steckel J, Tavoni M, Monjon S (2012) Time to act now? Assessing the costs of delaying climate measures and benefits of early action. Clim Change 114(1):79–99

Jayawardena AW (2015) Climate change: is it the cause or the effect? KSCE J Civil Eng 19(2):359–365

Jolliffe I (2002) Principal component analysis. Springer, Berlin

Jones N, Clark J, Tripidaki G (2012) Social risk assessment and social capital: a significant parameter for the formation of climate change policies. Soc Sci J 49(1):33–41

Kalamatianou AG (2000) Social statistics, methods of one—dimensional analysis. The Economic, Athens

Kim SY (2011) Public perceptions of climate change and support for climate policies in Asia: evidence from recent polls. J Asian Stud 70(2):319–331

Lorenzoni I, Hulme M (2009) Believing is seeing: Laypeople's views of future socio-economic and climate change in England and in Italy. Public Underst Sci 18(4):383–400

Lorenzoni I, Pidgeon NF (2006) Public views on climate change: European and USA perspectives. Clim Change 77(1–2):73–95

Lubell M, Zahran S, Vedlitz A (2007) Collective action and citizen responses to global warming. Polit Behav 29(3):391–413

Lupton R (1993) Statistics in theory and practice. Princeton University Press, Princeton

Matis K (2001) Forest sampling, 2nd edn. Democritus University of Thrace, Xanthi

McCright AM, Dunlap RE, Marquart-Pyatt ST (2016) Political ideology and views about climate change in the European Union. Environ Polit 25(2):338–358

Midden CJH, Huijts NMA (2009) The role of trust in the affective evaluation of novel risks: the case of CO_2 storage. Risk Anal 29(5):743–751

Murtagh F, Heck A (1987) Multivariate data analysis. D. Reidel Publishing Company, Holland

O'Neill SJ, Boykoff M, Niemeyer S, Day SA (2013) On the use of imagery for climate change engagement. Glob Environ Change 23(2):413–421

Obani PC, Gupta J (2016) The impact of economic recession on climate change: eight trends. Clim Dev 8(3):211–223

Pagano M, Gauvreau K (2000) Principles of biostatistics. Ellin, Athens

Pearce G, Cooper S (2011) Sub-national responses to climate change in England: evidence from local area agreements. Local Gov Stud 37(2):199–217

Persson J, Sahlin NE, Wallin A (2015) Climate change, values, and the cultural cognition thesis. Environ Sci Policy 52:1–5

Pielke RA Jr (2005) Attribution of disaster losses. Science 310(5754):1615–1616

Prins G, Galiana I, Green C, Grundmann R, Hulme M, Korhola A, Laird F, Nordhaus T, Pielke Jr, Rayner S, Shellenberger M, Stehr N, Hiroyuki T (2010) The Hartwell paper: a new direction for climate policy after the crash of 2009. Available: http://eprints.lse.ac.uk/27939. Accessed 20 Sept 2017

Shanahan M (2007) Talking about a revolution: climate change and the media. International Institute for Environment and Development (IIED). Available: http://www.iied.org/pubs/pdfs/17029IIED.pdf. Accessed 13 Feb 2013

Sharma S (1996) Applied multivariate techniques. Wiley, New York

Siardos G (1998) Multivariate statistical analysis methods. Part one, investigating relationships between variables. Ziti Publications, Thessaloniki

Siardos G (2005) Methodology of sociological research, 2nd edn. Enhanced and completed. Ziti Publications, Thessaloniki

Stern DI, Kaufmann RK (2014) Anthropogenic and natural causes of climate change. Clim Change 122(1–2):257–269

Tampakis S (2000) Forest fires in Greece from a forest policy point of view. Ph.D. thesis. Aristotle University of Thessalloniki

Terwel BW, Daamen DDL (2012) Initial public reactions to carbon capture and storage (CCS): differentiating general and local views. Clim Policy 12(3):288–300

Tjernström E, Tietenberg T (2008) Do differences in attitudes explain differences in national climate change policies? Ecol Econ 65(2):315–324

UNFCCC (2009) The Copenhagen Accord 2/CP.15.FCCC/CP/2009/11/Add.1, United Nations Framework convention on climate change (UNFCCC), Bonn, Germany. Available: http://unfccc.int/resource/docs/2009/cop15/eng/11a01.pdf. Accessed 20 Sept 2017

WCED (1987) Our common future. World Commission on environment and development. Oxford University Press

YPEKA (2016) National strategy for adaptation to climate change. Ministry of Environment and Energy, pp 1–115. Available: http://www.ypeka.gr/LinkClick.aspx?fileticket=crbjkiIcLlA%3D&tabid=303&language=el-GR. Accessed 22 Sept 2017

Zerva A, Tsantopoulos G, Grigoroudis E, Arabatzis G (2018) Perceived citizens' satisfaction with climate change stakeholders using a multicriteria decision analysis approach. Environ Sci Policy 82:60–70

Aikaterini Zerva has received an undergraduate degree in Forestry and Management of the Environment and Natural Resources from Democritus University of Thrace (2010), a Master of Science in Sustainable Management of Protected Areas from the University of Ioannina (in cooperation with the University of Patras and Aristotle University of Thessaloniki) (2012), a Master of Education and Disability from Universita degli Studi di Roma Tor Vergata (2015) and a Philosophy Doctorate also from Democritus University of Thrace (2018). She teaches in secondary education.

Georgios Tsantopoulos has received an undergraduate degree in Forestry and Natural Environment from the Aristotle University of Thessaloniki (1995) and a Philosophy Doctorate also from the Aristotle University of Thessaloniki (2000). He is Associate Professor in the Department of Forestry and Management of the Environment and Natural Resources at the Democritus University of Thrace.

Evangelos Manolas was born in Naxos, Greece, in 1961. He has received a Bachelor of Arts in Sociology from the University of Essex (1983), a Master of Arts in International Relations from the University of Kent at Canterbury (1985) and a Philosophy Doctorate from the University of Aberdeen (1989). He is Associate Professor in the Department of Forestry and Management of the Environment and Natural Resources at the Democritus University of Thrace.

Stilianos Tampakis was born in Thessaloniki, Greece, in 1967. He has received an undergraduate degree in Forestry and Natural Environment from the Aristotle University of Thessaloniki (1990) and a Philosophy Doctorate also from the Aristotle University of Thessaloniki (2000). He is Associate Professor of Forest Policy in the Department of Forestry and Management of the Environment and Natural Resources at the Democritus University of Thrace.

Treasuring Evaporation: The Radical Challenge of a Museum of Water

Amy Sharrocks

Abstract Museum of Water invites people to gather water: any water in any bottle, encouraging people to look closely at this extraordinary substance and detail the way it impacts on their life. It is both live artwork and museum, a chorus of voices from all ages, races and social backgrounds, influenced by each donor, changing shape with each new gift and comment. This is a museum that remembers your words; it moves beyond treating people as visitors or audience, instead making everyone donor, curator, protagonist and custodian. It seeks to engender a new relationship between people and the world, a responsibility for care, fostering the role of custodian in each of us. Museum of Water began in 2013 and has travelled to 50 sites worldwide, been visited by over 60,000 people, and was nominated for European Museum of the Year 2016. It has used its work to curate wide, cross-cultural programmes to explore questions of water, migration, fear, climate change and urbanisation through science, literature, ecology and anthropology, music and play. Water is the most important substance for human life, and is the substance at the front line of Climate Change. Questions of access, ownership and care will define the coming century: this Museum and its programmes help to equip people to play an active role in the situations and debates to come. This paper details the work of this unusual museum, exploring its spectacular public engagement and the radical challenge it presents to previous systems of collecting and to traditional processes and economies, how it treasures a substance and experience that we cannot hold onto, the process of evaporation itself. Museum of Water is an act of witness, providing a platform for different voices, and an instigating force for future care. This paper will look at the different systems and ways of listening it promotes, supporting responsibility and bravery for the coming century: how to look more closely and not look away.

A. Sharrocks (✉)
Museum of Water, 28 Commercial Street, London E1 6AB, UK
e-mail: amysharrocks@mac.com
URL: http://www.museumofwater.co.uk

© Springer Nature Switzerland AG 2019
W. Leal Filho et al. (eds.), *Addressing the Challenges in Communicating Climate Change Across Various Audiences*, Climate Change Management,
https://doi.org/10.1007/978-3-319-98294-6_29

481

1 Introduction

Museum of Water is a live artwork and museum that seeks to prompt people worldwide, individually and collectively, into deeper consideration of water and the world around us.

Museum of Water invites people to gather water: any water in any bottle, encouraging people to look closely and detail the way water impacts on their life. It is a collection of water wholly gifted by the public that treasures and honours each gift as it is conceived, gathered and donated.

The Museum began in London in 2013, and since travelled across the UK, into Europe and to Western Australia. The Museum has been commissioned over the last 4 years at 50 different sites worldwide, and has used these opportunities to curate wide, cross-discipline cultural programmes to explore questions of water: climate change, migration, urbanisation, fear, thirst, flood and drought, through science, literature, ecology, philosophy, anthropology, music and play.

This Museum develops entirely site specifically, in appreciation of the different cultural, geographic, social, aesthetic and philosophical histories each country enjoys. Through its process of careful listening, its recognition and enjoyment of delicate detail and difference, the Museum re-connects people with their environment, incites personal responsibility and cultivates a new economy of care.

This paper will detail the work of this unusual museum, exploring its spectacular public engagement and the radical challenge it presents to previous systems of collecting and to traditional processes and economies. We explore how important it is in a century of hoarding—where even the internet is now an Internet of Things—to treasure a substance and experience that we cannot hold onto, the process of evaporation itself.

Museum of Water is an instigating force for future care. As we encounter a warmer atmosphere and the problems presented by climate change heat up relationships and pressurise resources, this Museum promotes careful listening, initiates actions, supports responsibility and the bravery to look more closely and not look away.

2 "It's Public Commons!"[1]

> Water is the force that flows through birth, love, loss, home, holidays, danger, adventure and grief. It is life, death and all the washing up in between. Which is what you want from a museum – the whole of human life. Bottled.[2]

---

[1]Visitor's remark to Museum of Water Rotterdam.
[2]Frizzell (2014).

Museum of Water is a collection of water, any water, in any bottle. It asks you, if you could keep one water, what water would it be? It is an invitation to each one of us to spend time with water, to really consider it, and think for ourselves exactly what we treasure most about it, which will be different for everyone.

The Museum is a completely public collection, made in collaboration with everyone who brings water. This open offer matters: we don't change the donation in any way, we don't try to 'improve it' or re-shape it, channel it or re-bottle your gift in any way. We treasure it, and we learn from it. You are the expert—we are the gleaners.

This sets the framework for an utterly new relationship between a museum and the public. We safeguard and care, but you are the creative force behind each bottle and the responsibility rests with you for what is cared for and what gets remembered. We offer a challenge to the usual authority of museums, which in the past have been built on bequests from mainly rich, white, western men. The experience of this Museum is radically different: it is "not a repository of national patrimony".[3] There is no hierarchy of knowledge here, no imposing atmosphere. It can often be chatty and loud, and because of the many places we have been, there is a huge diversity of experience cherished.

This full collaboration is an unusual offer from any Museum—it's up to you what goes into it. Usually when you visit a museum you are a visitor, just that. Here you are protagonist, curator, artist, donor, co-creator. Here your journey makes a difference to the museum, here your words will be remembered, your presence counts: this begins a very different relationship. This Museum wholly relies on the public, so you influence and shape the collection, which means that it matters whether you come or not. I don't know other museums that do that.

Museum of Water is a declaration of inter-dependence.

3 Mosaic of the Universe

Museum of Water is a subtle and profound work, taking us on a journey that starts with sipping from a glass of water and ends with pondering our relationship with our planet.[4]

Museum of Water is an act of witness to our lives lived with water.

People are often aware of the metaphor they are making, and claim the connection themselves. For them this is not a new thought—they may make daily appreciations of water, and sometimes they even have their own collection. The invitation to bring water may be a new, safe home for a bottle cherished for many years, or held onto for one reason or another. For them, this Museum is a welcome return, where their intimate perceptions can find common understanding.

---

[3]Al-Qattan (2016).
[4]Perkovic (2017).

I have lived my life by the tides

Alfie, #267 in Museum of Water Collection.[5]

The alcohol I poured down into my body mirrored the waves that pounded on the shore

Amy Liptrot, # 395 in Museum of Water Collection (see Footnote 5).

The Pendle – that's ME

Anita Borrows, # 297 in Museum of Water Collection (see Footnote 5).

The collection traces our commonality and charts the democratisation of public history. The Museum and all its programmes are free to visit and it questions our economies, suggests ways to share. It questions our real water rates—how expensive is our current way of life?

Museum of Water has been called a 'mosaic of the universe'.[6] It charts the impact of water through our days, the performance of everyday life. It is the specific and particular, each one a galaxy of meaning and intent in its own right, that builds up to a spectacular and astonishing whole. The invitation of the Museum implicates each one of us in this universe, pinpoints our choice, reminds us of our position on earth and helps us to ponder our relationship with the planet. The global perspective is clear and urgent throughout: **What will you keep?** There is a cosmic perspective too, that comes from understanding that the water in the collection is many different, particular waters, and part of one whole. We are part of the timeline and geography of this earth and this planet system, which is both reassuring and dizzying. It serves to remind us both of our tiny scale in this universe, and also of our impact on our small part of it, how we can affect the shape and nature of our planet.

4   Beginnings

The Museum started on a street corner in Soho, part of the bi-centennial celebrations for John Snow, who in 1854 painstakingly plotted the journey of cholera from water pumps right into our stomachs, saving hundreds of thousands of lives. John Snow was a physician, a scientific detective, walking the streets of London looking for something in the water. His work reminds us to look carefully, to travel for our discoveries and to question people. His expeditions are echoed by the new journeys of people who have gathered water for the Museum, adding to the story of water here. This Museum is a mapping of our feeling for water—a cartography of sorts— of our experience of water in this century.

From the start this Museum has been a collaboration between art and science, time and geography, with people, journeying and conversation at its centre. At each

---

[5]Collection can be viewed at www.museumofwater.co.uk; not all audio files available for public use.

[6]A visitor to Museum of Water in Moseley Baths, Birmingham.

site we develop a rigorous cultural programme for the serious investigation of water. We know that this Museum operates in context with the world and the questions facing us, so we have partnered with anthropologists from Denmark, Oxford University and Lancaster, with water engineers and scientists from the UK, Holland and Australia, geographers, writers, poets, artists and ecologists from across the world, a catholic priest and the music ethnographer from Pitt Rivers Museum. We have built each iteration to curate wide, cross-discipline cultural programmes to explore questions of water, migration, fear, climate change and urbanisation through science, literature, ecology and anthropology, music and play.

We have explored the sound of water, the language of water, water law and the politics of water while travelling through the UK floods of 2013 and 2015, the saturated land of the Netherlands to the parched and salted plains of Australia, as Cape Town heads towards Day Zero.

5 Specific and Particular

In each country, Museum of Water is a two-year live artwork gathering stories and water from people across each land, and has grown to become a major collection now. From 2013-16 we travelled across the United Kingdom from Glasgow to Shoreham, from Brightlingsea to Swansea. We have spent 18 months in the Netherlands collecting water, and 2016–18 the Museum has been making its way around Western Australia.

This is an exhibition of moments from everyday life—the ritual of our mornings, the first summer rain, the glass of water by our bed, a secret swimming spot. Different collections can tell of different experiences, of joyous adventures, family and loved ones, of flood and leaks and rain, of loss and disaster, climate change, polar expeditions, desalination, irrigation and thirst.

Being site-specific means that the Museum engages each time with the local and global questions of each space. In Cumbria it was extraordinary to be there in 2016, just after terrible floods, just as people were putting their houses and infrastructure back into some kind of order. There was a palpable feeling of trauma and recovery. But we were in an 18th century boathouse on the banks of Windermere, the deepest lake in Britain, an extraordinary landscape carved by glacier, so the experience of shock and revival was balanced with different histories and pre-histories.

Each new collection grows organically in full collaboration with the each space. In Rotterdam, much of the city lies below the water table. The Netherlands has used water as an extremely successful flood defence system for centuries. This is the Hub of Europe; the harbour is so large it has its own mayor. The area of South Rotterdam where we were based is the most diverse neighbourhood in the whole of The Netherlands, with 143 different languages spoken. Here we dealt with the fear of water, urban design with water, questions of Water Law and the ancient practice of poldering, an ancient system of negotiation which evolved from the constant collaboration needed to support the polders, the complex system of water dykes that allowed people to reclaim land from the sea. Poldering comes to mean the constant social negotiation towards co-operation—to negotiate with other's interests taken into account—which only ends when all parties are satisfied.[7]

This Museum marks an end to empirical models that insist on hierarchy. This is not one precious collection that is shipped around the world, but begins with nothing in each new country, and appreciates what is there.

Museum of Water's WA edition began life traveling across the state in a mobile trailer, hitched to the back of a car. In Australia our whole notions of museum-hood and sense of self have had to be questioned and re-understood. European cabinetry suddenly seemed so… inappropriate. Our cabinets were out of date and place: they speak strongly of another geography, and were developed in social conditions that were very different, where cities and people were crammed together, jostled up against each other in small spaces. In many ways they were specifically un-Australian: they talk of joinery and small spaces, of nooks and crannies. Of hoarding too, and keeping safe, a conspicuous display of wealth, of Sunday bests, and carving out a room of one's own. In Western Australia we faced the vast open plains of an extraordinarily different landscape, the escarpments and plateaus of the Swann Coastal plain; a very different light, space, and a very different connection to country. Enclosing suddenly felt wrong—we had to be open to exposure. We had to change to meet the light and appreciate the vibrations of this country. We had to re-understand ourselves in its context. Museum of Water in WA is an exploration of resilience and an encounter with vulnerability.

6   Tool for Survival

> The authenticity of a thing is the essence of all that is transmissible from its beginning, ranging from its substantive duration to its testimony to the history which it has experienced.[8]

Museum of Water is a wondrous collection, funny, warm, full of joy and delight, adventurous ideas and philosophical concepts. It can be heartbreakingly sad one minute, and have you laughing in the aisles the next.

We have magic potions from 8 year old boys, tears of joy from a man on hearing he had been accepted to be ordained a priest, a bucketful of drips from a leaking roof, sweat from the walls of a disco after a long night of dancing and a memory of water from a dried up qanat in Iran.

---

[7]For further examination of poldering, please see http://one-europe.net/to-polder-or-not-to-polder-that-s-the-question.

[8]Benjamin (1969).

We have bath water and bad dream water, toothpaste spit carefully gathered from a whole family, condensation wiped from the windows of a damp student flat, or dripped from the instruments of a 5-piece brass band. We have water that is full of the hopes and dreams and ashes of our loved ones.

We have floodwater from the 2013 UK floods, light summer rain and a torrent of election day rain, we have water from England's largest spring tide for 30 years in Morecambe Bay, a huge wave from Half Moon Bay in Antigua, and a muddy puddle from an 8 year old girl in Birmingham.

We have the first cup of tea, the last bath of the day, disappearing wetlands, a child's first swimming lesson, icicle, hail, sweat, snowballs from a 3-year-old boy, the evaporation of grief, water under the bridge, water from the point of no return, and the breaking waters of a woman, and the water her baby was born into.

We have been told how the changing sandbeds of the Wyre estuary affected people and their professions. We know the best place to fish for mackerel in Brightlingsea, the tidal drop in Swansea bay, the watering habits of Glasgow botanical gardens, how Rotterdam Aquarium gets its water and what the hierarchy of animals is—*who gets it first*—we know the bathing customs of Victorian England, and how the ironwork infrastructure of incredible Victorian swimming pools echoes across Britain from Leamington Spa to Govan Hill Baths to Reading and London.

We have heard how people yearn for the sea in Birmingham.

We have been given water from refugees around the world and a Mediterranean now filled with bodies in numbers that beggar belief. Who notices? Who can bear to look at the repeated images of boats crammed with bodies? How can we affect change?

7 Privilege

Museum of Water is itinerant, which is one of our greatest qualities. We travel because we understand that not everyone else can, and we wouldn't want to miss an opportunity to hear a different voice. We want to widen the discussion, so we have made a huge effort to get around. There are barriers to stepping inside museums that are not just ones of access and ability. You need a certain sense of entitlement to cross a threshold sometimes, and some people just don't feel they can or want to do this. So we go to street corners, festivals, busy market squares, riverbanks and canal towpaths in remote villages, riverbanks, schools, palaces, an old people's home, botanical gardens, boats, markets and shopping malls.

Because of the many places we have been, there is a huge diversity of experience cherished here. There is no one dominating, curator voice here, but the shared heritage of many journeys in this Museum, present voices for future generations.

The spaces have looked very different, and each has had a different resonance. So in Birmingham, that most landlocked of British cities, we were in a derelict public slipper baths. Suddenly we began to chart the changing habits of public bathing in just one generation, as people came in and remembered their weekly family outing for a bath. Somehow the bottles people gave us here reflected a yearning for water, a sense of missing water: a disused canal, a closing swimming pool, a muddy puddle, water from a dehumidifier, sighs and smiles … whereas Brightlingsea for instance was full of beach huts, fishing, ponds and canoes, and Shoreham was sailing, seas, beaches and rivers, life by the tides.

For years we have worked to trace the impact of water across the British Isles, out into The Netherlands, then onto Western Australia, always on the trail of a whisper of water words made public: a hesitation, an opening of experience, a loss, a precious moment, tracing something in the water.

As we crossed Europe between Rotterdam and London, passing increasingly guarded checkpoints, with walls and barbed wire and lookout posts more reminiscent of WW2, the privilege of travel, the exclusivity and encumbrance of borders, became correspondingly apparent. Is our plan to build higher and higher walls? To fortify ourselves against *the other*—whether that is climate change or refugees—to live in gated communities, constantly privatising more and more land, paying ever higher taxes to secure our borders?

There has to be an alternative.

8   Careful Listening

We are a different kind of museum, which gives primacy to each person's voice. The dual role of Museum of Water—both live artwork and museum—gives it extraordinary creative force. Museum of Water custodians occupy the space, not invigilating, but attending to people and caring for the Museum collection daily: a Ministry of Water. Visitors have the chance every day to donate water to the growing collection. The accessioning process involves a conversation with a Museum of Water Custodian, either one-on-one or in small groups. Each conversation is audio recorded if the visitor agrees, forming a sonic archive of the collection. The accessioning is semi-private: a view onto it, without the conversation being overheard. The bottle is given a unique number and donors write a label to display with their water, so that the handwriting of the donor is displayed along with their water. Custodians look after the water, explore the bottles with you and remember the phrases of people who come. In this way it is a Museum of Words as much as Water, and is different every day: you can add to the collection even if you didn't know to bring water, and your words might get woven into a bottle's history. Because your words matter. The Museum is a lively, chatty space where people share stories, encouraging you to have your say. There are many ways to get involved. You can go on a journey to get our own water, leave your writing in *The Water We Would Have Brought,* or come and share your thoughts at our Water Bar: it's what you bring to it that matters.

Custodians have been integral to the experience of Museum of Water. We welcome you, we explore the bottles with you, we notice and listen. We remember your words, the phrases you use. Even if you haven't brought water to the Museum your comments might get woven into a bottle's history, or your words might spark a connection with a bottle which could take us all somewhere new. *No visit to the Museum is ever alike.*

The subtle care of the custodians, their delicate interventions, their warm welcome, have played a large role in the thoughtful, moving, extended conversations that the Museum has housed over the years. The accessioning process has always been open-ended: it takes as long as it needs to talk a bottle through, to explore together what is in it. The offer of this Museum is one of time and care.

The Museum gives primacy to each person's voice.

Here, the institution does the listening: a Museum of Now, that captures how we feel right here, right now, our language and our ideas, ducking and diving, bobbing and weaving, not stiff not dry, but fluid.

9 Australia

the water hole is the epicentre of social cohesion and harmony

the community of water[9]

lawn – what you got here is a very greedy event.[10]

Working in different sites across Western Australia for 18 months, we have explored the experience of being Australian, in all its multi-faceted, many cultured ways, the extraordinary understandings of traditional owners, as well as the experiences of more recent times: the hand harvesting, thirst, the European movement to Australia (the change of bathing habits, the loss of family, the severance of ancestral ties). The importance of the Australian generosity of water (when camping, you take enough for you and someone else) as well as the constant drought, the vulnerability of life when over 80% of your drinking water comes from de-salination.

Museum of Water is concerned with the stewardship of memory, and strongly correlates with aboriginal ideals of Custodianship. There have been series of conversations and talks around the repetition of stories, circuits of stories, song lines and dreamtime stories, about reading a landscape, our feeling for country, how we educate our children about where they are going and where they come from through oral history. We have explored the different weights given to voices or scripts on a piece of paper. How do we know things, and what does it feel like to put someone else's words in your mouth? How do we speak a landscape? The refrain of Noongar voices repeating the land and our time here has been paramount. We are joining

---

[9]Walley (2017).
[10]Walley (2017).

with the original voices to repeat the story of this land and our time here, and the chorus has to build up. It is a strong and ancient answer to the extractive economy of the West.

In WA, donors know that the environment sustains you:

I used to pretend that the ocean was my boyfriend

Mei Swan Lim, # 777 in Museum of Water Collection.[11]

The ocean helped me befriend myself

Hanny Handoyo, #800 in Museum of Water Collection (see Footnote 11). What do we do when we pass history down through our mouth? Why do we travel?

We each have a narrow corridor of knowledge, and it is our duty to go outside of it and bring different knowledge in. This ancestral knowledge, this connection to country, is our public commons. This Museum makes custodians of us all.

10   *A Museum of Whaaat?* The Radical Challenge of Water

In museums, water is usually the enemy, guarded against at all costs. Museum curators shudder and conservation experts tremble at the idea of bringing water into a collection.

We make no effort to conserve the water.

Every bottle is in dynamic process with the world. Because where does it go after all—in you? In me? Each bottle's borders can be crossed, are themselves in constant process of evaporation. This is the constant process of give and take in the Museum: every act of gathering and giving is an act of care, returned with kindness and attention in careful reciprocity. This is a fluid economy, an economy of care, and we all gain from it. Movement and plurality are at the heart of Museum of Water. We deal in permanence and impermanence. We question how you can give something that is already gone, that can't be held onto?

There is almost an absurdity in the effort of the Museum, a Sisyphan attempt to hold onto something that is always slipping away from you, to shore up (a flood) against a flood. There are parallels here with the gargantuan efforts to stem the tide of climate change, that can feel overwhelming.

Very early on in the Museum, it became clear we needed a space for the things that aren't there, that for one reason or another couldn't make it into the collection. *The Water We Would Have Brought* has always held a cherished place in the Museum, and has grown and developed over the years. In this separate collection there is water that doesn't exist, or was unreachable for one reason or another: water from the Sea of Tranquillity on the moon, or '*the water I almost drowned in off the*

---

[11]Bottle can be viewed at www.museumofwater.co.uk; audio not yet available publically.

coast of Dorset when I was 5', or *'the drip from Captain Zambra's nose'*. Or in that common drama, as one woman wrote in her poem:

The Water I Would Have Brought

Is in my Other Bag.[12]

Like the main collection, this has also become a marvelous space for creative acts. People have gone to great lengths to shape it: origami swans and boats, elaborate drawings, graphs and pages and pages of words on water. The huge and wild space of the conditional—all the glorious hypothetical possibility of that *'would'*—has left a Derridan differance of possibilities of meaning that people have filled with delight, joy, hilarity and aching sadness.

The Museum deals in things which are disappearing, and there is a huge effort to trace things that were only ever almost there to begin with: the Museum deals entirely in something that is evaporating before our eyes, is impossible to hold in your hands, and whose narrative is constantly changing.

Yes, the water in collection is evaporating and we make no effort to stop it.

Masai water, tears of joy, Apocalypse Water and precious Australian Wetlands, all going.

How can we bear that?

This is a compelling force behind Museum of Water. This collection moves and changes, the contents of the bottles in constant and elaborate motion. There is no fixity. It reminds us that we are all evaporating. Those small drops are gone from their bottles, they are now a memory of water, a few whispered words in your ear. They are a reminder of the world's constant motion and bring our focus back to the reality of a drier future, parched and thirsty. But where does the water go? In you? In me? The Museum is a reminder both of how precious things are, how we have to prize and look after our resources, but also that our boundaries are more fluid than we imagine.

Displacement is a central fact of life. This Museum connects the seas and oceans around us, connects Syria to Camden Lock or the condensation of a Falmouth flat. It's the same amount of water, the same water itself, that has existed since the beginning of the world. The world is composite, not exclusive, and this collection helps to remind us of that truth. This water is slipping through our fingers, in constant process with our bodies, the constructs of our days and our built environment. This Museum connects us to the movement of air in a room, and to the ups and downs in our life: we are all in a process of disappearance.

Sarah Maguire's poem *From Dublin to Ramallah* rages at the borders in this world and imagines the water that flows through the world connecting us all and breaking down all barriers of geography and understanding,

I ask for a liquid dissolution:

let borders dissolve, let words dissolve,

---

[12]Unattributed, part of Museum of Water UK collection of *The Water We Would Have Brought.*

let English absorb the fluency of Arabic, with ease,

let us speak in wet tongues.

Look, the Liffey is full of itself. So I post it

to Ramallah, to meet up with the Jordan,

As the Irish Sea swells into the Mediterranean,

letting the Liffey

dive down beneath bedrock

swelling the limestone aquifer from Hebron to Jenin.[13]

Museum of Water is Maguire's wild fantasy come true, and more. Here the Liffey does swell next to the Jordan, the Clyde, the Rhine, the Maas. Here we join Mecca with Lourdes, with Knock, with Walsingham and the Vatican. Here, the Thames flows to the Mediterranean—*and beyond*—to the Ganges, to Canadian glaciers, to poisoned rivers in Brazil and further on to the Antarctic, to the last ice age, and still further on and back in time, to the last warm period in the earth's history, 129,000 years ago. Maguire notices the similarities of words and water, the fluency of both, their ability to dissolve borders and difference. Here is a real sharing, a fervour to fight the frenzy of religious wars with a final coming together, this liquid dissolution, absorption. (*Water is the currency of England, all that rain. It's a cliché of life here, damp, soggy, drenched.*)

It is surprising we have not had a Museum of Water before. We have a Maritime Museum, in honour of men's exploits on the water, we have a Water & Steam Museum celebrating how we can use water to power our engines, but we haven't had a Museum to consider the thing itself. It is surprising, since we use water metaphors every day to describe our thoughts and feelings, and a lack or over-abundance consistently threatens our existence.

On islands we are used to seeing water as both barrier and conveyor. But our seas and oceans have taken on whole new meanings lately. The Museum began in a year of devastating floods in the UK, and is now operating at a time when our seas and oceans are the sites of unimaginable horrors. A water crisis in Syria was one of the beginnings for a war which brought a flood of refugees and migrants to Europe, stretching borders to breaking point and threatening the whole concept of a European Union. We were well aware, coming to Australia, of their nil by water laws in regard to refugees. Sharing and diverting water are political acts that affect the world. Whoever controls the water controls the power. We all have much to learn from the crucial Dutch skills of poldering.

---

[13]Maguire (2015).

11 Museum of Evanescence

There is a place for loss in the collection, for absent and disappearing things, for things that aren't there. Honouring and treasuring the things that made into the collection, but also keeping space things for that didn't make it: not the thing itself. *The Water We Would Have Brought* is an answer to the fetishism of objects. If the water itself is occasionally intangible (the dry qanat, the water under the bridge, point of no return, the last drop, John Doust's smear) *The Water We Would Have Brought* is manifestly virtual. It stops the collection prizing only objects, and is a delicate reminder that this is not about the end result, the end of the journey, some kind of solution, conclusion or thing: this is about the moments in between, the ungraspable, ineffable, elusive moments and meaning beyond our grasp. It is very clear that this stretches out to all the meanings that cannot be included in a bottle or gathered on a page. This work is about the attempt of trying, of sitting with unknowns and listening carefully. This offers everyone another chance to define and refine a meaning, in a different media. It allows more people to give, more people to make their mark.

Eventually, the collection in this Museum will vanish.

The water will completely evaporate from the bottles, and the collection will become a collection of bottles that used to hold water. All that will remain will be the recordings of the donors' voices, the whisper of water words that have drawn the Museum across the world. Oh, the power of words to make all this happen! The whole Museum created from a breath of words and a few molecules of water. The carefully edited recordings of 1000 voices will stand as witness to the Museum and the water, testament to people's understandings and care.

Which begs the question: what exactly does our collection consist of? Is it the water or the bottles? The voices, the words, the exchange or the conversations? The original journeys of the donors? Is it "the quivering tension of the in-between",[14] as Astrida Neimanis describes water, and our interconnectness? And is it the transformative event in each donor, custodian and visitor, the event that happens inside each one of us, and is a private experience within the confines of our own bodies, even if it is then taken out into the world to effect change?

12 The Water Bar

At each site there is a Water Bar, for the sharing of a free glass of our finest, a moment for people to pause in their busy days for refreshment and respite, in celebration of our access to fresh water here.

In Rotterdam, in the very multicultural area around the Afrikaanderplein, the fresh mint grown by the local cooperative group RotterdamseMunt flavoured the

---

[14]Neimanis (2012).

water we offered to visitors. In Australia, marri blossoms and bankshia flowers, peppermint leaves and lemon myrtle fill the water containers, a bright reminder of the spectacular natural world of Australia. This is a world that the Aboriginal peoples knew intimately, their seven seasons attuned to the movement and qualities of the winds, the annual fish and bird migrations and the blossom and fruiting of the landscape. It has been surprising how many Australians have come up to the Water Bar amazed at the flowers in our water, unaware that they could be used for anything apart from visual delight. The effect of the Water Bar should not be underestimated. It has been a huge way of re-connecting people to their country, delighting people with new understandings of the use and bounties of the natural world. There are books in the Museum library that can then support further investigation into the flora and fauna of Australia.

13   Conclusion

One might subsume the eliminated element in the term "aura" and go on to say: that which withers in the age of mechanical reproduction is the aura of the work of art. This is a symptomatic process whose significance points beyond the realm of art. One might generalize by saying: the technique of reproduction detaches the reproduced object from the domain of tradition. By making many reproductions it substitutes a plurality of copies for a unique existence. And in permitting the reproduction to meet the beholder or listener in his own particular situation, it reactivates the object reproduced.[15]

Museum of Water champions the particular, the authentic, the testimony of history.

We deal in essences, liquid distillations of memory. The authority of each of these objects in Museum of Water reinvigorates each bottle. The Museum is a battle against the loss of significance, what many have complained about as the meaninglessness of modern life, the reduction of the object in the age of mechanical reproduction, the ennui which begins not to care, which does not cherish.

"*Bring me the sweat of Rafar Nadal!*" A worker at Thames Water once excitedly exclaimed, as he understood the work of Museum of Water. It is a challenge, and an always open offer, that puts the impetus on you to decide what to save in this world. *What will you keep?*

The maps we make to navigate this world are being constantly re-drawn, the data we draw on changing fast. Here the perceived edge of each bottle is not in fact the fixed ending it at first seems. The collection moves constantly between the bottles, transfers between the glass and plastic boundaries, into the air around us, between us and through our very bodies. The work is about the transference of meaning, its transmission between us. We are part of the collection.

This Museum began as a way of beginning a conversation, but through the variety of places we've been, the programmes we have curated and the testimonies

---

[15]Benjamin (1969).

of people we have met, the museum has become many more things: a reason for gathering, a space for performance, an incitement to action, a museum of public history, a reed bed for experience, an archaeology of the future. The Museum operates in a context of water issues facing the world and the people in it. It has taken us all by surprise. It has become crucial, with huge relevance for our future. I am suspicious of civilisation—it doesn't seem to enact the kindness it supposes.

We are out of our depth, and there is a bewilderment at modern life. These are some of the phrases people used in The Swimmers' Manifesto to talk about why they swim:

A process of being ok with drowning

a suspension of all that country

This is the centre of our being

There's something about being in that salt water that connects you to place.[16]

Keeping the planet liveable for the billions of people who want to live on it is the most crucial work for all of us. So a marker for all of our museums in this conference should be, *how do we effect change?* We have to bring about a re-consideration of water, of our days, and our environmental impact. As Donna Haraway says, *"You have to keep doing positive things. You have to do this, you have to be here now, doing this."*[17] This takes stamina and we all need to join in.

In Museum of Water to date there are 1038 bottles. These are 1038 creative acts: these bottles are both the residue from each journey and a new artistic creation, both documenting artefact and original artwork. There are many sustaining dualities and dialogues in the Museum: it deals in absence and presence, invention and reiteration. Each bottle is rare and common, priceless and free, unique and identical, irreplaceable and dematerialising before your eyes.

Silence is the ocean of the unsaid, the unspeakable, the repressed, the erased, the unheard... If the right to speak, if having credibility, if bring heard is a kind of wealth, that wealth is now bring redistributed... By redefining whose voice is valued, we redefine our society and its values.[18]

This Museum fights silence and hierarchical awe, instead enables, encourages, listens. This Museum questions where the authority for every collection lies, to whom are we looking for our influence? It explores alternative systems of value, and how we can learn from each other.

Each gift talks of boundaries crossed, thresholds stepped over and out into the world, commemorates a point of contact with the world. It is the repeated reaching out of our bodies and minds to nature, a repeated moment of appreciation and

---

[16]The Swimmers' Manifesto (2017).
[17]Donna Haraway: Story Telling for Earthly Survival (2016).
[18]Solnit (2017).

prizing. And although we each take something away from the original site, there is no theft of nature, since we only hold it for a moment, we are always in a process of losing it.

Museum of Water offers a platform to people who might not be used to having their voices listened to with close attention. It makes space for different cultural viewpoints, the details of existence, intimate moments, undiscovered feelings. It makes room for things that more often go unnoticed, are usually discounted. It exists in direct response to the repeated phrases from others: "Don't worry about it"... "It doesn't matter"... the litany of those who choose for us what is important to our days. This Museum goes out of its way to hear different voices, not in a cacophony of voices where no one is listened to, but with considered care offers the equal presentation of difference. This Museum intentionally searches out people who might not usually have their say and offers a platform to hear them. Historically, whole swathes of society have been discounted, different races, women, the young, the old... Here we cherish the vulnerable understandings we have sometimes only half formed, and might hesitate to frame and phrase... we make space for careful attention to be paid to opinions hazarded, ideas reached for and attempted.

I hope that our work explores what a Museum can be and do; what we value and who can influence us; how we gather information, prizing different processes. It harks back to the original understanding of a museum as a forum of ideas, questions the idea of tingling life and the magic of intimacy as much as contamination, ruin, governance, 'civilisation', healing and regrowth. We all need a different future than the one that is currently being shaped for us. Museum of Water is a strident call to change, to talk, to listen and not look away.

Post Script

This paper was written to bring attention to the work of Museum of Water and to share the processes of our work, in case others could benefit.

Museum of Water is an unusual museum, presenting a radical challenge to previous systems of collecting and to traditional processes and economies, how it treasures a substance and experience that we cannot hold onto, the process of evaporation itself. It is a challenge to a capitalist society that thrives on things it can buy and sell: it bottles water but it doesn't commodify it. Museum of Water explores our borders. It understands water as a shared resource, and seeks to embolden our connectedness and care. Its focus on a mundane substance has occasionally meant it is underestimated as a museum, and its double status as live artwork is sometimes underappreciated, but the understandings from each different discipline intrinsically support its presence and importance. By inviting us to re-consider our relationship to water, and making space for each different opinion, this Museum makes us all protagonists in our water history. The Museum continues to travel to different sites around the world, collaborating with people to help us all to question our inheritance and explore the estate of water, as we come under increasing global pressure to share. Water questions will define the coming century, so empowering people to engage in the questions and to influence their surroundings is crucial work. Through its work and its programmes, Museum of Water helps to equip us, because we are all implicated. This Museum makes custodians of us all.

References

Al-Qattan O (2016) The Welfare Association (Taawon), The Guardian Newspaper, UK edition
Benjamin W (1969) The Work of art in the age of mechanical reproduction. In: Arendt H (ed) Illuminations (trans. Zohn H, from the 1935 essay). Schocken Books, New York, p 4
Donna Haraway: Story Telling for Earthly Survival (2016) Documentary film dir. Fabrizio Terranova, Belgium. http://earthlysurvival.org
Frizzell N (2014) The Guardian Newspaper, UK edition
http://one-europe.net/to-polder-or-not-to-polder-that-s-the-question
http://museumofwaterrotterdam.tumblr.com
https://www.perthfestival.com.au/event/museum-of-water
Maguire S (2015) From Dublin to Ramallah, in Almost the Equinox, Selected Poems. Chatto and Windus
Neimanis, A (2012) Hydrofeminism: or, on becoming a body of water. In: Gunkel H, Nigianni C, Söderbäck F (eds) Undutiful daughters: mobilizing future concepts, bodies and subjectivities in feminist thought and practice. Palgrave Macmillan, New York, p 108
Perkovic J (2017) Realtime. http://www.realtimearts.net/article/137/12538
Solnit R (2017) A short history of silence. In: The mother of all questions. Granta Books, Great Britain, p 23
The Swimmers' Manifesto (2017) Museum of Water at Perth Festival. Available as a podcast on Sonic Encounters and Conversations on Water. www.perthfestival.com.au/event/museum-of-water
Walley R and Walley T (2017) Part of the series Conversations on Water recorded for Perth Festival, available for listening on the Sound Archive at www.museumofwater.co.uk

Effectiveness of Communication Strategies in Confronting Climate Change: The Views of the Citizens of Greece

Aikaterini Zerva, Evangelos Manolas, Constantina Skanavis and Georgios Tsantopoulos

Abstract The aim of this paper is to identify groups of citizens and their characteristics so that effective climate change communication strategies can be designed. The research was carried out from January 2014 to June 2015, 1536 questionnaires were completed which were evaluated on the basis of the co-efficient α-Cronbach, descriptive statistics, Friedman's non-parametric criterion, factor analysis and cluster analysis. The research showed that Greek citizens think that the most important concerned parties for taking action against climate change are environmental organizations, scientists and local citizen environmental groups. In addition, two groups of citizens were identified. In the first group belong mostly citizens aged 31–40 and less citizens aged 41–50, married, mainly graduates of secondary and tertiary education, most are public servants or unemployed and satisfied from governmental activities regarding municipal projects concerning adaptation, energy saving and lifelong learning. In the second group of citizens also belong young to middle age citizens, unmarried, who mainly work in the public and private sector, who are satisfied both from the activities of non-governmental concerned parties as well as governmental activities regarding adaptation to extreme environmental phenomena, mitigation and waste management.

A. Zerva · E. Manolas (✉) · G. Tsantopoulos
Department of Forestry and Management of the Environment and Natural Resources,
School of Agricultural and Forestry Sciences, Democritus University of Thrace,
193 Pantazidou Street, 68200 Orestiada, Greece
e-mail: emanolas@fmenr.duth.gr

A. Zerva
e-mail: azerva@fmenr.duth.gr

G. Tsantopoulos
e-mail: tsantopo@fmenr.duth.gr

C. Skanavis
Department of Environment, University of the Aegean, University Hill,
"Xenia" Building, 81100 Mytilene, Lesvos, Greece
e-mail: cskanav@aegean.gr

© Springer Nature Switzerland AG 2019   499
W. Leal Filho et al. (eds.), *Addressing the Challenges in Communicating Climate
Change Across Various Audiences*, Climate Change Management,
https://doi.org/10.1007/978-3-319-98294-6_30

1 Introduction

Climate change constitutes a serious social issue which many citizens do not understand, do not deal with seriously and do not regard it a public threat (Dimendo and Doughman 2007). This is so because the nature of the phenomenon of climate change is not perceived as an immediate problem (Weber 2016). However, public understanding of the problem regarding perception of the danger, in the framework of environmental communication, plays a vital role in dealing with the challenges of global climate change (Hagen 2016). As a result, the impact of communication on the formation of citizen beliefs through communication models is integrative and complex (Eveland and Cooper 2013). The designing of environmental communication strategies characterized by timely and clear representation of concerned parties and citizens in activities will greatly help confront the problem of climate change and contribute to sustainable development. Of course, the use of environmental communication is influenced by many factors. One of these factors is the means of mass communication which can easily influence the behaviour, the attitudes and the views of the citizens. The means of mass communication can frame environmental issues and influence the way through which the local society considers such issues important (Marin and Berkes 2013). The result is that viewers with low environmental knowledge and judgment are influenced and led easily to the wrong conclusions. Therefore, the means of mass communication have great responsibility towards citizens since they constitute a source of information and views (Carvalho 2007) especially on issues which the public is not experienced enough. Scientific information may not influence directly the views of the public on climate change but, at least, this small intervention helps to increase the awareness of the public on the issue. In order to combat this matter, there are groups of citizens, ecologists and environmental organizations which have organized collective activities and which, at the same time, have created various organizations on climate change. In such activities various communication practices are used which go hand in hand with technology such as internet, email services and web pages, all of which are important for communication with their members but also for approaching new members. Their mission is to make the public aware with regard to the threats and consequences of climate change but also to inform the public. However, in order to make the public aware on climate change more sources of information are needed on climate change which in combination with information which is not difficult to understand will increase the awareness of the public (Toma et al. 2014). In addition, the participation of citizens in a frequent dialogue on climate change is very important.

However, the different meanings and interpretations which are given to terms and reports with regard to climate change may create tension (Lassen et al. 2011) which will make it difficult for citizens to understand the problem and the probable solutions for confronting it. The delay with regard to scientific information is mainly due to the gap between the scientific community and public opinion (Risbey 2008). The public may not read or may not easily understand scientific articles, but the percentage of anxiety of the public may increase as a result of publication in magazines which use simpler language and terms (Brulle et al. 2012). In this process, the biggest barrier scientists face, in order to take the right decisions or in order to properly describe the relevant concepts, is the lack of information with regard to the audience they address (Bruin and Bostrom 2013). The result is that a gap is created between scientists and public opinion which is mainly due to the use of language used by the two groups (Sterman 2011). Despite all this, there were several efforts in implementing strategies in cities in Germany and the United Kingdom whose action plans were mainly focused on the energy sector just like the Greek national strategy on adaptation (YPEKA 2016) which, however, was abandoned because of financial difficulties (Bulkeley and Kern 2006).

Environmental education may significantly contribute to the minimization of the impact of factors which block the application of methods of environmental communication. Environmental education aims at fundamental changes with regard to the attitudes, behaviour and values of individuals and social groups (Souchon et al. 1996; Rangou 2013) and can impact the level of knowledge of students through understanding of variables which influence climate change as well the necessary knowledge for action through understanding of behaviours (Kloser and Bofferding 2014). Thus, environmental education and communication can directly influence the pro-environmental behaviour of citizens so that they themselves can participate in activities on climate change and sustainable development. The incentive for the realization of this research has to do with the fact that the majority of the concerned parties implements activities against climate change without taking into account the views, attitudes and knowledge of the citizens on climate change, the satisfaction and trust of citizens have for the concerned parties as the activities which citizens are willing to implement. Therefore, this chapter will contribute to the advancement of the field and its enrichment with new knowledge, because it will produce important information for the designing of effective communication strategies and activities against climate change, the discovery and improvement of the activities of concerned parties and the selection of activities which citizens intend to realize. In addition, it will give the possibility for the exploitation of the results and conclusions of the research by the concerned parties for improvement, designing and implementation of appropriate communication strategies for the participation of citizens against climate change and sustainable development.

2 Review of Literature

In recent years, environmental problems were considered by Greeks and other European citizens as less important compared to economic problems (European Commission 2004, 2007, 2010, 2013). To this may contribute the fact that economic and material problems caused by environmental issues may not be easily incorporated in the economic budgets of states (Papoulis et al. 2015).

However, the economic problems of each country, in addition to the economic situation of each country, may also influence the protection of the natural environment. Thus, the lack of available economic sources of states lead to the reduction of effective governmental activities in confronting climate change which, in turn, contributes to the reduction of citizen interest in dealing with environmental problems (Scruggs and Benegal 2012; Obani and Gupta 2016).

The opinion Greek citizens have for environmental parties which are involved with climate change causes anxiety to citizens because these parties play a determining role in confronting problems which are caused by human interventions in the environment (Zerva et al. 2018). Therefore, in order for this to change and in order to improve the connection between government and public opinion the government may need to co-ordinate the various bodies which are involved and interact among themselves in confronting climate change. The government may also need to incorporate them in the framework of their regional and local action (House of Commons Environmental Audit Committee 2008). This way these bodies will have the possibility to improve their arguments or ideas in confronting climate change so that they can encourage the public to participate in activities (Pearce 2014) and, at the same time create the opportunity to increase the degree of trust of citizens towards the government and other bodies with the ultimate aim the minimization of the consequences of climate change in the future. It has been found that the taking of timely activities is very important for the stabilization of atmospheric concentrations even though the mitigation efforts in the rest of the states are realized with delay (Jakob et al. 2012). On the contrary, the delay for action will be more costly when compared with actions from mild policies on climate (Bosetti et al. 2009).

Therefore, in order to encourage direct action for the climate, it is necessary to combine factors which influence the perceptions of citizens on climate change and have the possibility to restrict the political, economic and social action in confronting dangers (Leiserowitz 2005). Thus, it is wise that politics focuses on social and institutional issues which restrict the participation of the public in activities designed to confront the problem (Bulkeley 2000). So, to start with, it is necessary that the concerned parties inform the citizens on climate change in order to encourage them to participate in activities against climate change. In fact, many of them use social marketing which, however, cannot by itself help significantly in encouraging the participation of the public against climate change (Corner and Randall 2011). Therefore, the understanding and trust of the public in this social process is important (Demeritt 2001) as this may make easier the taking of decisions by concerned parties (Earle and Cvetkovich 1995).

However, in order to achieve this, it will initially be necessary that the concerned parties, and mainly the state, which constitutes one of the most important bodies on climate, (Boehmer-Christiansen 1994), improve their activities on climate change so that they can contribute directly to the adaptation of citizens in the new changes. Therefore, the strategies for confronting the impacts of climate change will need to focus on climate change not only as an international but also as a national issue (Bulkeley and Kern 2006). One of the most important national concerned parties in every European state is the government (European Commission 2013) which plays a central role in the global management of climate since it contributes to the creation of international and national agreements and their implementation (Bulkeley and Betsill 2005). Next in importance were considered the following parties: businesses, European Union, citizens, local self-government and environmental organizations (European Commission 2013). Similarly, it was found that the majority of Greeks (67.0%) think that the government is more responsible in confronting climate change and less so businesses and industries (European Commission 2015).

In addition, the majority of citizens in global level think that governments will need to realize more activities of energy consumption in order to encourage citizens to use renewable energy sources more (World Bank 2010). Similarly, most European and Greek citizens think that it is very important that governments set targets for increasing the use of alternative energy sources (European Commission 2013, 2015). Since, as it was recently found, Greek citizens are adequately informed on renewable energy sources, they are sufficiently willing to invest in renewable energy sources (Tsantopoulos et al. 2014) and think it is very important that strategies of energy policy are implemented despite the economic crisis (Papoulis et al. 2015). Key to the effective designing of policies by parties which desire to influence public opinion is the designing of environmental communication programs based on the views and attitudes of the public. The role of communication experts is important here since these experts are called to facilitate or participate in discussions related to environmental policy with public opinion playing a significant role in the process (Waddell 2000). During this process, public communication and commitment should not be taken as a simple way of transmitting science, in order to convince the audience to see the discussions scientifically or perceive them as a scientist would do (Nisbet and Schenfele 2009), but the importance of their responsibility and their values should be thought of as intermediaries in order for the public to have correct, simple and objective information. However, the objectivity and understanding of information do not only depend on values which characterize scientists, communication experts and the means of mass communication but they also depend on the level of the public they address which has to do with the upbringing, the place of residence and its culture (Corbett 2006). For this reason the implementation of an interactive process is important with messages which come from the sender to the public and vice versa. This will lead to effective communication (Valenti 1995).

3 Methodology

3.1 Research Area

This research was conducted in Greece which as a Mediterranean country is one of the most vulnerable Mediterranean countries since it is impacted by four types of climate (Mariolopoulos 1938, 1982). These types are the marine Mediterranean type (Western Greece and Ionian islands), the terrestrial type (Mainland Greece, Eastern Peloponnesus, Central Aegean and Crete), the Continental type (Thrace, Macedonia, Epirus and Thessaly) and the mountainous type which includes all the mountainous regions.

3.2 Questionnaire and Research Sample

This research was carried out from January 2014 to June 2015. 1536 questionnaires were collected. The questionnaire comprised of closed questions and it was distributed with simple random sampling to adult Greek citizens. The process also involved the use of interviews which lasted about 30 min each. The questions mainly referred to the concerned parties which citizens trust in order to participate in activities on climate change and then on adaptation and mitigation activities realized by the state with regard to confronting climate change. The majority of the questions were devised on the basis of the Likert scale which is used to study patterns of behaviour of a social group (Siardos 2005).

For the evaluation of data we used the statistical package SPSS, the α-Cronbach co-efficient, descriptive statistics, Friedman's non-parametric criterion, factor analysis and cluster analysis. The α-Cronbach co-efficient was used to find the internal reliability of a questionnaire, i.e. if the data have the tendency to measure the same event. The α-Cronbach co-efficient expresses the square of the correlation between the (observed) marking a person gets in the given scale and the marking he would have gotten (real) if he had been asked on the totality of topics (Siardos 1998). Friedman's non-parametric test was used to compare the values of three or more correlated groups of variables with regard to questions Q1 and Q2. In this process the values of the variables for each criterion were classified separately. Also, we calculated the average class of classification values for each variable (Freund and Wilson 2003; Ho 2006). Next we used factor analysis in order to investigate the existence of common factors in a group of variables (Sharma 1996). In particular, we used the principal components method which is based on the spectrum analysis of the variance (correlation) table (Murtagh and Heck 1987; Lupton 1993; Siardos 1998; Djoufras and Karlis 2001; Jolliffe 2002). The criterion used for the significance of the principal components was the one suggested by Guttman and Kaiser (Cattell 1978; Frangos 2004), according to which the limit for obtaining the appropriate number of principal components is determined by the values of the characteristic roots which are equal or greater than one. In addition, for better results we conducted rotation of the matrix of the principal components through Kaiser's method of maximum variance rotation (Harman 1976). Finally, we applied cluster analysis on the loadings of the final factors in order to investigate the possibility of existence of types of citizens from the factors extracted from the secondary factor analysis.

3.3   Calculation of Sample Size

The size of the sample was calculated on the basis of the formulas of simple random sampling (Freese 1984; Kalamatianou 2000; Matis (2001). Although we used simple random sampling without replacement, the correction of finite population can be ignored because the size of the sample n is small compared to the size of population N (Pagano and Gauvreau 2000).

$$n = \frac{t^2 \cdot \bar{p} \cdot (1 - \bar{p})}{e^2} = \frac{1.96^2 \cdot 0.50 \cdot (1 - 0.50)}{0.025^2} \cong 1536$$

where t = the value of Student distribution for probability $(1 - \alpha)$ = 95.0% and n − 1 degrees of freedom. Since the size of the conducted pre-sampling is big (bigger than 50) the value t is taken from the probability tables of normal distribution for the desired probability.

In practice for probability 95% the value is 1.96 (Matis 2001).

p = estimation of analogy
e = maximum accepted difference between the sampling mean and the unknown mean of the population. We accept that it is 0.025, i.e. 2.5%.

In order to calculate the size of the sample it was necessary to conduct pre-sampling with the size of the sample being 50 individuals. Thus, for each variable we calculated the real analogy of the population. The use of the questionnaire is not restricted to the estimation of only one variable of the population but of more variables. Thus, it is necessary to estimate the size of the sample of each one of the variables. If the sample sizes estimated are similar and the size of all is within the financial possibilities of the sampling, then as size of the sample we choose the biggest. This way the most changing variable is calculated according to the desired accuracy while the rest with greater accuracy than from what was originally determined (Matis 2001).

However, the main problem, as in every sampling process, is the creation of the sampling framework. This is so because we would have to create a voluminous catalogue of all citizens per prefecture in alphabetical order and give every citizen a unique code number. It is obvious that the above was time-consuming, difficult and costly. But even if this was applied no-one could have guaranteed that the particular individual would live or be in the random spot at which he was recorded. Next, we had created the sampling framework in which every citizen was defined with a number which defines the region, the prefecture, the municipality and the local community in which it is located. In order to create this framework we used the results of the most recent census of the Hellenic Statistical Authority (2011). Of course, in combination with the number of deaths and births but also in combination with the number of population movements, the real population of municipalities is different from the one recorded in the census (Tampakis 2000).

4   Results

The results mainly focus on the socio-demographic features of the respondents and, also, on their views, knowledge and attitudes with regard to climate change. In addition, they include the factors which influence the activities of the concerned parties have undertaken in confronting the problem but, also, the activities in which the citizens themselves will take part in, through the concerned parties. Also, from the analyses we extract the features of the groups of citizens who themselves are willing to take part in, in confronting climate change.

4.1   Socio-demographic Features of the Sample

Most of the respondents, more than half of the sample, are adults of middle age, no more than fifty years old who have at least completed secondary education. Women are more than men while the majority of the sample is graduates of secondary education (35.4%) and tertiary education (45.9%). In addition, most of them are public servants (28.3%), private employees (16.4%) and free lance professionals (13.2%) while those who are involved in primary sector production (arming and livestock farming) or housework are fewer. A significant percentage, about 15.5% of the respondents, remains unemployed.

4.2   Citizen Satisfaction from the Concerned Parties and the Activities of Government Agencies in Confronting Climate Change

In order to find the party citizens are more satisfied with for activities it has undertaken in confronting climate change we used Friedman's analysis. This analysis showed that citizens are more satisfied from environmental organizations with average value 5.99. Also, citizens showed they were very satisfied with activities undertaken by scientists and local citizen environemntal groups with mean ranking values 5.36 and 5.22 respectively. Similar mean values were found for activities undertaken by non-governmental organizations and fellow citizens while, on the contrary, the rest of the concerned parties showed relatively lower mean values with the state being in the last position (Fig. 1).

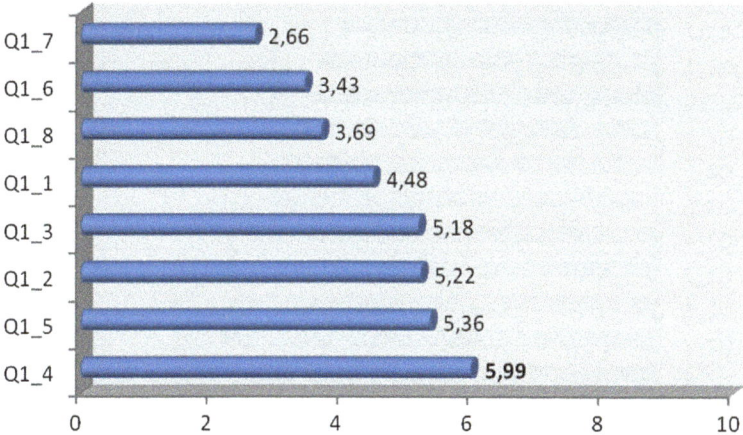

Fig. 1 Hierarchy of the concerned parties as a result of the evaluation of citizen satisfaction for their activities. Q1_1 = from simple citizens, Q1_2 = from local environemntal groups, Q1_3 = from non-governmental organizations, Q1_4 = from environmental organizations, Q1_5 = from scientists, Q1_6 = from local self-government, Q1_7 = from the state, Q1_8 = from the global community. N = 1536, Chi-Square = 3224.46, df = 7 Asymp. Sig. <0.001

4.3 Citizen Satisfaction from the Activities of Governmental Bodies

Next, we identified the activities of governmental bodies for which citizens are more satisfied. From the analysis it was found that citizens are more satisfied from activities realized by the state with regard to citizen servicing and the modernization of the means of mass transport (7.04) in order for citizens to move with greater safety but also contribute to the reduction of dangerous pollutants so that they can strengthen their activities in combating climate change. Also, the citizens seem to be relatively satisfied from state activities regarding the use of alternative energy sources and the construction and the subsidization of energy efficient housing with mean ranking values 6.72 and 6.99 respectively. Also, the citizens revealed significant mean ranking values with regard to activities on activities on checking the percentage of exhaust gases emitted by passenger vehicles and the management of waste. On the contrary, the rest of the activities reveal lower ranking values with the citizens being less satisfied from these activities. Last in the hierarchy is the training of adults in confronting climate change (Fig. 2).

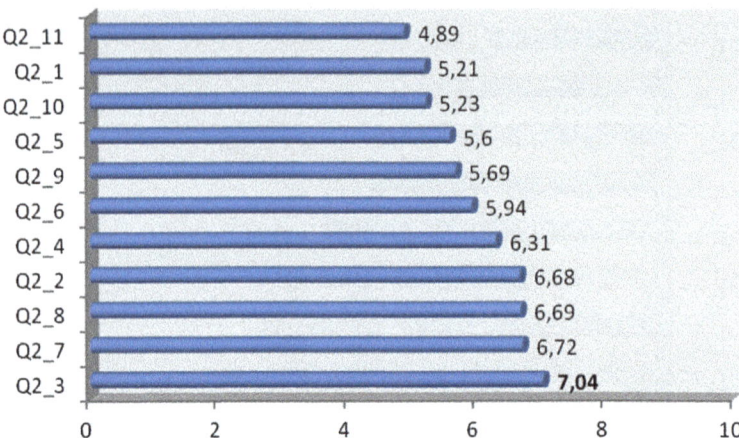

Fig. 2 Hierarchy of activities undertaken by the state through citizen satisfaction in confronting climate change. Q2_1 = regarding the reduction of greenhouse pollutants emitted by industries and businesses, Q2_2 = regarding checking the percentage of exhaust gases emitted by passenger vehicles, Q2_3 = regarding servicing and modernization of the means of mass transportation, Q2_4 = regarding the management of waste, Q2_5 = regarding political protection from extreme events, e.g. floods, typhoons, drought etc., Q2_6 = regarding public projects which support the effort of confronting climate change, e.g. bicycle lanes, green parks etc., Q2_7 = regarding the use of alternative energy sources, Q2_8 = regarding the consrtuction and the subsidization of energy efficient housing, Q2_9 = regarding provision of information to the public on climate change and its impact, Q2_10 = regarding training of the public via courses on climate change, Q2_11 = regarding training of adults on climate change. N = 1536, Chi-Square = 1161,27, df = 10 Asymp. Sig. <0.001

4.4   Satisfaction Factors Through the Activities of Concerned Parties

Next with regard to finding the views of the respondents on applying strategies on climate change activities the following was done: first class factor analysis and next second class factor analysis with orthogonal rotation. In particular, through first and second class factor analyses we analyzed citizen satisfaction from governmental and non-governmental concerned parties regarding their current activities in confronting climate change as well as citizen satisfaction from activities undertaken from the state exclusively.

On the basis of our analyses and Tables 1 and 2 the total variance of factor analyses was about 62.6%. It was also found that the women and men of the sample are more or less equally satisfied from the activities of governmental and non-governmental concerned parties (with emphasis on governmental activities). In particular, the results showed that there are two factors which impact the satisfaction of citizens from activities undertaken by governmental and non-governmental concerned parties and, specifically, activities undertaken by the government in confronting climate change.

Table 1 List of factors from the first factor analysis on the satisfaction of citizens from the activities of governmental and non-governmental concerned parties (with emphasis on state activities)

List of principal factors—results of the first factor analysis	Loadings	Eigenvalue	Variance (%)
Non-governmental concerned parties (Q1_A)		2.706	33.827
From single citizens	0.554		
From local citizen environmental groups	0.801		
From non-governmental organizations (NGOs)	0.837		
From environmental organizations	0.833		
From scientists	0.529		
Governmental concerned parties (Q1_B)		2.105	26.308
From the local self-government	0.827		
From the state	0.884		
From the global community	0.726		
Total variance (%)			60.463
Keiser–Meyer–Olkin = 0.770, Bartlett χ^2 = 4,123,977, df = 28, $p < 0.01$			
Governmental activities for municipal projects concerning adaptation, energy saving and life-long learning (Q2_A)		3.803	34.575
Regarding public projects which support the effort of confronting climate change (e.g. bicycle lanes, green parks etc.)	0.562		
Regarding the use of alternative energy sources	0.635		
Regarding the construction and the subsidization of energy efficient housing	0.643		
Regarding provision of information to the public on climate change and its impact	0.814		
Regarding training of the public via courses on climate change	0.863		
Regarding training of adults on climate change	0.856		
Governmental activities regarding adaptation to extreme environmental phenomena, mitigation and waste management (Q2_B)		3.085	28.047
Regarding checking the percentage of exhaust gasses emitted by passenger vehicles	0.813		
Regarding servicing and modernization of means of mass transportation	0.785		
Regarding reduction of greenhouse pollutants emitted by industries and businesses	0.738		
Regarding waste management	0.633		
Regarding political protection from extreme events, e.g. floods, typhoons, drought etc.	0.502		
Total variance (%)			62.622
Keiser–Meyer–Olkin = 0.904 Bartlett χ^2 = 9,169,795 df = 55, $p < 0.01$			

Table 2 List of factors from the second factor analysis on the satisfaction of citizens from activities of governmental and non-governmental concerned parties (with emphasis on governmental activities)

List of principal factors—results of second factor analysis	Loadings	Eigenvalue	Variance (%)
Satisfied citizens from the activities of governmental concerned parties (P1)		1.280	32.012
Governmental concerned parties (Q1_B)	0.793		
Governmental activities for municipal projects concerning adaptation, energy saving and life-long learning	0.769		
Satisfied citizens from activities of non-governmental concerned parties (P2)		1.138	28.451
Non-governmental concerned parties (Q1_A)	0.724		
Governmental activities regarding adaptation to extreme environmental phenomena, mitigation and waste management (Q2_B)	0.748		
Total variance (%)			60.463
Keiser–Meyer–Olkin = 0.453, Bartlett χ^2 = 271,899, df = 6, p < 0.01			

Of those factors, the first (P1) may be named "satisfied citizens from activities of governmental concerned parties" and refers to citizens who are more satisfied from activities undertaken by concerned parties and more satisfied from governmental activities for municipal projects concerning adaptation, energy saving and life-long learning. The second factor (P2) can be named "satisfied citizens from activities of non-governmental concerned parties" since it includes the satisfied citizens from activities undertaken by non-governmental concerned parties and who are more satisfied from activities undertaken by governmental activities regarding adaptation to extreme environmental phenomena, mitigation and waste management.

4.5   Clustering of Citizens Through Common Features

Of these factors we found that there are two types of citizens CL1 and CL2 which influence the application of strategic activities on climate change. Both exhibit similar percentages in each cluster (Table 3). Thus, in the first cluster mainly belong the citizens which are positively impacted by the first factor, i.e. they are satisfied from the activities of governmental concerned parties. Also, in the first factor belong the citizens which are negatively impacted by the second factor, i.e. they are not satisfied by the activities of non-governmental parties. On the contrary, in the second cluster belong the citizens which are negatively impacted by the first factor and positively by the second.

Table 3 Mean loadings of factors in each cluster of citizens regarding activities undertaken by governmental and non-governmental concerned parties regarding climate change (with emphasis on governmental activities)

	CL1 (51.7%)	CL2 (46.3%)
P1	0.24427	−0.26070
P2	-0.72527	0.77407

In addition, on the basis of factor loadings and the analysis of statistical correlation of the two clusters it was found that in the first cluster CL1 which constitutes the biggest cluster of the sample belong young citizens most of whom are married, graduates of secondary and tertiary education the majority of whom work for the public sector or are unemployed. In parallel, with regard to cluster CL2 we observe similar results since the percentages of the two clusters only differ slightly. Therefore, in the second cluster also belong young and middle age citizens who are mainly unmarried and most of whom work in the public and private sector (Table 4).

5 Discussion

The strategic planning of concerned parties will contribute to the participation of citizens on activities confronting climate change. This presupposes identifying the individual characteristics of citizens. This also presupposes identifying the most important parties and activities citizens desire to take part in. Of these, the most important concerned parties regarding the participation of citizens in activities with regard to confronting climate change were considered to be the government (Bulkeley and Betsill 2005, European Commission 2013, Zerva et al. 2018) and the local self-government and then the environmental groups and the rest of the non-governmental parties (European Commission 2015). However, the majority of Greek citizens is not particularly satisfied with the activities of parties involved in confronting climate change and this may be due to the minimum effort made by these parties concerning the realization of activities regarding the reduction of pollutants by industries, political protection and awareness, education and training of the public (Zerva et al. 2018). To this they have mainly contributed financial factors which influence directly the views of citizens and, by extension, their participation in activities, while according to Papoulis et al (2015) this may be due not only to the lack of available financial resources but also to the difficulty of incorporating environmental problems into Gross Domestic product as well as to the dissatisfaction of citizens from governmental activities in confronting climate change. Citizens are more interested in participating in the activities of

Table 4 Statistically significant differences of both clusters of citizens on activities undertaken by governmental and non-governmental concerned parties regarding climate change (with emphasis on governmental activities)

Variable	Scale	CL_1 (53.7%)	CL_2 (46.3%)
Age	18–31	31.5	37.0
	31–40	28.9	29.6
	41–50	23.2	20.3
	51–60	12.0	9.8
	>60	4.4	3.2
Marital status	Unmarried	43.6	49.9
	Married	51.8	44.5
	Divorced (separated)	3.7	4.4
	Widowed	0.9	1.1
Education	Elementary school graduate	3.7	2.0
	Graduate of lower secondary school	6.4	2.0
	Graduate of technical school	11.2	7.4
	Graduate of upper secondary school	27.5	24.5
	University graduate	25.2	35.9
	Graduate of technological educational institute	15.6	15.3
	With post-graduate degree	10.3	12.8
Profession	Public servant	24.2	32.6
	Private employee	15.5	18.3
	Free-lance professional	13.1	13.3
	Farmer-livestock breeder	9.0	3.1
	Unemployed	17.0	13.6
	Housework	6.2	4.6
	Pensioner	7.1	6.5
	Student–Pupil	7.9	8.1

non-governmental organizations but also in some of the activities of governmental parties which are mainly about alternative energy sources and the more frequent use of renewable energy sources. This compared to all other activities, probably comes from the fact that Greek citizens are more informed about activities concerning renewable energy sources (Tsantopoulos et al. 2014). In addition, governmental activities concerning mitigation and adaptation of citizens to the new changes were also considered important, probably because of the insecurity and dissatisfaction of citizens regarding environmental activities which refer to the future impacts of climate change.

6 Conclusions

In conclusion, despite the economic crisis in Greece, citizens are interested in participating mainly in actions of non-governmental organizations and less in those of governmental organizations. Citizens of young and middle age are willing to participate in activities which mainly concern municipal projects of adaptation to climate change such as bicycle lanes, parks and green projects and in some cases governmental activities which concern adaptation to extreme phenomena, mitigation activities and waste management. Through the use of these most important characteristics of citizens the concerned parties will have the possibility to encourage citizens more, so that they participate in relevant activities. Thus, the parties involved in confronting climate change will need to create new communication strategies which will come from the characteristics of the groups of citizens extracted from the results of this research so that they can directly contribute to the adaptation of citizens. This will create opportunities both for governmental and non-governmental parties to use them in the commencement of their strategic approaches in their various methods of environmental communication and environmental management in order to succeed in their aims.

References

Boehmer-Christiansen SA (1994) Global climate protection policy: the limits of scientific advice, part II. Glob Environ Change 4(3):185–200

Bosetti V, Carraro C, Sgobbi A, Tavoni M (2009) Delayed action and uncertain stabilisation targets. How much will the delay cost? Clim Change 96(3):299–312

Bruin W, Bostrom A (2013) Assessing what to address in science communication. Proc Natl Acad Sci USA 110(3):14062–14068

Brulle RJ, Carmichael J, Jenkins JG (2012) Shifting public opinion on climate change: an empirical assessment of factors influencing concern over climate change in US 2002-2010. Clim Change 114(2):169–188

Bulkeley H (2000) Common knowledge? Public understanding of climate change in Newcastle, Australia. Public Understand Sci 9(3):313–333

Bulkeley H, Betsill M (2005) Cities and climate change: urban sustainability and global environmental governance. Routledge, London

Bulkeley H, Kern K (2006) Local government and the governing of climate change in Germany and the UK. Urban Stud 43(12):2237–2259

Carvalho A (2007) Ideological cultures and media discourses on scientific knowledge: re-reading news on climate change. Public Understand Sci 16(2):223–243

Cattell RB (1978) The scientific use of factor analysis in behavioral and life sciences. Plenum, New York

Corbett JB (2006) Communicating nature: how we create and understand environmental messages. Island Press, Washington, DC

Corner A, Randall A (2011) Selling climate change? The limitations of social marketing as a strategy for climate change public engagement. Glob Environ Change 21(3):1005–1014

Demeritt D (2001) The construction of global warming and the politics of science. Ann Assoc Am Geogr 91(2):307

Dimendo JF, Doughman P (2007) Climate change: what it means for us, our children, and our grandchildren. MIT Press, Cambridge. http://books.google.gr/books?id=PXJIqCkb7YIC&lpg=PP1&hl=el&pg=PR4#v=onepage&q&f=false. Last Accessed 1 Nov 2014

Djoufras I, Karlis D (2001) Elements of multivariate data analysis. Notes on course "Data Analysis I". Department of Business Administration University of the Aegean, Chios (in Greek)

Earle TC, Cvetkovich GT (1995) Social trust: toward a Cosmopolitan Society. Praeger, Westport, Connecticut

European Commission (2004) The attitudes of European citizens towards the environment, Special Eurobarometer 217/Wave 62.1 European Commission: Brussels, Belgium, 2004. http://ec.europa.eu/commfrontoffice/publicopinion/archives/ebs/ebs_217_en.pdf. Last Accessed 20 Oct 2016

European Commission (2007) Attitudes of European citizens towards the environment, Special Eurobarometer 295/Wave 68.2; European Commission, Brussels, Belgium. http://ec.europa.eu/commfrontoffice/publicopinion/archives/ebs/ebs_295_en.pdf. Last Accessed 20 Oct 2016

European Commission (2010) National report-executive summary—Greece, Standard Eurobarometer 72; European Commission, Brussels, Belgium, 2010.Διαθέσιμο: http://ec.europa.eu/commfrontoffice/publicopinion/archives/eb/eb73/eb73_vol1_en.pdf. Last Accessed 20 Oct 2016

European Commission (2013) Climate change, special Eurobarometer 409/Wave EB80.2; European Commission, Brussels, Belgium. Διαθέσιμο: http://ec.europa.eu/commfrontoffice/publicopinion/archives/ebs/ebs_409_en.pdf . Last Accessed 27 Oct 2016

European Commission (2015) Citizen support for climate action, Special Eurobarometer 435/Wave EB83.4. Commission, Brussels, Belgium. Διαθέσιμο: https://ec.europa.eu/clima/sites/clima/files/support/docs/gr_climate_en.pdf. Last Accessed 27 Oct 2016

Eveland Jr WP, Cooper KE (2013) An integrated model of communication influence on beliefs. RNAS Proc Nat Acad Sci USA 10(3):14088–14095

Frangos CK (2004) Methodology of market research and data analysis with the use of the Statistical Package SPSS for Windows. Interbooks Publications, Athens

Freese F (1984) Forest sampling elements (trans. Karteris MA (ed) Aristotle University of Thessaloniki, Thessaloniki) (in Greek)

Freund R, Wilson W (2003) Statistical methods. Elsevier, New York

Hagen B (2016) Public perception of climate change: policy and communication: Routledge Studies in Environmental Communication and Media, pp 1–183

Harman H (1976) Modern factor analysis. The University of Chicago Press, Chicago

Ho R (2006) Handbook of univariate and multivariate data analysis and interpretation with SPSS. Chapman & Hall, New York

House of Commons Environmental Audit Committee (2008) Climate change and local, regional and devolved government, HC225. The Stationery Office, London

Jakob M, Luderer G, Steckel J, Tavoni M, Monjon S (2012) Time to act now? Assessing the costs of delaying climate measures and benefits of early action. Clim Change 114(1):79–99

Jolliffe I (2002) Principal component analysis, Wiley StatsRef: Statistics Reference Online

Kalamatianou AG (2000) Social statistics, methods of one-dimensional analysis. The Economic Publications, Athens (in Greek)

Kloser M, Bofferding L (2014) Middle and high school students' conceptions of climate change mitigation and adaptation strategies. Environ Educ Res 21(2):275–294

Lassen I, Horsbøl A, Bonnen K, Grethe A, Pedersen J (2011) Climate change discourses and citizen participations: a case study of the discursive construction of citizenship in two public events. Environ Commun J Nat Culture 5(4):411–427. http://dx.doi.org/10.1080/17524032.2011.610809. Last Accessed 25 Aug 2016

Leiserowitz AA (2005) American risk perceptions: is climate change dangerous? Risk Anal 25(6):1433–1442

Lupton R (1993) Statistics in theory and practice. Princeton University Press, Princeton

Marin A, Berkes F (2013) Local people's accounts of climate change: to what extent are they influenced by the media? Wiley Interdisc Rev Climate Change 4(1):1–8

Mariolopoulos EG (1938) The cimate of Greece. A.A. Papaspyrou Press, Athens

Mariolopoulos EG (1982) Climate epicenter of Greece. Center for Atmospheric Physics and Climatology Academy of Athens, Publication 7, Athens

Matis K (2001) Forest sampling. 2nd ed. Democritus University of Thrace, Xanthi

Murtagh F, Heck A (1987) Multivariate data analysis. D. Reidel Publishing Company, Holland

Nisbet MC, Scheufele DA (2009) What's next for science communication? Promising directions and lingering distractions. Am J Bot 96(10):1767–1778

Obani PC, Gupta J (2016) The impact of economic recession on climate change: eight trends. Climate Dev 8(3):211–223

Pagano M, Gauvreau K (2000) Principles of biostatistics. Ellin, Athens

Papoulis D, Kaika D, Bampatsou C, Zervas E (2015) Public perceptions of climate change in a period of economic crisis. Climate 3(3):715–726

Pearce W (2014) Scientific data and its limits: rethinking the use of evidence in local climate change policy. Evidence Policy 10(2):187–203

Rangou P (2013) Economy and environment: a framework for "re-reading" environmental education. Forestry and Management of Environment and Natural Resources Issues 5th Volume: International Environmental Policy, Confrontations with the Future. Democritus University of Thrace, pp 114–124

Risbey JS (2008) The new climate discourse: alarmist or alarming? Glob Environ Change 18 (1):26–37

Scruggs L, Benegal S (2012) Declining public concern about climate change: can we blame the great recession? Global Environmental Change. http://sp.uconn.edu/ ∼ scruggs/gec11.pdf. Last Accessed 15 Dec 2017

Sharma S (1996) Applied multivariate techniques. Wiley, New York

Siardos G (1998) Multivariate statistical analysis methods. Part one, investigating relationships between variables. Ziti Publications, Thessaloniki

Siardos G (2005) Methodology of sociological research. Enhanced and completed, 2nd edn. Ziti Publications, Thessaloniki

Souchon C, Raichvarg D, Goffin L (1996) Module d'Autoformation à Distance en Education pour l'Environnement (M.A.D.E.R.E.). Document de travail. Association Didactique, Innovation, Recherche en Education Scientifique (D.I.R.E.S.), Paris

Sterman JD (2011) Communicating climate change risks in a skeptical world. Clim Change 108 (4):811–826

Tampakis S (2000) Forest fires in Greece from a forest policy point of view. PhD thesis. Aristotle University of Thessalloniki

Toma L, Barnes A, Revoredo-Giha C, Tsitsoni V, Glenk K (2014) A behavioural economics analysis of the impact of information and knowledge on CO_2 capture and storage acceptance in the European Union. Proc Econ Finance 14(14):605–614

Tsantopoulos G, Arabatzis G, Tampakis S (2014) Public attitudes towards photovoltaic developments: case study from Greece. Energy Policy 71:94–106

Valenti JK (1995) Commentary how well do scientists communicate to media? Sci Commun 21 (2):172–178

Waddell P (2000) Towards a behavioural integration of land use and transportation modelling. International Association of Travel Behaviour Research Tri-annual conference, Gold Coast, Australia

Weber EU (2016) What shapes perceptions of climate change? New research since 2010. WIREs Climate Change 7(1):125–134

World Bank (2010) World development report 2010. Public attitudes towards climate change: finding from a multi-country poll. World Bank, Washington, DC. http://siteresources. worldbank.org/INTWDR2010/Resources/Background-report.pdf. Last Accessed 2 Sept 2015

YPEKA (2016) National strategy for adapatation to climate change. Ministry of Environment and Energy. http://www.ypeka.gr/LinkClick.aspx?fileticket=crbjkiIcLlA% 3D&tabid=303&language=el-GR. Last Accessed 15 Dec 2017
Zerva A, Tsantopoulos G, Grigoroudis E, Arabatzis G (2018) Perceived citizens' satisfaction with climate change stakeholders using a multicriteria decision analysis approach. Environ Sci Policy 82:60–70

Aikaterini Zerva has received an undergraduate degree in Forestry and Management of the Environment and Natural Resources from Democritus University of Thrace (2010), a Master of Science in Sustainable Management of Protected Areas from the University of Ioannina (in cooperation with the University of Patras and Aristotle University of Thessaloniki) (2012), a Master of Education and Disability from Universita degli Studi di Roma Tor Vergata (2015) and a Philosophy Doctorate also from Democritus University of Thrace (2018). She teaches in secondary education.

Evangelos Manolas was born in Naxos, Greece, in 1961. He has received a Bachelor of Arts in Sociology from the University of Essex (1983), a Master of Arts in International Relations from the University of Kent at Canterbury (1985) and a Philosophy Doctorate from the University of Aberdeen (1989). He is Associate Professor in the Department of Forestry and Management of the Environment and Natural Resources at the Democritus University of Thrace.

Constantina Skanavis is a Professor in Environmental Communication and Education at the Department of Environment, University of the Aegean, Mytilene, Greece. She is also the Head of the Research Centre of Environmental Education and Communication. She joined the University of the Aegean 15 years ago. Before that she was a Professor at California State University, Los Angeles; she has developed several courses on issues of environmental health and education. She currently teaches environmental education, environmental communication and environmental interpretation courses in undergraduate and postgraduate levels. She has numerous publications on an international basis and has given presentations all over the world.

Georgios Tsantopoulos has received an undergraduate degree in Forestry and Natural Environment from the Aristotle University of Thessaloniki (1995) and a Philosophy Doctorate also from the Aristotle University of Thessaloniki (2000). He is Associate Professor in the Department of Forestry and Management of the Environment and Natural Resources at the Democritus University of Thrace.

Climate Hack: Rapid Prototyping New Displays in Multi-disciplinary Museums

Charlotte Connelly

Abstract Ten years ago the Science Museum commissioned research in advance of the development of its new climate science gallery, finding that 'museums are high on the trust scale, but currently have a low profile in the climate change debate.' A decade later new research is needed into the role that museums can and should play in exploring the impacts of climate change. With eight diverse museums and a botanic garden, the University of Cambridge Museums are well placed to carry out research with a range of audiences and collections, not only those that are actively engaged with science and the environment. In January 2018 over a three-day 'climate hack', teams of climate experts, collections experts, storytellers, makers and hackers produced prototype exhibits that interpreted the links between museum collections, narratives and climate change. The event straddled the different types of collections held at the Museum of Zoology, Whipple Museum of the History of Science, Museum of Archaeology and Anthropology and the Polar Museum. It explored changes in animal diversity, the long history of scientific investigations, climate refugees and polar tourism. Visitor and participant feedback provides us with useful insights for developing future displays and developing a programme of audience research about how climate change might be interpreted across different types of collection and audience.

1 Introduction

How would you change one of our museums to tell stories about climate change? That was the question we posed to 24 people who participated in the University of Cambridge Museums' Climate Hack, an event that brought together makers, environment experts and museum professionals for an intensive long-weekend to build prototype museum experiences relating to climate change.

C. Connelly (✉)
The Polar Museum, Scott Polar Research Institute,
University of Cambridge, Lensfield Road, Cambridge CB2 1ER, UK
e-mail: charlotte.connelly@spri.cam.ac.uk

© Springer Nature Switzerland AG 2019
W. Leal Filho et al. (eds.), *Addressing the Challenges in Communicating Climate Change Across Various Audiences*, Climate Change Management,
https://doi.org/10.1007/978-3-319-98294-6_31

The University of Cambridge Museums and Botanic Garden (UCM) is a consortium of eight museums and collections across a range of subject disciplines. The different types of museums attract a range of different audiences, which puts the UCM in a strong position to investigate how museums can interpret and communicate content relating to climate change in multi-disciplinary museums. This is essential if climate change is to be understood and communicated as culturally embedded, as well as a set of processes that can be described and understood through scientific enquiry. This type of understanding of climate change, as a meeting of the natural world and our cultural activities within it, has been described by geographer Mike Hulme:

> The story of climate change… is a meeting of Nature and Culture, about how humans are central actors in both of these realms, and about how we are continually creating and re-creating both Nature and Culture. Climate change is not simply a 'fact' waiting to be discovered, proved or disproved using the tenets and methods of science. Neither is climate change a problem waiting for a solution, any more than clashes of political ideologies or the disputes between religious beliefs are problems waiting to be solved. (Hulme 2009: 20)

In seeking an approach to understanding climate change that cuts across our diverse museums and collections, the UCM has explored environmental change as something we respond to culturally as well as scientifically. In recent years the UCM has invested in exploring the 'green' content in its museums and collections, primarily through the 'Green Museums' project which took place from 2014–15. The project delivered schools workshops and events, and produced an online resource that presents 'green' material from across all of the UCM collections, grouping them under several themes including the history of environmental science, cultures under threat, and contemporary environmental science (University of Cambridge Museums 2015).

The Climate Hack was a continuation of the UCM's strategy of working across museum sites to explore climate and the environment using a variety of approaches, not just through science communication. The event was also a pilot for a model of co-curation across our sites, which the UCM is actively exploring as it works to become more representative of its audiences and democratise how our collections are interpreted and shared.

Co-developing content with audiences in this way is a well-established practice. The now fully assimilated 'Participatory Turn' in science centres and museums appeared in the literature well over a decade ago, with researchers including Sheila Jasanoff writing about it some fifteen years ago (Jasanoff 2003). The Participatory Turn was a response to the one-way 'deficit model' of science communication prevalent in the 1990s, and the later 'engagement' model which encouraged members of the public to take part in debates about the implications of science and research. Exhibitions reflected the shift towards engagement by asking questions of visitors, with the museums' role being to help visitors to find their own answers. The 'engagement' model gave way to a more participatory way of working with audiences not only to discuss, but also to develop interpretation of science in museums.

Having an arena of dialogue and debate is important, but it became increasingly clear that the follow up to those debates is as important as the opportunity to have them. Science communication happens not only between scientists and the public, but involves a complex network of stakeholders, all of which need to be involved in the conversation... European museums have demonstrated their capacity to act not only as forums for discussion, but also as brokers able to convey the public's ideas, opinions, desires and fears to a vast network of stakeholders. Museums have therefore become 'full players' in the governance of science. (Bandelli and Konijn 2015)

This type of practice was never limited to science topics. Nina Simon's influential book *The Participatory Museum* (2010) offered a framework for participatory practice which many museums took up. In the United Kingdom the rapid expansion of this type of practice was fuelled by the prevalence of Heritage Lottery Fund (HLF) support in the museum sector; in guidance published at the same time as Simon's book, the HLF stated one of their aims as to 'help more people, and a wider range of people, to take an active part in and make decisions about heritage (HLF 2010).'

Museums with anthropological collections have been particularly active in working with a wide range of communities to develop new collections and interpret existing collections. For example, in a 2014 collaboration between two museums, 'Rethinking Home: Climate Change in New York and Samoa', two diverse groups of about 15 people each participated in workshops and meetings to explore and interpret the meanings of climate change. In the wake of Hurricane Sandy and Cyclone Evan, both groups were aware of the impact of extreme weather on them as individuals, on their homes and on wider infrastructures. They could draw their own expertise into the project in which a new object was collected, and existing collections gained new interpretations (Newell 2017).

In the Climate Hack one of our goals was to explore climate change as both a cultural and scientific phenomenon, which means understanding both participatory practice and the democratisation of museum collections and the barriers to engagement with the science and technologies of climate change. In 2008 the Science Museum commissioned focus groups which revealed three factors that affected people's level of trust in sources of information about climate change:

- Hypocrisy—people did not want to be told what to do without seeing any evidence of others taking action;
- Profit—concerns were raised about (a) the altruistic act of combating climate change being championed by, for example, energy companies encouraging consumers to switch to renewable sources and (b) 'independent' scientists being funded by the Government or businesses;
- Inconsistency—sources are expected to have a consistent stance rather than changing their views to suit the latest trend.

The implications of these views were that the government, businesses and the media were least trusted by the focus group participants, while scientists, charities and non-profit public organisations such as museums were most trusted. In reporting these findings, Dillon and Hobson (2013) comment that:

The public's association of 'science' with 'truth' and 'facts' results in scientists being viewed primarily as independent truth-seekers. Confusion therefore arises over scientists' lack of agreement around climate change issues and this perceived 'inconsistency' is often cited as a reason for lack of belief in climate change.

One way to tackle the perceived need for consistency of voice in communicating climate change is to look beyond the scientific arguments. Cameron et al. (2013) argue for pluralistic responses to climate change in museum spaces, suggesting that artists and other well-informed publics can present climate change in new ways: 'museums need to rely less on presenting audiences with information and more on creating and designing richer experiences. The emotions they aim at should have range and balance, encompassing joy, wonder, and delight, rather than just pressing the buttons of fear and guilt' (Cameron et al. 2013: 18). Museums are in a unique position of being able to bring together artefacts, people with diverse expertise and audiences to interrogate climate change and its meanings. This is reflected in the three underlying goals of the Climate Hack:

(1) To explore environmental and climate change as a scientific and cultural phenomenon;
(2) To use a variety of approaches in our interpretation of objects and ideas relating to climate change in museums;
(3) To democratise how our collections are interpreted and shared.

2 An Introduction to the Climate Hack

The premise of the Climate Hack was simple. Over three days we challenged four teams to make a prototype exhibit or display to be installed in one of our museums: the Polar Museum, the Museum of Zoology, the Whipple Museum of the History of Science and the Museum of Archaeology and Anthropology (MAA). Participants applied by sending a brief outline of their skills and what they hoped to gain from participation. From the applicants we were able to put together four teams that each had a combination of skilled makers and artists; audience and interpretation expertise; and an understanding of environmental science or the impacts of climate change. The event was based at the Makespace, a community workshop in Cambridge for makers with a wide range of equipment for members to use including lathes, laser cutters, 3D printers, electronics workbenches and workshops for wood and metal work.

Participant recruitment targeted makers, environmental scientists and museum professionals. There was an emphasis on recruitment from within Cambridgeshire, however several participants travelled from further affield. The event was heavily oversubscribed and a selection process used the limited information we had asked for in the application form to form teams that had a range of complementary skills. As a pilot event we concentrated most of our planning efforts on the practicalities of running the event and ensuring teams had the skills they needed, however recruitment for future events should be more carefully designed to encourage a diverse range of applicants and participants. Diversity data was not gathered about the participants; however, one team member did comment that they were concerned 'by the (lack of) diversity in the participants' (anonymous post-event participant feedback). Given the underlying ambition of the event, to explore climate change as a cultural challenge as much as a scientific one, ensuring diversity and a range of viewpoints among the people invited to co-curate interpretations of climate change is extremely important.

The event itself took place over three full days. The decision was taken not to introduce team members to each other in advance for two reasons: (a) the event already required a substantial time commitment from participants which they had agreed to on application, and by sharing contact information some participants may have felt pressured to commit still more time in advance of the event; and (b) the aim of the event was for teams to respond to the museum collections, and there was concern that with time to correspond in advance ideas might generated and agreed by team members before they had encountered the museum or collection they would be working with. Participant feedback after the event revealed a mixed response to this approach. One participant suggested that we could establish a dialogue before the event, 'encourage tweeting by seeding questions, set up a Slack channel for chatting and sharing examples of good practice before and throughout the process.' Other participants valued the intensive and self-contained process that they had been involved with. Around a quarter of the participants suggested that the thing they would have valued most, if we could make changes, would be to have 'more time', however some respondents also recognised that a requirement to commit more time might have prevented their participation altogether.

On the first day of the Climate Hack all the participants met in the Makespace for a heavily structured day. Throughout the morning, facilitated team building activities were interspersed with presentations about our audiences and what audience research has shown about museum visitors and climate change in the past. Teams were invited to map out their knowledge about climate change and what it means to them. The majority of comments presented in feedback from the teams discussed the impacts of climate change, for example on sea levels or melting ice caps. A smaller proportion of comments reflected on changes that individuals can make in order to reduce their personal contribution to climate change, such as recycling or reducing air travel. There was also some discussion about how trustworthy information in the public domain is about climate change, mirroring the audience research carried out in 2008 for the Science Museum (Dillon and Hobson 2013). In this regard our participants, despite being skewed through self-selection towards taking an active interest in climate change, interacted with the topic of climate change in a similar way to focus groups with little or no specialist knowledge or interest. Other findings from the research carried out for the Science Museum were also shared (TW Research 2008), in particular the aspects of climate change that consistently gained interest among audiences which included:

- Impact on poor countries
- Potential impacts in the UK
- Human stories
- Climate change has been known about since the 1950s.

These four key findings were particularly relevant as each of the prototypes produced by the teams had one of these messages as its foundation.

On the afternoon of day one, the teams travelled to their individual museums where they were introduced to the collections and the particular challenges faced by each of the museums in engaging audiences with the topic of climate change. Following these encounters the teams had time to begin discussing ideas for the prototypes they would make. At this stage practical guidance about what was or was not possible in a prototype design was given. The suggested outcome of a 'prototype' was kept deliberately vague so that teams did not feel constrained or heavily steered by the UCM organisers. This also meant that we had no idea what the teams would produce. Handing over control of our museums' content was both a little frightening and very exciting, especially as we were inviting the public into the museums to see the teams' final creations on a topic that often proves to be controversial. Post-event participant feedback suggests that the guidance on what a prototype might be was a little too vague, and several of the teams would have benefited from a few concrete examples of expected outcomes to help them understand the type of product the different museums were expecting or hoping for. Nevertheless, on the second day of the Climate Hack, after a short period of time to prepare in the morning, all of the teams pitched their ideas to the rest of the group. The pitches were an opportunity to share constructive feedback and for the organisers to provide practical guidance or further support if needed before the teams began building their prototypes.

3 Museum of Archaeology and Anthropology

The Museum of Archaeology and Anthropology (MAA) introduced their team to developing plans for a new Pacific gallery. Many people living in the Pacific region live on islands that are already suffering the consequences of rising sea levels. The Climate Hack team based at MAA were introduced to the challenges faced by people living in Kiribati, where moderate projections suggest that 55% of the main island could be vulnerable to inundation or storm surges by 2050, and that Kiribati could face annual economic damages due to climate change of up to a third of its gross domestic product (Wyatt 2013). The team focused on the idea of flooding, and decided that their exhibit would share stories of flooding and migration in Kiribati, but also in the fens around Cambridgeshire. In learning about Pacific islanders the team encountered artworks created from 'ghost nets'—abandoned, lost or discarded fishing nets that continue to float around the oceans and cause damage to marine organisms (Wilcox et al. 2015). A number of indigenous artists living in the Pacific regions have begun making artworks from ghost nets, some of which were displayed in an exhibition called *Au Karem ira Lamar Lu* at the Asian Civilisations Museum in Singapore in 2017.

Inspired by artworks created from ghost nets and stories of flooding, the Climate Hack team built a life-size 'ghost boat' in which written stories and wooden 'story hooks' engraved with links to online stories of flooding from around the world were entangled. Visitors to the exhibit were also invited to listen to a storyteller, share their own stories, and make a paper boat in which they could write down what they would save in a flood. The project also has an online web presence which captures a number of stories, as well as describing the processes the team went through and some of the outcomes in much more detail than can be explored here (McKenzie et al. 2018).

The prototype display was extremely engaging, and some visitors spent a significant amount of time listening to stories, making a paper boat and choosing what they would save from a flood. All the visitor feedback that was left was positive, with around 30% of the feedback comments specifically mentioning storytelling as a positive part of the experience. Although this way of engaging audiences is very resource intensive—with 2–3 people needing to be present to facilitate visitors' experiences—the dwell time and level of engagement was deeper than in static displays. The focus on people and their experiences was also highlighted as valuable by several pieces of audience feedback, with comments including: 'It brings attention to the extreme challenges facing many populations and societies, that aren't always immediately present in the mind of us privileged groups' and 'Great to hear a story about how those who haven't caused the problem are nevertheless suffering the consequences of others' action and inaction.' These kinds of comments also revealed that the comparisons between fenland populations a few miles outside of Cambridge and remote communities of islanders in the South Pacific did not always register with visitors despite comparable weight being given to both types of narrative in the museum interpretation.

4 Museum of Zoology

The team based at the Museum of Zoology had a different challenge to the other three teams because the Museum was still closed for redevelopment when the Climate Hack took place. The only part of the Museum that was accessible was the whale hall—a large and mostly empty hall, apart from an impressive finback whale skeleton which hangs from the ceiling of the double height room. Following an introduction to the collections of the Museum of Zoology the team developed two ideas into exhibits.

Visitors first encountered an interactive activity that explored carbon capture. They were encouraged to grab models of molecules from a wall and bin them if they were greenhouse gases. A sign offered some explanation about the challenges of capturing carbon and other greenhouse gases, and one of the team was present throughout the public opening hours to explain more about this to visitors. As with the experience at MAA, having a person available to offer interpretation was a valuable part of the experience for visitors.

A second exhibit developed by the team, like the MAA experience, invited visitors to listen to and share stories. This time the storytelling was delivered through an audio exhibit. The physical form of the exhibit was inspired by the Holme Post, a cast-iron column installed in 1850 in the clay floor underlying the layer of peat that has built up on the fenlands of Whittlesey Mere in Huntingdonshire, just outside of Cambridge. The post provides a fixed-point against which peat wastage can be measured, showing how the environment has changed over time (Hutchinson 1980). The Holme Post inspired listening-post built by the Climate Hack team also offered a way to understand the passage of time: visitors were invited to plug headphones into the post in order to listen to stories of environmental change or sounds of the environment, for example now-extinct bird calls. Visitors were also encouraged to leave their own stories about environmental change which could be incorporated into the audio experience later. The idea of change over time was explored further through some of the visual interpretation near the post. Graphics showing sedimentary layers and ice cores explored the different ways and different time scales that we use to understand environmental and climate change.

Feedback from visitors was again generally positive, with several comments noting the interesting use of timelines as an aid to understanding and the 'unusual' use of audio to explore how our environments have changed over time. The use of carbon-capture as a way to explore greenhouse gas emissions provoked some less positive responses. Some comments wanted the exhibit to be embedded in more practical considerations, suggesting 'It'd be great to have an interactive display to allow children to 'stop the flow' of greenhouse gases (e.g. turning off lights, not going by car) to see how many emissions you've saved.' Another commenter thought it better not to highlight carbon capture at all:

> I do not agree on highlighting carbon capture in a public exhibition. I think the priority should be put in stopping adding carbon to the atmosphere. Therefore, public expositions should condemn the construction of new fossil fuel infrastructure above all.

Throughout the Climate Hack the audience feedback was predominantly positive, with no feedback indicating a position of climate-scepticism. Where there was negative feedback, as shown above and in further examples below, it was from a position of environmental advocacy. From the feedback it was evident that some visitors expect museums to take a clear position not only that climate change is happening, but also to communicate ways of preventing or mitigating further environmental change. The feedback indicated that for those visitors, museums should not be places to explore pragmatic approaches to living with climate change. It should be remembered that the Climate Hack and the public facing element of the event, the Climate Hack Showcase, were very clearly marketed as being about climate change, and that audiences to some extent will have self-selected. Potential visitors with ambivalence toward or an active dislike of the topic of climate change would be unlikely to attend. For a full picture of visitor expectations and responses to these interventions an event that was less explicitly concerned with climate change would be needed.

5 Whipple Museum of the History of Science

The Whipple Museum team developed an exhibit that used the Museum's collections as a starting point, selecting objects from the history of the environmental sciences for display. One of the objects they selected was a Campbell-Stokes sunshine recorder (object number: Wh.5173), a device that uses a glass globe to focus the light from the sun onto a detecting plate, where the focused sunlight burns a track in a measuring card. Although called a sunshine recorder, the Campbell-Stokes instrument really provides a measurement of what is between the sun and the recorder. If there is a lot of cloud cover, no track is burnt in the card; if there are some particles in the air the track will be discoloured and blackened but not burnt through; and on a clear day a clean gash is burnt along the length of the measuring card. This object inspired a physical interactive which enabled visitors to manipulate a large scale interpretation of the Campbell-Stokes device and see how it could be used to record atmospheric conditions. The exhibit also looked at contemporary exploration of the sun and its effects on the Earth, delivered through the display of a model satellite and a short animated film describing some research being undertaken today.

More than a third of the people who left feedback commented on the comparison of time periods, leaving remarks such as 'It was interesting to think about the long-term history of the way we have measured the climate' and 'The exhibit showed me the past and future of climate change and the science behind it. It was very informative.' The experience was particularly good at linking the collections on display with visitors' understandings of climate change, with visitors commenting on how it made concepts of climate change more tangible: 'Tactile sunlight recorder really helps to visualise how such a tactile instrument works in real life.' The display itself, more than the other prototypes, very consciously used found and recycled materials, which also appealed to some of the visitors. The team also presented a very high-quality illustration of what the prototype could be developed into with some investment.

All of the visitor feedback at the Whipple Museum was positive, and the exhibit scored highly on visitor voting slips leading to the installation winning the audience award. In many respects this installation was the most straightforward piece of science communication of any of the prototypes, and in the process it avoided engagement with any controversial or political topics. Nevertheless, the instrument and concepts it delivered were complex, and the visitor feedback demonstrated that many people who engaged with the installation enjoyed the opportunity to explore a single instrument in detail. Displays of climate science often focus on whole-Earth systems, and the focus on a single instrument and its tactility helped to make the observations of climate change more comprehensible and more human.

6   Polar Museum

The Polar Museum team learnt about the opening Northwest Passage, a sea route across the Arctic which until recent times has rarely been navigable because sea ice has blocked the route. In recent years the passage has become much more easily navigated as temperatures rise in the Arctic and sea ice lessens. The team were introduced to Arctic sea ice data published in the Hadley Centre Sea Ice and Sea Surface Temperature data set (HadISST.2); the data reveals the rapid changes in sea ice in recent years after a period of relative stability (Titchner and Rayner 2014). The team became interested in the impacts the reduction in sea ice is having on Inuit peoples living in the Arctic, and particularly the resilience of those communities. Their interests led them to work of researchers like Elizabeth Marino, who writes that indigenous Arctic communities are often described as 'vulnerable'. Yet, she argues: 'labelling certain groups as "vulnerable" can be stigmatizing and can result in the re-creation of outdated and racist stereotypes of indigenous peoples needing the help of white outsiders. The label can imply a lack of agency and competence' (Marino 2015: 29). Elsewhere in her book Marino describes how settlement, partly enforced by legislation requiring all school-age children to attend school, diminished patterns of nomadism that had made Arctic peoples resilient when living in harsh and changeable environmental conditions. As the team explored the effects of climate change on Arctic peoples the Climate Hack team uncovered both challenges and opportunities that arise from the rapid changes they are facing. Challenges to the communities stem in part from strong dependence on climate sensitive resources, such as fish, wildlife and plants, while rising sea levels, coastal erosion and thawing permafrost damage settlements and compromise infrastructure. Despite these substantial challenges, there have been some benefits to communities including opportunities for economic development, growing tourism and the potential for new commercial fisheries (Stern and Gaden 2015: 404).

The team developed a prototype that would demonstrate the changes in sea ice and provide a pluralistic interpretation of the impacts of climate change on the Arctic, offering a range of perspectives from different types of people affected. The first part of their exhibit was an interactive display; a digital projector was used to overlay Arctic sea ice extent onto a map of the Arctic that is painted on the ceiling of the Polar Museum, using data from the HadISST.2 dataset. Visitors were invited to turn two dials, the first dial altered the month that was being displayed, while a second dial altered the year that was being displayed—this made it very easy to visualise both changes in sea ice throughout the year, and how in recent years there has been a noticeable decline in sea ice in the Arctic. The interactive display prepared visitors for the second part of the exhibit, a family friendly museum trail which took visitors on a tour of the Northwest Passage. As they followed the trail, finding clues and being directed towards objects in the museum display, visitors also encountered three characters: a scientist, a tourist and an Inuk. Each of the characters was introduced in an information panel which explained why the Arctic is important to them, and some of the challenges and opportunities they have encountered as a result of climate change.

As with the other three prototype displays, visitors were very engaged with the exhibit and the majority of the feedback was positive, with one piece of feedback particularly noting the balance of positive and negative that the team had sought to present: 'this was my favourite climate hack as it really embraced the museum as a whole and showed multiple sides, both positive and negative to the opening up of the north-western passage.' Also in line with the other displays, where there was negative feedback it was from climate advocates arguing that the viewpoints we were presenting were unethical or irresponsible: 'Western capitalist socioeconomic logic has wrecked ecosystems, landscapes and communities in the Arctic. You may see a solution in tourism, but I just find it an immoral perversion to impose them in a drastic livelihood change in the basis of yet more aggressive search for "profit".' There were also concerns raised that we were speaking on behalf of indigenous communities without their input. Although the messages being communicated were derived from recent research by and with Arctic communities, querying the role of those communities in the development of an exhibit is reasonable. This was acknowledged by the team during the event, and in post-event feedback one of the team commented: 'With more time it would have been nice if we could contact and interview Inuit people to find out their actual views on tourism in the Arctic, but obviously within the time constraints I believe we did what we could.' In a longer term project the Polar Museum would seek to consult or co-create displays with Arctic peoples in line with its usual practices.

7 Evaluation

The climate hack was a new venture for the University of Cambridge Museums, and one that yielded results that were both excellent and unexpected. Twenty-four participants engaged very deeply with collections and climate change, and a much larger number of visitors had the opportunity to interact with the prototype exhibits. Of those visitors who chose to leave feedback, 88% agreed that the exhibits had 'inspired them to think about differently about climate change'. The hack format allowed rapid development of new interpretations of UCM collections and their relationships with climate change, and enabled a range of people to choose the narratives shared by the museums with their visitors.

The event also provided an opportunity to find out more about what visitors to UCM museums expect from narratives about climate change. The public element of the event was marketed as the 'Climate Hack Showcase', which may have led to a degree of self-selection among visitors. The total absence of feedback from visitors expressing sceptical views about climate change is therefore not surprising, although it could also indicate shifting public opinion towards the topic—more audience research is required to understand this in detail. Instead negative feedback came from visitors and participants with a high level of pre-existing knowledge about climate change, who felt that the museums should be acting as advocates for particular ways of living with our changing world. Perhaps the most explicit articulation of the high expectations people hold for the museum was in a feedback comment from one of the Climate Hack participants. In discussing the prototype at the Polar Museum they argued that the choice to share not only challenges but also opportunities for people living in the warming Arctic, the exhibit presented 'totally the wrong climate change message, that shouldn't be in any museum.' The implication of feedback like this is that for climate advocates, museums are seen as spaces that should engage with the topic of climate change primarily to inform and educate visitors about how to mitigate climate change. However, the University of Cambridge Museums strive to be a space for sharing current research and enabling debate and discovery not only about hoped for futures, but also about the often messy and complex realities that people are and have been living with. By taking on a strict advocacy role, museums would have to limit the types of research and narratives they could share.

8   Conclusion

Museums are in the fortunate position that they are seen by many people to be removed from marketplace and governmental influences, and as a result they benefit from high levels of respect and trust. The perception of museums' impartiality which partly supports the degree of trust our visitors place in us, makes it vital that museums enable a range of voices to be heard and experienced. As Bunning et al. (2015) have described, 'a key value ascribed to participatory work in much of the [museum studies] literature is the shift in power from 'experts' to 'non-experts' that enables museums to foreground new and diverse voices.' Although the language of expertise is problematic, this statement comes close to describing the third of the stated ambitions of the Climate Hack. In sharing the opportunity to develop new displays with a group of participants, the event was also able to achieve the first and second of its goals:

(1) To explore environmental and climate change as a scientific and cultural phenomenon;
(2) To use a variety of approaches in our interpretation of objects and ideas relating to climate change in museums;
(3) To democratise how our collections are interpreted and shared.

Overall then, the Climate Hack was a success. The learning points outlined below will feed into future participatory practice and explorations of climate change across all of our museums and collections.

- Visitor feedback showed that their experience was overwhelmingly positive and that they enjoyed encountering prototype exhibits that varied from their usual museum experience and offered new perspectives on the topic of climate change.
- With a relatively small budget and a short time scale it is possible to enable meaningful engagement with a significant number of people on a topic as complex and sometimes controversial as climate change.
- Although quantitative data was not gathered, anecdotally it appeared that our participants and visitors self-selected and were already engaged in the topic of climate change. To reach a more diverse audience the focus on climate should be less overt when recruiting for and promoting similar events in the future.
- In recruiting team members, more efforts should be made to encourage and enable participation from a more diverse group of people. Care is also needed to ensure that as far as possible the timing and level of commitment required for the event does not prevent participation from key groups of people.
- Concrete examples of what a 'prototype' could be would have been beneficial to the teams, while still allowing them freedom to develop their own ideas.
- Some participants expected the museums to take on the role of climate activist. An open discussion about the museums' approaches to communicating climate change, and other contested and complicated topics, could have been productive and helped to build a stronger mutual understanding between participants and the museums.

The findings of this paper are necessarily limited to the relatively small group of 24 participants and around 100 visitors who experienced the outcomes of the Climate Hack. Future projects and audience research will provide a greater understanding of how the topic of climate change can be woven into narratives across a wide range of different subjects and types of museum. Meanwhile continued investment in participatory practice will continue to diversify the voices and perspectives present in our museums. In Cambridge, the 'hack' model of inviting audiences into the UCM's spaces to reinterpret collections will continue to develop, and it is hoped that some of the prototypes will be developed further in consultation with the teams—meaning that the Climate Hack will leave a real legacy. With only the small number of exceptions described above, feedback from visitors and participants was overwhelmingly positive. The greatest achievement of the Climate Hack, though, was that through the use of storytelling, audio installations and interactive experiences, the enormous and often very abstract topic of climate change was transformed into something that was both tangible and personal for participants, visitors and museum staff.

References

Bandelli A, Konijn EA (2015) Museums as brokers of participation. Sci Museum Group J 3
Bunning K, Kavanagh J, Mcsweeney K, Sandell R (2015) Embedding plurality: exploring participatory. Sci Museum Group J 3
Cameron F, Hodge B, Salazar JF (2013) Representing climate change in museum space and places. WIREs Clim Change 4:9–21
Dillon J, Hobson M (2013) Communicating global climate change: issues and dilemas. In: Gilbert JK, Stocklmayer S (eds) Communication and engagement with science and technology: issues and dilemas. Routledge, London
Heritage Lottery Fund (2010) Thinking about... community participation. Available at https://www.hlf.org.uk/community-participation. Accessed 17 Mar 2018
Hulme M (2009) Why we disagree about climate change: understanding controversy, inaction and opportunity. Cambridge University Press, Cambridge
Hutchinson JN (1980) The record of peat wastage in the East Anglian Fenlands at Holme Post, 1848–1978 A.D. J Ecol 68(1):229–249
Jasanoff S (2003) Technologies of humility: citizen participation in governing science. Minerva 41 (3):223–244
Marino E (2015) Fierce climate, sacred ground: an ethnography of climate change in Shishmaref, Alaska. University of Alaska Press, Fairbanks
McKenzie B, Ash J, Brooks R, Leeper M, Marshall K, Velez Vago M (2018) Ngaru: the heart of the world. Available at https://ngarughostboat.wordpress.com. Accessed 19 Feb 2018
Newell J (2017) Talking around objects: stories for living with climate change. In: Newell J, Robin L, Wehner K (eds) Curating the future: museums, communities and climate change. Routledge, London
Stern GA, Gaden A (2015) From science to policy in the western and central Canadian arctic: an integrated regional impact study (IRIS) of climate change and modernization. Available at www.arcticnet.ulaval.ca. Accessed 20 Feb 2018
Titchner HA, Rayner NA (2014) The Met Office Hadley Centre sea ice and sea surface temperature data set, version 2: 1. Sea ice concentrations. J Geophys Res Atmos 119:2864–2889
TW Research (2008) A climate change gallery at the Science Museum (unpublished)
University of Cambridge Museums (2015) Green Museums. Available at https://www.museums.cam.ac.uk/working-together/green-museums. Accessed 16 Feb 2018
Wilcox C, Heathcote G, Goldberg J, Gunn R, Peel D, Hardesty BD (2015) Understanding the sources and effects of abandoned, lost, and discarded fishing gear on marine turtles in northern Australia. Conserv Biol 29:198–206
Wyatt K (2013) Escaping a rising tide: sea level rise and migration in Kiribati. Asia Pacific Policy Stud 1:171–185

Planning a Life Cycle Analysis Library and Beta Tool for Sustainable Cultural Heritage Preservation and Exhibition Practices

Sarah Nunberg, Sarah Sutton and Matthew Eckelman

Abstract Custodians of cultural heritage have begun to employ Life Cycle Assessment (LCA) to evaluate the environmental impact of materials and actions that curators, conservators, registrars and art handlers employ. LCA is a popular systems modeling tool useful for quantifying the total resource inputs and environmental burdens of a particular product or process. This paper discusses a US-based, federally-funded project established to create a free online LCA library of materials and methods relevant to preservation of cultural heritage. The project will also produce an LCA beta tool that provides collections care professionals with guidance to achieve sustainable goals through informed choices. The paper reports on two of the three LCAs, both addressing different aspects of maintaining cultural heritage: (1) cleaning methods and their environmental and human health impacts, and (2) cradle-to-gate impact of manufacturing, using and displaying three seventeenth and eighteenth century silver objects. The lists and categories gathered to populate the beta LCA tool are also discussed. The specific scoping and classification challenges that were encountered during the initial project phase are presented, and the methods for fulfilling the grant, the specific LCAs and materials lists are described. The issues learned and information acquired during the project will be useful in defining the terms for the next project phase where the full tool and

S. Nunberg (✉)
The Objects Conservation Studio, LLC, Brooklyn, NY, USA
e-mail: snunberg@aol.com

S. Nunberg
Department of Math and Science, Pratt Institute, Brooklyn, NY, USA

S. Sutton (✉)
Sustainable Museums, Waialua, HI, USA
e-mail: sarah@sustainablemuseums.net

M. Eckelman (✉)
Department of Civil and Environmental Engineering,
Northeastern University, Boston, MA, USA
e-mail: m.eckelman@neu.edu

© Springer Nature Switzerland AG 2019
W. Leal Filho et al. (eds.), *Addressing the Challenges in Communicating Climate Change Across Various Audiences*, Climate Change Management,
https://doi.org/10.1007/978-3-319-98294-6_32

library are realized. The final project will be freely available to users worldwide, and will support further research in preventive conservation, treatment, and exhibition through conducting material analysis, organizing knowledge and sharing it openly.

1 Introduction

Through implementation of a National Endowment for the Humanities (NEH) Tier 1 Research and Development Grant, this project addresses the important question: *at what cost to the environment and to human health is cultural heritage collected, preserved, and exhibited?*

Custodians of cultural heritage use known carcinogens in conservation treatments, curate exhibitions that require long distance transportation of artifacts world wide, and advise energy intensive environmental management, often with little consideration of potential impacts to the environment and health. Many custodians of cultural heritage believe that the quantities of solvents, amount of air travel, frequency of construction for exhibition and storage, and use of heating and cooling are minimal compared with industry and most business uses, justifying their lack of concern and reluctance to change work habits. Yet there are 726,000 museum sector jobs that involve transport, care, and display of objects in the United States (AAM 2017), United States art museums currently house 16,943,955 collection objects (AAM 2016), and in 2004 American museums recorded housing over a billion objects in collections (IMLS 2016). The reluctance to recognize personal impact exists not only in cultural heritage but is a problem throughout society, as individuals and many professions readily recognize the consequences of actions from other industries, not their own (Marshall 2014).

An additional significant barrier to sustainable practices is the lack of clear information concerning the consequences of our actions, which requires analyzing not just what we do in our own practices, but also how our resources are manufactured, transported, and disposed. This Tier 1 grant strives to create a specialized tool that will help custodians of cultural heritage choose best work practices with redused environmental and human health impact. Some major museums have begun to dedicate exhibition time towards educating the public about sustainable practices, but a tremendous gap in sustainable practices and education remains. Consequently, many institutions forgo teaching opportunities and modes for sustaining their relevance towards twenty-first century concerns (Janes 2011; Anderson 2012).

The National Endowment for the Humanities is an independent grant-making agency of the United States government that was established to address major challenges in humanities research, education, preservation of humanities collections, and public programs. One such project granted in 2016 to the Foundation for the American Institute for Conservation of Art and Historic Artifacts (FAIC), explores expanding resources for devising sustainable, non-toxic approaches to preserving cultural heritage. This grant, titled "Planning a Life Cycle Analysis Library of Preventive Conservation Methods," funds a Tier 1 research project to create metrics that will allow cultural heritage professionals to evaluate the health and environmental impacts of sustaining humanities collections. The FAIC project envisions a Life Cycle Assessment (LCA) library that will house LCA research selected to serve the collections care field, and an LCA beta tool using sets or lists of materials and actions that professionals can customize according to their own projects. The LCA beta tool and LCA library will be explored and tested during this Tier 1 project, with the goal of further development in a Tier 2 implementation project and eventual release to the community.

The users of the proposed LCA library and LCA tool will be professionals associated with the care and exhibition of cultural heritage collections: facilities managers, registrars and collections managers, curators and conservators, librarians and archivists, exhibit designers and preparators, and shippers and handlers. Through familiarizing themselves with LCA and the use of the proposed library and tool, custodians of cultural heritage will be able to manipulate components of the LCAs in the library to examine their own choices and practices (LCAs are described in detail below). Work this far, two-thirds of the way through the grant, has shown the evolution of a research project with constant exploration of new and unexpected findings. The Tier 1 research and development phase has provided the basis for a strong Tier 2 implementation grant application to complete the project.

The tool will be easily accessible field-wide, allowing professionals to evaluate their work plans and make effective and educated changes towards sustainable, non-toxic practices. In comments on the project, the proposal reviewers noted the potential for this project to be successful both with "the difficult task of creating concrete yet widely applicable" tools, and one that would reveal "new information of value to decision-making."

2 Sustainability in Collection Care

For more than a decade, new and renovated "green buildings" under the auspices of the United States Green Building Council (USGBC) Leadership in Energy and Environmental Design (LEED) rating system, have been the museum profession's icons of environmental sustainability and response to climate change. The California Academy of Sciences, renovations to systems at the National Gallery of Art in Washington, DC, and the development of the collections facility at the Museum of Northern Arizona are widely recognizable examples. Much of the efforts have focused on the physical impact of the structures that house collections, specifically gallery lighting and environmental management, generally exploring how to change those spaces and systems to save energy and therefore money. The payoffs have been measured by tangible returns on investment, while removing a key imagined barrier to building and operating more environmentally responsible museums.

Cultural heritage institutions often overlook the well being of both the people who maintain collections and the objects in the collections as directors and curators continue to acquire materials and conservators carry out extensive, complex treatments. Goals to achieve sustainable practices opens a great opportunity to challenge working methods. The products that professionals select, whether they are chemicals or containers, application tools or packaging materials, all have environmental and human health impacts from production through disposal. Actions carried out including international loans, or large scale renovations, increase energy use and result in a myriad of environmental impacts. LCA offers the opportunity to identify worst and best practices and has already been helpful in designing sustainable treatments and evaluating energy use in environmental management in six studies. Three of these studies have been published to date (Nunberg et al. 2016).

3 Life Cycle Assessment (LCA)

3.1 Development and Implementation

The development of life cycle assessment began in the 1960s and 1970s and initially focused on quantifying energy use and emissions (Curran 1993), in order to compare different materials and product designs on a life cycle basis. LCA methods have since undergone several waves of development and formalization, and new data and scientific models are continually being incorporated to improve the accuracy and relevance of LCA results. LCA tools are now used around the world by companies, governmental agencies, NGOs, and scientific researchers, and have been applied to thousands of products and processes across a wide range of sectors. While the goals and context of each LCA study may differ, many common protocols and standards have been codified to provide a common methodological platform for all LCA practice. The most important of these consensus standards is the series promulgated by the International Organization for Standardization (ISO 2010).

3.2 Standards

The series of standards documents ISO 14040–14049 describes the principles and framework for LCA including: definition of the goal and scope of the LCA, the life cycle inventory analysis phase, the life cycle impact assessment phase, the life cycle interpretation phase. LCA should follow the principles and guidelines of the ISO 14040. ISO 14040 sets standards for the LCA, which each LCA practitioner must follow (ISO 2010).

3.3 Components

Whereas many analyses focus just on museum energy use or direct emissions during object conservation, an LCA is so-named because it examines a product or process over its entire life cycle and quantifies resource use (energy, water, materials) and emissions in each phase. A product that is 'emissions-free' during use may be dangerous or toxic to manufacture or have deleterious effects on the environment when it is disposed, but this information is not typically available to consumers. Choices made with good intentions but without considering life cycle effects may end up doing more harm than good. LCAs are typically scoped in one of two ways. *Cradle to grave* analysis encompasses the lifespan of an object or action from first manufacture to use and final disposal. *Cradle to gate* analyses include only up to the 'gate' of the manufacturing facility. For example, examining a piece of paper from cradle to gate considers growing and processing the raw materials, including fertilizers and other agricultural inputs, harvesting, transport, and all of the materials including water and energy that go into making the paper. Cradle to grave would examine everything in the cradle to gate study along with sale, using the paper and disposal after use. Storage after use could also be explored. Whether a study encompasses cradle to gate or cradle to grave of an object or action depends on the extent of the *system boundary* defined for the study.

When analyzing the act of exhibiting an object of art, the system boundary may include: art transport and the courier, crate construction and packing materials; art storage and the associated energy for environmental management (HVAC system); gallery preparation including exhibition cases, vitrines, gallery wall construction and paint finishes; environmental management of the gallery; administration processes involving report writing, travel arrangements and associated energy through computer use. One might choose not to include the energy and resources expended in making the art object or the impact of building the museum that houses the objects.

Through establishing system boundaries, LCA helps to track performance of a product or action by structuring or organizing the relevant parts. Definition of the system boundaries will depend on the goals of the study. Because the scope of an LCA must be designated by the practitioner and the client, LCAs are often product or project-specific (the practitioners or the person who commissions the LCA product selects which direct impacts to focus on in the LCA).

Because the field of cultural heritage is so material specific, and many actions and materials are adapted from other industries, it is difficult to draw only from the established databases. To support effective, educated decision-making, this project aims to populate the library of an online tool with materials and methods used by cultural heritage professionals. Access to this database will allow professionals to examine specific scenarios to understand the ramifications of his or her actions. One goal of running the database will be to identify aspects of a group of materials or actions that are the least sustainable, either producing the most waste or resulting in the highest energy use or toxic output. These identified areas are called hot spots.

For example, a museum may hope to study the impact of replacing halogen lights with LED housings in a specific gallery or throughout the museum. LCA has been used at the Museum of Fine Arts Boston (MFA Boston) to explore parameters of their choices, understand the ramifications of using the different bulbs and housings, and make appropriate choices (Nunberg et al. 2016). With knowledge of hot spots, a conservator can evaluate and alter aspects of a system to result in more sustainable practices, such as choosing local sources to reduce the impact of transportation, or design the system with the least toxic solvents. Instead of depending on decisions based on habit, familiarity or incomplete information, cultural heritage professionals can make their own informed decisions without requiring expertise in industrial hygiene or environmental engineering.

3.4 Categories of Environmental Impact

After quantifying resource use and emissions over the life cycle, the LCA practitioner then considers what effects these might have on the environment and public health. There are many types of harmful impacts that we would like to avoid, and so LCA typically considers a range of impact categories in order to avoid unintended consequences and to make the best product choices overall. Common categories of environmental impact include but are not limited to global warming/climate change potential, ozone depletion potential, smog formation, acidification, eutrophication, cultural eutrophication, human health toxicity, cancer/non-cancer potential, and ecotoxicity. Additional impact categories include: radiation, abiotic resource depletion, fossil fuel depletion, biotic resource depletion, energy demand, water use, land use, nuisance-related effects (noise, odor), and indoor air quality. The selection of impact categories to include in an LCA is up to the practitioner, but is commonly guided by industry recommendations or assessment model developers such as the US Environmental Protection Agency.

4   Research Process

Three institutions, differing in size and mission, were identified as sites for LCA case studies; the MFA Boston, The Princeton University Art Museum, and The Gibson House Museum. The MFA and the Gibson House Museum, among other institutions, provided materials lists for the beta tool. Acquiring data for the beta tool and case studies for the LCA library required site visits to the Gibson House Museum and the MFA by the project co-directors, Sarah Nunberg and Sarah Sutton, the environmental engineer, Matthew Eckelman, and two consultants, Michael Henry, architect and engineer, and Pamela Hatchfield, head objects conservator from the MFA Boston. The goal was to review with curators from each institution the daily, seasonal, and annual practices to begin to develop basic materials lists, processes, and practices, and to help define the outside boundaries of the workplace needs and practice for future users of the LCA Library and the beta tool.

4.1 Cultural Heritage LCA Tool for Sustainable Practices

4.1.1 Developing the Beta Tool

To populate the LCA library, the project directors and peers developed lists of materials and actions that are relevant to collections care, attempting to identify as many items as possible. The project co-directors guided peer reviewers to collect data that would support a beta version of an end-user LCA decision-making tool available through the AIC website.

The first step was to acquire lists of materials used and actions carried out by conservators, registrars, curators, art handlers, and exhibition management. This list would include preventive conservation, exhibition related activities, all aspects of traveling art for loans, and treatment materials and methods. The project directors asked a range of art museums, historic houses, archives, and private conservators to provide lists. However, many institutions did not have access to lists of all the materials purchased in a year, or many museums found list-gathering difficult due to department overlap in product ordering. Out of fifteen museums, historic houses, library archives, and private conservation studios contacted for this project, three were able to easily provide lists, two assigned interns to assemble the lists, and three took the opportunity to assemble the list themselves (one conservation lab head and two private conservators).

Educating colleagues about the role of the lists for this project, their proper format for inclusion, and the relevant information was a challenge. Smaller institutions and private conservators provided lists with specific focus such as materials for textiles treatments or surface cleaning solutions. These lists were helpful to populate the LCA tool, but were not complete. Larger institutions that provided lists, some of which were gathered specifically for the NEH LCA grant project, tended to include items that were not relevant to this project and had to be culled. Specifically, general household cleaning supplies, books, "hardware-brass fixtures," or highly specific items such as twenty Epson printer cartridges-one for each color. In some cases the use for materials was not clear, so they could not be included due to uncertainty concerning their appropriate category. Some items were not specific enough, such as "magnifiers" and some were so specific they were not included due to irrelevance to general use. Some items, though useful did not belong to a specific category. Another challenge related to selecting items mostly related to research, such as X-ray units, microscopes, and other highly specialized equipment. The relevance of understanding the environmental impact of examining an object under a sterobinocular microscope seemed uncertain. Combining the lists, flagging the repeat items on the lists for importance according to repetition among lists, was also a time-consuming challenge. The staff at all the institutions that did not have lists immediately acknowledged that assembling lists were a priority for them and their institutions. Lists would help them avoid waste in purchasing products, reduce costs of unneeded purchases, streamline activities, and help them practice good house-keeping. Much like the preferred-purchases list for sustainable options in an office or a household, lists of preferred and commonly purchased items would be a valuable tool for collections care professionals.

As expected, much was learned from the list gathering phase and the Tier 2 proposal will include provisions for a robust plan to accommodate this difficult task and result in a complete library. A successful library will house the largest possible amount of relevant items, organized into recognizable categories.

4.1.2 LCA Beta Tool Categories

After assembling lists, general categories were defined for an initial beta tool. These categories provide two types of information for the LCA tool—the background and support needed to collect product information from existing sources, and the basic components or special activities to plan individualized LCAs. The list categories include:

Treatment: solvent cleaning and dry cleaning; inpainting; varnishes; adhesives; consolidants; glues; tools; repairs; loss compensation

Materials: tools; storage bottles; paper

Documentation: photography and computer equipment required for photo documentation and written reports

Preventive: pest control; art storage; environmental monitoring and management

Equipment: personal protection equipment; water purification system; waste disposal

4.1.3 LCA Beta Tool Use

To use the tool, the collections care professional would first need to understand the parameters of LCA and choose a specific system to analyze. For example, a conservator might question the environmental and health impacts of a solvent such as cyclomethicone D4, a silicone solvent that sublimes after application (Cremonesi 2017). The conservator would open the LCA tool, similar to an Excel spreadsheet, select the components of the solvent system, identify amounts of solvent, and select the relevant types of environmental impact. From this data, the user will be able to produce graphs and charts that explain the types of impact and highlight areas of greatest concern. The tool will be able to evaluate one product or to compare any number of products.

4.2 LCA Library: Three Representative Case Studies

The project team identified three LCAs to exemplify the capabilities of LCA and act as case studies in an LCA library. To choose among the many pressing sustainability questions facing cultural heritage professionals, the team discussed the goals and scopes of possible topics. Through these discussions, the team eventually settled on investigating aspects of acquisition, care, and exhibition of three silver objects, each in different size institutions with a range of resources: the Museum of Fine Arts, Boston the Princeton University Art Museum, and the Gibson House Museum in Boston.

The MFA Boston is one of the largest museums in the United States with more than 450,000 works of art. One of its more well-known objects is the Sons of Liberty Bowl by Paul Revere 1768. The bowl measures 14 × 27.9 cm and was acquired by the Museum in 1949. It has never traveled outside the commonwealth and has never been loaned out from the MFA, always on display in the Americas collection. It is housed in a Goppion exhibition case with RHapid Silica Gel and Zorflex to stabilize relative humidity. The gallery is kept according to museum standards at controlled relative humidity and temperature.

The Gibson House Museum was completed in 1860 as one of the first of Back Bay Boston's classic four story brownstones, housing a collection of a wide array of artifacts. In 1957 after the last family occupant died, the house was dedicated as a museum and in 2001 as a National Historic Landmark. The house has no specific environmental management and all windows are kept shut year round to reduce pollution infiltration. The house is maintained with three part-time staff. Objects are displayed with minimal security and the trustees work to just sustain the building. The collection features a silver bowl that was an engagement gift to Rosamond Warren Gibson, dating 1871.

The Princeton University Art Museum houses 97,000 objects. The American Decorative Arts department is curating an exhibit, *Nature's Nation,* that explores the environmental implications of an exhibit, encompassing the impact of a work of art from creation to museum acquisition. One object in this exhibition is a silver 1790 Joseph Lowell sugar urn, a third object compared in the pilot life cycle assessment.

Gorham Bowl, Gibson House
Exhibition: no environmental management, little security
History: Gibson House/Boston Treatment: polish regularly

Revere Bowl MFA Boston
Exhibition: microchamber/ gallery standards, seasonal shift
History: Rhode Island, NYC, MFA
Treatment: no treatment Since 1949

Lowell Silver Urn PUAM
Exhibition: gallery standards
History: PUAM, Morvan Museum, loans
Treatment: polish then coat with incralac

The Pilot LCA, or LCA 1 examined the *cradle to now* life cycle of the three seventeenth and eighteenth century silver vessels housed in the three case study institutions (Huan et al. 2017). This LCA is unique in its examination of the human impacts behind creating a piece of art.

For each vessel, the LCA explored the impact of mining silver on the indigenous communities where mines were located, refining and processing the silver, including the impact of the mercury involved in production, transporting the silver, making the object, transporting and housing the object before and after acquisition, exhibiting and preserving the object in the institution. The Gorham silver bowl is exhibited in the Gibson House Museum dining room without any environmental management and polished every few years. In sharp contrast are the Princeton Lowell Urn and the MFA Sons of Liberty Bowl. The Lowell Urn underwent conservation treatment only in 2006, involving polishing and coating to prevent further tarnish. The Sons of Liberty Bowl has not been treated since it entered the MFA collection in 1949. Both are housed in Goppion cases with RHapid silica gel and Zorflex fabric, in galleries maintained to museum standards.

The LCA 1 results challenge many treatment standards and exemplify how LCA can uncover unexpected hot spots. Current preservation approaches focus on environmental management practiced by many cultural heritage institutions such as the MFA and PUAM. Best practices for many types of objects, including silver, first consider preventive conservation, which typically attempts controlling environmental factors before exploring treatment. The Gibson House Museum, without the funds to support costly environmental management systems, polishes their bowl regularly, reducing the lifespan of the bowl because each polishing removes silver, often leaves residue, easily scratches the surface, and reduces surface decoration. Given the best intentions, many treatments are not reversible, and introduce new materials with their own degradation properties, ultimately further damaging the objects. Treatments are also often toxic to the conservator who generally works in extremely close proximity to the object, different from industry uses and increasing exposure. Often environmental management provides the least invasive and most beneficial practice as in the case with the silver objects. However, this LCA indicates that environmental management based on energy from nonrenewable resources presents significantly greater impact to the environment and is significantly less sustainable than treatment. The environmental impact of the Gorham bowl is negligible compared to the MFA and PUAM pieces due to the energy use for exhibition in the larger, more established and more intensely climate-controlled institutions.

LCA 2 examines forty cleaning agents such as solvents, water, and detergents. It also compares five specific systems involving solvents, buffered water and gel systems. The LCA 2 final recommendations were based on environmental impact, excluding the human health impact, so solvents that were found acceptable in the production phase, such as benzene, are highly toxic and not acceptable for the use phase (Chen et al. 2017). This LCA provided helpful information for a Tier 2 project, where information concerning human health will require additional analysis and considerations. List categories and material selection for the lists was also a challenge and the importance of simplifying material comparisons and justifying each item was realized.

LCA 3 looks at the environmental impact of three types of storage systems and the energy required to establish and maintain each system. The data from LCA 1 was applied as case studies for this third LCA. The most complex system examined is the display case for the Sons of Liberty bowl at the Museum of Fine Arts, Boston. The bowl is housed in a glass and powder coated steel case with interior lighting and the relative humidity controlled with desiccated silica gel. The piece is exhibited in the museum gallery with temperature gallery standards with seasonal shifts. The second storage system examined a sealed plastic container holding an object similar in size to the Sons of Liberty Bowl with enough silica gel to condition the container to 35% RH. The third storage system examines a storage closet of a specific size (approximately 3.0 m by 3.0 m) maintained at 35% RH. At the time of publication, these results are not complete.

5   Discussion

Fulfilling the goal to populate an online library with LCA case studies and to collect lists that would support a beta tool presented unexpected challenges. Through this study it became clear that institution-wide materials lists are uncommon, cumbersome to create, and when they do exist they require extensive editing and organizing before they can be applied to LCA. Disseminating the materials and compiling lists required much time and many discussions before a satisfactory goal was achieved. Similarly, identifying three LCA case studies to populate the online library proved challenging and involved multiple discussions between the conservators, engineers, and project directors to identify relevant products and actions that appropriately compared "apples to apples." A first attempt to create an LCA comparing environmental management in microchambers with controlled storage rooms proved too complex with unlimited variables not appropriate for LCA. Only after considering one complex proposal, was a simpler format devised and an LCA examining the impact of controlled storage and exhibition environments established. With the AIC LCA tool, guidelines for use will be established to reduce trial and error for expeditious, useful results.

The Tier 2 proposal will assign sufficient time for establishing lists, modes for full input from colleagues and related professionals, increase focus on educating colleagues about the requirements for LCA, and will establish workshops, classes, publications and webinars to effectively teach tool use.

6 Conclusions

This planning project explored a range of approaches to LCA for custodians of cultural heritage and has revealed areas that require alterations and redirection to develop a tool that is useful, readily accessed, and easily evaluated. The importance of connecting the human health impact and toxicity of materials and actions with environmental impact has become clear through this project. The disconnect between our work and its consequences leads to the lack of incentive to change habits. Even knowledge of direct toxicity and carcinogenic risk of products often does not prompt reconsideration of materials use and practices. Overcoming the lack of environmental concern regarding the impact from cultural heritage preservation and exhibition is one of the greatest challenges of this project and was realized during the Tier 1 phase. Connecting the human health impact with the environmental impact has become the shifted focus for the Tier 2 project, aiming to provide the most useful information and lead to the greatest change in preservation practices, towards more sustainable means.

References

American Alliance of Museums, "Museum Facts" (2017) http://www.aam-us.org/about-museums/museum-facts. Last Accessed 20 Mar 2018
American Alliance of Museums (2016) Museums as economic engines. http://www.aam-us.org/about-museums/museums-as-economic-engines-a-national-report. Last Accessed 20 Mar 2018
Anderson G (2012) Reinventing the museum. AltaMira Press, Maryland
Chen S, Kinnaly S, Qi Z (2017) Life cycle of cleaning systems for art conservation. CIVE 5275 Project Report, Northeastern University, Boston, MA

Cremonesi P (2017) Solvents, toxicity. In: Gels conference, London. https://www.youtube.com/watch?v=pID6rVME6qk. Last Accessed 20 Feb 2018

Curran M (1993) Life cycle assessment: inventory guidelines and principles. U.S. Environmental Protection Agency, Cincinnati, Ohio

Huan W, Que Q, Varela K, Wilborn M, Yang T (2017) Life cycle assessment of historical silver objects. CIVE 5275 Project Report, Northeastern University, Boston, MA

Institute of Museums and Library Services (2016) American heritage health index. https://www.imls.gov/publications/heritage-health-index-full-report. Last Accessed 27 Jan 2018

International Organization for Standardization (2010) 14044:2006—environmental management-life cycle assessment—requirements and guidelines. International Organization for Standardization, Geneva. http://iso.org/iso/home/standards_development.htm. Last Accessed 10 April 2017

Janes R (2011) Museums in a troubled world. Routledge, Oxon

Marshall G (2014) Don't even think about it: why our brains are wired to ignore climate change. Bloomsbury USA, New York

Nunberg S, Eckelman M, Hatchfield P (2016) Life cycle assessments of loans and exhibitions: three case studies at the Museum of Fine Arts, Boston. J Am Inst Conserv 55(1):2–11

Sarah Nunberg, Fine Art Conservator, PA AIC focuses on conservation and preservation of cultural heritage objects made of wood, stone, ceramic, and metal. She incorporates sustainable practices into treatments and preventive methods. Along with her private practice in Brooklyn, NY, Sarah teaches and researches polymer degradation at Pratt Institute, is fulfilling a fellowship in the Math and Science Department, and is working closely with Pratt chemists to archive and preserve a collection of 1960s sculptures made from polyester resin. Sarah has spearheaded introducing life cycle assessment to the cultural heritage field and has worked with Northeastern University Professor, Dr. Matthew Eckelman and his students in carrying out six LCAs.

Sarah W. Sutton, LEED-AP, Principal, Sustainable Museums was a founding member of AAM's PIC Green and a past co-chair, and is a founding co-chair of AASLH's Task Force on Environmental Sustainability. She trained originally in administration of cultural heritage sites at The Colonial Williamsburg Foundation and The College of William & Mary. She is happy to leave the science of LCA to co-director Sarah Nunberg, and focus on the strategy and policy of caring for collections and professionals sustainably. Her consultancy, Sustainable Museums, is committed to world-wide efforts to align the good work of museums, zoos, gardens, historic sites, and other cultural and natural resource organizations in creating a just, verdant, and healthy world. She is the co-author, with Elizabeth Wylie, of two editions of *The Green Museum, a primer on environmental practice,* and author of *Environmental Sustainability at Historic Sites and Museums.*

Matthew Eckelman, Ph.D., Environmental Engineering, Assistant Professor at Northeastern University in Civil and Environmental Engineering, with secondary appointments in Chemical Engineering and Public Policy. His research focuses on large-scale modeling of industrial resource use and emissions and subsequent impacts on the environment and public health. He consults regularly on sustainability-related projects with a range of businesses, institutions, and government agencies. Dr. Eckelman previously worked in the Massachusetts State Executive Office of Environmental Affairs and was a Peace Corps science instructor in Nepal. He received a doctorate in chemical and environmental engineering from Yale University.

Moving Forward in Climate Change Communication: Recommendations for Rethinking Strategies and Frames

Annemarie Körfgen, Alina Kuthe, Sybille Chiari, Andrea Prutsch, Lars Keller and Johann Stötter

Abstract Climate change communication tailored to different target groups aims at creating awareness of climate change and willingness to engage in climate change mitigation, adaptation, and transformation, thus fostering a development towards a low-carbon society. As there is a long way to reach these goals, the question remains how to create effective communication strategies and topics. This paper identifies the *status quo* of climate change communication including existing challenges by taking the example of Austria. 101 stakeholders from science, civil society, public administration, media, and economy, who are actively communicating climate change, were questioned on their goals, target groups, strategies, and topics of communication. Further, they were asked to assess existing challenges in climate change communication from their experience as communicators. In order to derive recommendations for a practical application, the results were discussed in expert interviews. Recommendations include a stronger institutionalisation of climate change communication as well as target-group-specific communication strategies. The findings presented in this paper contribute to a greater understanding of climate change communication in its practical dimension. They can support

A. Körfgen (✉) · A. Kuthe · L. Keller · J. Stötter
Institute of Geography, University of Innsbruck, Innrain 52f, 6020 Innsbruck, Austria
e-mail: annemarie.koerfgen@uibk.ac.at

A. Kuthe
e-mail: alina.kuthe@uibk.ac.at

L. Keller
e-mail: lars.keller@uibk.ac.at

J. Stötter
e-mail: hans.stoetter@uibk.ac.at

S. Chiari
Center for Global Change and Sustainability, University of Natural Resources and Life Sciences, Borkowskigasse 4, 1190 Vienna, Austria
e-mail: sybille.chiari@boku.ac.at

A. Prutsch
Umweltbundesamt, Spittelauer Lände 5, 1090 Vienna, Austria
e-mail: andrea.prutsch@umweltbundesamt.at

© Springer Nature Switzerland AG 2019
W. Leal Filho et al. (eds.), *Addressing the Challenges in Communicating Climate Change Across Various Audiences*, Climate Change Management,
https://doi.org/10.1007/978-3-319-98294-6_33

stakeholders in applying climate change communication approaches aiming towards sustainable development, by contributing to the implementation of climate change mitigation and adaptation strategies.

1 Introduction

When climate change (=CC) entered the public agenda in the 1980s, it were scientists who communicated the issue, i.e. through initiatives such as the Intergovernmental Panel on Climate Change (Bodansky 2001). Thus, scientific framings of causes and effects of CC dominated climate change communication (=CCC). As scientific consensus on the anthropogenic causes of CC grew (cf. Oreskes 2004; Maibach et al. 2014), it was soon realised that CC mitigation and adaptation measures had to be implemented urgently (Moser 2010). Communicators from the non-scientific field, such as policy makers, journalists, civil society organisations, and stakeholders from economy, entered the stage of CCC. These communicators provoked that—beyond scientific frames—also political, cultural, social, and economic frames were applied in CCC (Nerlich et al. 2010). Adapted to different target groups, these 'new' frames were expected to foster public engagement in CC mitigation and adaptation (Nisbet 2009). The growing research field of CCC has been contributing to practice with insights on public perceptions of CC, framing and messaging, channels and formats as well as different target audiences (Moser 2016). However, communication is often observed to be uncoordinated and discontinuous, because synergies between stakeholders communicating CC are not exploited. Perspectives and experiences of stakeholders from CCC practice and research are not exchanged sufficiently (Moser 2016). Fostering more tailored and continuous CCC strategies for different target groups could provide a vital contribution to a development towards a low-carbon and climate-resilient society by increasing awareness and willingness to engage in CC mitigation, adaptation, and transformation activities (Ockwell et al. 2009).

In order to take advantage of these potentials, the networking project *C4Austria —Connecting Climate Change Communicators in Austria* aims at building up an interdisciplinary, transdisciplinary CCC network. The project facilitates exchange between communicators from practice and research and gives impulses for the enhancement of communication strategies. Through better networking between scientific, political, and societal realms, existing synergies are exploited and challenges in CCC addressed. Besides several networking activities and stakeholder workshops, the project included the identification of the *status quo* of CCC in Austria. This analysis creates a discussion base for the exchange between communicators and will be presented in this paper. By addressing the following questions, experience from CCC practice are gathered in order to contribute to a greater understanding of climate change communication and to foster its improvement:

- RQ1: What is the *status quo* of CCC targets, prevailing topics, strategies, and target audiences addressed by communicators, by taking the example of Austria?
- RQ2: What are existing challenges in CCC?
- RQ3: How can CCC be improved in future?

2 Methodology

In order to answer RQs 1–3, two sets of empirical data were generated and interpreted in a multimethod approach (Fig. 1). Figure 1 illustrates the workflow from data generation to interpretation to dissemination. The question of the *status quo* of CCC (RQ1) was addressed via an online survey among CC communicators in the frame of the networking project *C4Austria—Connecting Climate Change Communicators in Austria*. Challenges in CCC (RQ2) were identified in a meta-analysis and discussion of the results through expert interviews. Based on these interviews, recommendations for rethinking CCC strategies and topics were developed and reflected with findings from literature (RQ3). These recommendations were also informed by the outcomes of an iterative dialogue with stakeholders throughout the project period during two workshops (Fig. 1).

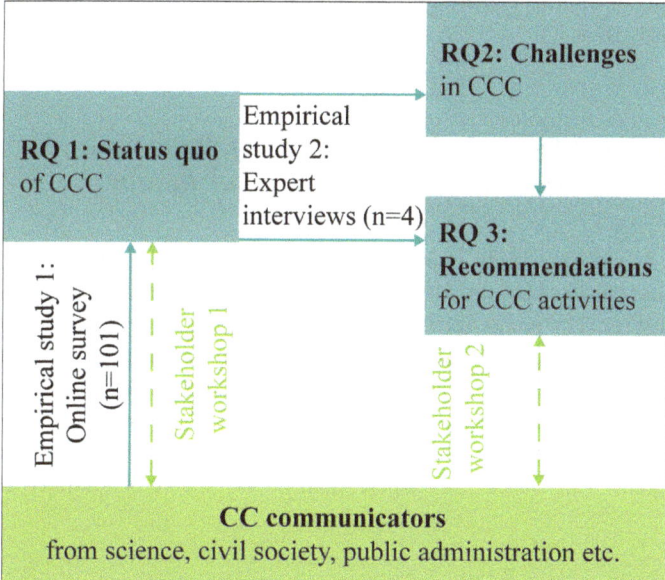

Fig. 1 Workflow diagram with work steps including an iterative transdisciplinary dissemination of results during two stakeholder workshops

2.1 Empirical Study 1: Online Survey with CC Communicators

An online survey containing open-ended and close-ended questions delivered first insights into the *status quo* of CCC in Austria and helped to identify current challenges (Fig. 1). The target group of this survey were stakeholders from science, civil society, public administration, churches, media, and economy, all communicating CC in Austria. During a stakeholder workshop (in the frame of the *17. Österreichischer Klimatag,*[1] Graz 2016) contacts of respective CC communicators were collected on a map. As a next step, these stakeholders were contacted via phone and asked to fill out the online questionnaire and additionally to forward the questionnaire within their respective networks in order to reach a higher number of stakeholders. The list of contacts was complemented by online research. By applying this 'snowball principle', 101 questionnaires were completed during two sampling periods in May–July 2016 and January–February 2017. In total, 65 different Austrian institutions are represented.

In a mixed-methods approach, quantitative and qualitative survey data were analysed. The closed-ended questions were evaluated and presented as absolute values. The open-ended questions were coded applying structuring content analysis (Mayring 2003; Krippendorff 2013). The code system was developed inductively in a first step, and then applied to the open-ended questions by means of a deductive approach in a second step. The results were synthesised into hypotheses on the *status quo* of CCC and disseminated and discussed with CCC stakeholders in an iterative transdisciplinary procedure during a first stakeholder workshop in October 2016, Vienna (Fig. 1).

2.2 Empirical Study 2: Meta-analysis of status quo in Expert Interviews and Synthesis of Results

In a second empirical study, the hypotheses derived from the online-survey were analysed and discussed with four CCC experts from science and media in interviews (Table 1). The semi-structured interview guide was developed based on the results of the online-survey and the derived hypotheses according to up-to-date standards on the development of interview guidelines (cf. Helfferich 2014; Mex and Mruck 2011). The interviews were conducted in April 2017 partly via phone and in live conversations, were fully transcribed using *F4transkript*, and analysed applying *MAXQDA 12*. Based on expert interviews, recommendations addressing

---

[1]Yearly plenary assembly of Austrian CC research community with the goal of fostering interdisciplinary and transdisciplinary exchange.

Table 1 Background of the interviewed experts

Interview no.	Type of interview	Organisation	Background	Region
I1	Phone (29/03/2017)	Climate Service Centre Germany	Science	Hamburg, Germany
I2	Phone (30/03/2017)	Research and Transfer Centre "Sustainability and CC Management", University of Hamburg	Science	Hamburg, Germany
I3	Live (04/04/2017)	Austrian Public Broadcasting Corporation (ORF)	Media	Vienna, Austria
I4	Phone (06/04/2017)	klimafakten.de—information platform on CC and CCC	Media	Berlin, Germany

existing gaps in CCC and giving impulses for new climate topics were derived. These recommendations were disseminated and discussed in a second stakeholder workshop in May 2017, Vienna (Fig. 1).

3 Results: *Status quo* of CCC Taking the Example of Austria

In the following, the results of the analysis of the *status quo* are presented according to the different items of the online-survey (Fig. 1).

3.1 CC Communicators

From the 101 CC communicators participating in the survey, nearly half of the attendees mentioned to work within the scientific field. Figure 2 illustrates the organisational background of all survey respondents. The second strongest group were NGOs (e.g. *Greenpeace*), followed by governmental institutions (e.g. *klima: aktiv* programme commissioned by the Austrian Federal Ministry of Agriculture, Forestry, Environment, and Water Management). Only very little response was received by education and consulting stakeholders (e.g. consumer advice centres and environmental consultancies dealing with sustainability questions for individuals and companies), the economic sector, and churches. Contacted media and policy makers did not respond to the survey.

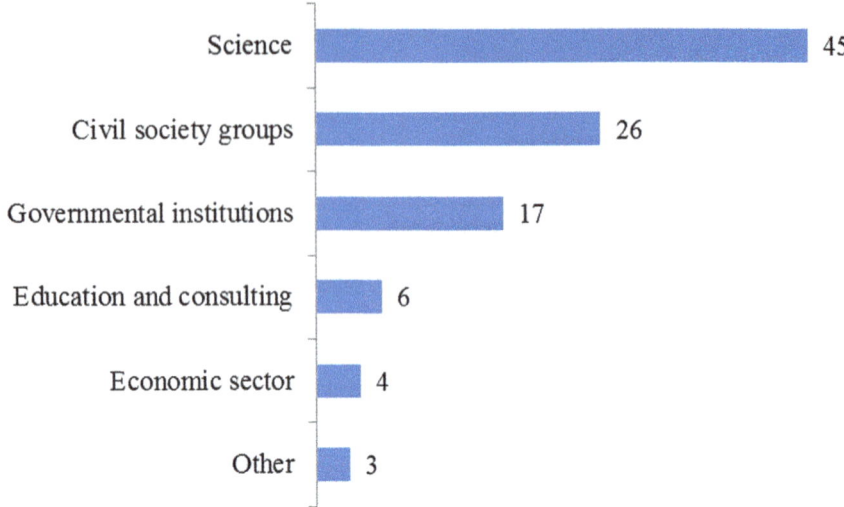

Fig. 2 Organisational background of the participants of the online-survey (n = 101, in absolute values, closed-ended question)

3.2 Target Groups

The analysis shows that CCC in Austria already addresses a variety of different target groups (Fig. 3). Mainly, public administration was mentioned by the respondents, followed by young people, interest groups, and the broad public. The target group of multipliers, such as teachers, is addressed only very randomly, science and the media are even less.

3.3 Targets of Communicators

The analysis of the open-ended survey item *focus and target of communication* resulted in 192 codings (Table 2). These roughly comprise the categories (i) *knowledge and awareness* (n = 143), (ii) *climate action* (n = 40) and (iii) *other* (n = 9).

(i) About three quarters of the communication targets named by the respondents of the online-survey belong to the category of *knowledge and awareness* (Table 2). Within this category, the sub-category *educational work* prevails, comprising codings such as awareness-raising by giving impulses for "critically questioning established attitudes and ways of life". About one quarter of the respondents is involved in research that leads to the generation of 'new' CC knowledge (sub-category *knowledge generation*). These also include a strengthening of climate research. A small fraction of respondents

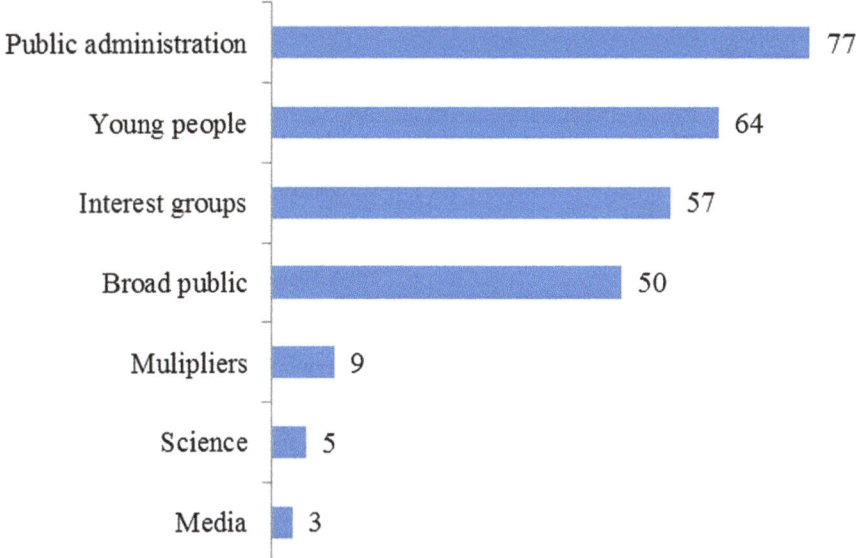

Fig. 3 Target groups mainly reached by CCC in Austria (n = 265, in absolute values, open question with multiple responses possible)

is active in *knowledge and awareness*-raising through consulting activities, e.g. with companies (sub-category *consulting*).

(ii) About one fifth of the codings in the category *climate action* focuses on the concrete implementation of climate mitigation and adaptation measures (Table 2). About half of these codings belong to the sub-category *CC mitigation*, contributing mainly to energy revolution and decarbonisation. Further, codings within the category *climate action* comprise the sub-categories *CC adaptation* as well as *sustainability*, while the latter includes "the transformation of society including the economic and financial systems and (...) the energy systems". Few respondents name networking and the building of cooperation of one target of their work (sub-category *networking*).

(iii) *Other* codings comprise general goals that are not used for the evaluation of CCC in this study.

3.4 Climate Topics

Communicated climate topics comprise six main categories (Table 3). Most topics communicated so far are *CC mitigation* topics. Within this category, the most communicated topics belong to the sub-category *sustainable production and*

Table 2 Targets of CCC as stated by the respondents of the online-survey (n = 192, in absolute values, open question)

Category	Sub-category	Description	Examples	No. of codes
Knowledge and awareness (n = 143)	Educational work	Knowledge transfer, awareness-raising	"To give incentives to critically reflect on consolidated attitudes and concepts of life"	102
	Knowledge generation	Research, analysis	"Develop CC adaptation strategies for agriculture"	32
	Consulting	Consulting with various stakeholders	"Advising for communities, education institutes and enterprises— support, accompany and motivate them"	9
Climate action (n = 40)	CC mitigation	Activities fostering reduction of GHG emissions	"Drive local energy transition"	19
	CC adaptation	Activities fostering adaptation measures	"Adapt housing to changing climatic conditions"	6
	Sustainability	Activities fostering socio-ecological transformation and sustainability	"Support transformation towards sustainable lifestyles and high quality of life"	9
	Networking	Activities supporting the networking amongst stakeholders from various fields	"Build up a networking platform for the energy industry"	6
Other (n = 9)		Items not directly relating to CCC	"Advocating the rights of indigenous partners in the Amazon region"	9

consumption, followed by *mobility and communication, structure, living, spatial planning* and *food and agriculture* (cf. WBGU 2011). Codings in the category *CC adaptation* are considerably rarer as well as *CC policy and economics*, meaning political and economic control mechanisms. Only few communicate on *CC causes*. The communicated topics depend strongly on the background of the individual communicator. Surveyed scientists, for example, predominantly communicate on *CC consequences*. Communicators from civil society groups, governmental institutions, and from the education and consulting sector, however, have their focus on *CC mitigation* topics, especially in the sub-category *sustainable production and consumption*.

Table 3 CCC topics as stated by the respondents of the online-survey (n = 213, in absolute values, open question with multiple responses possible)

Category	Description	Examples	No. of codes
CC mitigation	Mitigation measures in the sub-categories of (a) Production and consumption, (b) Mobility and communication, (c) Structure, living, regional development, (d) Food and agriculture (WBGU 2011)	"Strategies for a low-carbon Austria"	96
CC adaptation	Adaptation measures to cope with CC consequences	"CC adaptation in agriculture and forestry"	21
CC consequences	Processes and effects directly or indirectly caused by CC	"CC impact on vector borne diseases"	30
CC causes	System knowledge on CC causes	"Greenhouse gas emissions"	15
CC policy and economics	Policy and economic measures to drive and regulate mitigation and adaptation measures	"World climate report and its implementation on national scale"	17
Systemic interferences	Topics integrating different interferences of CC with other environmental and anthropogenic systems	"Systemic coherences of the climate system with diverse spheres of the globe as well as anthropogenic societal and economic systems"	11

3.5 CCC Formats

For the analysis of the most commonly applied communication formats, it was distinguished between *one-way* and *interactive* formats (Table 4). About three fifth of the communication formats are assigned to the category *one-way*, thus represent the dominant format chosen, independently of the background of the communicator and

Table 4 CCC formats as stated by respondents of the online survey (n = 268, in absolute values, open question)

Category	Sub-category	No. of codes
One-way (n = 153)	Media	59
	Publications	51
	Presentations	38
	Exhibitions	5
Inter-active (n = 115)	Workshops/ seminars	51
	Consulting	25
	Events	22
	Interactive/ hands-on	7
	Other	10

the addressed target group. It should be considered that boundaries between these formats and their labelling are often fluent. For example, exhibitions or presentations, which are assigned to the category *one-way*, could be designed quite interactively, whereas seminars or workshops could also include one-way elements and mono-directional implementations. Nevertheless, as the set of data does not provide more insights into single formats, this simplified categorisation system was chosen.

4 Results: Challenges in CCC

In the online-survey as well as the expert interviews, different challenges in CCC were identified (Table 5). The online-survey focused on challenges given by target groups and topics neglected so far in CCC as well as deficits in communication strategies. In the expert interviews, additional challenges were discussed, critically reflecting on the framing of CCC and the handling of uncertainties.

4.1 C1: Neglected Target Audiences

The respondents' estimations on possibly neglected target audiences varied depending on their specific background (Table 6). Stakeholders from the scientific field estimated that young people as well as policy makers and decision makers should be addressed more in CCC. Socio-civilian groups named socially

Table 5 List of challenges in CCC derived from the online-survey and expert interviews

#	Challenges
C1	Neglected target audiences
C2	Neglected climate topics
C3	Challenges in messaging
C4	Dealing with uncertainties

Table 6 Assessment of target groups neglected in CCC as stated by respondents of the online-survey (n = 157, in absolute values, open question with multiple responses possible)

Neglected target audiences	No. of codes
Citizens	26
Economists	21
Policy/decision makers	21
Socially disadvantaged people	21
Young people	18
Interest groups	15
Seniors	13
Multiplicators	6
"Hard-to-reach-audiences"	2
Other	14

disadvantaged individuals as target group that should be reached more. Communicators from public institutions stated that especially the economy should be taken more into account.

4.2   C2: Neglected Climate Topics

Among the climate topics not communicated sufficiently the socio-ecological transformation was mentioned more often than all other topics (Table 7). First, respondents stated that chances of socio-ecological changes could get more weight and recommend addressing "that CC means change, but changes can be utilised in a positive way". Second, CCC often neglects that "a transformation process is needed in order to reach the goals of the Paris Agreement" and that there is an urgency to do so, because in order "to reach the 2 °C goal and thus prevent dangerous changes of the climate system, we (=all stakeholders) cannot lose more time. We have to act resolutely now!". Other topics not sufficiently communicated are self-efficacy and opportunities for action, "showing that each individual's behaviour has consequences for the system".

Communicating a complex issue such as CC also involves the challenge to address systemic interactions. The respondents of the online-survey estimate that a systemic view is often missed in communication. Rather, it should be stressed that "CC is connected with all areas (agriculture, energy politics, trade, industry, mobility, living), and without a change it will be impossible to prevent the crisis". In this context, it is important to recognise that "climate protection is one part of a problem that has to be solved holistically".

According to the online-survey, "CC consequences not connected to extreme events" are often not addressed, more concretely consequences affecting

Table 7 Assessment of climate topics neglected in CCC as stated by the respondents of the online-survey (n = 157, in absolute values, open question)

Neglected climate topics	No. of codes
Socio-ecological transformation	25
CC mitigation	19
Self-efficacy	18
CC consequences	17
Systemic interactions	14
CC adaptation	10
Values/responsibility	10
Scientific basics	8
Everyday life topics	4
Political/economic control mechanisms	3
Other	18
No indication	11

every-day-life for specific target sectors and areas, because "people are concerned, if they realise what an unabated CC means for them as individuals, their surrounding and the environment they are living in".

This view was shared in the expert interviews, where interviewees addressed the problem that CCC usually deals with "unpleasant CC consequences, which, I would assume, have a rather paralysing than activating effect, because they make the issue big and global and the path towards single activities of individuals or policy action is very long" (I.4). Especially, the discrepancy between single measures with only slight potential to reduce greenhouse gas emissions, such as turning off the light when leaving the room, in contrast to significant CC impacts, such as the melting of Antarctic ice shelves, "leads to doubts about the credibility of these measures" (I.4).

Further, CC consequences for societies of the global south, resulting migration processes and consequences for future generations are topics not communicated enough so far according to the respondents of the online-survey. These topics are bound to the question of "how (in which society and which environment) we want to live".

4.3   C3: Challenges in Messaging

Respondents of the online-survey noted that one deficit in current communication strategies is that "information from the political level is transferred in a very abstract way. (…) that is why the topic gets vague and difficult to understand". This leads to the question of how to break down scientifically proven knowledge to key messages that are relevant for different stakeholders.

This problem is also identified in the expert interviews: "at the moment (…) it is communicated very broadly on the topic of CC and people think that it is too abstract" (I.2). Experts assume that this broad CCC in a „scattergun approach" (I.1) does not reach each target group. Further, communicators often apply "obsolete communication strategies (…) or communication strategies are not reflected upon" (I.4).

4.4   C4: Dealing with Uncertainty

One further challenge identified in the expert interviews is the handling of uncertainties in CCC. With a focus on uncertainties, arguments often seem not to be strong enough and an impression of not sufficient scientific consensus on CC evolves. Therefore, interviewees stated that "it is urgently necessary to place a stronger focus on more secured knowledge, so that CC communicators can rely on more facts—also in order to understand and counter arguments from the opposite sides" (I.3). It poses a difficult task to find the balance between stressing secured knowledge on the one hand and communicating uncertainties inherent in climate modelling on the other hand.

5 Discussion and Derivation of Recommendations for Future CCC Activities

In the following section, nine recommendations (R1–R9) for future CCC activities as derived from the results of the online-survey and the expert interviews are discussed (Fig. 4). These recommendations go beyond the case study of Austria and deliver impulses for international stakeholders active in CCC. Three fields of action are covered by the recommendations, illustrated in Fig. 4. The first three recommendations R1–R3 comprise issues concerning communicators (left column, Fig. 4). Next, recommendations R4–R6 on target groups are given (middle column, Fig. 4). The last section is designated to climate topics (R7–R9, right column, Fig. 4).

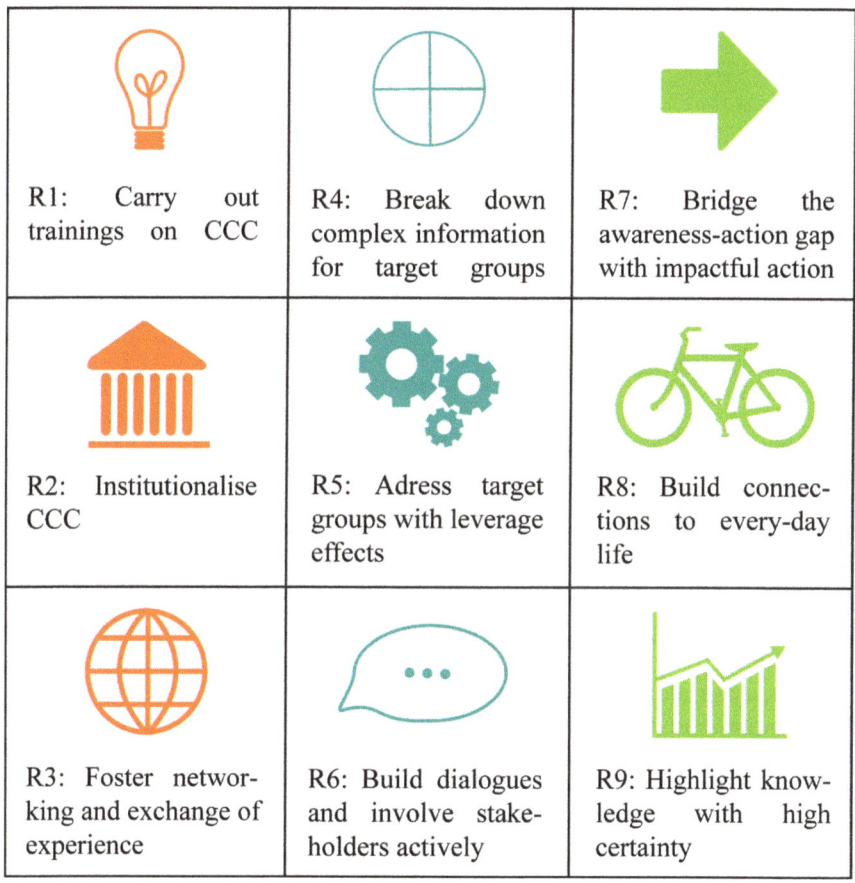

Fig. 4 List of recommendations derived from the expert interviews

5.1   Carry Out Trainings on CCC (R1)

Viewing the analysis in the online-survey of communicators' backgrounds (Fig. 2), interviewed experts think it is "not surprising [authors' remark: that science plays a dominant role,], because the topic contains a great deal of complexity, so many people do not dare to talk about climate topics, because they think they do not understand it properly" (I.2). Other reasons for this dominant participation of stakeholders from the scientific field could be that they are more likely to be reached by the study conducted within a scientific environment. Further, the scientific community might be particularly reflective on the issue of CCC and thus more likely to participate in such surveys and networking activities. In non-scientific institutions, however, CC is often only one topic amongst a broad spectrum of topics treated; in many cases even just a side issue.

In order to support potential communicators from the non-scientific field, e.g. the media, these should get better access to scientific knowledge that is prepared clearly and simply. Trainings for journalists, allowing a dialogue with scientists, should go beyond providing facts on CC, but "help to make journalists feel more comfortable with the complex issue of CCC" (I.2). On the other side, media trainings for scientists could give "insights into the work of journalists, the information they need for profound reporting of scientific facts" (I.1). Further, education on CCC is needed for multipliers such as teachers, e.g. through "teacher trainings, so that they can communicate the topic in school with certainty" (I.2).

5.2   Institutionalise CCC (R2)

Like climate research was institutionalised by the IPCC in the 1980s, interviewed experts stated that "maybe it needs someone who takes care of these communicative aspects, an institution at the interface (...) some kind of agency, which intensely deals with knowledge and also transfers it" (I.1). Such institutions with the explicit task of CCC—as a 'collecting point' between science and different fields of practice - could close the "science-action gap" (Moser and Dilling 2011: 162) by synthesising scientific knowledge with communication know-how. Science would be relieved from the conflict to provide profound research and target-group-specific communication at the same time and instead of "(letting) the scientist alone with the question of how to communicate the topic, an interdisciplinary and holistic dialogue could be fostered" (I.3). For example, such institutions could be realised through public relations offices, where "someone operating as a press officer, who knows the information and also transfers it" (I.3) could take the role of CCC.

Such an institutionalisation would further address the challenge of discontinuity in CCC: "there is a whole range of climate projects, often with very short-dated financing, which always somehow include communication aspects. But this does not work at all" (I.4). Due to short project periods, communication approaches are often random and reach only selected groups for a short period. In order to address this shortcoming, one of the experts suggests "continuity for all, regardless of information for policy makers, students or consumers" (I.2). Communication efforts should be focused and synergies between different communicators exploited through long-term networks. An institutionalisation would also enable more continuous, long-term communication.

5.3 Foster Networking and Exchange of Experience (R3)

Networking and exchanging experiences on good and bad practice between communicators with various backgrounds and interests is vital to facilitate mutual learning and enhance communication strategies. This also involves the analysis of historical experiences in science communication to identify strategies that already proved to work. Of course, the issue of CC has some characteristics that make it quite complex to communicate, e.g. its invisible risks, or a temporal and geographic divergence between causes and effects (Moser 2010). However, insights from other fields of study (Gammelgaard Ballantyne 2016), such as environmental communication, as well as looking at past communication campaigns that have led to changes, could be inspiring.

As for a long time, CCC has taken place intuitively without reflecting on the factors that determine successful communication, a stronger emphasis on the evaluation of processes is recommended: "a good communicator always evaluates his message, and then he will realise that messages trigger different reactions for different target groups, (...) leading to the insight that different messages should be developed for different target groups" (I.4).

5.4 Break Down Complex Information for Relevant Target Groups and Sectors (R4)

According to the interviews, communication contents should be adapted to the respective target group, which goes in line with current recommendations of CCC (Moser 2010). A sectoral communication approach would "consider and break down complex interconnections (...) with regards to different societal sectors" (I.2) and provide the information this target group really needs, thus, making the issue more comprehensible. Interviewed experts identified a lack of such approaches, as "at the moment, there is not such a system, but communication on CC is very broad —so people think it is too abstract" (I.2). Farmers as a target group, for example, need information that increases their resilience to deal with extreme weather events, while communication with young people rather focuses on increasing awareness on CC. Therefore, sectoral targets for all relevant target groups and their specific requirements for CCC should be defined thoroughly as first step in the communication process.

Such approaches, however, should balance between breaking down information for sectors and considering that "issues like CC, resources, sustainability, quality-of life are strongly interconnected and should be communicated as such" (I.3). Communicating interactions of different fields in a way they are easy to grasp helps to contribute to the development of a holistic understanding of CC (Kagawa and Selby 2013). As CC interferes with many different areas, such as agriculture, health etc., experts recommend that "anywhere decisions are made, CC has to be included" (I.1).

5.5   Address Target Groups with Leverage Effect (R5)

The analysis of neglected target groups showed very heterogeneous results in the online survey (Table 5). Because all sectors that are "part of the problem are part of its solution at the same time" (I.4), interviewed experts recommend that stakeholders with leverage effect are to be engaged more deeply in CCC. This includes policy makers, public opinion leaders, but also stakeholders from education, media or consumer advice centres. Policy and decision makers should be involved, as they have a great discretion to act, so strategies to sensitise them (not) to take certain decisions are needed. Very high potential is lying in the economic sector, as "the causes of anthropogenically induced CC are deeply embedded in the socio-economic fabric of modern society" (Lidskog and Elander 2009: 32).

5.6   Build Dialogues and Involve Stakeholders Actively (R6)

Viewing the results of the online-survey (Table 4) monologic formats are dominant in Austria. According to the interviewed experts, these "are not completely wrong, but not necessarily what we need" (I.4). Instead, interactive and participative communication strategies should be fostered (supported by Moser and Dilling 2011): "I think that dialogues are incredibly important; when together with people from the community or with companies something is worked out. (…) And this is very effective, because the people really get a lot of in-depth knowledge" (I.1). This dialogue should also allow debates including opposing views, because "it is not about just placing [information] in front of somebody's feet, but getting oneself into an exchange" (I.4). In this context, experts considered as most constructive the exchange between groups that represent different values and "people, one usually does not speak with, in a discursive way" (I.4) (supported by Corner et al. 2014). Participative communication on a community level with stakeholder involvement is estimated to be very promising, because "when involving the people who are concerned, all different areas of society are reached" (I.1) (supported by

Otto-Banaszak et al. 2011). All interviewed experts agreed that "the open format is the better one, definitively the interaction with citizens" (I.3), who should be integrated into a dialogue with science and decision-making from beginning (Nisbet and Scheufele 2009).

5.7 Bridge the Awareness-Action Gap with Impactful Climate Action (R7)

Knowledge alone, such as suggested by the information deficit model, will not be sufficient to trigger engagement in CC mitigation and adaptation, because "the idea that if scientists just communicate clearer and simpler, everything will be different [authors' remark: in terms of the public taking action], is a misunderstanding" (I.4). (supported by Gammelgaard Ballantyne 2016). Fostering awareness is a necessary, but not sufficient condition for engagement. Therefore, "beyond awareness-raising, we need intelligent ideas how to motivate people to change their behaviour. Most people are motivated to take action if they see the advantages" (I.2). This gap is also referred to as the awareness-action gap (Kollmuss and Agyeman 2002). As knowledge on the effectiveness of (climate) actions (Frick et al. 2004) has the most impact on engagement, it should be targeted more in CCC. Therefore, the communication on mitigation measures (action-related knowledge) should always include its effectiveness (effectiveness knowledge), "because only few are able to assess the weight of single mitigation measures—if they have a big impact or are in a small range" (I.4). Knowledge transfer, however, is not sufficient to get people to take action, but systemic intervention strategies including factors such as emotions and feeling of self-efficiency (Kollmuss and Agyeman 2002; Nicholson-Cole 2005) combined with infrastructural, organisational and policy measures (Grothmann et al. 2017) significantly influence if climate-friendly behaviour is taken.

5.8 Build Connections to Every-Day Life (R8)

In order to increase feeling of self-efficiency, frames that are connected to every-day-life, such as economic issues or food and agriculture, induce that "everybody listens, because it affects all of us. An ideal communicator can build the bridge between these regional aspects with the global context" (I.2) (supported by Wibeck 2014). Mobility and sustainable consumption were items named in this context in the online-survey (Table 7) and expert interviews. These make "the topic of CC something more concrete and real, showing that it affects all of us" (I.2) (supported by Nisbet and Scheufele 2009). Especially, local stakeholders are authentic if they communicate "on issues they can influence or shape. And these are often local and concrete, so lots of things are implemented on this level" (I.4) (supported by Lorenz 2010). Communicating 'local' topics can contribute to a feeling of self-efficiency, which, "experienced in cooperation and amicably relationships at the place where they live, is what finally motivates people" (I.4). When communicating everyday life topics, however, the broader context of CC should not be neglected, because "the intensity of communication of climate protection on a small-scale often hides the bigger context, such as decarbonisation (…) and creates the impression that lots of measures are taken already. But on a systemic level, few measures are communicated" (I.4).

5.9 Highlight Knowledge with High Certainty (R9)

The determination of uncertainties as one challenge in CCC as stated in the expert interviews (see C4) corresponds to scientific literature, where uncertainties in climate research have been identified as a driving force for climate scepticism and counter arguments (Oreskes and Conway 2010). Communicating these uncertainties is vital part of scientific progress and discussion (Weingart et al. 2000), but can be problematic, when it comes to decision-making: "noting a common problem among scientists [authors' remark: regarding communication with politicians]: the tendency to emphasise uncertainties rather than settled knowledge" (Oreskes and Conway 2010: 75).

Interviewed experts think "it would be good to stick to facts and to talk about the hard facts and for the soft facts say that they are not ascertained yet" (I.3), corresponding to findings from CCC research (cf. Corner et al. 2015). Interviewed experts further suggest the development of "tools how to deal with these ranges, with uncertainties, which at the moment are still part of climate modelling. This cannot just be solved through better models, because it is inherent in the system. Rather, visualisation tools presenting these uncertainties could help here" (I.1). Also, strengthening the scientific community by networking helps to signalise consensus amongst science on knowledge with high certainty (Maibach et al. 2014).

6 Conclusions and Outlook

In this paper the *status quo* and challenges of CCC in Austria were presented and discussed. Nine recommendations tailored to the need of CC communicators for how to improve CCC were provided. The analysis of CCC in Austria shows that there is already diverse action taken in CCC, but communication could get more voice by strengthening institutions at the interface between science and practice (R1–R3). Within the action field of target groups (R4–R6), there is potential to improve communication by addressing different stakeholders with target-group-specific information via interactive and participative communication formats.

The communication and demonstration of successful and meaningful climate actions, which are relevant in every-day life, enhances the acceptance, credibility, and impact of CC communicators (R7–R9).

The study provides general recommendations for how to improve CCC among different societal and scientific groups. Nevertheless, the work is constrained to the communicators' evaluation. Questionnaires or interviews with respective target groups were not subject of this study. Insights to the communication among specific stakeholders, e.g. scientists and policy makers, will require further transdisciplinary research activities involving these groups. For example, further in-depth interviews can provide insights into the communication requirements of specific target groups. As the study is limited to Austrian stakeholders, a repetition of the online survey on an international level can deliver valuable results regarding differences between countries.

Next steps should tackle the recommendations that seem realisable, and keep in view the ones that comprise more complex factors for implementation. As there is a great willingness amongst stakeholders to work together on these topics, these developments give reason to be optimistic that the nexus between science, policy, and the public regarding CC mitigation, adaptation, and transformation could be considerably strengthened by pursuing the initiated networking activities amongst CC communicators.

Acknowledgements We would like to thank the Climate Change Centre Austria (CCCA) for financing the networking project *C4Austria—Connecting Climate Change Communicators in Austria*. The paper has additionally been prepared in the framework of ACRP financing—special thanks go to the Austrian Climate and Energy Fund. Further, we would like to thank everyone who has contributed to the success of this study, especially the climate change communicators involved in the project, the participants of the online-survey, and the interviewed experts.

References

Bodansky D (2001) The History of the global climate change regime. In: Luterbacher U, Sprinz DF (eds) International relations and global climate change. MIT Press, Cambridge, pp 23–40

Corner A, Markowitz E, Pidgeon N (2014) Public engagement with climate change. The role of human values. WIREs Clim Change 5(3):411–422

Corner A, Lewandowsky S, Phillips M, Robert O (2015) The uncertainty handbook. University of Bristol, Bristol

Frick J, Kaiser FG, Wilson M (2004) Environmental knowledge and conservation behavior. Exploring prevalence and structure in a representative sample. Personality Individ Differ 37 (8):1597–1613

Gammelgaard Ballantyne A (2016) Climate change communication. What can we learn from communication theory? WIREs Clim Change 2:329–344

Grothmann T, Leitner M, Glas N, Prutsch A (2017) A five-steps methodology to design communication formats that can contribute to behavior change. SAGE Open 7(1):1–15

Helfferich C (2014) Leitfaden- und Experteninterviews. In: Baur N, Blasius J (eds) Handbuch Methoden der empirischen Sozialforschung. Springer Fachmedien, Wiesbaden, pp 559–574

Kagawa F, Selby D (2013) Ready for the storm: education for disaster risk reduction and climate change adaptation and mitigation. J Educ Sustain Develop 6(2):207–217

Kollmuss A, Agyeman J (2002) Mind the Gap: Why do people act environmentally and what are the barriers to pro-environmental behavior? Environ Educ Res 8(3):239–260

Krippendorff K (2013) Content analysis. An introduction to its methodology. SAGE Publications, Los Angeles

Lidskog R, Elander I (2009) Addressing climate change democratically. Multi-level governance, transnational networks and governmental structures. Sustain Dev 18:32–41

Lorenz S (2010) Das Klima erkennen, verhandeln, prozessieren - Ein Einblick und Vorschlag zur transdisziplinären Diskussion. In: Voss M (Ed) Der Klimawandel. Sozialwissenschaftliche Perspektiven. VS Verlag für Sozialwissenschaften, Wiesbaden, pp 61–73

Maibach E, Myers T, Leiserowitz A (2014) Climate scientists need to set the record straight: There is a scientific consensus that human-caused climate change is happening. Earth's Future 2 (5):295–298

Mayring P (2003) Qualitative Inhaltsanalyse: Grundlagen und Techniken. Beltz, Weinheim

Mex G, Mruck K (2011) Qualitative Interviews. In: Naderer G, Balzer E (eds) Qualitative Marktforschung in Theorie und Praxis. Springer Fachmedien, Wiesbaden, pp 258–288

Moser SC (2010) Communicating climate change: history, challenges, process and future directions. WIREs Clim Change 1(1):31–53

Moser SC (2016) Reflections on climate change communication research and practice in the second decade of the 21st century. What more is there to say? WIREs. Clim Change 113:92

Moser SC, Dilling L (2011) Communicating climate change: closing the science-action gap. In: Dryzek JS, Norgaard RB, Schlosberg D (eds) The Oxford handbook of climate change and society. Oxford University Press, Oxford, pp 161–174

Nerlich B, Koteyko N, Brown B (2010) Theory and language of climate change communication. WIREs Clim Chang 1(1):97–110

Nicholson-Cole SA (2005) Representing climate change futures. A critique on the use of images for visual communication. Comput Environ Urban Syst 29(3):255–273

Nisbet MC (2009) Communicating climate change: why frames matter for public engagement. Environ Sci Policy Sustain Develop 51(2):12–23

Nisbet MC, Scheufele DA (2009) What's next for science communication? Promising directions and lingering distractions. Am J Bot 96(10):1767–1778

Ockwell D, Whitmarsh L, O'Neill S (2009) Reorienting climate change communication for effective mitigation. Sci Commun 30(3):305–327

Oreskes N (2004) Beyond the ivory tower. The scientific consensus on climate change. Science 306(5702):369–383

Oreskes N, Conway E (2010) Merchants of doubt. How a handful of scientists obscured the truth on issues from tobacco smoke to global warming. Bloomsbury Press, New York

Otto-Banaszak I, Matczak P, Wesseler J, Wechsung F (2011) Different perceptions of adaptation to climate change: a mental model approach applied to the evidence from expert interviews. Reg Environ Change 11(2):217–228

WBGU (2011) Welt im Wandel. Gesellschaftsvertrag für eine Große Transformation. WBGU, Berlin

Weingart P, Engels A, Pansegrau P (2000) Risks of communication: discourses on climate change in science, politics, and the mass media. Public Understand Sci 9(3):261–283

Wibeck V (2014) Social representations of climate change in Swedish lay focus groups: local or distant, gradual or catastrophic? Public Understand Sci 2(23):204–219

A Quest for Green: An Analysis of Environmental and Other Appeals in Pakistani Ads

Khansa Tarar and Rabia Qusien

Abstract With the emergence of Climate Change as a grave issue across the globe, media houses are using the environmental appeal in advertisements frequently and hence the media advertisements are major contributor towards public awareness about the significance of green lifestyle by highlighting aspects of climate change communication. In underdeveloped countries with low literacy rate, environmental appeal in ads is still not very common and the consumers' awareness about the environmental issues is slim. This study offers a content analysis of advertisements of automobiles and housing schemes appeared in Pakistani Media exploring the frequency and kind of appeals (Green or Non-Green), the style in which they have been appeared and also the depth of green appeal specifically used in these ads during the past one year. Both categories (Automobiles and Housing Schemes) were selected carefully with an understanding that auto industry is a major pollutant and housing schemes are growing on the expense of forests and agricultural land and with the new wave of urbanization in Pakistan both of these sectors are touching new heights of production and expansion. Results offer a depiction of how shallow and deceptive the use of environmental appeal in Pakistani advertisements which resulted in inadequate climate change communication. Also the slender use of climate change communication in ads, when compared to the pressing need of communication about the subject, the results offer nothing as promising. The final outcome is a consumer mind-set who has little or no clue about the grave issue of climate change Pakistan has been facing for years.

K. Tarar · R. Qusien (✉)
School of Creative Arts, University of Lahore, Lahore, Pakistan
e-mail: rabiaqusien@gmail.com

K. Tarar
e-mail: khansa.tarar@soca.uol.edu.pk; tarar_honey@hotmail.com

© Springer Nature Switzerland AG 2019
W. Leal Filho et al. (eds.), *Addressing the Challenges in Communicating Climate Change Across Various Audiences*, Climate Change Management,
https://doi.org/10.1007/978-3-319-98294-6_34

1 Introduction

Since long environmental issues are taking the attention of research scholars, scientists, authorities and public across the globe. As Priestley (2017) noted that "average global temperature is increasing, icecaps are melting, sea level is rising and extreme weather events are becoming frequent and main reason is the increase in carbon dioxide pumped into the atmosphere by the fossil fuels we've been burning since the start of industrial revolution". It also suggests that "Climate change involves the entire world, and is bound up with issues of poverty, economic development and population growth" (Bird et al. 2008).

Many efforts have been done such as Kyoto Protocol and likes of Paris Agreement—to curb human activities contributing to climate change and many other are underway to counter the menace of Climate Change by bringing nations to idea of realizing it as the threat and to bring forward the honest efforts of countering it.

1.1 Climate Change Communication

Media also constitutes key influences among a set of complex dynamics which are shaping information dissemination in this politicized environment. Media coverage of climate change is not simply a random amalgam of newspaper articles and television segments; rather, it is a social relationship and also a corporate social responsibility among scientists, policy actors and the public that is mediated through various media forms (Hou and Reber 2011). The presence of green/ environmental appeal in advertising and persuasive messages is one of many ways through which the information about this pertinent issues of global climate change can be transferred to the audience (D'Souza et al. 2006).

1.2 Climate Change Communication and Advertising

Climate change specifically and the environment more generally are becoming increasingly central features in much of contemporary persuasive messages. Meaningful academic attention to environmental cues in advertising at a global arena of media can be seen of as occurring in two waves (Rahim et al. 2012). First, was more concerned with the advertising containing environmental appeal and questions about green washing—deceptive environmental claims were emerged. In second wave, issues of environmental preservation and conservation dominated the climate change communication in ads. In late 2000s it has also brought to the focus not only the environment but also how audiences understood the ads during this wave (Fernand et al. 2014).

1.3 Green Appeal

According to Banerjee et al. (1995), defining green advertising or green appeal in advertising is complex but to say is that green appeal tells the relationship between the product and its environment friendliness, and advocating an environment friendly life style through those features.

The past few decades have seen a subtle increase in concern along few particular dimensions (Bremner 1989). Most significant is environmental appeals are now appearing more frequently in advertising, but all green ads are not created equal. Particularly at early stage of green advertising, companies quickly issued green claims, which sometimes included deceptive or confusing truths or even false promises. These misleading or exaggerated appeals are called "green washing" which dominated the green/environmental appeal appeared in the ads generally (Baum 2012).

1.4 Pakistan and Climate Change Communication

Now the Asian region is also embracing the significance and power of 'going-green' as the environmental threats are multiplying (Lee 2008). Pakistan has been ranked seventh among the top ten countries most affected by the climate change. Many Asian Countries are focusing on advocating about the issue through climate change communication (Lee 2008). Pakistan is suffering intensely and it needs profound attention towards climatic changes (Mustafa 2006). Hudiara drain is a highly polluted tributary which flows through India and Pakistan. An example of going green in Pakistan is the revival of this drain. The people are now displaying more eagerness towards green resolve and the concern about the environmental issues is aggregating (Mustafa 2006).

In order for public to be more responsive towards environmental issues more effective climate change communication is needed through different mediums (Xue and Muralidharan 2015). Living life in an eco-friendly and environmentally responsible manner is an attempt to minimize the harms mankind has done to the planet. To obtain this environmental awareness seems pivotal. Learning an attitude to respond in favourable or unfavourable manner consistently with respect to the environment is required at the moment along with a permanent environmentally responsible behaviour on the part of public (Rashid 2009).

2 Rational

Global Climate Risk Index places Pakistan at seventh number in top ten most climate affected countries. Due to lack of responsibility and little awareness among masses regarding the issue of climate change, developing nations are more vulnerable to climatic changes (Kreft et al. 2016). UNESCO highlighted the importance of media in climate change communication in book titled, 'Media as partners in education for sustainable development'. It raises the questions on role of Media in increasing the awareness over the critical issue of climate change. Advertising is most important of all due to its persuasive nature. The presence of green or any environmental appeal in advertising can be one of the many ways through which the climate change communication can be done along with persuading the audience in most effective manner (D'Souza et al. 2006). Existing literature on the subject has suggested that very few research studies have been done in the area of climate change communication let alone investigating climate change communication in ads. This provides a dire need for conducting this study. Also due to the alarming situation highlighted by Global Climate Risk Index (2017), Pakistani advertisers have to practise more responsibility and vigilance in their messages. This study has explored the aspects of climate change communication present in the ads of housing schemes and automobiles as these two sectors are directly responsible for de-forestation and environmental pollution in Pakistan.

3   Research Questions

1. Which appeal (green and other) has been appeared most in the TVCs?
2. Which level of green appeal is most frequent in TVCs of both housing schemes and automobiles?
3. What are the commonly used styles in green appeal?

4   Review of Literature

Climate change has appeared to be the most important matter on diverse global forums like UN etc. Climate change is acritical issue which is leaving its footprints on both societies and individuals and at the universe at large. The implications are ranging from daily choices about lifestyle, the way to adjust into swift change in the climate, to role of an individual in debating and enacting related social alterations (Whitmarsh et al. 2013). Climate change has engulfed both developed and under-developed countries and seems to affecting sustainable development around the globe. European Union insight (2009) noted that "Rising temperatures, altered precipitation patterns, increased sea level, altered bushfire dynamics and extreme weather events will retard economic development and pose increasingly severe threats to agriculture, international security, infrastructure, water security and energy systems. The poorest are most at risk".

Hepburn (1997) noted that mass media including advertising is agent of socialization which depicts behaviours and actions which are widely accepted in the social system. Furthermore, "media are embedded in the textures and routines of everyday life" and gives social identity to its consumers which touches them both consciously and subconsciously (p. 250).

Research by Harwood Group showed the acknowledgment of Americans about their consumption patterns but they are unaware of the strategies which are mandatory to prevent this. Their research concluded that there is need of changing the lifestyle of Americans to conserve the environment. Americans are more responsible for the change in environment because America is utilizing more resources and its market is showing higher consumption rate (1995). Brower and Leon (1999) highlighted that every purchase matters because "[w]e vote with our dollars when we choose to buy particular products". Researchers also maintained that food, transportation and products like cosmetics and cleaning brands together are responsible for 80% of the environmental change due to consumers (pp. 14–15).

Media is considered as one of those forums which voices this issue and generates awareness in masses. Bigger chunk of public is still unaware; so persistent and accurate coverage of issues like global warming and climate change can positively affect the public engagement across the world (Thaker et al. 2017). Some scholars view this as tricky because of heterogeneity of the target audience and message construction techniques, context of communication, etc.

Mercado (2012) found that issue of climate change is of international concern for media in Argentina. News related to climate change appeared in international section of the newspaper instead of national or science section. Generic framing of this issue in Argentinean media is "conflict" which is linked with conflicting points in international negotiations. Studies offered that one of major role media plays is to raise eco-literacy by enabling audience to interpret the information or environmental cues offered in the media content.

Hou and Reber (2011) conducted content analysis of nine media companies of the United States with reference to their CSR (corporate social responsibility) activities and unveiled that seven out of nine media groups are into environmental activities. "These media companies have engaged in various programs related to the major focus areas of environment policy, such as building high-performance green office buildings, measuring and analysing energy used, and engaging stakeholders with environmental issues".

4.1 Research Methodology

Methodology selected for this study was qualitative content analysis because of lack of literature on analysis of media content with special emphasis to green/environmental appeal.

4.1.1 Universe

All types of advertisements of housing schemes and automobiles which yielded the consumerism in the society were universe of this study.

4.1.2 Sample

Purposive sampling was employed to select the right sample.

Four housing schemes were selected on the basis of their sound brand image in the country and their nationwide operations. Two TVCs of each housing society were selected for content analysis.

(a) *Bahria town*, (b) *Fazaia Hopsuing Scheme*, (c) *Royal residencia*, (d) *DHA housing authority.*

Automobile industry is comprised of many different vehicles but mostly car and motorbikes are advertised through TVCs in Pakistan. There are three big names in cars that have monopoly on car market in Pakistan and widely accepted in the public and few brand names in motorbikes as well. Three cars brands with two TVCs each and 1 motorbike brand with two motorbikes TVCs were selected for this study.

(a) *Honda*, (b) *Suzuki* and (c) *Toyota* (*cars*), (d) *Yamaha* (*motorbike*).

4.1.3 Unit of Analysis

Unit of analysis for this study was every single shot of the TVC.

4.2 Operationalization of Variables

4.2.1 Levels of Green Appeal

As highlighted by Corbett (2006), green appeal has been measured on four levels:

1. Nature as Backdrop (where greenery/nature are used in the background of any shot)
2. Green Product Attributes (does the product is environment friendly or there are facilities/systems which provide green or natural environment and does not disturb the nature)
3. Green Image (in this category researchers have looked into the organizational/ brand image as green organizations. Their names, logos and taglines are analysed)

4. Environmental Advocacy (does the TVC contain proper environmental advocacy that product is environment/nature friendly, it will not affect the nature rather it will help in protecting it etc.).

4.2.2 Frequency of Green Appeal

The number of times each level of green appeal highlighted by Corbett (2006) has been used in each TVC.

4.2.3 Non-Green Appeal

Researchers have found non-green appeals and divided them into rational and non-rational appeals. Researchers came up with most commonly appeared themes across the sample during pilot testing. Rational appeal for housing schemes comprises of Security, education, health, location, pricing, power supply, lifestyle, and international recognition.

Non-rational appeals were emotional appeal, musical appeal and religious appeal. For automobiles rational appeal comprises of innovation & technology, comfortability/luxury, engine power and body design. Non-rational appeals were celebrity endorsement, emotional appeal, social appraisal, musical appeal, women objectification (Fig. 1).

5   Results

Researchers have analysed the appeals used in the ads of housing schemes and automobiles under two major categories, i.e. green appeal and non-green appeal. The green appeals analysed were same for the ads of housing schemes and automobiles as highlighted by Corbett (2006) and they were Nature as Back Drop, Green Product Attribute, Green Image and Environmental Advocacy. The non-green appeal was further categorized as rational and non-rational appeal which contained various items for both ads of housing schemes and automobiles keeping in view the complete different nature of the products. An elaborated overview of green and non- green appeal is given below (Tables 1 and 2) with the detailed analysis and the interpretation of results by the researchers comes afterwards.

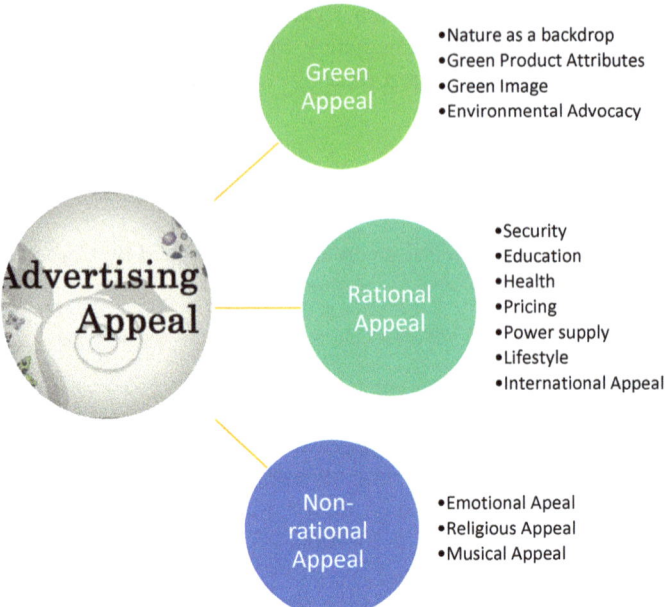

Fig. 1 Model for advertising appeal in housing schemes

5.1   Analysis of Ads of Housing Schemes

Findings of (Table 1) housing schemes'ads have been discussed by researchers according to the research questions.

1. Which appeal (green and other) has been appeared most in the TVCs?
   The content analysis of TVC's of housing schemes has shown that the Green appeal has been used for (63) times and the non-green appeal appeared for (85) times. The most frequently used appeal among the non-green appeals was the rational appeal i.e. (64) times. The rational appeal included several appeals like security, education, health, pricing, power supply, life style and international appeal. Among all of these the most used theme was international appeal which appeared (17) times in the form of claims, recognition, international media acknowledgement, certification and awards. The lifestyle appeal comes after the international appeal appearing (15) times offering consumers with luxurious, comfortable, executive living. The non-rational appeal was also part of non-green appeals appeared (21) times of which the most commonly repeated was the emotional appeal that appeared (13) showing happy families with kids and grandparents. The other non-rational appeals were religious and musical.
2. Which level of green appeal is most frequent in TVCs of both housing schemes and automobiles?

Table 1 Housing schemes TVC's analysis

Appeals			
Levels of green appeal		Frequency	Style
Nature as back drop		31	Greenery, blue sky, freshness, trees, green paths, green grounds and trees, green patches
Green product attribute		18	Royal garden, royal green house, royal wildlife, farmhouse, wind farm, parks, green play-ground, lush green grounds, green golf course, night safari, green landscape, water ponds, theme parks
Green image		12	Logo of petals in different shades of green, names of societies, eagle has been shown flying-a symbol of co-existence with nature, sun rays in logo, green colour in logo
Environmental advocacy		2	A green place to live, nature friendly residence
Non green appeals			
Rational	a. Security	9	Security cameras. Security assured by Pak air-force, international certified security system, safe city
	b. Education	4	Royal library, educational district, best schools, international study hall
	c. Health	5	International standard hospital, health facilitation
	d. Pricing	5	Affordability, according to budget, easy instalments
	e. Power supply	9	No load shedding, continuous power supply, 0% load shedding, power grid
	f. Lifestyle	15	Classy, luxury, executive sports facilities, extraordinary luxurious lifestyle, Luxury lifestyle, comfortable living, facilitated living
	g. International appeal	17	International standard residence, international standard sports, international standard health facilities, international standard stadium, International awards and claims, international media acknowledgement, international consultancy, built by Asia's largest developer, international level security, standard theme park, world renowned consultancies
Non-rational appeals	Emotional appeal	13	Happy family, joyful kids, happy grandparents, friends, A complete family
	Religious appeal	4	Grand mosque, Mosque, magnificent mosque, Bahria grand mosque Islamic architecture
	Musical appeal	4	background Music with VO

Table 2 Automobiles TVC's analysis

Appeals			
Levels of green appeal		Frequency	Style (element shown)
Nature as back drop		67	Trees, mountain area, green fields, green ground, water, snow fall, desert, mountain roads, theme park Seashore, rocky mountain area, sun set
Green product attribute		Nil	None
Green image		1	Showing horses
Environmental advocacy		Nil	None
Non-green appeals			
Rational	a. Innovation and technology	18	Push button start, headlights, comfortable handling, good head lights Navigation system, LED lights, automatic transmission, alloy rim, key less entry, push button start, paddle shift transmission for manual control of transmission, Sports mood, updated sound system, attractive speedo meter and RPM, beautiful rear view mirror, stylish fog lights, ABS brakes sporty steering, Cruise control, paddle shifting, sun roof, for ac control they give touch panel
	b. Comfortability/ luxury	14	Comfortable seats, good quality tyres, good suspension, beautiful speedo meter and RPM + fuel gage, adjustable rear suspension, Comfortable leather seats, executive, classy lifestyle, AC, Very good suspension, music system, sun roof, unique transmission change panel, comfortable while off road drive Executive, classy lifestyle, power windows
	c. Engine power	18	Drive on road, 125 cc, metaphor of athlete, high speed drive, engine power, metaphor of racing horses, off road drive, drive on water and on rough mountain show engine power, 1.3 cc engine, engine power show by high RPM, off road drive, adventure
	d. Body designing	24	Stylish body design, suitable body size, car design and features are highlighted with a metaphor of a woman, different life experience, focus on car design

(continued)

Table 2   (continued)

Appeals			
Levels of green appeal		Frequency	Style (element shown)
Non-rational appeals	Celebrity endorsement	2	Ali Zafar, Fawad Khan
	Emotional appeal	6	Family ride, friendship zone, family travelling
	Social appraisal	5	Women is getting impressed by Ali Zafar, I look good on the bike, showing fencing sport, women is getting impressed, you are always ahead
	Musical appeal	8	Ali Zafar song, simple music, music + musical concert, english song, fun & exciting music theme, english song
	Women objectification	11	Unnecessary women image, unnecessary women image, showing women holding lamp, texture seats are compared with women body, magnificence of women and elegance is used to show elegance and magnificence of car design, unnecessary women image
	Adventure	15	Motion rides, ride on rocky mountains, major focus on off—road drive and thrill

Green Appeal in TVC's of Housing schemes

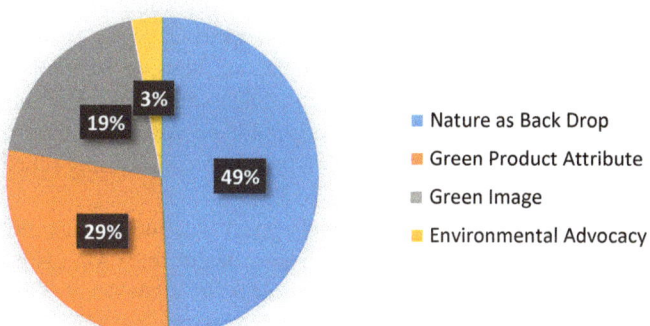

* Percentage of sub-categories of green appeal in TVC's of housing schemes

Fig. 2   Green appeal in the TVC's of housing schemes *Percentage of sub-categories of green appeal in TVC's of housing schemes

Table 1 has shown that most dominant green appeal was 'Nature as Backdrop' which appeared (31) times almost half the number of total number of green appeal been used in ads. The second most appeared level of green-appeal was

Rational Appeal in TVC's of Housing schemes

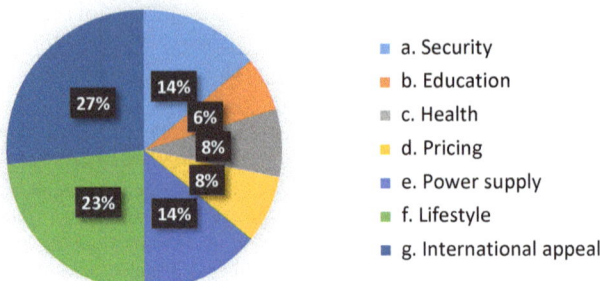

* Percentage of sub-categories of rational appeal in TVC's of housing schemes

Fig. 3 Rational appeal in the TVC's of housing schemes *Percentage of sub-categories of rational appeal in TVC's of housing schemes

the 'Green Product Attribute' appearing (18) times and then the 'Green Image' used (12) times in the ads. The fourth; last and most important one appeared only (02) times of the total number (see Fig. 2 for percentages).

3. What are the commonly used styles in the green/Non-Green Appeal?

The styles or elements used depended largely on the kind of given appeal. While showing 'Nature as Back Drop' the visuals mostly were trees, green lanes, lush grounds, skies and waters. 'Green Product Attribute' was depicted through gardens, wildlife, farmhouses, green golf courses, safaris, water ponds and theme parks. In case of 'Green Image' of the corporation, the green was shown in form of embedded green or natural appeal in logos, names of housing schemes and their tag lines, for example one housing scheme carries a logo which entails 'flower petals in different shades of green'. The most little used level was 'environmental advocacy' depended on voice over than visuals telling potential consumers that how choosing their brand of housing schemes will lead them to live a green lifestyle (Fig. 3).

Among non-green appeals the rational appeal was further categorized into different appeals such as security, education, health, etc. Almost every housing scheme ad mentioned the feature of security in great detail. Claims like Pak Airforce security or international standard equipment were also present. Education appeal focuses on school and libraries. As Load shedding has emerged as a big issue during recent past in Pakistan, so, almost every housing scheme mentioned non-stop power supply as a special feature. Health appeal focused on health facilities. Pricing focused on offering easy instalments and affordable deal to the customers. Life style was shown through the luxury and comfortability. International claims, acknowledgements, awards, recognition, certification, architecture and standard were central to the rational appeal of housing schemes ads.

Non-Rational Appeal in TVC's of Housing schemes

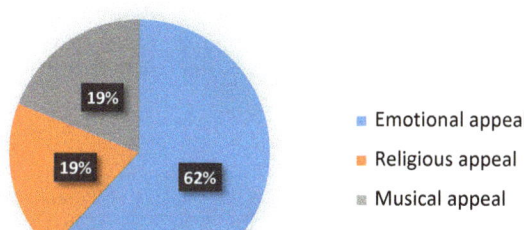

* Percentage of sub-categories of non-rational appeal in TVC's of housing schemes

Fig. 4 Rational appeal in the TVC's of housing schemes *Percentage of sub-categories of non-rational appeal in TVC's of housing schemes

The non-rational appeal features emotional, religious and musical. Emotional appeal focused on showing family and friends with kids playing around or family dining. Religious appeal featured magnificent mosques, grand mosques and claims of Islamic architecture. The musical appeal provided background music complementing the visuals (see Fig. 4).

5.2 Analysis of Ads of Automobiles

Findings of (Table 2) automobiles ads have been discussed by researchers according to the research questions.

1. Which appeal (green and other) has been appeared most in the TVCs?
   From the above table it is evident that green appeal has been used (68 times) in all TVCs of automobiles. When comes to other appeals analysis showed that major focus of all automobiles TVC in Pakistan was centred on rational appeal. The sub-categories of rational appeal innovation & technology (18), comfortability/luxury (14), engine power (18) and body design (24) are dominating other appeals in the TVCs. Adventure is another important appeal (15) which used is in the TVCs of automobiles followed by women objectification (11). Music is also used in the background of all TVCs. Emotional appeal and social appraisal are also important appeals which are frequently used by the major car and motorbikes brands in Pakistan. Celebrity endorsement by any car/ bike brand remained marginal in the sample (as only two advertisement featured celebrities) (Fig. 5).

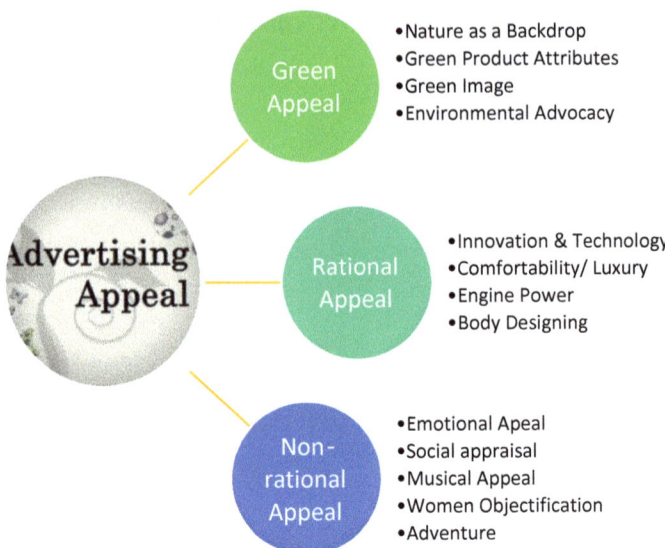

Fig. 5 Model for advertising appeal of automobiles

2. Which level of green appeal is most frequent in TVCs of both housing schemes and automobiles?
   Table 2 showed that 'Nature as Backdrop'has been widely used by the all automobiles. Nature has been used 967) times in background through depiction of green fields, trees, water, rocky mountains, hilly areas, and grass etc. 'Green Image' is shown only once where speed of the car engine is compared with speed of the horses. Not a single TVC mentioned 'Eco-friendly Attributes' of the products neither any use of 'Environmental Advocacy' was observed (see Fig. 6).
3. What are the commonly used styles in the green/Non-Green Appeal?
   Most common appeared styles or approaches which were used in the TVCs of the automobiles comprised of both green appeal and other appeals. Analysis showed that to show 'Nature as Backdrop' shots of trees, mountain area, green fields, green ground, Water, Snow fall, desert, mountain roads, theme park, seashore, rocky mountain area, sun set etc. were used. In the 'Green Image' horses were used as a metaphor to show the efficiency of the engine. Other appeals contained rational appeals which were shown with multiple attractions: Innovation & technology "Push button start, headlights, comfortable handling, navigation system, LED lights, automatic transmission, alloy rim, key less entry, push button start, Sports mood, updated sound system, speedo meter and RPM, beautiful rear view mirror, stylish fog lights, ABS brakes sporty steering, Cruise control, paddle shifting, sun roof, AC Control touch panel".

Green Appeal in TVC's of Automobiles

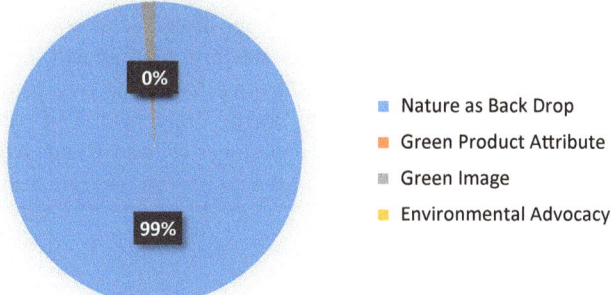

Legend:
- ■ Nature as Back Drop
- ■ Green Product Attribute
- ▨ Green Image
- ■ Environmental Advocacy

0%

99%

* Percentage of sub-categories of green appeal in TVC's of automobiles

Fig. 6 Green appeal in the TVC's of automobiles *Percentage of sub-categories of green appeal in TVC's of automobiles

Comfortability/luxury: "Comfortable leather seats, good quality tyres, good suspension, adjustable rear suspension, executive, classy lifestyle, AC, music system, sun roof, comfortable while off road drive, power windows".

Engine power: "125 cc, Metaphor of Athlete, High speed drive, Engine power, Metaphor of racing horses, off road drive, Drive on water and on rough mountain show engine power, 1.3 cc engine, Engine power show by high RPM, off road drive, adventure" (Fig. 7).

Body design: "Focus on stylish body design, suitable body size, focus on logo of the car brand".

Adventure was shown through drifting, off-road drive, drive on Rocky Mountains and seashore, metaphor of motion ride etc. Women Objectification

Rational Appeal in TVC's of Automobiles

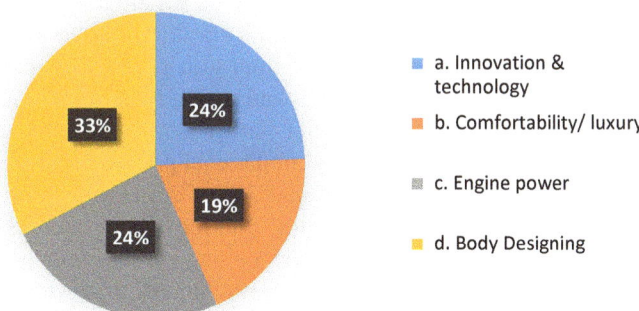

Legend:
- ■ a. Innovation & technology
- ■ b. Comfortability/ luxury
- ▨ c. Engine power
- ■ d. Body Designing

24%
19%
24%
33%

* Percentage of sub-categories of rational appeal in TVC's of automobiles

Fig. 7 Rational appeal in the TVC's of automobiles *Percentage of sub-categories of rational appeal in TVC's of automobiles

was depicted by using women as a metaphor to show the softness of leather seats, her swirls were compared with spin of the wheels etc. Music was used in the background. In one TVC where a singer turned actor was featured and his song was played in the TVC. For emotional appeal family and friends were used and social appraisal was depicted by showing that women are getting impressed by the cars and bikes passing by them. Only two TVC held celebrity endorsement featuring famous actor and singer (Fawad Khan, Ali Zafar).

6 Discussion

An analysis of findings of both the ads of housing schemes and automobiles showed that the use of environmental appeal in ads was evident but shallow. The aspects of climate change communication were present in ads but the depth of climate change communication can be questioned through the level of appeal being used as offered by Corbett (2006). For instance, Hansen and Machin (2008) concluded that "the more notable change appears to be the commercial appropriation of this discourse, and therefore the effect of promoting greater consumption..." instead of focusing on promotion of environmental literacy or green lifestyle among masses.

The ads of both sectors relied heavily on using green appeal in form of 'Nature as Backdrop' which showed the maximum green in the background. They limited the use of nature as 'pleasure for the eye' at most places and as 'complementary' at some places—adding something to the larger picture on the screen (for example, showing a tree in front of every house or showing green lanes on both sides of roads or greenery while showing a bird eye view of the architecture and design of housing scheme). It was evident from analysis that Green was used in background extensively which has built an overall aura of the product as a green product in viewers' eyes though it does not offer anything to the nature, climate or environment in reality.

The 'Green Product Attribute' was used in the ads of housing schemes depended upon the nature of product offered. It mostly featured all the possible distinctive attributes a housing scheme can offer—from lush green play grounds, green golf courses, and night safaris to theme parks and ponds. None of them really focused on how it may prevent environment from further deterioration or how much these acclaimed attributes were truly offered in the final product or how can they help in fighting back the danger of climate change. The 'Green Product Attribute' appeared more 'persuasive' and eye-catching than offering 'nature as backdrop'. In automobile ads the focus was simply more on alluring viewers with the technology and innovation offered by the launched product rather on what green attribute a particular gadget can offer.

Talking about 'Green Image' of the product the ads of housing schemes strived to portray their product as 'green product' by embedding a natural attribute in either the logo or name of their product (Hartmann et al. 2016). An 'illuminated sun' in the logo of a housing scheme or 'flowers petals in shades of green' offered nothing in actual to the nature but considered natural and familiar to the viewer's eye as it helps them identify the product as nature friendly product. One housing scheme had used the metaphor of 'eagle's flight' as how bigger the goals of that housing scheme or for how elevated lifestyle that housing scheme will offer. The 'Eagle's flight' is also a symbol of Pakistan Air Forces (PAF) and by using it, the particular housing scheme was basically trying to cash the trust of their potential buyers which they have over the institution of PAF. Similarly, at one place an automobile ad has used the metaphor of 'racing horses' to that of the speed of car. Such need based uses of natural symbols appeared more exploitative than nature-friendly in essence.

Interestingly despite significant use of nature and green imagery as backdrop in the ads of both housing schemes and automobiles both of their ads were failed to offer 'environmental advocacy' of any sort. Only a one liner given by an ad of a housing scheme claimed to provide a green life style (*sarsabz-o-shadaab zindagi*— an urdu phrase for green life style) was identified as 'environmental advocacy'. The gap between use of nature (pleasure, complement, persuasion, symbolic, etc.) and advocacy was blatant. This glaring gap had shown how much the nature of products was in contrast to their own visuals. The makers were ready to use the nature and green to freshen or beautify the visuals but when it came to 'environmental advocacy'—it had been ignored deliberately. It showed how greenwashing had been done by using nature merely as ploy for the prospective customers.

On other hand, the use of non-green appeals outnumbered the green appeal was another interesting dimension to this analysis which reduced the use of nature in the ads to a prop only. Non-Green appeals were the actual perks these ads were offering —may it be the security, international claims/certifications, comfortable life style, health, education or emotional appeal in case of ads of housing schemes—or may it be the technology/innovation, engine power, body designing, comfortability or social appraisal in case of automobiles ads. Among non-green attributes the heavy use of rational appeal over non-rational appeal displayed what side of viewers' perceptions manufacturers wanted to hit at. As highlighted by Thaker et al. (2017) that there is responsibility of the media to inform and educate the public about climate change issue, the analysis showed that this concern is undermined by the advertisers and they used nature as a commodity used to heighten the saleability of product. Though Blouin (2016) highlighted that "as spectators of media constructions of climate change, we need to begin the arduous task of scrutinizing ideological traps associated with the problem: the numerous ways in which consumerism overshadows the environmental advocacy".

7 Future Prospects

In order to mitigate the causes for climate change a more issue-sensitive and responsive audience is needed. This study was intended to fill the gap between research studies on climate change and climate change communication in Pakistan.

Very few studies have investigated and analysed the aspects of climate change communication in advertising. This study has offered a detailed content analysis of ads of two major sectors of Pakistan, i.e. automobiles and housing schemes. It served various purposes. At one hand it added to the existing literature present on the subject. Secondly, it highlighted the aspects of climate change communication in advertising by analysing the green/non-green appeals present in the ads. Findings have not only suggested the extent to which advertisers haven been practising climate change communication and to what level but has also endorsed how important it is for this persuasive medium to use climate change communication. It has also maintained that climate change communication is real/operational need and it can only be dealt by adopting a more responsible advocacy approach in ads. It will also serve as a guideline for future researches to look up to the subject through various other dimensions.

8 Limitations

- This study has only investigated the presence of green appeal in advertisements of two sectors only i.e. Housing Schemes and Automobiles.
- A more exhaustive study can be done by also bringing ads of other sectors into investigation which will give an exclusive picture of presence climate change communication in advertisement.
- There was lack of relevant literature related to Pakistan on the topic. Only a few studies investigate climate change communication. More literature is available on climate change.
- Being self-financed researchers, the study was limited by the non-availability/ shortening of resources.
- The sample size was relatively small due to time-constraint. The ads of only 2016 and 2017 of top housing schemes and automobiles were analysed for the study.

9 Conclusion

In a developing country like Pakistan with a low literacy rate, masses are striving hard to afford a reasonable living. Their definition of 'reasonable' is largely fashioned by the media industries with private media thriving in the country since 2002. Due to media meagre coverage of environmental issues—public awareness over the issue is either limited or null. In this bigger picture, the advertising usually focuses on what attracts the public rather what benefits the public. In order to present their product in most persuasive manner both sectors (housing schemes and automobiles) have not shown any reluctance even at one point to use the nature as an object of pleasure, a prop, or as mere symbol. Rather focusing on nature friendliness of their product and advocating about the environment through their ads (climate change communication)—they reduced the use of nature for deception only. Noting the fact that in Pakistan agriculture sector is considered as backbone of economy there is a need for corporate social responsibility of presenting an even bigger picture of how housing schemes are expanding on the expense of agricultural land around the country no matter how skilfully they present the green attributes of their product. On other hand automobiles ads scarcely failed to highlight any nature friendly feature of their product reducing the consumer motive for choice to rational appeals only which can give rise to consumerism only. Being ranked on seventh on the list of countries most vulnerable to climate change Pakistan's marketing sector needs to practise corporate social responsibility approach in their advertising not through greenwashing but actually offering nature friendly products and also by not reducing the use of nature for pleasure only but also for environmental advocacy which is much needed at the moment.

References

Banerjee S, Gulas S, Iyer E (1995) Shades of green: a multidimensional analysis of environmental advertising. J Adv 23(Summer):21–31

Baum LM (2012) It's not easy being green… Or is it? A content analysis of environmental claims in magazine advertisements from the United States and United Kingdom. Environ Commun J Nat Cult 6(4):423–440

Bird E, Lutz R, Warwick C (2008) Media as partners in education for sustainable development: a training and resource kit. United Nations Educational, Scientific and Cultural Organisation, Paris, France

Blouin MJ (2016) Climate change and the "greenwashing" of hollywood fantasies. In: Magical thinking, fantastic film, and the illusions of neoliberalism. Palgrave Macmillan US, p 109

Bremner B (1989) The new sales pitch: the environment, business week, (July 24), 50

Brower M, Leon W (1999) The consumer's guide to effective environmental choices: practical advice from the Union of Concerned Scientists. Harmony

Corbett JB (2006) Communicating nature: how we create and understand environmental messages. Island Press

D'Souza C, Taghian M, Lamb P, Peretiatkos R (2006) Green products and corporate strategy: an empirical investigation. Soc Bus Rev 1(2):144–157

Fernando AG, Sivakumaran B, Suganthi L (2014) Nature of green advertisements in India: are they greenwashed? Asian J Commun 24(3):222–241

Fund MF (1995) Yearning for balance: views of Americans on consumption, materialism and the environment. Author, Milton, MA

Global Climate Risk Index (2017) Available at http://germanwatch.org/de/download/13503.pdf

Hansen A, Machin D (2008) Visually branding the environment: climate change as a marketing opportunity. Disc Stud 10(6):777–794

Hartmann P, Apaolaza V, Eisend M (2016) Nature imagery in non-green advertising: the effects of emotion, autobiographical memory, and consumer's green traits. J Adv 45(4):427–440

Hepburn MA (1997) TV violence: a medium effects under scrutiny. Soc Edu 61(5):244–249

Hou J, Reber BH (2011) Dimensions of disclosures: corporate social responsibility (CSR) reporting by media companies. Public Relat Rev 37(2):166–168

Insight E (2009) Climate change: a global problem requiring a global solution. Delegation of the European Commission to Australia and New Zealand, Yarralumla

Kreft S, Eckstein D, Melchior I (2016) Global climate risk index 2017: who suffers most from extreme weather events? Weather-related loss events in 2015 and 1996 to 2015. Germanwatch Nord-Sud Initiative eV

Lee K (2008) Opportunities for green marketing: young consumers. Market Intell Plann 26 (6):573–586

Mercado MT (2012) Media representations of climate change in the argentinean press. J Stud 13 (2):193–209

Mustafa (2006) Pakistan, India join hands to clear Hudiara Drain

Priestley R (2017) Communicating science to climate change deniers Retrieved 1 June, 2017, from http://www.victoria.ac.nz/news/2017/03/communicating-science-to-climate-change-deniers

Rahim MHA, Zukni RZJA, Ahmad F, Lyndon N (2012) Green advertising and environmentally responsible consumer behavior: the level of awareness and perception of Malaysian youth. Asian Soc Sci 8(5):46

Rashid NA (2009) Awareness of eco-label in Malaysia's green marketing initiative. Int J Bus Manag 4(8):132–141

Thaker J, Zhao X, Leiserowitz A (2017) Media use and public perceptions of global warming in India. Environ Commun 11(3):353–369

Whitmarsh L, O'Neill S, Lorenzoni I (2013) Public engagement with climate change: what do we know, and where do we go from here? Int J Media Cult Polit 9(1):7–25

Xue F, Muralidharan S (2015) A green picture is worth a thousand words?: effects of visual and textual environmental appeals in advertising and the moderating role of product involvement. J Promot Manag 21(1):82–106

Environmental Entrepreneurship: Adapting Our Museums for a Greener Future

Elliot Goodger

Abstract Our museums are not only a space for promoting education concerning climate change, but they also must function as the forward-thinking green institutions of the future. Taking measures to educate about climate change is our moral duty, but we must go a step further and set the example by creating and implementing greener business strategies that enable us to operate sustainably, whilst creating and maintaining a dialogue surrounding climate change with our staff and visitors. By following the research and hard work of green museums around the world, much can be learned regarding the most efficient methods for communicating to visitors and delivering the sustainability message of change. We must make haste to combat climate change, not only in our museum narratives, but in our entrepreneurial activities too. This paper challenges the common perception that business development must be compromised in order to pursue greener business strategies for our organisations. By detailing examples of more suitable workspaces, the importance of green policies and practices, better communication strategies, and energy-saving installations, materials and technology, this paper will provide the tools for museums to adapt their management approach in order to pursue an energy efficient, greener and more cost-effective future.

If left unchecked, climate change will result in food and water scarcity, exacerbate the risk of severe weather conditions, contribute to the rise in sea levels, and ultimately be a grave cost to all of us in the long-term, both economically and environmentally. It is the undoubted moral responsibility of all museums to reflect the contemporary issues that impact society in a changing world, and it has become an extension of the role of the museums to include environmental issues. It is our duty to be accountable for adding sustained public benefit to our communities and the wider world through our greener choices. It is not only about adjusting our attitudes and our narratives but this also involves practicing what we preach by

E. Goodger (✉)
Nantwich Museum, Pillory St., Nantwich, Cheshire CW5 5BQ, UK
e-mail: exg006@alumni.bham.ac.uk

© Springer Nature Switzerland AG 2019
W. Leal Filho et al. (eds.), *Addressing the Challenges in Communicating Climate Change Across Various Audiences*, Climate Change Management,
https://doi.org/10.1007/978-3-319-98294-6_35

introducing clean energy technology and reducing carbon emissions in our institutions. We must be both internally focused and externally mindful, by promoting both education and action on climate change and indeed a growing number of museums are starting to do so.

Many museums have adapted their function and outlook through the course of several decades or even centuries and have been intuitively adaptive to the current social and scientific climate. One of the UN sustainable development goals is to ensure that by 2030, there is 'adequate access to clean energy research and technology', whilst promoting investment in clean energy technology and energy infrastructure (UN 2015a). This paper aims to help address that knowledge gap for the museums sector. Currently there are no national schemes in the UK that are focused on helping museums to go green, but there are resources available and methods that can be employed, as will be explored.

Greener solutions may not be an immediately apparent aspect of many museums functions and missions but if we consider that environmentalism is related to saving money, increasing public value and promoting sustainability, then of course green solutions are just another extension of resource and funding efficiency. How many ticket sales does it take to cover your museum electricity bill and think what those annual savings could be spent on instead? If not sufficiently motivated by our moral obligations to consider climate change in our museums, then underneath that, there are huge savings to be made on our utility bills and our equipment costs by simply switching to more environmentally-friendly and power-saving installations, mind-sets and methods.

Some changes are as simple as monitoring your systems, examining water and energy usage, recycling more whilst reducing wastage, and other solutions are about locating and building partnerships with greener companies when concerning products and services. The green economy is significantly more advanced and accessible than a decade ago and it is a myth to think that implementing green methods are always the more expensive option. Green building now matches the cost of traditional building, and many projects have the additional benefit of saving money in the long-term (Brophy and Wylie 2013: 4).

Through research and participation in museum management in the UK and around Europe, and through reviewing museum overheads to see where savings can be made, I have identified a common pattern of mistakes in management decisions that can be costly in the long-term. I have compiled four key areas of consideration in museum management and spending decisions that not only consider the commercial direction of the museum, but the environment too. By having transparent and permanent green materials, policies, practices and installations in the museum space, as outlined in this paper, it allows museums to function as permanent reminders and educators on climate change for their staff and for their visitors.

The following figure is a summary of the four areas of focus for environmental entrepreneurship if we are to take a holistic approach to entrepreneurship through our approach to workspaces, the policies and mission of our museums, informative communication with staff and visitors, energy and material consumption, our operation practices and funding, and technology usage and building design for our

ENVIRONMENTAL ENTREPRENEURSHIP			
1	**2**	**3**	**4**
Suitable Workspace	**Green Technology and Material Selection**	**Energy Consumption and Management Practices**	**Communicating Environmentalism**
Comfortable and healthy work environment	Energy saving technology	Waste and cleaning considerations	Engaging with local markets
Plant-based foods	Building design and retrofits	Combined water and energy usage methods	On-going staff and visitor engagement
Ensuring Employee wellbeing	Product and material selection	Responsible Funding Sources	Making museum policies greener

Fig. 1 The four sections of environmental entrepreneurship *Source* the Author

museums. Such an approach is necessary to ensure that environmental entrepreneurship is sustainably reflected in our museums. Each of the following four areas will be explored in greater depth to detail how and why our museums should adopt an ethos of environmental entrepreneurship (Fig. 1).

1 Suitable Workspace

The first area of consideration creating a mindset of environmental entrepreneurship is to ensure that it firstly reflected in our workspaces, through a healthy working environment. Its perhaps of no surprise to discover that there is a direct correlation between higher wellbeing amongst employees and higher economic performance overall as an organisation, as well as longer retention of workers (Wright 2006: 118–120). Encouraging a healthy lifestyle amongst staff is the best way to achieve this, as well as acknowledging a duty to provide a healthy work environment, in a suitable green museum environment that includes good ventilation, clean air, ergonomic workspaces and natural lighting where possible in the communal spaces (Brophy and Wylie 2013: 5–6).

Large amounts of natural lighting in rooms that do not house collections are essential in saving on the electricity bill, but these types of rooms are not appropriate for the storage or display of objects due to the damage caused by UV light. Of course, relocation of buildings that house museums is a laborious and costly process, unrealistic for most buildings in the current economic climate, but for new buildings and extensions, an environmental policy and eye for detail comes into play, which I discuss more later.

The UN has stated that climate change is unequivocally human produced and a UN study in 2006 concluded that animal agriculture is responsible for 18% of all global CO_2 emissions, more than all forms of transport put together (FAO 2006). The global food industry is devastating rainforests as huge areas of land are being cleared and large amounts of water is being wasted, to make way for livestock and the vast quantities of crops required to feed them. This system creates further environmental impacts as the products are then shipped across the globe.

We are increasingly seeing museums and galleries including vegan options in their cafes, as well as dedicated plant-based menus. Leading restaurants across Europe and the US are introducing vegan options and plant-based menus, with a huge spike in their popularity over the last three years. Veganism has increased by 600% in the US since 2014, meaning that now 6% of the US population eats a plant based diet, and similarly in the UK, veganism has risen by 360% in the last decade (Chiorando 2017). Cattle require approximately 7 kg of grain in order to generate a 1 kg of beef, and pigs require 4 kg of grain for 1 kg of pork (White 2000: 145–153). Consequently, staff and visitors that select the vegan option, use an average of only 20% of the land required for the meat option, and in the process, their meal requires dramatically less water and pollutants for its production.

Eating plant-based foods also helps towards achieving several of the UN's sustainable development goals including promoting good health and wellbeing and ensuring food security and sustainable agriculture. For those museums that rent out their café space to third parties, renting to a plant-based organisation is not only better for the environment, but also likely to boost revenue and footfall to the museum space, especially if the trend towards plant-based spending and revenue for sales of vegan products continues at its current rate.

2   Green Technology and Material Selection

Next, we can explore the way in which greener technologies, buildings and services contribute to environmental entrepreneurship. This should help direct towards products and installation selection that will help to make savings through reduced energy consumption in the museum space in the long-term. A huge 80% of the Greenhouse gas emissions responsible for global warming come from cities, and roughly half of that can be attributed to energy in-use in buildings and running appliances (Williams 2007). We have a responsible for running our museums in a way which minimalises our energy consumption.

Collections care is an area where we should consider low impact technology and more sophisticated museum storage and displays, as sensible green investments. Preserving our objects requires a well-controlled microclimate and environmentally unsuitable buildings and practices speeds up the degradation of our objects in storage.

Display lighting used for collections care is considerably wasteful to the environment. One alternative that is both longer lasting and more energy-efficient (as it operates at a lower temperate) is adjustable-strength laser lighting. It also does not contain the mercury and other toxic materials usually found in ordinary lamps.

The V&A museum use display cases that have a low rate of air exchange between the display case and the external air. This means that reasonable changes in temperature will have less effect on the degree of stability for relative humidity, as well as resulting in reduced absorption of pollutants from the air, and consequently require less frequent maintenance (Cassar and Martin 1994: 171). Simply selecting equipment that slow down the rate of decay for objects without consuming any energy saves on conservation costs and resources, and in addition, does not rely as heavily on the use of expensive and high-energy external temperature-regulating equipment.

One way to assess the environmental impact of new technologies and products is to look at life-cycle assessments and life-cycle analysis. This is an examination of the whole lifecycle of a product or technology and it assesses the environment impact and financial cost of the product at each stage of its life, from the extraction of the material, through to processing, manufacturing, distribution, maintenance and disposal. Before changing to a new energy-saving product or service, analysing the new projected energy usage versus the existing one, will allow for an estimated time in which we can predict that the new product will pay for itself, and everything after this will be savings. Manufacturers often publish this information on demand, and for green companies, they will usually be even more transparent about their green and energy saving credentials.

What about technology in regards to communicating with our visitors? Increasingly, research suggests that museum visitors are not ordinarily visiting alone and it is the free conversations between visitors that guides them to evoke and shape their views on an exhibition theme. There are user interfaces and screens available that are built into café tables and help to shape visitor discussion and influence behaviour by using images and keywords related to the exhibition content (Zancanaro et al. 2011: 396). This same technology can be used to relate exhibition objects to environmentally-friendly products available in the shop or café, although it is not necessary for this process to be hi-tech. Signage in the café that links museum objects and narratives to available products, also serves the same environmental and commercial purpose as an interactive display.

For sustained cost-effective management, greener buildings and infrastructure should be central to our planning. It is far easier to incorporate money-saving green ideas into a completely new building, extension or renovation project, as the Grand Rapids Art Museum in Michigan found when their new green museum building was completed in 2004. It became the first museum in the world to receive the LEED Gold certification, featuring heat-recovery ventilators and a grey water re-use system on-site. LEED, or Leadership in Energy and Environmental Design is a worldwide green building rating system that in practice reduces operating costs by saving on energy, water, resources and waste products whilst simultaneously creating a more comfortable environment for the occupants. LEED publishes purchasable reference guides online for project planners.

The Nature Research Center at the North Carolina Museum of Natural Sciences is the location of the Green Square Complex, and a 7400 m² LEED certified extension to the museum completed in 2012 (Garafalo 2013: 75). It is an environmentally friendly building made from locally sourced granite and houses an exhibit to educate the public on data collection methods for natural resources preservation and sustainability projects. The building is designed to transfer rainwater to underground cisterns to save on water bills, and the focus of the exterior was on ensuring that it had an airtight envelope (Garafalo 2013: 76).

Adapting existing buildings is also crucial, and the simplest way to control heat and humidity is to control the natural ventilation in a building and make improvements to the way buildings are insulated. This is in order to avoid the worst of the effects of the local climatic conditions on a museum collection and the accompanying fluctuations in temperature and humidity that are the most detrimental to the longevity of the collection. Retrofitting existing buildings internally with new technology is a very costly process, and in the current climate of funding for museums, it is difficult for museums to justify the prioritisation of such changes. Therefore, it is more sensible to better manage and improve the building itself in which a collection is stored.

Passivhaus or 'Passive House' is a growing worldwide energy performance standard which helps to ensure high thermal performance, mechanical ventilation and outstanding airtightness on the creation of new buildings, storehouses, or extensions. Having these high performance factors helps to reduce the necessity for internal energy consumption and traditional heating systems, and a Passivhaus reduces fuel costs by 5–10 times, saving significant sums of money annually and reduces dependency on the consumption of dwindling and increasingly costly fossil fuels (Passivhaus 2017).

There is no standard environmental safeguard or climate control mechanism as it very much dependent on the nature of the collection and on the climate of the country in which the collection is stored. An alternative approach has been adopted by the Central Institute for Conservation in Belgrade, Serbia, as only four out of 140 museums were purpose-built museums, and with very little money directed towards conservation, consequently, they lack central environmental control mechanisms, HVAC, active temperature, relative humidity and pollution control systems (Zivkovic and Dzikic 2015: 117).

Due to a chronic lack of funding across the sector in Serbia, the alternative to the installation of more costly mechanical equipment has been a focus on cost-effective installations and small interventions on building envelope. Thus, the focus was on maintaining stable climatic conditions for storage by controlling ventilation through windows, and focussing on external changes, investing in building maintenance and improving the integrity and air-tightness of buildings. This helps to ensure climate control without the need for huge and increasingly expensive energy usage.

3 Energy Consumption and Management Practices

Some environmental methods are much simpler, such as methods for cleaning and waste management. Waste is defined as simply something for which the museum no longer has a use for, and the best way to reduce wastage is to generate less waste to begin with and the museum ethos should be to Reduce, Reuse, and Recycle. Waste that goes to landfill produces high levels of damaging methane, which is far worse for the environment than CO_2 emissions. General cleaning products can be purchased in bulk, and some green suppliers reject the use of plastic containers, distributing refills in large bags. This is more economical and the sources of green cleaning product have a less detrimental effect on the environment. More specialist environmentally friendly conservation cleaning equipment is much more widely available than a decade ago and more competitively priced. The use of green alternatives in the museum space also reduces the risk of respiratory problems for staff and visitors.

It is essential for museums to learn to care for collections without the total dependence on energy consumption and museum staff must adapt their current technology and practices accordingly. There is an undisputed link between higher temperatures for collections storage and the deterioration of objects. An increase of just five degrees in temperature can result in the acceleration of water and air through stable particulates increasing by 30% and can accelerate the chemical process of the deterioration of objects even more dramatically (Pavlogeorgatos 2003: 1460). Temperature and relative humidity fluctuation is responsible for the expansion and contraction of exhibition objects, and the drying of materials such as wood and paper, is highly damaging. Reducing the amount of money spent on energy, releases money that could be spent on other development projects. It is clear that in responding adequately to needs in collections care, we continue to control the internal environment of our museums and collections stores.

The free Green Museum step-by-step guide produced by the Museums Association details a guidance framework for a participating museum to fill in the annual total water, gas, electricity, heating, oil and fuel usage alongside detailed recommendations for landfill, boiler status and usage, insulation and ventilation, equipment usage, recycling and disposal (MA 2008: 13–14). There are a number of self-assessed targets to help the museum save money through reduced energy consumption, there is an action plan which looks at wastage, supplies, transportation and staff awareness. There is also an accompanying spreadsheet which calculates the environmental impact summary based on the data inputted, and the formulae converts the energy usage to a CO_2 emissions value.

The Chicago Museum of Science and Industry secured funding in 2003 to install a cogeneration system in its museum, which provides the buildings electricity, hot water, cooling and dehumidification system. This greener system was projected to save the museum around $200,000 a year based on current gas prices, but savings change as gas prices fluctuate (Blankinship 2003: 128). Additionally, the new cogen plant saves the museum around $16,000 a year on the hot water system for the food court alone (Blankinship 2003: 128). Integrated energy systems are a sensible investment for a sustainable future, and as this example is based at a science museum, it has become the basis for a scientific exhibit which helps to educate visitors and the wider public about the logistics and long-term benefits of a cogen system.

Ever the frontrunner in progressive developments in the UK museums sector, the V&A reduced their carbon emissions attributed to energy usage by 20% over a two year period via the installation of a combined heating and power system and by utilising low energy lighting (V&A). Additionally, studies have shown that sales increased by 40% with the introduction of natural sunlight in retail environments, so by simply adjusting the lighting in the retail space, we can shape customer behaviour (Brophy and Wylie 2013: xviii).

A museum must also consider the use of paper, wastage, materials used in exhibition planning and building materials. There are a number of European companies that sell competitively priced green and recycled office supplies. Setting up a group for a coalition of green museums could be used as a collective force for utilising their combined buying power to convince popular manufacturers to implement more green products and processes in regards to the supplies sold to the museums, and to negotiate a lower price for the members of the collective.

Shortages of fresh water are deemed to be one of the most pressing environmental concerns over the next half a century, and water is only renewable if it is properly managed. Two-thirds of the world is estimated to be water stressed by 2025 (UN 2015b). Water rates across Europe are in perpetual climb. If our concern is cutting our water bills (and energy bills if hot water is used) then there are a number of straightforward ways in which to achieve that. Firstly, ensuring all staff and volunteers share this environmental mind-set and disseminate information about water usage via emails and newsletters. Low-flow valves and fixtures in restrooms reduce the amount of water used by taps and flushing toilets. Reducing water for outdoor areas can be achieved by harvesting rainwater and replacing plants and flowers with indigenous and less water-intensive species. For museums who have a particularly high water bill, it is worth investing in a water audit to identify leaks and potential repairs and some companies offer this service for free. Simple signage in museum restroom facilities is a cheap way to inform and educate on water usage.

Museums are increasingly pressured to think about the environment through their themes, through the running of their institutions and even through their funding. Funding is a clear conflict for some museums and in some instances they are forced to play a balancing act as many are extremely underfunded in this current economic climate. Consequently, they must consider whether to accept funding from individuals and organisations that have a questionable environmental impact. Increasingly, these decisions have faced considerable public scrutiny and negative media attention that is potentially detrimental to the image of the museum under fire and detrimental to the standing of museums more generally.

Greenpeace let their disapproval of the British Museum's sponsorship by BP be known, as the museum was forced to close temporarily for several hours on the

16th May 2016, as environmental activists hung huge banners from the columns outside the BP-sponsored Sunken Cities Egyptian exhibition (Vaughan 2016). The banners read 'Sinking Cities', in reference to rising sea levels. BP also backs the National Portrait Gallery and the Royal Opera House. It made the news in Europe and North America in January 2018 when more than a hundred climate change scientists and curators wrote an open letter to call for the resignation of Rebekah Mercer from her trustee position at the American Museum of Natural History (Milner 2018). Mercer is one of Donald Trump's top donors and has given huge funding to organisations such at the Heartland Institute that host regular conferences, and critics claim that these events dismiss any human involvement in climate change.

4 Communicating Environmentalism

It is the responsibility to not only educate on climate change and environmental protection through museum narratives and interpretation, but also, the museum space should be used as a platform to encourage dialogue between visitors, in a similar style that the Happy Museum Project in the UK encourages: part of exhibitions should involve the museum asking an open question about an environmental issue and then allowing visitors to write their responses to the ways in which visitors can work as a collective in order to help resolve these issues.

40% of the Greenhouse gas emissions responsible for global warming come from transportation (Williams 2007). Therefore, sourcing food and trading with businesses locally helps to reduce the environmental impacts of transporting goods and helps to strengthen local partnerships. Multisensory food programs can place a new community emphasis on local food production and local businesses, and bring heightened awareness to the local food system. As a result of a board member driving a new policy which addresses the issue of food scarcity, a seasonal farmers market was setup based at Parksville Museum, Vancouver Island, Canada (Bell and Clover 2017: 22). The farmers market then evolved to include gardening classes, as well as other skills such as pottery and weaving. There are multiple benefits to this approach: it creates an attractive platform in which staff at the museum can discuss environmental issues with visitors and the wider public, it brings a new revenue stream to the museum, engages with audiences that might not otherwise occupy the museum space and additionally, stimulates the local economy and food market, building new partnerships with local businesses in the process.

A recent study into pro-social and pro-environmental behaviour in the museum space showed that encouraging visitors to decrease waste, save water, recycle, eat local foods, walk more and use public transport, were all received positively by the audience and were influential in shaping decision making (Han and Hyun 2017: 1251). The study concludes that it is consequently worthwhile to reinforce these behavioural patterns and educate on the consequences of environmental neglect in a local setting and is therefore valuable to incorporate environmentalism into our exhibitions and events.

Artist Tiffany Holmes pushes the concept of eco-visualization: artworks that function to re-interpret environmental data into meaningful visual narratives for visitors and staff to understand, in order to encourage stewardship and make visitors aware of environmental risks, some of which are site-specific or unique to that area (Holmes 2007). The data from these visualisations will be especially interesting when two artworks compare the environmental footprint of a museum before and after the incorporation of an environmental plan. Engaging staff is also crucially important, the V&A museum have environmental forums for environmental sustainability for staff, a cycle to work scheme, and have also launched a scheme to allow staff to borrow a Wattson home energy monitor, to more closely understand energy consumption (V&A 2016).

Aside from communicating with audiences in regards to environmental exhibition design, content, conservation and planning, as the Climate Change Museum in New York City demonstrates, this should also be reflected in the range of retail products available. Emphasis should be on locally crafted and sourced goods, and recycled materials. Signage in the retail space should also emphasise current wastage and the positive impact of environmentally conscious spending, a retail trend that is growing rapidly, in which environmentally marketed products have demonstrated to be more profitable (Han 2015: 164–177). Ecotourism is now increasingly shaping the choice of destination and reflected in the purchase choices of an increasingly environmentally conscious public.

One important step that all museums must do is to ensure that environmental sustainability is reflected in the policy documents of our museums, and that a separate environmental sustainability document is created in addition to the revision of our collections development and conservation policy documents. This ensures that all business decisions consider the environment and money-saving investments, and that in the running of our museums we consider reducing, reusing and recycling materials. Crucial in connecting environmental sustainability with a museum mission statement, is identifying the link with three aspects: research, public education and public service. Environmental sustainability is just a natural extension to the sustainability of the museum and its collection that all museums strive for, and ensures that heritage is protected for future generations as well as for the foreseeable future.

Simbarashe Chitima states that the Zimbabwe Medical Museum was taking too long to employ green practices in their museum due to the fact that no green practice existed in any sort of policy document or governance strategy for the museum (2015: 231). Chitima describes the 'haphazard implementation' of a number of green practices, as a result of them being treated as an 'after-thought' and states that according to tour guides and curators at the museum, such a lack of policy has resulted in a failure of communication between management, employees and visitors and a failure to make environmentally-conscious decisions a priority (2015: 231). This is then more costly for museums such as the Zimbabwe Medical

Museum who will have to deal with more costly bills and a higher cost in order to retrofit greener and more economical energy and utility strategies, rather than incorporating these into new plans in the first place.

It is now a requirement for accreditation in the UK that a museum has an environmental sustainability policy statement, and 'meets all relevant legal, ethical, safety, equality, environmental and planning requirements' (Arts Council 2014). As such, the sustainability policy should be used as a means in which to consider money-saving, energy efficient heating, lighting and monitoring systems. It should contain an action plan and pinpoint the individuals that are responsible for implementing it. The policy should be reviewed annually and should be signed and approved by trustees, displayed on the museum website and on public display in the museum space. Being transparent with environmental concerns is essential for spreading the message of climate change and should feature as part of regular staff meetings to ensure it remains an active consideration for the workforce. A draft environmental policy is available from the Museums Association website (MA 2008: 41).

A transparent and prominent environmentally friendly museum space acts as justification for lenders and grant-funding bodies, and demonstrates that a museum is sufficiently prepared to accommodate new collections or exhibitions. By investing in long-term energy-saving equipment, we should be open about their usage in order to help justify the acquisition of long-term funding solutions. A number of large and small grant funding bodies in the UK including the Esmèe Fairbairn Foundation, the Bridge House Estates Trust and others, invest millions of pounds into environmental projects and designated grants for environment-focused projects form a significant part of their funding activities. Sourcing grant funding to invest in infrastructure and green technology in order to save money on energy bills and wastage is undoubtedly a viable option.

5 Conclusions

To understand how the activities of our museums impact climate change, we have to make the link between how we manage the museum space, and this effect on the food and transport system, as well as the detrimental environmental impacts of excessive energy and water usage in our buildings. As stated in Sects. 1 and 2, a comfortable work environment will help with the retention of workers and on saving on energy and bills, and so the central focus of all new building and installation programs should be on systems that reduce the need for costly temperature regulation. This doesn't have to mean exceptionally costly installations as demonstrated with the external installations in Belgrade. Simple changes in the workspace like food selection in cafes in order to include plant-based alternatives helps to widen the demographic and introduce a more environmentally-friendly option.

Section 3 looked at calculating and reducing our carbon footprint in the museum space, through reviewing our internal energy systems. We should integrate our systems into our exhibitions in order to bring the topic of climate change and energy-saving methods to our audiences. It is vital for all museums to maintain a good public image through wise investments and ethical funding bodies. We should consider the implementation of the methods outlined in Sect. 3 that allow us to review and restrict our water usage and get cheaper rates on environmentally-friendly museum supplies through collective purchasing power.

Section 4 reminds us that transparent environmental policies and processes in the museum space are attractive to investors and that environmental processes are now a crucial part of the accreditation process. Implementing green policies as an afterthought risks costly retrofits. Switching to local trade can improve the standing of the museum in the community and can boost local business partnerships, especially important for bolstering museum events calendars. It is now economically sensible to incorporate environmentally friendly products into the museum shop.

The Bizot Group, a group comprised of the world's leading museums devised the following guiding principles to characterise the Green Protocol in 2015, which has been subsequently adopted by a number of other organisations including the UK National Museum Directors' Council (NMDC 2015). The NMDC have adopted the appropriate operational mind-set and environmental sustainability priorities for museums moving forward, something we should adopt into our management ethos:

- Environmental standards should become more intelligent and better tailored to specific needs. Blanket conditions should no longer apply. Instead conditions should be determined by the requirements of individual objects or groups of objects and the climate in the part of the world in which the museum is located.
- Where appropriate, care of collections should be achieved in a way that does not assume air conditioning or other high energy cost solutions. Passive methods, simple technology that is easy to maintain, and lower energy solutions should be considered.
- Natural and sustainable environmental controls should be explored and exploited fully.
- When designing and constructing new buildings or renovating old ones, architects and engineers should be guided significantly to reduce the building's carbon footprint as a key objective.
- The design and build of exhibitions should be managed to mimimise waste and recycle where possible.

Undoubtedly, museums should review both their policy and practice in regards to environmentalism, in order to ensure sustainable profitability is prioritised in the running and maintenance of the buildings and collection, just as much as it is in the retail and café product selection and in their scheduled events and exhibitions programmes. Environmental sustainability alongside cost and resource saving initiatives should be regarded as companions to any entrepreneurial activities. As demonstrated, these environmental entrepreneurship activities and communications should be holistic: encompassing attitudinal, legislative, functional, and technological aspects of museum management and every effort should be made to ensure that they are deeply entrenched into the ethos of our museums and workforces to ensure greater success, more flexible budgets and greener, more sustainable museums.

References

Arts Council (2014) Accreditation guidance: an introduction. Arts Council. Available at: artscouncil.org.uk/sites/default/files/download-file/FINAL_201406_GuidanceIntroduction_PrintFriendly.pdf. Accessed: 21 Dec 2017

Bell L, Clover D (2017) Critical culture: environmental adult education in public museums. New Dir Adult Cont Edu 153:17–29

Blankinship S (2003) Chicago museum goes green with cogen system. Power Eng 107(11):128–131

Brophy S, Wylie E (2013) The green museum: a primer on environmental practice, 2nd edn. Rowman & Littlefield, Maryland

Cassar M, Martin G (1994) The environmental performance of museum display cases. Stud Conserv 39(2):171–173

Chiorando M (2017) Veganism skyrockets by 600% in America to 6% of population. Plant Based News. Available at: plantbasednews.org/post/veganism-skyrockets-by-600-in-america-over-3-years-to-6-of-population. Accessed: 20 Dec 2017

Chitima SS (2015) Developing sustainable museums through 'greening': a case study of the Zimbabwe military museum. In: Mawere M, Chiwaura H, Thondhlana TP (eds) African museums in the making: reflections on the politics of material and public culture in Zimbabwe, 1st edn. Cameroon, Langaa RPCIG

FAO: Food and Agriculture Organisation of the United Nations (2006) Livestock a major threat to environment. FAO. Available at: fao.org/newsroom/en/news/2006/1000448/index.html. Accessed: 1 Mar 2018

Garafalo S (2013) Sustainability serves as inspiration for museums. Stone World 30(10):74–77

Han H (2015) Travelers' pro-environmental behavior in a green lodging context: converging value-belief-norm theory and the theory of planned behavior. Tour Manag 47:164–177

Han H, Hyun SS (2017) Fostering customers' pro-environmental behavior at a Museum. J Sustain Tour 25(9):1240–1256

Holmes T (2007) Place-based education and the museum. J Museum Edu 32(3):275–285

Museums Association (2008) Green museums: a step by step guide. Museums Association. Available at: museumsassociation.org/download?id=282631. Accessed: 20 Feb 2018

Milner O (2018) Museum of natural history urged to cut ties with 'anti-science propagandist' Rebekah Mercer. Guardian. Available at: theguardian.com/us-news/2018/jan/25/american-museum-of-natural-history-rebekah-mercer. Accessed: 26 Jan 2018

NMDC (2015) Environmental sustainability—reducing museums' carbon footprint. NMDC. Available at: nationalmuseums.org.uk/what-we-do/contributing-sector/environmental-conditions. Accessed: 10 Dec 2017

Passivhaus (2017) The Passivhaus standard. Passivhaus. Available at: passivhaus.org.uk/standard.jsp?id=122. Accessed: 10 Dec 2017

Pavlogeorgatos G (2003) Environmental parameters in museums. Build Environ 38(12):1457–1462

UN (2015a) Transforming our world: the 2030 Agenda for sustainable development. UN. Available at: sustainabledevelopment.un.org/post2015/transformingourworld. Accessed: 15 Dec 2017

UN (2015b) Water and sustainable development. UN. Available at: un.org/waterforlifedecade/water_and_sustainable_development.shtml. Accessed: 15 Mar 2018

V&A (2016) Sustainability at the V&A. V&A. Available at: vam.ac.uk/content/articles/s/v-and-a-sustainability/. Accessed: 15 Dec 2017

Vaughan A (2016) Greenpeace activists scale British museum to protest BP sponsorship. Guardian. Available at: theguardian.com/environment/2016/may/19/greenpeace-activists-scale-british-museum-to-protest-bp-sponsorship. Accessed 29 Dec 2017

Williams B (2007) Climate change statement at the UN commission on sustainable development, 15th session. UN. Available at: sustainabledevelopment.un.org/content/documents/habitat_2may_cc.pdf. Accessed: 1 Mar 2018

White T (2000) Diet and the distribution of environmental impact. Ecol Econ 34:145–153

Wright T (2006) To be or not to be [happy]: the role of employee well-being. Acad Manag Perspect 20(3):118–120

Zivkovic V, Dzikic V (2015) Return to basics-environmental management for museum collections and historic houses. Energy Build 95:116–123

Zancanaro M, Stock O, Tomasini D, Pianesi F (2011) A socially aware persuasive system for supporting conversations at the museum café. In: Proceedings of the 16th international conference on intelligent user interfaces, Palo Alto, CA, USA, 13–16 Feb 2011, pp 395–398

Communicating Climate Change: Reactions to Adapt and Survive Exhibition and Visitors' Thoughts About Climate Change in the Pacific Islands Region

Sarah L. Hemstock and Stuart Capstick

Abstract This paper examines the content and responses to an art installation addressing climate change in the Pacific, collected at the Adapt and Survive exhibition held at the University of the South Pacific Oceania Centre Gallery in 2014. The artist statement on the exhibition emphasised that it sought to explore the causes and effects of climate change, and to raise awareness of its wider impacts for cultural loss and societal change. As well as conducting a series of interviews with the artist, visitors to the exhibition were invited to complete a short survey concerning their thoughts about climate change and reactions to the exhibition, both before and after they viewed the artworks. The artist's perspectives emphasised the significance of climate change for the region, in the context of traditional responses to environmental problems. The audience survey results suggest that there were high levels of agreement among visitors that the place where they live is being affected by climate change. While emphasising both negative and positive emotional reactions to the artworks, people for the most part expressed confidence and hope that climate change can be effectively addressed, although there was uncertainty on whether or not Pacific islands had the resources to do so. Our study is limited by the small sample size available, but points to directions for future research in this under-developed field.

S. L. Hemstock (✉)
Geography, School of Humanities, Bishop Grosseteste University,
Longdales Rd, Lincoln LN1 3DY, UK
e-mail: sarah.hemstock@bishopg.ac.uk

S. Capstick
Low-Carbon Lifestyles and Behavioural Spillover Project (CASPI),
School of Psychology and Tyndall Centre for Climate Change Research,
Cardiff University, Park Place, Cardiff CF10 3AT, UK
e-mail: capsticksb@cardiff.ac.uk

© Springer Nature Switzerland AG 2019
W. Leal Filho et al. (eds.), *Addressing the Challenges in Communicating Climate Change Across Various Audiences*, Climate Change Management,
https://doi.org/10.1007/978-3-319-98294-6_36

1 Introduction

Environmental degradation—including, until recently, using the Pacific region as a nuclear testing ground—can be seen as the Pacific region's greatest contemporary challenge. Climate change impacts on Pacific societies and cultures are far reaching and rapid. Geographic isolation, ecological uniqueness and fragility, human population pressures and associated waste disposal problems, limited land resources, depleted marine resources, exposure to damaging natural disasters, and global changes in climate; all contribute to the increasing vulnerability of small island developing states in the Pacific Islands region (Woods et al. 2006; Weir et al. 2016; Taylor and Kumar 2016). As outlined by Smith and Hemstock (2011, p. 67) the Pacific Islands "have come to represent the 'front-line' or the 'canary in the coalmine' in raising awareness regarding the potential negative consequences of climate change and impacts on land use, livelihoods, food and energy security." The more pessimistic forecasts of climate change impacts state that low-lying island nations such as Tuvalu and Kiribati could turn out to be uninhabitable inside a generation (Maclellan 2009). Various responses have been put forward by Pacific small island states—including discussion of mass migration of the affected population, with Australia and New Zealand governments already looking into policy options for accepting entire displaced populations (Shen and Gemenne 2011). Kiribati is considering migration as an "adaptation strategy". This abandonment of territory and culture appears to represent the most dramatic and fatalistic approach to tackling the encroachment of the ocean. It is the option of last resort. There is resistance to implementing this policy, both from the host governments and the islanders themselves who do not wish to leave their homeland. Therefore the dominant policy strategy, both within the island states and among the international aid donor community, is to continue to focus on in-country adaptation and literal survival (Smith and Hemstock 2011).

Against the enormous scale of the contemporary challenge of climate change impacts in the Pacific, the UNDP (2013) concluded that the extent of climate change awareness of most Pacific island populations and community participation in appropriate adaptation strategies and agendas has been negligible. This is partially due to a lack of awareness and engagement which leaves communities powerless to make informed choices about adaptation to climate change impacts affecting their livelihoods and resources (Nunn 2012)—both now and in the future. Furthermore, if this region is to access available funding for climate change adaptation and be an active participant in international climate change debate, then it is essential for the general population to have not only an awareness but also an opinion in order to take part in decision making and inform leadership on these issues—leadership at community, national, regional and international levels. Likewise, there has been little practical guidance available on how to effectively communicate climate change in ways that increase community resilience and capacity to adapt (McNaught et al. 2014).

With this lack of awareness in mind, artists seeking to promote engagement with climate change impacts and adaptation in the Pacific cannot be sure that the approaches commonly used are effective. But, in contrast to this, there is a growing literature which points to appropriate and effective ways to engage people in climate change and environmental issues (Whitmarsh and Corner 2017); including research that has sought to understand the most appropriate way to frame messages about climate change, drawing on metaphors, and the use of emotion and imagery (Moser 2010). While there remains an important role for conveying factual and scientific information about the topic, this can be alienating for those without specialist knowledge (Duxbury 2010). It is important therefore that climate change communication resonates with an audience through being meaningful to people's own lives, values and social aspirations (Moser 2010). Visual arts can therefore be an effective medium to communicate climate change and environmental issues since both can engage people directly on an emotional level, allowing the viewer to explore, reflect and respond based on their own personal experiences.

It has been suggested that the use of visual arts and artistic approaches may be effective in stimulating changes in attitudes regarding environmental issues, often more so than can be achieved through scientific communication (Robinson et al. 2014). For these reasons and others, the role of the arts and use of visualization in climate change communication has been identified as a priority area for future research (Moser 2010, 2014; O'Neill and Smith 2014).

The objective of this research was to evaluate the potential for the visual arts to communicate and enhance public engagement with climate change issues in a Pacific island context. The research focuses on the audience perspective, in addition to those of the artist.

2 Background to the Research and Research Methods

To integrate with a Pacific cultural context it is important to consider capacity building in arts as well as sciences. From the researchers' experience with regional community EcoArts projects (Capstick et al. 2018) and with projects at the University of the South Pacific and the NGO Alofa Tuvalu, it is apparent that traditional/local knowledge/wisdom and artistic/social activities such as drama, visual arts, and dance have been amongst the most successful methods of communicating climate change awareness. However, there has been very little research on the potential for 'the arts' to facilitate public engagement with climate change in the context of the Pacific.

The focus of the research is on an exhibition titled 'Adapt and Survive' held at the University of the South Pacific Oceania Centre Gallery in 2014. The artist statement with respect to the work presented read:

Our climate has always changed, but now those changes are happening over a human lifespan. In order for our species to survive in a way and in a world we recognise, we have to adapt.... The works in this exhibition are an attempt to explore the causes and effects of climate change and human imposed environmental degradation and seek to bring an awareness of how these issues reverberate around cultural loss and societal change within the Pacific islands region. The works reference current threats from climate change and parallel this with the region's past use as a testing ground for destructive and constructive technologies and how it has lived with and survived these uses.

As part of our research, a series of 2 semi-structured interviews were undertaken with the artist. One interview with the artist was undertaken while they were completing work for the exhibition and a second interview was undertaken during the exhibition. We set out to understand the motivation behind the artist's work, as well as the anticipated or hoped-for responses that it might engender. A more detailed depiction of the content of the exhibition is also given below (Table 1).

Table 1 Description of artworks in the exhibition

Name	Media	Size	Description
Burn with me	Found objects (driftwood, children's toys, animal bones, red and black paint)	2.5 m × 0.5 m × 0.5 m	Installation—3D sculpture on the floor of the gallery
Footprint 1: here there be monsters	Black acrylic paint, soot, food colouring, mahogany, PVA glue, canvas	2 m × 2 m	Abstract image on canvas
Footprint 2: waveform	Black acrylic paint, soot, food colouring, mahogany, PVA glue, canvas	0.3 m × 0.3 m	Abstract image on canvas
Footprint 3: monsters Marquette	Black acrylic paint, soot, food colouring, mahogany, PVA glue, canvas	0.25 m × 0.15 m	Abstract image on canvas
Are you my mummy? (Fig. 1)	Found objects: (bamboo driftwood, drinks bottles and cans), coconut fibre string, gypsum, hessian, family photos and indentured servitude shipping lists of family names	1.5 m × 2.5 m × 1.2 m	Installation—3D sculpture on the floor of the gallery (see illustration 1 below)
Photographs	Photographs of various adaptation projects from across the Pacific mounted in brightly coloured plastic frames	2.2 m × 1.3 m	Framed and wall mounted photographs of various sizes
Soul ages neon carbon	Acrylic paint, food colouring, mahogany, PVA glue, canvas	1 m × 0.3 m	Abstract image on canvas

(continued)

Table 1 (continued)

Name	Media	Size	Description
Footprint 4: straw in the wind	Black acrylic paint, soot, food colouring, mahogany, PVA glue, canvas	0.3 m × 0.3 m	Abstract image on Canvas
Dream a little green with me	Acrylic paint, food colouring, mahogany, canvas	1.2 m × 0.3 m	Image on canvas
Pure pandanus: carbon neutral trousers	Pandanus	36 Regular	Woven Pandanus trousers
Cheesy what sits—homage to Henry Moore's Atom	Acrylic media, soot, gypsum, toilet tissue, canvas	0.15 m × 0.15 m	A series of four relief images on canvas
Hollow cost harvest henge	Found objects (driftwood, flip-flops) fishing line, concrete.	3 m × 2 m × 3 m	3D installation set on the floor and ceiling
Footprint 5: end of days	Black acrylic paint, soot, food colouring, mahogany, PVA glue, canvas	0.6 m × 0.3 m	Abstract image on canvas
Colograph: death	Watercolour paper, found materials (tuna tin, leaves, plastic, moss), PVA glue, oil paints	0.35 m × 0.2 m	A series of 4 colographs plus 1 printing plate
Colograph: mask	Watercolour paper, found materials (tuna tin, leaves, plastic, moss), PVA glue, oil paints	0.35 m × 0.2 m	A series of 6 colographs plus 1 printing plate
Colograph: herbarium botanicals	Watercolour paper, found materials (Plants), PVA glue, oil paints	0.15 m × 0.12 m	A series of 9 colographs plus 2 printing plates
The causeway (Tuvalu) and seascape	Watercolour paper, watercolour paints, plastic, salt	0.15 m × 0.12 m	A series of 7 increasingly abstracted images
Adapt and survive	Animation using stock footage from Government Public Information Films and stop-motion animation	20 min	Animated film

In addition to inviting exhibition audiences to comment in a visitors' book, we separately invited them to complete a short survey concerning their thoughts about climate change and reactions to the exhibition, both before and after they viewed the artworks. Nineteen people returned surveys prior to entering, with seventeen of these visitors then completing a second survey having viewed the exhibition.

3 The Exhibition

The Adapt and Survive exhibition was held between June and July, 2014. It was a solo exhibition by the artist who prefers to be known by the name "Ex-Isle". Outlining the aims of the exhibition and influences prior to the exhibition, the artist provided a written statement which specified the following:

> The works reference current threats from climate change and parallel this with the region's past use as a testing ground for destructive and constructive technologies and how it has lived with and survived these uses. The artist's personal experience growing up in the cold war is also reflected.

> The artworks raise many questions – not all of which can be answered. For example, we find out about history by examining what nature, civilisations and cultures leave behind… pottery, footprints, documents, images, clothing, tools, artworks, oratory, stories… Due to climate change and environmental degradation, many of the low lying atoll nations of the South Pacific will soon be history – in a human lifetime it is likely that land and cultures will belong to the past – what will be left behind and why? How can these "traces" of what once was be anticipated and presented? How are the people of the region going to respond to these challenges? What are people doing in order to adapt and survive?

> The artwork is influenced by the nexus of the modern human induced environmental degradation and the ancient, possibly even pre-historic, natural resource based strategies that are now seen as a pre-requisite to adapt and survive. Traditional adaptation strategies were, and still are in some cases, tied into spiritual/supernatural beliefs. Additionally, from the artist's personal experience with communities in Tuvalu, spiritual beliefs are also linked to the climate change phenomenon and community responses to its challenges. For example, the process of materials collection is important in terms of resource availability, social activity and beliefs. As knowledge, belief and culture are lost, the artefacts they produced are likely to survive, be exhibited in museums and researched… as the physical traces of once existent cultures and beliefs. I am hoping that the exhibition reflects the reverence given to artefacts by giving a theatrical experience to the presentation of the work.

A major influence on the artist was revealed to be their ongoing development work. Their work has always dealt with communities, promoting sustainable development and climate change adaptation via use of local natural resources. The artist also stated that "Sustainable development and "traditional" forms of climate change adaptation intrinsically involve working with natural renewable resources such as wood, biomass and found materials, hence the basis for working with found materials and natural forms."

Both researchers into climate change adaptation and development agencies working directly with communities recognise traditional/local knowledge and the management of the natural environment for ecosystem services as the best bet for a sustainable future for most Pacific island countries (McNamara and McNamara 2012; Hemstock 2012; Rosillo-Calle et al. 2015). According to the artist, these behaviours and activities have been passed on from generation to generation and represent the traces of previous generations.

These adaptation strategies were, and still are in some cases (Hemstock 2012), tied into spiritual/supernatural beliefs—with obvious links to the same beliefs that are communicated through traditional ethnographic art. Gender issues are also important in this context since women have a valuable role to play where natural resource-based strategies for what we now term climate change adaptation are concerned (livelihoods; food and medicine collection, preparation, and storage, food collection; etc.), but women have a diminished role where these resources are both portrayed and linked with spiritual/supernatural beliefs (Hemstock 2012; Teilhet 1983). These differences in gender roles are something the artist also tries to explore—particularly within the context of the "standard" definition of ethnographic art as "a system of communication which manifests the ideologies and beliefs that bring order and definition to a person's culture" (Teilhet 1983; Firth 1973). Additionally, spiritual beliefs are also linked to the climate change phenomenon and how some communities respond to its challenges (Havea et al. 2017). Within the context of the nexus outlined above, the process of materials collection is important in terms of resource availability, social activity and belief. In respect of this, the artist reviled that it was important to them that the works in the exhibition encompassed found objects and natural raw materials (Table 1).

Traditional knowledge relating to climate change will be lost as a "meme" (a unit of cultural transmission) e.g. religious belief, adaptation strategy, behavioural response, etc. (Dawkins 1976) as their transmitters become extinct. However, the artefacts produced by that knowledge, and in response to belief are likely to survive, be exhibited in museums and researched… as the physical "traces" of once existent cultures and beliefs.

To some extent, this has already happened as many authors on the subject of Oceanic Art do not focus on or even acknowledge the role of the artists' personal contribution to the production of the work. Oceanic art tends to be defined by culture and belief (Forge 1973; Wingert 1953)—it appears to be defined by the "meme". There are a multitude of books and catalogues depicting dance, costume, objects and "art" of the Pacific islands that are in fact memes—thus shifting the focus from the artist to the product. This is in contrast to the "creative" role ascribed to western artists—for example, in Price (1980) an artist in a photograph is described as "Fijian potter begins shaping the first cylindrical form", whereas Picasso is unlikely to be described as "Spanish potter" in a similar pose.

4 Image-Making and Animation

The interview with the artist during the exhibition revealed that the denotative and connotative images in the exhibition were made using a variety of aesthetic perspectives and techniques. In many cases, photographs formed the initial starting point for image-making (e.g. The Causeway series was described by the artist as depicting the Causeway in Tuvalu in a series of watercolour sketches moving from "photographic representations to abstractions of shapes and colours"). Sketch books allowed the artist to strip images down to the artist's view of their core communicative value by considering the essence of the object and distilling the elements of the object/subject being portrayed. According to the artist:

> Images were broken down into recognisable segments, and sometimes digital effects were used to render different values or accentuate existing values of the images – this can be seen in the animation Adapt and Survive.

Denotative images also formed an intrinsic aspect of the 3D work. For example, the installation "Are You My Mummy" used photocopies of original documents and photographs. Collating documents and photographs for this installation involved carrying out research at the UK and Fiji National Archives on the indentured servitude programme and other subject areas.

Experimentation and process, with particular reference to the generative image production of colographs and watercolours provided an unexpected element to the work making it on the one hand complex, but on the other loose and gestural.

Generative denotative image-making was used for Causeway and the Herbarium Botanicals series so that the images can still be read as the object the image is meant to portray—although it may be abstracted to some extent, it is a representation of what the object it is supposed to be. In the case of the series Herbarium Botanicals it is a direct print from a branch of a plant.

The artist's references for the Footprint series, which depict textured black shapes, some of which appear to look like mushroom clouds, are works by the artists Pierre Soulages and Richard Wilson.

The animation Adapt and Survive used traditional stop-motion techniques combined with digital editing using Sony Vegas Pro. Stop-motion images were overlayed on a series of copyright-free UK Government public information films made in the 1970–80s centred on what to do in the event of a nuclear attack. This series of UK Government information films is called Protect and Survive and can be accessed from the UK National Archives website.

5 Audience/Exhibition Viewers' Perceptions

Spontaneous thoughts about climate change—before and after the exhibition

Visitors were first of all asked to provide open-ended responses to the question "What is the first thing that comes to mind when you hear the phrase 'climate change' or 'global warming'?" based on an image association question commonly used in research that has examined public understanding of climate change (e.g. Tvinnereim and Fløttum 2015).

Before people had viewed the artworks, responses to this question mainly stressed the effects of climate change on the weather and environment, such as changes to weather patterns and rising sea levels. References were also made to causes of climate change such as 'pollution' and 'greenhouse gas emissions'.

After people had visited the exhibition, responses to the question were markedly different in tone. Several people stressed negative emotional reactions, including 'pain', 'unhappy people', and 'pollution of man's mind'. By contrast, other visitors stressed positive emotional reactions, with comments noting the potential for climate change art to be connected to hope and inspiration. Examples include the comments 'there is hope through exhibitions to raise awareness' and 'climate change can... inspire some quality artwork'.

All responses obtained before and after people had visited the exhibition are shown in Table 2.

Thoughts and feelings about the exhibition

After they had viewed the exhibition, visitors were asked about their responses to it, via the question "*What thoughts or feelings did the 'Adapt and Survive' exhibition bring to mind for you?*"

Again, visitors emphasised both negative and positive emotional reactions. Of the seventeen responses obtained, five of these stressed a negative emotional response. Two visitors noted simply that 'I felt sorry for myself', another that seeing the human causes of climate change had 'made me feel depressed'. Reference was also made to the 'sad effects of human activity' and to the 'death of a world'.

By contrast, others emphasised positive emotions in response to the exhibition, and the importance of acting to protect the environment. One visitor remarked that 'it's important to prevent climate change', with another commenting 'I am so pleased to learn a lot on climate change'. Two further visitors referred to the importance of conserving resources and taking personal responsibility on climate change. One response emphasised the beauty of the natural environment and the importance of its preservation for future generations.

Visitors' comments on their thoughts and feelings about the exhibition are shown in Table 3.

Memories from the exhibition

Visitors were asked whether they thought they would remember anything (image, sculpture, phrase, feeling) from the exhibition in one year's time.

The majority of visitors indicated that they would remember aspects of the exhibition. The sculpture titled 'Are You My Mummy?' was referred to by several visitors.

Table 2 Spontaneous thoughts about climate change before and after the exhibition

'First thoughts'—before exhibition	'First thoughts'—after exhibition
Sea level rise, GHG emissions	Pain
Heaps of people coming to TU8 [Tuvalu] for a climate change related purpose	Apparently climate change can also inspire some quality artwork
Soil erosion and disappearing islands	Unhappy people
Destruction that is occurring due to the increase of pollution being let loose into the atmosphere	This is becoming very serious and accelerating at a very fast pace
A serious problem that no one is taking seriously	As an artist myself I know that there is hope through exhibitions to raise awareness
Sea level rise, change of the world as we know it	(No response)
Climate change—the change of environment [caused] by the weather pattern	The change in surrounding—erosion [and] rise in sea level
Climate change is when the weather pattern is not always normal like what I used to experience before when I was on land	The change in weather pattern
Changing the weather pattern	Changing the weather pattern
Climate change is the change in the weather pattern and the effects on nature	(No response)
Disaster, fraud	Pieces of plastic
Atmospheric catastrophe, over-pollution and misuse of natural environment	Pollution of man's mind
Changes in the climate, sea level rising	It affects our nature as climate change occurs
The planet is getting hotter and some parts of the world are sinking because the speed of sea level change is faster than we expected	Protecting the environment and balance up nature
Climate change is change in weather and it does affect our nature	(No response)
Global warming and change in weather average over a period of time from humans using fuels/fishing/cutting down rainforests and trees	Change in the weather and increase in natural disasters and a reduction or extinction of resources by humans using Earth's natural resources at a rate they can't naturally replenish and from fuel consumption polluting the atmosphere
The first thing is change of weather and how the environment has changed	I'm amazed to find out what we can create rubbish to something valuable
Climate change is the effects of the ozone layer that effects the natural environment	Climate change is the change in weather pattern in a particular place/country
Climate change is the change in the weather pattern and the effects on the nature	Climate change is the change in the weather pattern and the effects on the nature

Table 3 Thoughts and feelings about the exhibition

Awe at [the artist's] process
Memories of 10 years in TU8 [Tuvalu] in which climate change was a high-frequency topic
It made me think of the sad effects of human activity on the environment
Made me feel depressed to see that the main cause of climate change is the increase in industrialisation around the world
A shame that exhibitions of this type are not more accessible and available to inspire the general public
The death of a world and what it contains
I've learned rubbish can be created into something useful
I felt sorry for myself
I felt sorry for myself
To utilise all material, no waste, express the feelings
Kind of explosive in a sense of creative medias
It's important to prevent climate change
Make a better world
That it is a personal responsibility that everyone changes their lifestyle to prevent further climate change
Yes, I'm so pleased to learn a lot on climate change. As I said collect something to [make] something valuable
The first thing that pop in my mind was that the natural environment is very beautiful thing and it should be preserved for the future generation
It brings out the human's behaviour and its causes on human nature

Visitors' comments on what they would remember about the exhibition are shown in Table 4.

Perceptions of climate change—before and after the exhibition

Visitors were asked to indicate the extent to which they agreed or disagreed with a series of statements about climate change. These related to beliefs about adaptation and prevention, climate change causation, and the degree to which people felt 'worried' or 'hopeful' about the subject. Visitors were asked for their views before and after entering the exhibition.

The findings from this section of the survey suggest that there were very high levels of agreement among visitors that the place where they live is being affected by climate change, with 18 of 19 respondents either agreeing or strongly agreeing with this statement prior to viewing the artworks. Responses overall suggest that people were worried about climate change, with all but two of 19 respondents *disagreeing* with the statement "*Climate change is not something that I am worried about*" before entering the exhibition. More visitors agreed with the statement "*It is not possible to stop climate change from happening*" (9 of 17 responses) than disagreed (five responses; a further three were unsure).

Table 4 Memories of the exhibition

The contrast between light and dark
I'll remember a diverse oeuvre and the carbon neutral trousers
Traces left behind—flip flops, trousers, footprint-waveform, Freddie's painting
The sculpture of the eroded tree [burn with me] which is caused by the seawater eroding more into the inland which causes the destruction of other plants and increase in soil erosion
Definitely the installations
The "are you my mummy" sculpture [Fig. 1]
The creation of the sculpture titled "are you my mummy"
"Burn with me" and I hope I'm not going to be like that
"Are you my mummy" [Fig. 1]
Slippers hanging [Hollow cost harvest henge]
The artist
Everything—image, sculpture, phrase and feelings
The flip-flops [Hollow cost harvest henge]
Hollow cost harvest henge
The log and it being displayed is truly unique where it once was a tree but got destroyed [burn with me]
Yes the Doctor who, the nature and Artfestoo way

Despite expressing concern in these ways, nevertheless people for the most part expressed confidence and hope that climate change can be effectively addressed. Before viewing the exhibition, a majority of visitors (14 of 18) agreed or strongly agreed that "*I am hopeful that we can act to address climate change*" (although a further four visitors disagreed). Most of those completing the survey beforehand also were of the view that "*I can do something to protect myself and my family from the effects of climate change*" (14 of 18 responses).

Visitors were more evenly split on the question of whether Pacific Island nations have the knowledge and resources to adapt and survive in response to climate change. Whilst a majority (10 of 17) expressed this view at the 'before' stage, the remaining seven responses either disagreed or were unsure. There was also substantial variability in whether climate change was considered to be mostly caused by natural processes. Six visitors agreed that this was the case; whereas a further nine visitors disagreed.

Based on a comparison of these measures before and after the exhibition, people's views did not appear to alter overall having viewed the artworks (though it is possible that small effect sizes might have been masked by our limited sample). Very similar scores were obtained at the before/after stage and differences are not statistically significant for any of these survey statements.

Figure 2 shows average scores for each of the survey measures at the before/after stage, as well as a combined average of before/after scores for all survey measures.

Fig. 1 Are you my mummy?

6   Limitations of the Research

The study reports the perspectives of the artist-practitioner and the exhibition audience. The study discusses the potential use of visual arts in communicating climate change issues in a Pacific islands context. Inferences are made regarding opinion and conjecture of those participating in the study. However, conclusions cannot be drawn from the findings regarding the effectiveness of specific artworks, or of arts communication as a general approach. This is particularly the case given the small opportunity sample available to us, based on those who chose to attend the exhibition. Furthermore, conclusions cannot be made about the wider and more general potential of the arts to communicate in the Pacific context.

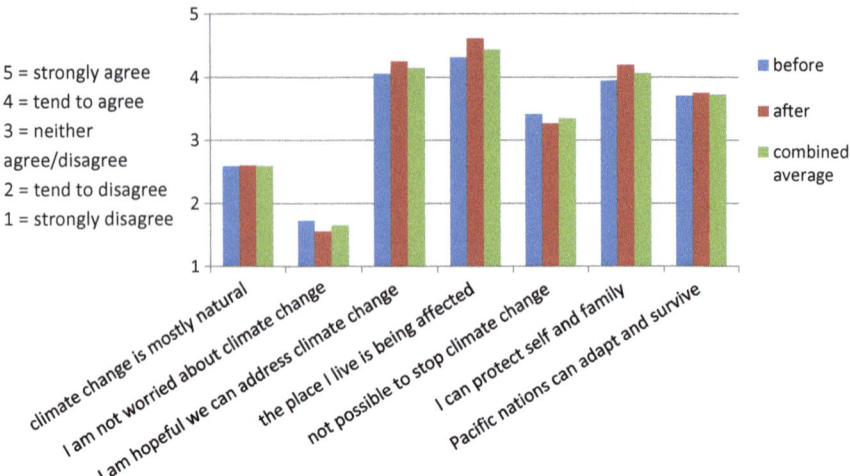

Legend text within figure:
5 = strongly agree
4 = tend to agree
3 = neither agree/disagree
2 = tend to disagree
1 = strongly disagree

before
after
combined average

X-axis labels:
climate change is mostly natural
I am not worried about climate change
I am hopeful we can address climate change
the place I live is being affected
not possible to stop climate change
I can protect self and family
Pacific nations can adapt and survive

Fig. 2 Attitudes towards climate change before and after the exhibition

7 Conclusions

A unifying feature across audience participant responses was the way in which climate change is recognised to be a part of the contemporary reality of life in the Pacific islands region. Additionally, participants appear to have found the exhibition thought provoking around climate change issues, with both negative and positive feelings and emotions referenced.

Previous research has suggested that many people better understand problems or issues by the clear use of illustrations, paintings, sculptures, photography and through other media of communication. According to McNaught et al. (2014), research on climate change communication in the context of adaptation is very limited in developing countries. One may well ask the question as to what role/s do artists have in the climate change debate. Considering the continuing adverse effects of climate change on the Pacific islands region now is the time to consider the importance of art and how it can be used to inform about climate change and adaptation for survival. In the build up towards the COP21 conference in Paris 2015, several art groups have taken the initiative to inform people about the effects of climate change using various art exhibitions. From these exhibitions have sprung fundraising activities to assist those severely affected by climate change including desertification problems in Africa, Asia, Latin America and the Middle-East. Known as the Artists 4 Paris Climate 2015, this initiative brought together thirteen bankable artists to exhibit in public spaces on various climate change issues.

Connections between people's 'experience' of climate change and their subsequent beliefs and actions has now been observed across a range of research studies: for example, where personal experience of flooding prompts adaptation intentions

(Demski et al. 2017) or where perceived changes to the weather predicts future risk perceptions (Akerlof et al. 2013). This said, the overwhelming focus of these studies thus far has been upon behavioural responses (such as saving energy; Spence et al. 2011) or beliefs (such as in the reality of climate change; Myers et al. 2013).

Audience participants provided diverse thoughts and feelings about the exhibition. Around half of the responses reflected values such as altruism, concern for others and for the natural world. The types of values embodied in such rationales and aspirations are in line with a wide range of literature which has demonstrated them to be linked to attitudes towards climate change and the taking of action to address it (Corner et al. 2014). Altruistic and pro-environmental values, broadly construed as 'self-transcending' in the psychological literature (Corner et al. 2014) are also considered to be important to convey and assert where seeking to engage and influence others (Crompton 2010). From a consideration of the work of many artists and arts projects in Australia, Curtis (2009) has concluded that arts projects are able to engender a connectedness and emotional relationship with the natural environment, as well as to instruct in a more educational sense.

The current European Union Pacific Vocational Education and Training in Sustainable Energy and Climate Change Adaptation project (EU PacTVET) has developed regional qualifications in "Sustainable Energy" and "Resilience—Climate Change Adaptation and Disaster Risk Management" (Hemstock et al. 2017). The inclusion of traditional/local knowledge in the learning outcomes of these courses as well as the development and use of novel teaching and learning resources to ensure qualifications were fully accessible to Pacific communities (including those with low levels of literacy and educational participation), was highlighted by stakeholders (Hemstock et al. 2017). Building on the observed connection between people's 'experience' of climate change and their subsequent beliefs, behavioral responses and future risk perceptions (Akerlof et al. 2013; Demski et al. 2017; Myers et al. 2013; Spence et al. 2011), the EU PacTVET project in partnership with the Coping with Climate Change in the Pacific Islands Region Project (CCCPIR) project have selected and developed a series of teaching resources which reference visual arts in a Pacific community context. This is as a result of lessons learned from this research and similar Eco-Arts projects in the Pacific region (Capstick et al. 2018). In addition, the EU PacTVET project is building on this research in the development of teaching resources and the use of methodologies which aims to explore and use the perceived personal impacts and experiences of learners to climate change, their behavioural responses (such as installing water tanks for rainwater collection) or beliefs (such as is climate change a human induced or super natural phenomenon). The current research does not make strong claims regarding a direct causal effect of experience in informing creative approaches and communication of climate change. However, for the EU PacTVET and CCCPIR initiatives, it is hoped that the inclusion of creative approaches to climate change communication in a formal educational environment may yield positive results.

The extent to which this field can develop is likely to be limited by the practicalities of creating and presenting material by and for a wider public. Nevertheless, arts communication has the potential to engender new forms of understanding and responses about climate change. We conclude that the activities carried out during the pursuit of this research offer an exciting and innovative way of raising awareness and communicating information concerning environmental issues and climate change to various communities and new audiences—from school children to Ph.D. level researchers.

References

Akerlof K, Maibach EW, Fitzgerald D, Cedeno AY, Neuman A (2013) Do people 'personally experience' global warming, and if so how, and does it matter? Glob Environ Change 23 (1):81–91

Capstick S, Hemstock SL, Senikula R (2018) Perspectives of artist-practitioners on the communication of climate change in the Pacific. Int J Clim Change Strat Manag 10/2. ISSN 1756–8692. https://doi.org/10.1108/IJCCSM-03-2017-0058

Corner A, Markowitz E, Pidgeon N (2014) Public engagement with climate change: the role of human values. Wiley Interdisc Rev Clim Change 5(3):411–422

Crompton T (2010) Common cause: the case for working with our cultural values, WWF, Surrey, UK. Accessed 15 Mar 2018. https://assets.wwf.org.uk/downloads/common_cause_report.pdf

Curtis D (2009) Creating inspiration: the role of the arts in creating empathy for ecological restoration. Ecol Manag Restor 10(3):174–184

Dawkins R (1976) The selfish gene. Oxford University Press, Oxford. ISBN019857519X

Demski C, Capstick S, Pidgeon N, Sposato R, Spence A (2017) Experience of extreme weather affects climate change mitigation and adaptation responses. Clim Change 140(2):149–164

Duxbury L (2010) A change in the climate: new interpretations and perceptions of climate change through artistic interventions and representations. Weather Clim Soc 2(4):294–299

Firth R (1973) Tikopia art and society. In: Primitive art and society. Oxford University Press, Oxford, pp 25–48. ISBN 0192129538

Forge A (1973) Primitive art and society. Oxford University Press, Oxford. ISBN 0192129538

Havea PH, Hemstock SL, Jacot des Combes H, Luetz J (2017) God and Tonga are my inheritance!—climate change impact on perceived spirituality, adaptation and lessons learnt from Kanokupolu, 'Ahau, Tukutonga, Popua and Manuka in Tongatapu, Tonga. In: Climate change impacts and adaptation strategies for coastal communities. Springer International Publishing, Berlin. https://doi.org/10.1007/978-3-319-70703-7_9. ISBN: 9783319707020; ISSN: 1610-2010

Hemstock SL (2012) The potential of coconut toddy for use as a feedstock for bioethanol production in Tuvalu. Biomass Bioenerg 49:323–332

Hemstock SL, Jacot Des Combes H, Martin T, Vaike FL, Maitava K, Buliruarua L-A, Satiki V, Kua N, Marawa T (2017) A case for formal education in the technical, vocational education and training (TVET) sector for climate change adaptation and disaster risk reduction in the Pacific Islands Region. In: Climate change adaptation in Pacific Countries: fostering resilience and improving the quality of life. Climate Change management. Springer International Publishing, Berlin, pp 309–324. ISBN 978-3-319-50093-5

Maclellan N (2009) The future is here: climate change in the Pacific, Auckland, Oxfam briefing paper, Oxfam New Zealand. Accessed 18 Mar 2018. https://www.oxfam.org.nz/sites/default/files/reports/The%20future%20is%20here-Oxfam%20report-July09.pdf

McNamara KE, McNamara JP (2012) Using participatory action research to share knowledge of the local environment and climate change: case study of Erub Island, Torres Strait. Aust J Indigenous Educ 40(1):1–10

McNaught R, Warrick O, Cooper A (2014) Communicating climate change for adaptation in rural communities: a Pacific study. Reg Environ Change 14:1491–1503

Moser SC (2010) Communicating climate change: history, challenges, process and future directions. Wiley Interdisc Rev Clim Change 1(1):31–53

Moser SC (2014) Communicating adaptation to climate change: the art and science of public engagement when climate change comes home. Wiley Interdisc Rev Clim Change 5(3):337–358

Myers TA, Maibach EW, Roser-Renouf C, Akerlof K, Leiserowitz AA (2013) The relationship between personal experience and belief in the reality of global warming. Nat Clim Change 3 (4):343–347

National Archives. Protect and survive. Accessed 15 Mar 2018. http://www.nationalarchives.gov. uk/films/1964to1979/filmpage_warnings.htm

Nunn PD (2012) Climate change and Pacific Island Countries. Asia Pacific Human Development Report, Background Papers Series 2012/07, United Nations Development Programme. Accessed 18 Mar 2018. https://www.uncclearn.org/sites/default/files/inventory/undp303.pdf

O'Neill SJ, Smith N (2014) Climate change and visual imagery. Wiley Interdisc Rev Clim Change 5(1):73–87

Price C (1980) Made in the South Pacific—arts of the sea people. The Bodley Head Ltd. ISBN0370302311

Robinson PA, MacNaghten P, Banks S, Bickersteth J, Kennedy A et al (2014) Responsible scientists and a citizens' panel: new storylines for creative engagement between science and the public. Geogr J 180(1):83–88

Rosillo-Calle F, De Groot P, Hemstock SL, Woods J (eds) (2015) Biomass assessment handbook: bioenergy for sustainable development, 2nd edn. Earthscan, London, 336 pp. ISBN-10: 1138019658

Shen S, Gemenne F (2011) Contrasted views on environmental change and migration: the case of Tuvaluan migration to New Zealand. Int Migr 49(1):224–242

Smith R, Hemstock SL (2011) An analysis of the effectiveness of funding for climate change adaptation using Tuvalu as a case study. J Clim Change Impacts Responses 3(1):67–78

Spence A, Poortinga W, Butler C, Pidgeon NF (2011) Perceptions of climate change and willingness to save energy related to flood experience. Nat Clim Change 1(1):46–49

Taylor S, Kumar L (2016) Global climate change impacts on pacific islands terrestrial biodiversity: a review. Trop Conserv Sci 9(1):203–223

Teilhet J (1983). The role of women artists in Polynesia and Melanesia. Chap. 3 in Art and artists of Oceania. Dunmore Press Ltd, NZ, pp 45–56. ISBN 0908564856

Tvinnereim E, Fløttum K (2015) Explaining topic prevalence in answers to open-ended survey questions about climate change. Nat Clim Change 5:744–747

UNDP (2013) Climate change and Pacific Island Countries. Asia Pacific Human Development Report. Background Papers Series 2012/07. United Nations Development programme. HDR 2013-APHDR-TBP-07

Weir T, Dovey L, Orcherton D (2016) Social and cultural issues raised by climate change in Pacific Island countries: an overview. Reg Environ Change (online first)

Whitmarsh L, Corner A (2017) Tools for a new climate conversation: a mixed-methods study of language for public engagement across the political spectrum. Glob Environ Change 42:122–135

Wingert PS (1953) Art of the South Pacific Islands, 1st edn. Thanes and Hudson, London

Woods J, Hemstock SL, Bunyeat J (2006) Bio-energy systems at the community level in the South Pacific: impacts and monitoring. Greenhouse gas emissions and abrupt climate change: positive options and robust policy. J Mitig Adapt Strat Glob Change 11(2):461–492

Disaster Risk Reduction Begins at School: Research in Bangladesh Highlights Education as a Key Success Factor for Building Disaster Ready and Resilient Communities—A Manifesto for Mainstreaming Disaster Risk Education

Johannes M. Luetz and Nahid Sultana

Abstract In many countries of the world the dream of achieving education, free and compulsory for all, remains elusive for large parts of the population. Bangladesh is a case in point. Drawing on field research conducted in Bangladesh in 2008, 2011 and 2012, including in conjunction with the international development organisation World Vision, this chapter discusses some of the linkages between education, extreme levels of poverty, forced human migration, environmental change, and disaster readiness. The study identifies protracted poverty as the predominant impediment to schooling in Bangladesh. It extends previous research by expressly inviting the participation of respondents in coastal villages in the Bhola and Satkhira districts, as well as in urban slum communities in the country's two largest cities Dhaka and Chittagong. The findings show that severe poverty forces school age children to work in low-paid jobs as garbage collectors, recyclers, domestic workers, servants, street vendors, hotel boys, burden bearers, couriers, etc., thereby thwarting their education and perpetuating the cycle of poverty. The research recommends a holistic portfolio of educational strategies comprising formal, non-formal and

Preamble A background video documentary on aspects of this research was published by UNSW, Sydney on 18 February 2015 and is publicly available at https://youtu.be/PBJeelgnadU.

J. M. Luetz (✉)
CHC Higher Education, Carindale, QLD 4152, Australia
e-mail: jluetz@chc.edu.au; j.luetz@unsw.edu.au

J. M. Luetz · N. Sultana
University of New South Wales (UNSW), Sydney, Australia
e-mail: n.sultana@unsw.edu.au

informal learning approaches that are integrated at the community level. Multi-stakeholder strategies seem to be best suited to Bangladesh's dynamic environmental, geodemographic and socioeconomic context. Disaster risk education offers auspicious benefits for resilience and disaster preparedness.

> Education is the most powerful weapon we can use to change the world.
> (Nelson Mandela 2003, para. 24; cf. Strauss 2013, para. 3)
> "Educate those children in the slum, and you will break that vicious cycle of poverty in which they find themselves." (Research Participant No. 23, interviewed in Dhaka on 5 December 2011 cited in Luetz 2013, p. 481)
> The creation of a thousand forests is in one acorn.
> (Ralph Waldo Emerson 2010, p. 2)

1 Introduction: Education Is the Sine Qua Non for Human Flourishing

The following quote, attributed to Salvano Briceño, Director of the United Nations Office for Disaster Risk Reduction, reflects a growing awareness within the disaster management community to leverage the education of children for the purposes of disaster risk reduction.

> When you have only a few minutes, it is important to know the actions you must take to reduce your risk, such as running to higher ground to avoid flood water. Many children have learnt to live with natural hazards in countries such as [...] Bangladesh. Everybody should have this basic knowledge [...] We need to work together to reduce the impact of natural hazards on children [...] *If we educate our children, there is hope that we can build a culture of prevention for future generations.* (United Nations Office for Disaster Risk Reduction [UNISDR] 2005, para. 7, 9, emphasis added).

Educating children to be disaster aware, safe and responsible is not a novel concept but one that dates back millennia.

> The power of educating children for the build-up of societal resilience has been known for millennia. More than 2300 years ago, the Greek philosopher Aristotle (384-322 BC), student of Plato and one of the most influential teachers of all times, identified education as the kingpin of societal transformation: "All who have meditated on the art of governing mankind have been convinced that the fate of empires depends on the education of youth. [...] Those who educate children well are more to be honored than parents, for these only gave life, those the art of living well [...] Education is an ornament in prosperity and a *refuge in adversity.*" [emphasis added] – In the context of disaster preparedness, the knowledge of what to do can be a literal 'refuge in adversity.' – According to the Oxford Dictionary, the verb 'to educate' is derived from Latin 'educere,' meaning 'to lead out.' [McKean 2005, p. 539]. This meaning is significant because education not only 'leads out' into a life of opportunity, but can 'lead out' of precarious disaster situations to safety. Educated people can create smart policies to govern themselves out of poverty, and identify smart escape routes to guide themselves out of danger. Children are change agents.

Promoting a culture of disaster readiness in children is pivotal to raising up a new generation of resilient people who are ready, responsible and response-able. (Luetz 2008b,
p. 79)

In short, educating children in areas of disaster preparedness is increasingly
recognised as an opportunity not to be missed, including by international organisations such as UNISDR and UNICEF, which have even "produced a board game
called Riskland to help educate children on practical actions to take when disaster
strike [while continuing efforts] to integrate disaster risk reduction into school
curricula." (UNISDR 2005, para. 7–8).

Given a definitive increase over recent decades in the number of people around
the world affected annually by disasters (Guha-Sapir et al. 2017), the rationale for
leveraging education for disaster risk reduction, resilience, and awareness raising is
quite straightforward and hence widely accepted by the international community.
This is noted, among numerous other examples, by the "World Conference on
Disaster Reduction Hyogo Framework for Action 2005–2015" (Priority 3): "Use
knowledge, innovation and education to build a culture of safety and resilience at
all levels." (United Nations Educational, Scientific and Cultural Organization
[UNESCO] and United Nations Children's Fund [UNICEF] 2009, p. 4).

Several years ago, the power of disaster education was famously demonstrated
by a typical yet heretofore unknown schoolgirl:

At the tender age of ten, British schoolgirl Tilly Smith can say in her résumé what few
people can claim even in the autumn of their lives – that she is credited with saving nearly
100 tourists at Maikhao Beach (Thailand) by raising the alarm minutes before the deadly
tsunami waves crashed into their hotel. On 26 December 2004, while on the beach in
Phuket with her parents, Tilly sensed something was wrong. As the sea receded and had
'froth on it like you get on the top of a beer' [Owen 2005, para. 2], Tilly immediately
recognised the tell-tale signs of an impending tsunami and pleaded with her parents to flee
from the beach. Remembering the words of her geography teacher, Andrew Kearney, who
showed the class a video of a tsunami in Hawaii, Tilly quickly connected the dots: 'I was
hysterical. I was screaming, I didn't want to leave my mom [...] I said, 'Seriously, there is
definitely going to be a tsunami.' [British Broadcasting Corporation (BBC) 2005, para. 13–
15]. Tilly's adamant warnings alerted her parents, who warned others, including the hotel
staff. The beach was evacuated before the tsunami reached shore, and was one of the few
beaches on Phuket with no reported casualties.[1] Tilly's mother Penny (43) says she is proud
of her daughter's quick thinking, as she herself did not recognise the danger signs: 'She was
screaming at us to get off the beach [...] I didn't know what a tsunami was, but seeing my
daughter so frightened made me think something serious must be going on.' [Owen 2005,
para. 4, 6]. Tilly received numerous awards[2] and was given the honour of closing the First

---

[1]In contrast, the Hollywood disaster drama "The Impossible", directed by Bayona (2012) and
starring Naomi Watts and Ewan McGregor, depicts the fate of tourists caught in a Thailand hotel
that was *not* evacuated in time before the 2004 Indian Ocean tsunami waves reached the shore.
[2]Selected honours include Child of the Year award (Randall and Berger 2005), Thomas Gray
Special Award from Second Sea Lord, Vice-Admiral Sir James Burnell-Nugent (BBC 2005), and
having Asteroid 20002 Tillysmith named after her by the National Aeronautics and Space
Administration (NASA) (2007) for "alerting beachgoers [... and saving] many lives on the island
of Phuket" (n.p.).

Anniversary Tsunami Commemorations in Khao Lak, Thailand, on 26 December with a poem before thousands of spectators. Former U.S. President Bill Clinton observed: 'Tilly's story tells us about the importance of teaching young people about natural hazards. *All children should be taught disaster reduction so they know what to do when natural hazards strike. Tilly's story is a simple reminder that education can make a difference between life and death.*' (UNISDR 2005, para. 3 cited in Luetz 2008b, p. 80, emphasis added)

Over recent years, integrating disaster risk reduction in school curricula has garnered growing interest among global development and education stakeholders, as noted by both publications (UNESCO and UNICEF 2009, 2012) and awareness campaigns (UNISDR n.d.): "UNISDR is promoting a global culture of safety and resilience through the integration of disaster risk reduction in school curricula and the continuous involvement of children and youth in the decision-making process for disaster risk reduction." (para. 1).

In addition to potentially making a difference between life and death, education is also almost universally embraced by humanity as a pathway to opportunity, peace, human flourishing and economic wellbeing, as borne out by a plethora of research studies. For example, education is understood to be beneficial for global conflict reduction: "Each year of education reduces the risk of conflict by around 20%." (Collier 1999, p. 5). Further, schooling heightens economic output in fiduciary terms: "Each additional year of schooling raises average annual gross domestic product (GDP) growth by 0.37%." (UNESCO 2011, p. 6). At the individual level, "[o]ne extra year of schooling increases an individual's earnings by up to 10%." (UNESCO 2011, p. 7). In short, investing in education reaps returns on investment that accrue back to the investor: "A dollar invested in an additional year of schooling, particularly for girls, generates earnings and health benefits of $10 in low-income countries and nearly $4 in lower-middle income countries." (International Commission on Financing Global Education Opportunity [ICFGEO] n.d., p. 14).

Given such benefits, it is unsurprising that the international community has been increasing efforts globally to ensure "Education for All (EFA) [...] an international initiative first launched in 1990 to bring the benefits of education to '*every citizen in every society.*'" (World Bank 2014, para. 1, emphasis added). This global aspiration was expressly enshrined as Goal 2 ("ACHIEVE UNIVERSAL PRIMARY EDUCATION") in the "United Nations Millennium Development Goals (MDGs),"[3] which spanned the years 2000–2015 (Luetz 2007, pp. 21–25). More recently, it was reiterated, re-emphasised and recast as Goal 4 ("ENSURE INCLUSIVE AND EQUITABLE QUALITY EDUCATION AND PROMOTE LIFELONG LEARNING OPPORTUNITIES FOR ALL") in the "United Nations Sustainable Development Goals (SDGs),"[4] which replaced the MDGs, and now span the operative time horizon 2015–2030.

---

[3]United Nations. (n. d.).
[4]Division for Sustainable Development, UN-DESA (2017).

In summary, it can be said that the international community remains deeply and quasi-universally committed to the cause of education (UNESCO 2011; UNESCO and UNICEF 2009, 2012), broadly agreeing that "[e]ducation and skills are essential for realizing individual potential, enhancing national economic growth and social development, and fostering global citizenship. In the coming decades, as technology, demographic change, and globalization reshape the world we live in, they will become ever more important." (ICFGEO n.d., p. 29). In short, education is "[t]he best investment the world can make" (ICFGEO n.d, p. 2; cf. Luetz 2007, pp. 33–36).

Even so, in many countries of the world the dream of achieving education, free and compulsory for all, remains elusive for large parts of the population. Worldwide, millions of children do not attend school:

> the total number of out-of-school children, adolescents and youth has remained nearly the same at around 264 million for the past three years [...] Some 61 million, or 23% of the total, are children of primary school age (about 6 to 11 years), 62 million, or 23% of the total, are adolescents of lower secondary school age (about 12 to 14 years), and 141 million, or 53% of the total, are youth of upper secondary school age (about 15 to 17 years) (herein children, adolescents and youth, respectively). (United Nations Educational, Scientific and Cultural Organization Institute for Statistics [UIS] 2017, p. 1)

Significantly for this discussion, a large proportion of non-attending school age children in the world live in Central Asia and Southern Asia, one of three regions in the world that collectively account for "nine out of ten out-of-school adolescents: sub-Saharan Africa (26 million), Central Asia and Southern Asia (20 million) and Eastern and South-eastern Asia (8.5 million)." (UIS 2017, p. 4).

Situated in this regional context, the nation of Bangladesh exemplifies global trends in education and offers important insights into some of the reasons for protracted school absenteeism. The country's geographic, demographic, environmental and natural disaster context makes the nation a useful case study for research into both the impediments to schooling and the consequences of absenteeism.

2 Bangladesh and Education: Research Rationale and Intended Contribution

Bangladesh is a densely-settled country with approximately 163 million citizens (World Bank 2018). Discounting city states, it has the highest population density in the world. With an average 1229 people living together on each available square kilometre of land, Bangladesh is home to more people than live in all of Russia[5] (Belt 2011, p. 64; United Nations Development Programme [UNDP] 2016, pp. 222, 224). The United Nations considers Bangladesh a Medium Human Development country with a rank of 139 (out of 188 countries) according to the latest Human Development Index ranking (UNDP 2016, p. 271).

---

[5]Russia, the world's largest country by size, is more than 100 times bigger than Bangladesh.

In Bangladesh, the "[m]ean years of schooling [is] 5.2 years" (UNDP 2016, p. 200), and the population with at least some secondary education is between 42% (female) and 44.3% (male) (p. 216). Importantly, the primary school dropout rate, defined as the "[p]ercentage of students [...] who have enrolled in primary school but who drop out before reaching the last grade of primary education" (UNDP 2016, p. 233) is 33.8% (p. 232). These data are significant: They imply that despite significant improvements made over recent years to raise the rate of education, more than 66% of children in Bangladesh still do not finish primary school, and more than 55% of adults do not have "at least some secondary education" (UIS 2017, p. 233).

Relatedly, UNICEF Bangladesh (2010) estimates that there are approximately 4.7 million working children (aged 5–14) in that country, and that "half of all child labourers do not attend school at all [...] As a result, working children get stuck in low paying, low skilled jobs, thereby perpetuating the cycle of poverty." (p. 3, attributed to International Labour Organization 2006).

Further, Bangladesh is also one of the most disaster-prone countries in the world (Luetz 2008a). With nearly 5% of the nation's citizens affected annually by disasters (UNDP 2011, p. 152), and more than 80% of the population fighting for survival on less than two dollars a day (p. 144), natural disasters can cause significant shocks to long-term human development prospects. Three disaster types stand out: (1) windstorms; (2) flooding; (3) erosion. These disaster types are discussed in detail in Luetz (2018, pp. 64–74) and will not be recapitulated here beyond the following brief mention, which highlights pivotal ripple effects on education.

2.1 *Windstorms*

Bangladesh is "affected by major cyclones on average 16 times a decade" (U.S. Department of State. Bureau of South and Central Asian Affairs 2008, "Geography" para. 1), and a number of studies indicate that continued warming of Indian Ocean surface waters as projected to occur under climate change "could spawn even stronger cyclones in the future, a scenario which could be particularly difficult for coastal communities if increases in windstorm intensity are accompanied by rises in sea level" (German Advisory Council on Global Change 2006, pp. 38–44; see also Allison et al. 2009, p. 19; Emanuel 2005; Intergovernmental Panel on Climate Change 2007, pp. 13, 53; Sánchez-Arcilla and Jimenez 1997; Webster et al. 2005 cited and elaborated in Luetz 2018, p. 65). As may be expected, windstorms can have a significant knock-on effect on education, as exemplified by Cyclone Sidr,[6]

---

[6]"According to the situation report released by the Bangladesh Disaster Management Information Center, Very Severe Cyclonic Storm Sidr killed 3292 people, injured 52,808, fully destroyed 563,877 households, and partially damaged 939,675. It affected 8,669,789 people, 2,000,848 families, and 30 of the 64 districts in Bangladesh. Moreover, '[c]rops on 596,516 acres of land were fully damaged while crops on 1,480,712 acres of land were partially damaged, [...] 2400 educational institutes were fully damaged while 12,399 more were partially damaged, [...]

which made landfall in Bangladesh on 15 November 2007: "2400 educational facilities were fully damaged while 12,399 more were partially damaged" (Ascension 2007, p. 4 cited in Luetz 2018, pp. 65–67).

2.2 Flooding

Constituting the world's second largest river delta system "comprising 100,000 km^2 of riverine flood plain and deltaic plain" (Sarker et al. 2011, p. 203), Bangladesh is among the most flood-prone regions in the world. "Up to one third of low-lying Bangladesh floods annually during the monsoon season" (Luetz 2008a, p. 4), and severe monsoon rains may cause flooding across "more than one-third of Bangladesh" (ReliefWeb 2017, para. 6). According to Oxfam, in 2017 "two-thirds of the country was under water" (Bennett 2017, para. 2). Geospatial analysis prepared by the United Nations Institute for Training and Research (UNITAR 2007) reflects what may perhaps be classified as a typical flood (2–5 August 2007): It covered 42.41% of Bangladesh (UNITAR 2007),[7] left 848 people dead, affected 11.4 million, and caused US$100 million in economic damages (EM-DAT data cited in Luetz 2008a, pp. 5, 7). Significantly for this discussion, "[a]ccording to the Network for Information, Response and Preparedness Activities on Disaster, the 2007 floods saw 332 schools in Bangladesh destroyed and 4893 damaged." (ReliefWeb 2007, para. 13 cited in Luetz 2008b, p. 85). Importantly and relatedly, "[a]s a result, students in flooded areas could not attend classes for months." (ReliefWeb 2007, para. 1).

2.3 Erosion

Criss-crossed by 230 rivers, Bangladesh's problems of river erosion are perennial (Haque and Hossain 1988). "Every year anywhere between 66,500 (BSS [Bangladesh Sangbad Sangstha] 2012) and 100,000 people (Shamsuddoha 2007) become homeless due to the effects of river erosion." (cited in Luetz 2018, p. 67). According to satellite-based geospatial data compiled and analysed by Sarker and colleagues (2011) at the Dhaka-based Centre for Environment and Geographic Information Services (CEGIS), erosion claims approximately 100 km^2 of land

---

1714 km of roads were fully damaged, […] while 5409 more kilometres were partially damaged'" (Ascension 2007, p. 4 cited in Luetz 2018, pp. 65–67).
[7]UNITAR (2007).

annually (CEGIS 2009). In places along the northeast coast of Bhola Island where some interviews for this research took place, erosion decimated coastal lands by as much as "six kilometres"[8] (see CEGIS 2009, p. 41). According to research participants interviewed in situ for this study in northeast Bhola in areas affected by the eroding coastline, about 35 km^2 of land were lost to erosion, and "more than 40,000 people made homeless."[9] As may be deduced, erosion on such a scale can have a detrimental impact on education. This was demonstrably exemplified by an eroded school depicted in a video documentary published by UNSW Sydney on 18 February 2015 from footage filmed by these researchers during field research. A former student can be seen standing at the site of his eroded school, explaining how erosion impeded the education of 250–300 students.[10]

In view of the compelling case for education highlighted above, and its notable absence in cross-sections of the Bangladeshi society, three independent field research studies were conducted in Bangladesh to better understand the impediments to education, as well as the interrelationships between levels of education, socioeconomic conditions, and success factors for managing slow- and rapid onset natural disasters and environmental changes.

The research extends previous studies by expressly inviting the views of research participants in coastal areas, as well as in urban slum communities in the country's two largest cities, Dhaka and Chittagong. In soliciting these unique grassroots perspectives this research aims to support more congenial human development outcomes. It also seeks to engender more concrete policy maker support so that the manifold benefits of learning may also be extended to those poor communities that are presently eclipsed.

3 Research Design and Methodology

Findings presented in this research chapter (Sect. 4) are based on a synthesis of field research conducted in Bangladesh as part of three separate studies undertaken in April 2008 (Luetz 2008b), November and December 2011 (Luetz 2013, 2018), and during mid-2011 to early 2012 (Sultana 2015).

As such, this chapter extends previous research by means of a three-pronged research approach that combines quantitative and qualitative data analysis with an element of longitudinal field site revisitation. Methodological approaches pertaining to each of the three studies are briefly and consecutively outlined below.

---

[8]See https://youtu.be/PBJeelgnadU @ 21:42–24:15 min.
[9]See https://youtu.be/PBJeelgnadU @ 11:30–11:45 min.
[10]See https://youtu.be/PBJeelgnadU @ 3:20–8:20 min.

3.1 World Vision Research Study: Heightening Community Level Disaster Preparedness

The first study was undertaken as part of research carried out for the World Vision Asia Pacific Annual Disaster Report entitled "Planet Prepare" (Luetz 2008b). In addition to facilitating the inception and design of subsequent Ph.d. field research (Sect. 3.2 below), this initial research visit also enhanced familiarity with cultural, environmental, and socioeconomic issues, which later constituted the foundation for strategic partnerships in areas of translation, interpretation, logistics and overall research support.

3.2 Doctoral Case Study Research into Rural-Urban Human Migration

The second study arose from Ph.D. research conducted in both rural areas of out-migration (communities of origin) as well as urban areas of in-migration (communities of destination). More specifically, semi-structured interviews were conducted in villages on the northeast coast of Bhola, Bangladesh's biggest island, where environmental changes have caused the continued and ongoing displacement of thousands of coastal dwellers.[11] In Dalalkandi, Tajumuddin, community members consulted in 2008 as part of the first study (Sect. 3.1 above; see also Luetz 2008b, pp. 26–28) were revisited for additional key commentaries on erosion and community displacement, and confirmed that "the whole area" where interviews were conducted in 2008 had since "disappeared". Incidentally, even the land on which interviews for this research study were conducted in 2011 has also since disappeared because of erosion.[12] In addition to Bhola Island, semi-structured data collection for this study also took place in slums in Dhaka and Chittagong, Bangladesh's two biggest urban catchments (Baker 2007, pp. xi, xiii; see also Muriel 2012).

A total of 49 semi-structured interviews took place, of which 48 were carried out on-site face-to-face. Eight interviews were held in Bhola Island, and 40 in the two urban conglomerates Chittagong (17) and Dhaka (23). One key informant interview with a Member of Parliament was conducted via skype. Of all respondents, 96% had Bangladeshi nationality, 86% were Muslim (4% Hindu, 10% Christian), and 53% were female. Precise ethnic background was inconsistently provided by respondents, albeit 92% of respondents can be described by the catch-all ethnicity "Bengali". Eleven key informant interviews (22% of the sample) were conducted

---

[11]"According to CEGIS (2009), erosion has caused the coastline to shift by about six kilometres, thereby displacing thousands of coastal dwellers." (p. 41).
[12]See geospatial data in https://youtu.be/PBJeelgnadU?t=3m20s and http://goo.gl/maps/1huUJ.

with 'experts', including a researcher/morphologist, local government officials, water resources and migration experts, development project officers and managers, and both local and international disaster management professionals with expertise in disaster risk reduction and community resilience. Of all interviews conducted, 38 were carried out in the country's lingua franca Bengali with the help of local guides and interpreters. The remainder were conducted in English. Respondent ages ranged from 18 to 82 years, with 37.5 years as the average age.

With a minimum of 14 focus group discussions taking place and between nine to 35 respondents participating in each conversation the total number of respondents in this field research is estimated by this researcher to be 289. An interview questionnaire guided conversations into key areas of interest, which are elaborated in Luetz (2013, pp. 78–80; 2018). Testing of the interview questionnaire occurred during a pilot study in Bougainville/Papua New Guinea during October and November 2010 (see Luetz and Havea 2018) and led to the incorporation of simplifying features into the final questionnaire design. Validity was ensured by accommodating feedback and input from the research, ethics, and pilot communities.

Broadly speaking this research followed a 'mixed method' approach and an 'exploratory design' paradigm that disembogued in a 'case study' write-up weighted heavily on qualitative analysis rather than quantitative study (Creswell 2013; Creswell and Plano Clark 2011; Punch 2014).

3.3   Doctoral Case Study Research into Coastal Zone Management

The third study also arose from Ph.D. research and delineated two place-based case studies within the context of wider research, which were supported by key informant interviews, surveys and focus groups. This approach drew on the theory that carefully selected diverse case studies at the commencement of research can provide the basis for the development of general theories through observation, thus constituting a fertile research approach for qualitative analysis (Bryman and Burgess 1999; Cameron 2000; Dunn 2005; McGuirk and O'Neill 2005; Morgan and Krueger 1993). Further, Benbasat et al. (1987) identify three strengths of case study research in information systems: (1) the researcher can study local social parameters in-depth in a natural setting, learn about current practices, and generate theories from practice; (2) the method allows the researcher/s to understand the nature and complexity of the process taking place; and (3) valuable insights can be gained into new topics emerging in rapidly changing local contexts.

The locations of the two coastal case study areas were selected from among the delineated 19 coastal districts of Bangladesh by the Integrated Coastal Zone Management (ICZM) Plan project started in 2005. The different locations contributed different perspectives, and the research sought to test the robustness of

ICZM as a theory and practice in coastal areas of Bangladesh. The comparative study between Bhola Island and the Satkhira coastal estuary context provided useful insights into the implementation of ICZM in different sociocultural and biophysical contexts in Bangladesh.

In each district a structured survey was conducted with 60 local community group members. In addition, focus group discussions were carried out with farmers (12), fishers (12), women householders (12), local businessmen (12) and local professionals (12) for each village, thus resulting in a total 120 local community participants from each coastal district. Interviews and surveys took place during mid-2011 to early 2012. The questions invited new ideas on how ICZM,[13] NAPA[14] and BCCSAP[15] might be improved for better coastal protection in the future. The focus groups were held in the following year at the case study villages of the coastal districts and involved 20–25 groups of people comprising the same participants who had also participated in the semi-structured interviews and surveys conducted before in the two coastal districts. As mentioned, methodology and design issues of the three independent studies are detailed in Luetz (2008b, 2013, 2018) and Sultana (2015), and will not be recapitulated beyond the overview provided above.

4 Results and Key Research Findings

From the research several key findings arose, which are synthesised under two themed subheadings below. Quantitative results are presented first (Sect. 4.1), followed by qualitative results (Sect. 4.2).

4.1 A Significant Proportion of Bangladeshis Has Had Limited Access to Education

As reflected by Figs. 1, 2, 3 and 4, quantitative data confirmed what may be categorised as an overall low level of educational attainment on the part of research participants encountered in Bangladesh. Figure 1 (reproduced with permission of the publishers from Luetz 2018, p. 83) provides a one-page overview of all non-expert research participants encountered in Bhola, Chittagong, and Dhaka. As

---

[13]Definitional approaches to the concept of ICZM are elaborated by Schernewski (2014).

[14]"National adaptation programmes of action (NAPAs) provide a process for Least Developed Countries (LDCs) to identify priority activities that respond to their urgent and immediate needs to adapt to climate change—those for which further delay would increase vulnerability and/or costs at a later stage." (United Nations Climate Change 2014, para. 1).

[15]Bangladesh Climate Change Strategy and Action Plan (BCCSAP).

Interview location	Age	Sex	Years	Highest level of education
Orig/Bhola	70	m	0	No school
Orig/Bhola	82	m	0	No school
Orig/Bhola	18	m	4	Primary
Orig/Bhola	55	m	5	Primary
Orig/Bhola	32	m	9	Secondary*
Orig/Bhola	32	m	9	Secondary*
Dest/Chittagong	20	f	5	Primary
Dest/Chittagong	20	f	5	Primary
Dest/Chittagong	30	f	0	No school
Dest/Chittagong	45	f	9	Secondary
Dest/Chittagong	49	m	0	No school
Dest/Chittagong	60	m	2	Primary
Dest/Chittagong	35	f	0	No school
Dest/Chittagong	45	f	0	No school
Dest/Chittagong	25	m	0	No school, neighbour taught
Dest/Chittagong	40	f	5	Primary
Dest/Chittagong	28	f	5	Primary, unaffordable school
Dest/Chittagong	26	m	7	Secondary
Dest/Chittagong	50	f	0	No school
Dest/Dhaka	57	f	2	Primary
Dest/Dhaka	26	m	1	Primary
Dest/Dhaka	34	f	9	Secondary*
Dest/Dhaka	35	f	0	No school
Dest/Dhaka	35	f	0	No school
Dest/Dhaka	25	f	1	Primary
Dest/Dhaka	19	f	3	Primary
Dest/Dhaka	20	f	2	Primary
Dest/Dhaka	20	f	2	Primary
Dest/Dhaka	30	f	0	No school
Dest/Dhaka	28	f	0	No school
Dest/Dhaka	31	f	3	Primary
Dest/Dhaka	50	f	0	No school
Dest/Dhaka	20	f	2	Primary
Dest/Dhaka	34	f	10	Secondary School Cert (SSC)
Dest/Dhaka	35	f	3	Primary
Dest/Dhaka	30	f	0	No school
Dest/Dhaka	30	f	1	Primary
Average Age	**26**		**2.8**	**Average years of schooling**

* NB: Precise grade unknown, grade 9 assumed. Figure reproduced with permission of the publishers from (Luetz 2018, p. 83).

Fig. 1 Years of schooling completed by 37 interviewed non-expert adult migrants

Name	Age	Children	Monthly Income	Schooling	Reason for migration to slum
Samsunnar	35	3	3,300 Taka[a]	no school	income
Surma	25	2	3,400 Taka	grade 1	can only write
Rabea	19	1	2,500 Taka	grade 3	was born here
Honufa	20	2	1,500 Taka[b]	grade 2	lost parents in childhood, came to Dhaka as 10-year-old maidservant
Josna	20	1	6,300 Taka	grade 2	came here with mother for better income, to escape poverty
Jorina	30	5	4,000 Taka	no school	poverty, better income
Rashida	28	4	2,400 Taka	no school	poverty, better income
Fatema	31	4	8,500 Taka[c]	grade 3	poverty, better income
Razia	50	2	not working	no school	river erosion, moved 3 times
Averages	**28.7**	**2.7**	**3,544 Taka**	**1.2 years**	**poverty implicated**

NB: Focus group interview (No 27) conducted in Dhaka 6 Dec 2011.
At the time of this research 1,000 Taka were equivalent in value to approximately US$13.
[a]Husband earns 2,400 Taka per month (p.m.), her son 600 Taka, and she 300 Taka. After paying 1,200 Taka house rent, the household has approximately 2,000 Taka p.m. remaining to meet all household expenses.
[b]Needs to pay 1,000 Taka p.m. on rent, which leaves her 500 Taka to live on (but with food provided).
[c]Husband earns 4,000 Taka p.m. as rickshaw puller, elder daughter 2,500 Taka p.m. as garment worker, and she earns 2,000 Taka p.m. as garment worker.

Fig. 2 Focus group exemplar A reflects nexus between education and income

shown, a definitive majority of adult respondents indicated "primary school" or "no schooling" as their highest level of educational attainment. On average, adult respondents had completed less than three years of schooling each.

The same situation was recorded during interactive focus group discussions. Indicating their age, number of children, monthly income, level of schooling, and their reason/s for rural-urban migration into the slum context, focus group discussions seemed to reveal a nexus between levels of education and available employment opportunities. Data from two sample focus groups, numbering five and nine participants respectively (Figs. 2 and 3), reflect what may broadly be interpreted as an appreciable interrelationship between low levels of education and arising corresponding socioeconomic situations (elaborated in Sect. 5).

Further, a comparative analysis based on data collected from 240 research participants in the Bhola and Satkhira districts of Bangladesh (Sultana 2015, p. 241), additionally confirmed that virtually half of all rural survey respondents had never attended school (Fig. 4). Correspondingly, these generally low levels of

Name	Age	Children	Monthly Income	Schooling	Reason for migration to slum
Lucky	20	1 son	4,200 Taka[a]	grade 2	From Borhanuddin, Bhola. Came 6 months ago to join her factory worker husband. Married 4 years ago.
Hasina	3	0	unknown[b]	Secondary School Cert. (SSC) = 10 yrs. [c]	From Chandpur. Came 10 years ago because of poverty. Husband is a garment employee and runs small grocery shop.
Anowara	35	2	3,000 Taka[d]	grade 3	From Brammonbaria. Came 15 years ago because of poverty. Got married here. Husband only earning family member. Rent = 1,000 Taka p.m., rest 2,000 Taka p.m. for household.
Rina	30	3	8,000 Taka	no school	Husband microbus driver @ 7,000 Taka / month; She maidservant @ 1,000 Taka / month. Gov't property: works in 2 houses, washing dishes, clothes, sweeping floors; works 4 hrs p. day @ 8 Taka p. hr. ~ 32 Taka p. day
Roxana	30	2	She earns 800 Taka as a maidservant per month	grade 1	Collects mango 30 days / month @ 8 Taka / hour. Two years ago, husband went to Egypt: borrowed 4 lakh = 400,000 Taka; went through broker – only tourist visa – now illegal – until now hasn't sent money.
Averages	29.8	1.6	4,000 Taka	3.2 years	poverty implicated

NB: Focus group interview (No 30) conducted in Dhaka 6 Dec 2011.
At the time of this research 1,000 Taka were equivalent in value to approximately US$13
[a]Husband's earnings.
[b]Involved with World Vision for four years. Has sewing machine, sometimes tailors.
[c]High School Certificate (HSC) = 12 years.
[d]Husband's monthly earnings as rickshaw puller.

Fig. 3 Focus group exemplar B reflects nexus between education and income

Highest level of education completed	Bhola district (%)	Satkhira district (%)
No Education	49.5	48.0
Class 1-10	43.0	35.5
SSC and HSC pass	7.5	9.0
University	-	7.5
Total	100.0	100.0
Sample size (Participants)	120	120

Fig. 4 Education rate of Bhola and Satkhira Districts at Village level [N = 240]

education also point to the limited livelihood options of rural people, who are overwhelmingly dependent on natural resources. In the absence of long-term improvements in these areas, there is a high risk for such communities to remain exposed to the ongoing impacts brought on by natural disasters and environmental change.

In addition to levels of education and corresponding socioeconomic situations, research results also highlighted important factors impinging on community resilience. In coastal areas most people are either farmers or fishermen. Given that these districts are situated within the Bay of Bengal, community member skill sets are for the most part restricted to farming, fishing and work related to marine resources. However, if people choose to move inland due to rises in sea level, disaster impacts, and/or erosion and corresponding land loss, they might have to find other kinds of work for which they have no skills, education and/or prior professional experience. For this reason people may have difficulty finding work. Hence people in both districts expressed "extreme" to "very extreme" concern about the lack of skills to switch occupations and the impact this might have on their socioeconomic situation (Fig. 5). People who migrated to Dhaka typically ended up working in garment factories, as daily wage labourers, or pulling rickshaws. However, if they failed in these endeavours, they commonly ended up begging or resorting to criminal activities and/or illegal work.

Further, there was overwhelming empirical evidence that respondents in both districts are either "very" or even "extremely" interested in participating in adaptation and disaster education and training programs as may be offered by government departments and NGOs (Figs. 6 and 7).

In summary, empirical perspectives arising from the quantitative data suggest that large parts of society do not enjoy the benefits of continuing access to education. This was also recurrently confirmed by what seemed to be a pervasive presence of slum dwelling children not attending school but instead supporting their parents as income generators.

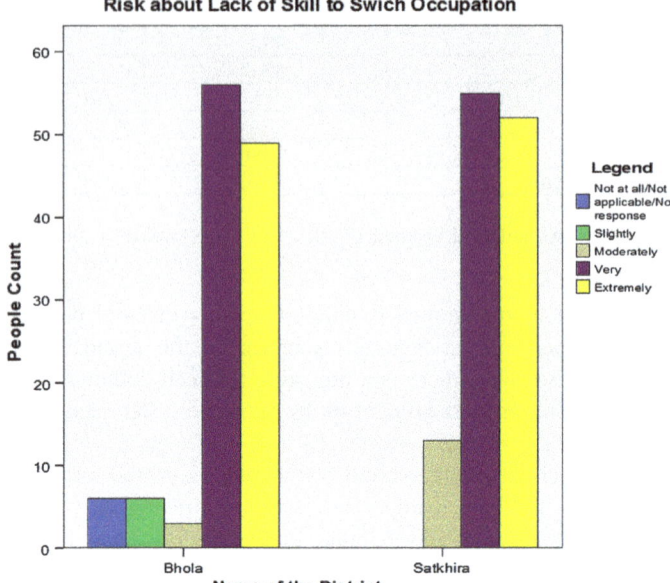

Respondents with lower levels of education are more risk averse
when considering new livelihoods.

Fig. 5 Participants' perception of risk from lack of skill to switch occupation (by district)

4.2 Qualitative Results Revealed a Significant Interest in Disaster Risk Education

As already highlighted by the quantitative data above (Sect. 4.1), qualitative results further confirmed both a pervasive absence of education in cross-sections of society, as well as broad interest in formal and non-formal education on disaster risk reduction. The following sample perspectives are provided as exemplars. One farmer encountered in a village in Bhola district recounted that for a range of reasons remote schools often remain closed for long periods of time:

> Due to lack of monitoring and supervision, and understanding between the teachers and the supervising authority, many schools in remote areas remain closed for indefinite periods from time to time, hampering proper education. (Local community level interview and survey, November 2011, Muslim para village, Char Kukri Mukri Union of Bhola district, Md. Jakir Hossain)

Furthermore, there were numerous suggestions of widespread interest in disaster risk education, as exemplified by workshops and initiatives promoting community based capacity building (Fig. 7).

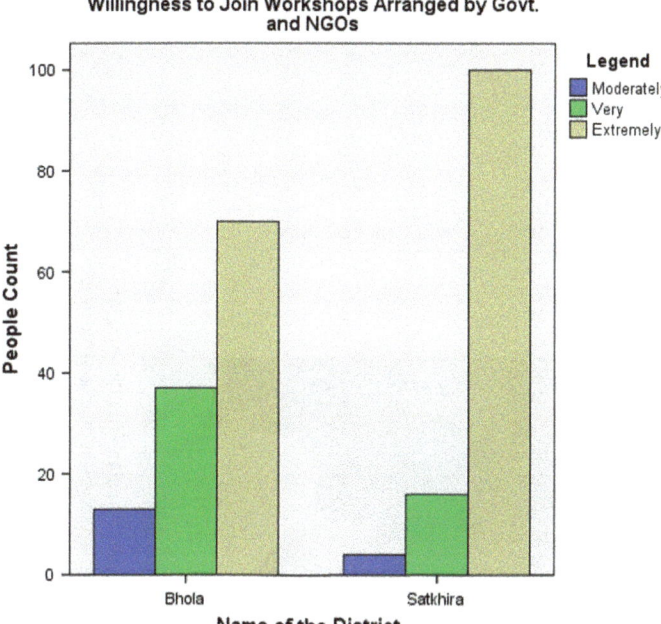

Respondents in both districts are eager to learn about adaptation and disaster risk education.

Fig. 6 Participants' perception of willingness of local people to join workshops (by District)

In summary, quantitative (Sect. 4.1) and qualitative results (Sect. 4.2) point to both deep-seated school absenteeism and related arising vulnerabilities and socioeconomic problems, and significant opportunities for formal, informal and non-formal learning.[16]

5 Discussion: Causes and Consequences of Non-education

As reflected in Sect. 4, research results suggest several impediments to education, which are explored in this section. Two areas are shortlisted for discussion: Impediments to education (Sect. 5.1), and consequences arising from these impediments (Sect. 5.2). A short synthesis concludes the section (Sect. 5.3).

---

[16]See UNESCO (2010) and Walid and Luetz (2018) for definitional discussions relating to formal, informal and non-formal education.

The main aim of this workshop was to strengthen local capacity on "Sustainability Planning of DUS for Mainstreaming Disability" on Bhola Island. This workshop serves as an exemplar of the far-reaching popularity of grassroots community based education initiatives on adaptation and disaster risk reduction.

Fig. 7 Workshop organised by NGO Dip Unnayan Society (Lalmohan Upazila, Bhola district), Nov 2011. *Photo* Sultana

5.1 Impediments to Education

Impediments to primary and secondary education seem to arise predominantly from a state of protracted poverty and are influenced by the sociocultural mindset of the people of Bangladesh. The coastal zone of Bangladesh has 19 coastal districts, which seem to have remained below the national poverty line in part because of their geographically derived exposure and vulnerability to intense and frequent natural disasters and a changing climate. People living in these areas largely subsist on a day-to-day basis by collecting food and ensuring shelter. In these circumstances education is generally regarded as a "luxury" that can only be afforded *after* livelihood security has been ensured. Hence parents typically involve their children in income generation activities from an early age. Moreover, many schools are not easily accessible by roads due to tidal effects and the unavailability of culverts over many canals, making infrastructure a related contributing problem (Fig. 8). Hence poor infrastructure tends to further compound challenges of access to education overall. The primary school in Char Kukri Mukri Union of Char Fesson Upazila[17] (sub-district of Bhola) exemplifies these issues: There is no transportation available except for some bicycles. In short, girls and boys need to walk long distances to reach the school every day. Therefore, lack of available schools and access challenges represent two key impediments to education in that locality (see Fig. 8). Although children in urban areas are not similarly hindered by an absence of transportation infrastructure, distance from the school building—or traversal through unsafe areas—may represent an unacceptable risk, as one mother disclosed in a Dhaka slum: "[my husband and I] have three children. None attends school. The school is far and there is no one to accompany my young daughters on the way to school". (Participant No. 31, Dhaka, 7 December 2011).

Other impediments to education are more multi-causally complex and seem to be related also to issues of ethnicity and religion: Almost 100% of the people living in the aforementioned Upazila are Muslims, for whom education has not been a priority while conditions of extreme poverty persist. Further, there was a sense during interviews that culturally, people residing in these areas would prefer to attend Madrasas (Islamic Schools) rather than conventional government schools. By contrast, the Munshigang Union of Shyamnagar Upazila of Satkhira district offers a nuanced perspective as it reflects a slightly higher rate of schooling than encountered in Bhola district (see Fig. 4). The Upazila has more schools and the transportation network is superior. Further, the ethnoreligious composition of the Upazila reflects mostly a Hindu faith background. For many of these respondents it seemed, no matter how much poverty they were facing, they would rather send their children to school than see them work for a living. Hence research in these Upazilas has raised faith persuasions as a fertile area and opportunity for further research into impediments and aspirations related to education. Notwithstanding, despite regional

---

[17]Bangladesh is administratively organised into seven Divisions, 64 Districts, hundreds of Sub-Districts or Upazilas, and thousands of Unions comprising tens of thousands of Villages.

Bhola District (Left Column)	Satkhira District (Right Column)

Primary School used as Cyclone shelter, Sahbazpur Village, Char Kukri Mukri Union, Char Fesson Upazila, Bhola district.	Primary School used as Cyclone shelter, Mathurapur Village, Munshigang Union, Shyamnagar Upazila, Satkhira district.

This primary school (Muslim para village of Char Kukri Mukri Union, Char Fesson Upazila of Bhola district) remains closed for long periods of time due to the frequency of tidal surges.	Primary school students walk long distances every day to and from school by using such tidal affected unfeasible roads (Gabura Village of Munshiganj Union, Shyamnagar Upazila, Satkhira district).
	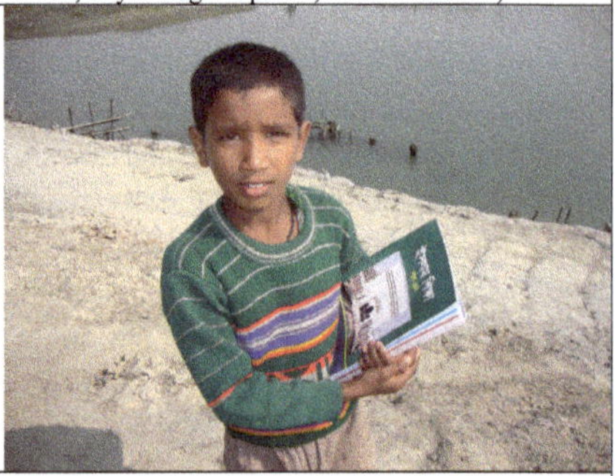
Children and elderly people in villages of Bhola district cross canals daily by using rickety culverts (branches of trees), including for walks to schools.	A primary student of Assasuni Upazila of Satkhira district returns home from school with high hopes for the future.

Fig. 8 Bhola District (Left Column) Satkhira District (Right Column). *Photos* Sultana

sociocultural and religious differences, overall the predominant impediment to education seems to be linked to recurring natural disasters such as cyclonic storm surges, which also represent a significant contributing factor to rural-urban migration, including into informal settlements and slum contexts in Khulna, Chittagong, and Dhaka (Luetz 2018). During and in the wake of natural disasters schools are often closed for a long time because of their dual use function as cyclone shelters. This situation can see children regularly and lengthily displaced from their classrooms (Fig. 8).

Finally, schools are frequently operating in two shifts. This practice has the tendency to limit the teaching exposure of children to *either* the morning (9 am to 12 pm) *or* the afternoon (12 pm to 3 pm) periods (Participant No 5, Bhola, 29 November 2011; Participant No 19, Chittagong, 3 December 2011).

5.2 Consequences of Non-education

The consequences of non-education were quite clearly revealed during this research. The most detrimental consequence seemed to be the inability of participants to find jobs in their own home districts. Relatedly, respondents with low levels of education and skills were then typically hesitant or even afraid to migrate to other districts, which reveals education to be an important factor in enabling (or disabling) mobility and adaptive capacity. While education as a factor in migration is not newly discovered (Belt 2011), it is yet another reminder that education raises options, lack of education forecloses them, and this includes mobility options, both locally, regionally, and internationally.

Further, according to a primary school teacher encountered in Bhola district, numerous students have such low levels of education that they remain unable to even sign their own names. Relatedly and importantly, this then has important knock-on effects as it hampers such students' literary capacity to comprehend available literature on the causes and consequences of natural disasters and how one might prepare for or protect oneself during disaster periods (see UNESCO and UNICEF 2009).

There are also equity problems. Although several NGOs are offering climate change and disaster awareness and training programs in these districts, program participants are not generally identified and recruited on an equitable basis. It seems that beneficiaries of such programs are usually close to local leaders and are therefore selected to participate on a preferential basis. Hence nepotism, favouritism, and/or corruption may constrain equitable access to education on disaster preparedness, response, recovery, and overall livelihood resilience. Moreover, available programs are also insufficient in number.

A related problem was observed in Satkhira district. One of the farmers encountered during field research recounted that middleclass people can sometimes remain particularly vulnerable as they are not identified as prospective trainees for disaster education. While NGOs tend to focus on the extremely poor as preferred beneficiaries, and government programs may identify rich and influential opinion leaders, people from the middleclass may be left entirely unidentified, unselected and untrained in areas of education on disaster awareness, relief, response and recovery.

Overall, the above consequences of non-education can be summarised as leaving people less resilient, and limiting their long-term ability to rebound and recover from recurrent disaster impacts. Additionally, and importantly, low levels of education also tend to render people more helpless economically and less well equipped and qualified to acquire good jobs. Notwithstanding, even secondary education is no automatic guarantee for employment. One respondent with a High School Certificate (HSC) elaborated severe difficulties finding a job in his district, and did not have the capacity to study more. Therefore, he remained unemployed for a long time and was contemplating migration to the capital city Dhaka to pursue a Class Four category government department job.

5.3   Towards Synthesis: Education Raises Options for Disaster Preparedness

In summary, if education raises options, lack of education forecloses them. According to recent interagency research, "[o]ver eight percent of 7–17 year-olds, almost three million [Bangladeshi] children in absolute terms, have never attended school. [...] Children's employment in Bangladesh appears incompatible with schooling, underscoring the importance of child labour as a barrier to achieving Education For All" (Understanding Children's Work 2011, p. ii). Hence, preparing for a better tomorrow (through education), while meeting the needs of today (livelihood security), seems to emerge as the predominant challenge facing development organisations and policy makers today.

This was poignantly raised by one research participant who made the following observation:

> The two priorities are education and livelihoods. But livelihoods is more important than education for the destitute person because it feeds them today. Education is about tomorrow. And they don't have the option of thinking about tomorrow. And it's back to the old poverty trap again. This is where policymakers and NGOs need to push harder on the things that provide for today but not ignoring the future. (Research Participant No 22, interviewed in Dhaka on 4 December 2011 cited in Luetz 2013, p. 208)

In view of the limited opportunities for formal education in Bangladesh, and the seemingly far-reaching societal interest in non-formal education (Sect. 4.2, Fig. 7), community based education initiatives on adaptation and disaster risk reduction are identified in this research as an auspicious complementary strategy for mainstreaming disaster risk education. This holds the dual promise of promoting human flourishing and fostering more disaster-ready and resilient communities. As such, holistic approaches involving formal, non-formal and informal education seem to be best suited to Bangladesh's geographic, demographic and socioeconomic context (Walid and Luetz 2018, pp. 818–822; also see definitional discussion in UNESCO 2010). Finally, improving the overall level of adaptive capacity is essentially an interdisciplinary endeavour that will require ongoing community-level research, training and education that will need to be sustained long-term.

6 Conclusion: Promoting a Culture of Disaster Readiness in Children

In Bangladesh school non-attendance can be explained on the grounds of present-day livelihood pressures, which are perceived to be so severe as to force many school age children to work and to contribute to family income as low-paid daily wage labourers. The study identifies a range of impediments to schooling in Bangladesh and explores the nexus between education and disaster preparedness and resilience. It extends previous research by expressly inviting the participation of respondents in coastal villages of Bhola and Satkhira district, as well as urban slum communities in the country's two largest cities Dhaka and Chittagong. In soliciting these unique grassroots perspectives, this research recommends more support from education stakeholders to extend the manifold benefits of learning also to those poor communities that are presently eclipsed. It further recommends a holistic portfolio of educational strategies that create more formal, non-formal and informal learning opportunities that are integrated at the community level. This offers auspicious benefits for disaster risk reduction.

7 Research Limitations and Opportunities for Further Research

The research findings presented in this chapter are subject to the following limitations. With 78% of interviews (second study, Sect. 3.2) requiring Bangla interpreters, it is conceivable that information could have been either lost in translation or lack thereof. Notwithstanding, the third study from which findings were synthesised for this chapter (Sect. 3.3), was not constrained by limitations of language, given that the author of that study is a national from Bangladesh who speaks Bangla as her mother tongue. Hence there is internal validity pertaining to the data presented in this research chapter (Punch 2014). Given one of the researchers' association with World Vision, and both researchers' affiliation with an Australian public research university, it is conceivable that some participant responses may have been coloured by a possible hope to gain certain benefits or influence future research or development programming commitments.

As mentioned in Sect. 5.2, this research raised faith persuasions as a fertile opportunity for further research into impediments and aspirations related to education. For instance, religious affiliation is indicated as a possible factor determining social outcomes for Bangladeshi people and as an area that may benefit from further exploration. While the multicausality in climate and disaster related human migration is well researched and documented (Luetz 2013, 2017, 2018; Luetz and Havea 2018), there are opportunities to investigate the role of religion in forced migration, as noted, for instance, in the European Refugee Crisis (Schmiedel and Smith 2018). Relatedly, there are opportunities to explore the nexus between limited access to education and the human experience of certain disaster types (e.g., soil erosion), whether/how these are related to each other, and whether factors may even mutually reinforce each other.

Additional opportunities for further research include comparative analyses between countries, both within and without the South Asian context that these studies (Sects. 3.1–3.3) were situated within. Finally, there are opportunities to research the role of social factors in development, especially vis-à-vis economic factors (Luetz and Walid 2019), and how businesses and organisations can operate in more sustainable and socially responsible ways (Leal Filho 2019).

8 Epilogue: Dignity Through Inclusivity and 'Bi-Directional Learning'

Global Perspectives: Applications of this research stretch beyond the borders of Bangladesh. In a world which is made up of multiple faith traditions and people who practice such traditions (or no tradition at all), it is pivotal to promote disaster education in a multi-faith context characterised by understanding, engagement, respect, and interreligious dialogue. The manifold benefits of such engagement have been theorised by Fernandez (2011, 2017), Kujawa-Holbrook (2014), Seib (2013), Moyaert (2011), and Wielzen and Ter Avest (2017). Even the UN has drawn attention to the benefits of harnessing theology for sustainable development in its landmark Human Development Report "Fighting Climate Change: Human Solidarity in a Divided World":

> Belief in the values of stewardship, cross-generational justice and shared responsibility for a shared environment underpin a wide range of religious and ethical systems. Religions have a major role to play in highlighting the issues raised by climate change. They also have the potential to act as agents of change, mobilizing millions of people on the basis of shared values to take action on an issue of fundamental moral concern. While religions vary in their theological or spiritual interpretation of stewardship, they share a common commitment to the core principles of cross-generational justice and concern for the vulnerable. (UNDP 2007, p. 61)

The report then cites examples of how sustainability is underpinned in theological or spiritual interpretations of stewardship found in Buddhism, Christianity, Hinduism, Islam, and Judaism (Climate Institute 2006; Islamic Foundation for

Ecology and Environmental Sciences 2006[18]; Krznaric 2007 cited in UNDP 2007, p. 61).

Even so, the authors of this research chapter emphatically advocate that poor people with little formal education are not to be merely viewed as "passive 're-cipients' and 'beneficiaries' (of benevolent concern) [but rather] active 'stake-holders' and 'partners' (of a common sustainability agenda)" (Luetz et al. 2019, p. 19). Recent research suggests that:

> the poor and marginalised are a valuable, although largely under-utilised and under-appreciated source for 'bi-directional learning' about sustainability. 'Reversals of learning' need to be standardised and normalised. Spirituality plays a significant role in the environmental perspectives of the poor, and the potential it brings for sustainability benefits should not remain under-utilised. (Luetz et al. 2019, p. 18)

In summary, living harmoniously in a globalising and interconnected world characterised by growing diversity is an issue that is of pivotal importance all over the world, and especially for people involved in education. Hence, promoting disaster readiness and resilience through mainstreaming disaster education is not an agenda that is somehow exclusive to Bangladesh (or any other country or faith tradition). Rather it emerges from a growing realisation that "we share a common planet, that we are all affected, and that our neighbor's suffering is not unlike our own [...] Only by concerted action that establishes a collective sense of affiliation with the entire biosphere will we have a chance to ensure our future. This will require biosphere consciousness." (Rifkin 2009, p. 616).

Acknowledgements The authors wish to thank Kirsty Andersen for her copy-editorial support, Balaram Chandra Tapader for his research assistance in Bangladesh, and Syed Abu Shoaib for his constructive comments and for field research support during visits in remote villages of the coastal districts. Grateful acknowledgment for relevant Ph.D. research support is also made to John Merson, Daniel Robinson, Eileen Pittaway, Russell Wise, Richard Rumsey, Geoff Shepherd, and to the international development organisation World Vision. Further, the authors wish to thank the Commonwealth Scientific and Industrial Research Organisation (CSIRO), the Centre for Environment and Geographic Information Services (CEGIS), the United Nations Institute for Training and Research (UNITAR), and its Operational Satellite Applications Programme (UNOSAT). Finally, the authors wish to thank the people of Bangladesh for generously sharing their stories, struggles, experiences and perspectives.

References

Allison I, Bindoff NL, Bindschadler R, Cox PM, de Noblet N, England MH, ... Weaver AJ (2009) The Copenhagen diagnosis: updating the world on the latest climate science. Climate Change Research Centre, The University of New South Wales, Sydney, NSW, Australia

Ascension AJ (2007) Loss in the wake of Sidr. World Vision Bangladesh communications, news and features packet 2007-522: 7 December 2007, Scribe 3.0. In: Lenssen H (ed) Input by

---

[18]This source (referenced in UNDP 2007, pp. 61, 211) is no longer available online at the indicated link.

Leeanne Grima. Input Date 5 December 2007, modification Date 8 December 2007. Document on file with researcher

Baker J (2007) Dhaka: improving living conditions for the urban poor. Bangladesh Development Series, Paper No. 17. Dhaka, Bangladesh, India: The World Bank. http://siteresources. worldbank.org/BANGLADESHEXTN/Resources/295759-1182963268987/dhakaurbanreport. pdf. Accessed 16 Jan 2018

Bangladesh Sangbad Sangstha (BSS) (2012) River erosion may make 66,550 homeless this year. BSS News. http://www1.bssnews.net/newsDetails.php?cat=0&id=269479&date=2012-08-02. Accessed 3 Aug 2012

Bayona JA (Producer) (2012) The impossible. (Lo imposible, original title). Motion picture. Warner Bros and USA/International: Summit Entertainment, Spain

Belt D (2011) The coming storm: the people of Bangladesh have much to teach us about how a crowded planet can best adapt to rising sea levels. For them, that future is now. Nat Geogr 58–83

Benbasat I, Goldstein DK, Mead M (1987) The case study research strategy in studies of information systems. MIS Q 11(3):369–386. https://doi.org/10.2307/248684

Bennett J (2017) South Asia floods: appeals for help as monsoon rains cause havoc in India, Nepal, Bangladesh. ABC News. http://www.abc.net.au/news/2017-08-31/india-nepal-bangladesh-floods-monsoon-rains/8858858. Accessed 15 Jan 2018

British Broadcasting Corporation (BBC) (2005) Award for tsunami warning pupil. BBC News. http://news.bbc.co.uk/2/hi/uk_news/4229392.stm. Accessed 28 Nov 2017

Bryman A, Burgess RG (1999) Qualitative research: vol. 1. Qualitative research methodology: a review. Sage, London

Cameron J (2000) Focussing on the focus group. In: Hay I (ed) Qualitative research methods in human geography. Oxford University Press, Oxford, pp 83–102

Center for Environmental and Geographic Information Services (CEGIS) (2009) Final report on monitoring planform developments in the EDP area using remote sensing. Government of the People's Republic of Bangladesh, Ministry of Water Resources, Bangladesh Water Development Board, Dhaka, India

Climate Institute (2006) Common belief. Australia's faith communities on climate change. Australia, Sydney

Collier P (1999) Doing well out of war. Paper prepared for the conference on economic agendas in civil wars, London, 26–27 Apr 1999. Paper 28137. The World Bank. http://siteresources. worldbank.org/INTKNOWLEDGEFORCHANGE/Resources/491519-1199818447826/28137. pdf. Accessed 14 Jan 2018

Creswell JW (2013) Qualitative inquiry and research design: choosing among five approaches, 3rd edn. Sage, Thousand Oaks

Creswell JW, Plano Clark VL (2011) Designing and conducting mixed methods research, 2nd edn. Sage, London

Division for Sustainable Development, UN-DESA (2017) Sustainable development knowledge platform. Sustainable development goals. Resource document. United Nations. https:// sustainabledevelopment.un.org/sdgs. Accessed 12 March 2018

Dunn K (2005) Interviewing. In: Hay I (ed) Qualitative reserach methods in human geography. Oxford University Press, Oxford, pp 50–82

Emanuel K (2005) Increasing destructiveness of tropical cyclones over the past 30 years. Nature 436:686–688. https://doi.org/10.1038/nature03906

Emerson RW (2010) Essays and poems by Ralph Waldo Emerson. Classic Books International, New York

Fernandez ES (2011) Burning center, porous borders: the Church in a globalized world. Wipf & Stock, Eugene

Fernandez ES (2017) Teaching for a multifaith world. Wipf & Stock, Eugene

German Advisory Council on Global Change (2006) The future oceans: warming up, rising high, turning sour. Special report. Berlin, Germany: Author. http://www.wbgu.de/fileadmin/user_

upload/wbgu.de/templates/dateien/veroeffentlichungen/sondergutachten/sn2006/wbgu_
sn2006_en.pdf. Accessed 15 Jan 2018
Guha-Sapir D, Hoyois P, Wallemacq P, Below R (2017) Annual disaster statistical review 2016:
the numbers and trends. Centre for Research on the Epidemiology of Disasters, Brussels. http://
emdat.be/sites/default/files/adsr_2016.pdf. Accessed 13 Jan 2018
Haque C, Hossain MZ (1988) Riverbank erosion in Bangladesh. Geogr Rev 78(1):20–31. https://
doi.org/10.2307/214303
International Commission on Financing Global Education Opportunity (ICFGEO) (n. d.) The
learning generation: Investing in education for a changing world. A report by the International
Commission on Financing Global Education Opportunity. Author, New York. http://report.
educationcommission.org/download/891/. Accessed 14 Jan 2018
Intergovernmental Panel on Climate Change (2007) Climate change 2007: synthesis report.
Contribution of working groups I, II and III to the fourth assessment report of the
Intergovernmental Panel on Climate Change. Author, Geneva, https://www.ipcc.ch/pdf/
assessment-report/ar4/syr/ar4_syr_full_report.pdf. Accessed 16 Apr 2016
International Labour Organization (2006) Baseline survey on child domestic labour (CDL) in
Bangladesh. International Labour Office, Dhaka, India. http://www.ilo.org/ipecinfo/product/
download.do?type=document&id=4647. Accessed 18 Jan 2017
Islamic Foundation for Ecology and Environmental Sciences (2006) EcoIslam. Newsletter, 02.
http://ifees.org.uk/newsletter_2_small.pdf
Krznaric R (2007) For God's sake, do something! How religions can find unexpected unity around
climate change. Human development occasional paper 2007/29. United Nations Development
Programme, New York
Kujawa-Holbrook SA (2014) God beyond borders: interreligious learning among faith commu-
nities. Pickwick, Eugene
Leal Filho W (ed) (2019) Social responsibility and sustainability: how businesses and
organizations can operate in a sustainable and socially responsible way. Springer, Cham,
Switzerland
Luetz JM (2007) Opportunities for global poverty reduction in the 21st century: the role of policy
makers, corporations, NGOs, and individuals. WDL-Verlag, Berlin
Luetz JM (2008a) Bangladesh: disaster monitor, 1. Asia Pacific fact sheets. World Vision Asia
Pacific, Dhaka, India. http://luetz.com/docs/dmfs1_bangladesh.pdf. Accessed 21 March 2013
Luetz JM (2008b) Planet prepare: preparing coastal communities in Asia for future catastrophes,
Asia Pacific disaster report. World Vision International, Bangkok, Thailand. http://luetz.com/
docs/planet-prepare.pdf. Accessed 16 Apr 2016
Luetz JM (2013) Climate migration: Preparedness informed policy opportunities identified during
field research in Bolivia, Bangladesh and Maldives. Doctoral dissertation, University of New
South Wales, Sydney, Australia. http://handle.unsw.edu.au/1959.4/52944. Accessed 31 May
2016
Luetz JM (2017) Climate change and migration in the Maldives: some lessons for policy makers.
In: Leal Filho W (ed) Climate change adaptation in pacific countries. Climate change
management. Springer, Cham, Switzerland, pp 35–69
Luetz JM (2018) Climate change and migration in Bangladesh: empirically derived lessons and
opportunities for policy makers and practitioners. In: Leal Filho W, Nalau J (eds) Limits to
climate change adaptation. Climate change management. Springer, Cham, Switzerland,
pp. 59–105
Luetz JM, Bergsma C, Hills K (2019) The poor just might be the educators we need for global
sustainability—a manifesto for consulting the unconsulted. In: Leal Filho W, Consorte-McCrea
(eds) Handbook of sustainability and humanities. Springer, Cham, Switzerland
Luetz JM, Havea PH (2018) We're not refugees, we'll stay here until we die!"—Climate change
adaptation and migration experiences gathered from the Tulun and Nissan Atolls of
Bougainville, Papua New Guinea. In: Leal Filho W (ed) Climate change impacts and
adaptation strategies for coastal communities. Climate change management. Springer, Cham,
Switzerland, pp 3–29

Luetz JM, Walid M (2019) Social Responsibility versus Sustainable Development in United Nations Policy Documents: A Meta-Analytical Review of Key Terms in Human Development Reports. In: Leal Filho W (ed) Social responsibility and sustainability: how businesses and organizations can operate in a sustainable and socially responsible way. Springer, Cham, Switzerland

Mandela N (2003) Lighting your way to a better future. Speech delivered by Mr N R Mandela at launch of Mindset Network. Speech transcript. Nelson Mandela Foundation. http://db. nelsonmandela.org/speeches/pub_view.asp?pg=item&ItemID=NMS909&txtstr=education%20 is%20the%20most%20powerful. Accessed 12 March 2018

McGuirk PM, O'Neill P (2005) Using questionnaires in qualitative human geography. In: Hay I (ed) Qualitative research methods in human geography. Oxford University Press, Don Mills, Canada, pp 246–273

McKean E (ed) (2005) The new Oxford American dictionary, 2nd edn. Oxford University Press, New York

Morgan DL, Krueger RA (1993) When to use focus groups and why. In: Morgan DL (ed) Successful focus groups: advancing the state of the art. Sage, London, pp 3–20

Moyaert M (2011) Fragile identities: towards a theology of interreligious hospitality. Rodopi, Amsterdam

Muriel S (2012) The fearless ferrymen of Dhakas Buriganga river: rush hour in the Bangladeshi capital sees thousands of Dhakas commuters boarding small wooden boats to cross the busy waters of the Buriganga river, one of the most dangerous waterways on earth, especially for the Ferrymen. BBC News Magazine. http://www.bbc.co.uk/news/magazine-19349949. Accessed 16 Jan 2018

National Aeronautics and Space Administration (NASA) (2007) JPL small-body database browser: 20002 Tillysmith (1991 EM). Reference: 20060809/MPCPages.arc. https://ssd.jpl. nasa.gov/sbdb.cgi?sstr=20002+Tillysmith. Accessed 28 Nov 2017

Owen J (2005) Tsunami family saved by schoolgirl's geography lesson. National Geographic News. https://news.nationalgeographic.com/news/2005/01/0118_050118_tsunami_geography_ lesson.html. Accessed 28 Nov 2017

Punch KF (2014) Introduction to social research: quantitative & qualitative approaches, 3rd edn. Sage, London

Randall C, Berger S (2005) Honour for young girl who saved tourists from tsunami. The Telegraph. http://www.telegraph.co.uk/news/uknews/1506286/Honour-for-young-girl-who-saved-tourists-from-tsunami.html. Accessed 28 Nov 2017

ReliefWeb (2007) Bangladesh: flood 2007 SitRep No. 12. Resource document. Author. https:// reliefweb.int/report/bangladesh/bangladesh-flood-2007-sitrep-no-12. Accessed 15 Jan 2018

ReliefWeb (2017) Monsoon floods: Bangladesh Humanitarian Coordination Task Team (HCTT) —situation report No. 1 (as of 28 August 2017). Resource document. Author. https://reliefweb. int/report/bangladesh/monsoon-floods-bangladesh-humanitarian-coordination-task-team-hctt-situation. Accessed 15 Jan 2018

Rifkin J (2009) The empathic civilization: the race to global consciousness in a world in crisis. Polity Press, Cambridge

Sánchez-Arcilla A, Jiménez JA (1997) Physical impacts of climate change on deltaic coastal systems (II): driving terms. Clim Change 35(1):95–118

Sarker MH, Akter J, Ferdous MR, Noor F (2011) Sediment dispersal processes and management in coping with climate change in the Meghna Estuary, Bangladesh. In: Proceedings of the workshop held at Hyderabad, India, September 2009, vol. 349. IAHS Press, IAHS, pp 203–217

Schernewski G (2014) Integrated coastal zone management. In: Harff J, Meschede M, Petersen S, Thiede J (eds) Encyclopaedia of marine geosciences. Springer, Dordrecht

Schmiedel U, Smith G (eds) (2018) Religion in the European refugee crisis. Palgrave Macmillan, New York

Seib P (ed) (2013) Religion and public diplomacy. Palgrave Macmillan, New York

Shamsuddoha M (2007) Climate change would intensify river erosion in Bangladesh. Campaign brief 6, coast trust fact sheet (M. Shamsuddoha, Ed.). Hardcopy on file with researcher

Strauss V (2013) Nelson Mandela on the power of education. The Washington Post. https://www. washingtonpost.com/news/answer-sheet/wp/2013/12/05/nelson-mandelas-famous-quote-on-education/?utm_term=.efc2d1e07f6b. Accessed 10 Mar 2018

Sultana N (2015) Adaptation to climate change impacts and coastal zone management in Bangladesh. Doctoral dissertation, University of New South Wales, Sydney, Australia. http://handle.unsw.edu.au/1959.4/55191. Accessed 12 March 2018

Understanding Children's Work (2011) Understanding children's work in Bangladesh. Author, Rome. http://www.ucw-project.org/attachment/Bangladesh_child_labour_report20111125_09 4656.pdf. Accessed 22 Sept 2012

United Nations (n. d.) We can end poverty. millennium development goals and beyond 2015. Resource document. Author. http://www.un.org/millenniumgoals/. Accessed 12 Mar 2018

United Nations Children's Fund (UNICEF) Bangladesh (2010) Child labour in Bangladesh: Key statistics. Resource document. Author. www.unicef.org/bangladesh/Child_labour.pdf. Accessed 28 Nov 2017

United Nations Climate Change (2014) National adaptation programmes of action (NAPAs). United Nations framework convention on climate change. Resource document. Author. http://unfccc.int/national_reports/napa/items/2719.php. Accessed 19 Jan 2018

United Nations Development Programme (UNDP) (2007) Human development report 2007/8: Fighting climate change: human solidarity in a divided world. Palgrave Macmillan, New York. http://hdr.undp.org/sites/default/files/reports/268/hdr_20072008_en_complete.pdf

United Nations Development Programme (UNDP) (2011) Human development report 2011. Sustainability and equity: a better future for all. Author, New York. http://hdr.undp.org/sites/default/files/reports/271/hdr_2011_en_complete.pdf. Accessed 27 Sept 2016

United Nations Development Programme (UNDP) (2016) Human development report 2016: human development for everyone. Author, New York. http://hdr.undp.org/sites/default/files/2016_human_development_report.pdf. Accessed 28 Nov 2017

United Nations Educational, Scientific and Cultural Organization (UNESCO) (2010) Guidelines for TVET policy review (Draft). Resource document. Author. http://unesdoc.unesco.org/images/0018/001874/187487e.pdf. Accessed 23 Jan 2018

United Nations Educational, Scientific and Cultural Organization (UNESCO) (2011) Education counts: towards the millennium development goals. Education for all (EFA) global monitoring report (GMR). Author, Paris, France, http://unesdoc.unesco.org/images/0019/001902/190214e. pdf. Accessed 14 Jan 2018

United Nations Educational, Scientific and Cultural Organization Institute for Statistics (UIS) (2017) Reducing global poverty through universal primary and secondary education. Policy paper 32/Fact sheet 44. Resource document. Author. http://uis.unesco.org/sites/default/files/documents/reducing-global-poverty-through-universal-primary-secondary-education.pdf. Accessed 15 Jan 2018

United Nations Educational, Scientific and Cultural Organization (UNESCO), & United Nations Children's Fund (UNICEF) (2009) Towards a learning culture of safety and resilience: technical guidance for integrating disaster risk reduction in the school curriculum, PILOT VERSION. Author, Geneva, Switzerland http://unesdoc.unesco.org/images/0021/002194/219412e.pdf. Accessed 13 Jan 2018

United Nations Educational, Scientific and Cultural Organization (UNESCO), & United Nations Children's Fund (UNICEF) (2012) Disaster risk reduction in school curricula: case studies from thirty countries. United Nations Children Fund, Geneva, Switzerland. http://unesdoc. unesco.org/images/0021/002170/217036e.pdf. Accessed 13 Jan 2018

United Nations Institute for Training and Research (UNITAR) (2007) Map of flood water over Bangladesh. UNOSAT. http://www.unitar.org/unosat/node/44/957. Accessed 12 Oct 2016

United Nations Office for Disaster Risk Reduction (UNISDR) (n.d.) Education. Resource document. Author. https://www.unisdr.org/we/advocate/education. Accessed 13 Jan 2018

United Nations Office for Disaster Risk Reduction (UNISDR) (2005) British schoolgirl hero meets President Clinton: "all children should know what a tsunami is... and how to react," says Tilly

Smith. Resource document. Author. https://www.unisdr.org/archive/5635. Accessed 28 Nov 2017

U.S. Department of State. Bureau of South and Central Asian Affairs (2008) Background note: Bangladesh. Resource document. Author. https://2001-2009.state.gov/r/pa/ei/bgn/3452.htm. Accessed 13 Aug 2012

Walid M, Luetz JM (2018) From education for sustainable development to education for environmental sustainability: reconnecting the disconnected SDGs. In: Leal Filho W (ed) Handbook of sustainability science and research. Cham, Springer, Switzerland, pp 803–826

Webster PJ, Holland GJ, Curry JA, Chang H-R (2005) Changes in tropical cyclone number, duration, and intensity in a warming environment. Science 309(5742):1844–1846. https://doi. org/10.1126/science.1116448

Wielzen D, Ter Avest I (eds) (2017) Interfaith education for all: theoretical perspectives and best practices for transformative action. Sense, Rotterdam

World Bank (2014) Education for all. Brief. Resource document. Author. http://www.worldbank. org/en/topic/education/brief/education-for-all. Accessed 14 Jan 2018

World Bank (2018) Bangladesh. Fact sheet. Author. https://data.worldbank.org/country/ bangladesh. Accessed 12 Mar 2018

Dr. Johannes M. Luetz is Senior Lecturer, Postgraduate Coordinator and Research Chair at CHC Higher Education in Brisbane, Australia, and Adjunct Lecturer at the University of New South Wales (UNSW) in Sydney in the School of Social Sciences, where he also completed his Ph.D. in Environmental Policy and Management with a thesis on forced human migration. Dr. Luetz has worked extensively with World Vision International on research projects raising awareness of the growing effects of climate change on poor and vulnerable communities in Asia, Africa and Latin America, and the need to meaningfully address vulnerabilities through praxis-informed education approaches that work in the real world. He has previously worked as a Lecturer in Development Studies at UNSW, Sydney.

Dr. Nahid Sultana is a Casual Academic and Tutor of the Masters of Environmental Management (MEM) program of the University of New South Wales (UNSW), Sydney. She has been involved in several projects on Disaster Management and Climate Change Adaptation in the South Asia Region as a Research Associate of School of Social Science, UNSW. Dr. Sultana has completed her Ph.D. in Environmental Management from the same university in 2015, and her research mainly focused on the decision-making process of coastal climate adaptation of Bangladesh. Before commencing her Ph.D. study, she was employed as a Scientific Officer of the Water Resources Planning Organisation (WARPO), Ministry of Water Resources of Bangladesh.

Stirring up Trouble: Museums as Provocateurs and Change Agents in Polycentric Alliances for Climate Change Action

Fiona R. Cameron

Abstract The aim of this chapter is to discuss key findings from the Australian Research Council funded international project, "Hot Science Global Citizens: The Agency of the Museum Sector in Climate Change Interventions" (2008–2012). The project looked to the museum sector—natural history, science museums and science centres—to play a role as resource, catalyst and change agent in climate change debates and decision-making locally and globally. The discussion focuses on a section of the research findings relating to the current and potential roles and agencies of natural history, science museums and science centres in polycentric climate change governing assemblages within Australian and US contexts. Through the analysis, eight strategic positions and role changes emerge for the different forms of the museum with a greater emphasis on collective action, networking and building more critical information on climate change as a complex issue and governing subject alongside activism in community and political contexts. In addition, re-working the relations between nature and culture across museum practice through a series of ecologizing experimentations was identified as a key role change in advancing sustainable practices in the long term.

1 Introduction

Climate change is an immense, complex phenomena of unknown magnitude dramatically impinging on life itself from the conduct of capitalist economies and the institutions of politics to the habits and cultures of the everyday, and at the same time stretching disciplines beyond current limits (Hodge 2011) and cultural

F. R. Cameron (✉)
Culture and Society (TemaQ), Department of Societal Development
and Culture Studies (ISAK), Linköping University (LIU), Linköping, Sweden
e-mail: f.cameron@westersydney.edu.ac

F. R. Cameron
Institute for Culture and Society, Western Sydney University,
Locked Bag, 1797, Penrith, NSW 2751, Australia

© Springer Nature Switzerland AG 2019   647
W. Leal Filho et al. (eds.), *Addressing the Challenges in Communicating Climate Change Across Various Audiences*, Climate Change Management,
https://doi.org/10.1007/978-3-319-98294-6_38

institutions such as museums further from their comfort zone. All these things invite a critical examination of and a renewed consideration of the roles and capacities of museums and science centres in these complex ecologies.

The Paris Agreement made at the Conference of the Parties in 2015 and signed by 195 nations was a significant step forward in climate change mitigation and adaptation on a global scale. The agreement seeks to strengthen global efforts to reduce greenhouse gases, setting ambitious emission reduction targets for those nations party to the agreement (UNFCCC 2016). Importantly, the agreement also detailed the important roles non-party stakeholders including cities, subnational authorities, civil society, the private sector and cultural institutions have in supporting actions to reduce emissions, vulnerability, build resilience and promote cooperation.

> Article 12: Parties shall cooperate in taking measures, as appropriate, to enhance climate change education, training, public awareness, public participation and public access to information, recognizing the importance of these steps with respect to enhancing actions under this Agreement.

Accordingly, the Paris agreement promotes polycentric modes of governance as a model for collective action. In doing so the agreement recognizes that global agreements not enough because while warming effects are global, they are often the result of actions across many sectors and scales. To bring about change, the day-to-day activities of individuals, families, firms, communities, and governments at multiple levels must change substantially (Orstrom 2010). Further the impacts of climate change vary dramatically according to their geographic location, ecological and economic conditions and prior policies and investments leading to damaging practices (Orstrom 2010). Further the United Nations Development Programme Sustainability Development Goals (2015) work alongside the Paris agreement to promote an equitable and habitable planet across a wider range of concerns and promoting education for sustainable development and sustainable lifestyles, human rights and cultural diversity, the promotion of peace and non-violence, global citizenship and of culture's contribution to environmentally sustainable development.

Building polycentric governance strategies to support the Paris agreement and the SDG's goals acknowledges the sheer diversity of approaches and solutions offered by diverse actors beyond global agreements on emissions reduction and as well as the importance of engaging the fourth sector, public sphere citizens and communities in promoting a viable future for the planet. Contemporary societies embody institutional diversity reflected in multi-level, multi-purpose, multi-sectoral, and multi-functional units of governance that can be mobilised in collective actions (Orstrom 2010). Polycentric systems tend to activate and engage pluralist, novel and imaginative climate change strategies that enhance innovation, learning, adaptation, trustworthiness, levels of cooperation of participants, and ultimately more effective, equitable, and sustainable outcomes at multiple scales (Hulme 2009; Orstrom 2010, 552). Polycentric governance therefore emerges as a networked, fluid system, an assemblage involving multisector cooperation characterized by mutual influence across the sectors and a willingness to undertake collective actions that take into account of the concerns and capacities of many parties. These processes of networks, pluralism, multi-level governance and adaptation strongly resonate with the idea of a 'political ecology'.

Museums can play a central role in climate change in polycentric climate change governance strategies however they are heavily influenced by a potent mix of highly charged information, conflicting political interests and affect which has resulted in constraining institution's capacities to take action. Against indecision and immobilisation, this chapter expounds the potential for museums and science centres to develop agential capacities for living creatively with the opportunities and threats posed by climate change through multi-sector cooperation (Cameron 2012). This paper presents key findings from the Australian Research Council funded international project, "Hot Science Global Citizens: The Agency of the Museum Sector in Climate Change Interventions" (2008–2012).[1]

2 Hot Science, Global Citizens: The Agencies of the Museum Sector in Climate Change Interventions

Hot Science, the pioneering international research project on museums and climate change looked to the museum sector—natural history, science museums and science centres—to play a role as resource, catalyst and change agent in climate change debates and decision-making locally and globally.[2]

The Project called on museums to take action on climate change but more importantly conducted research to detail how we could go about building a community of engaged institutions with the capacity to develop initiatives and build global alliances for action. The project used an interdisciplinary approach to develop new knowledge about what constitutes effective action around climate change, and how it can be represented and debated in local and global public spheres (Cameron 2011). For these purposes the Project was divided into three strands:

---

[1]Australian Research Council International Linkage Project, Hot Science, Global Citizens: the Agencies of the Museum Sector in Climate Change Interventions (2008–2011). The lead chief investigator was Dr Fiona Cameron from the Institute for Culture and Society (formerly the Centre for Cultural Research) with chief investigators, Professor Bob Hodge; Professor Brett Neilson (Institute for Culture and Society); Dr Juan Salazar; Professor Jann Conroy (Western Sydney University); Professor David Karoly (University of Melbourne, Earth Sciences) with the Australian Museum (Dr Lynda Kelly); Powerhouse Museum (Seb Chan); Museum Victoria (Carolyn Meehan); Questacon (Professor Graham Durant); Liberty Science Center, New York (Wayne LaBar); University of Leicester, Museum Studies (Professor Richard Sandell).
[2]Hot Science, Global Citizens: The Agencies of the Museum Sector in Climate Change Interventions http://ics.westernsydney.edu.au/hotscience/.

Strand 1: Museum interventions and institutional forms
Strand 2: Science-humanities interfaces in programming
Strand 3: Citizenship and media.

The research findings alongside presentations, discussions and debates at our international symposium, Hot Science, Global Citizens: The Agencies of Museums in Climate Change Interventions held at the Powerhouse Museum and the Australian Museum in Sydney on the 5th and 6th May 2011 were collated to produce a manifesto. The manifesto details the potential roles of institutions in local, national and global networks and complex governing arrangements providing pointers on how institutions might re-evaluate and re-frame their operations to more meaningfully act on climate change. Recommendations challenge museums and science centres to:

- adapt their operations and responses rapidly across different scales and to institute polycentric responses forming new cross-sector allies;
- build new relationships with many stakeholders and audiences who have differing views and responses;
- extend networks bringing together disparate people, ideas and institutions across social and geographical distances;
- institute new complex, thick modes of communication and interpretative practices;
- deal more effectively with complexity, dissent and conflict crossing sectors and scales in trans-national and cosmopolitan formations involving many people and ideas;
- re-consider ways of producing knowledge beyond disciplinary frames that acknowledge the entanglements and agency of both human and non-human forces;
- bring the past-present and future together as a focus for concern and for formulating creative thought and action;
- use art to communicate climate change issues, affects and solutions and to re-work relations between nature and culture.

The discussion in this chapter draws on a section of the quantitative and qualitative research relating to 12 positioning statements framed as a way of articulating both current and potential roles and agencies for natural history, science museums and science centres in climate change. The analysis of the 12 positioning statements sought to test perceived institutional performance detailing role enhancements and changes for natural history, science museums and science centres in Australian and US contexts.

Agencies tested ranged from more established roles as places to communicate up-to-date science and the refashioning individuals' behaviour, research and resource provision to ones that have a stronger activism and advocacy flavour involving collective action, decision-making and policy critique, lobbying, the formation of global networks to promote discussions and decisions (Cameron 2012, 318). To do this the Hot Science study employed a mixed method approach utilising both qualitative and quantitative research methods. This included a general demographic survey involving 2100 participants (1500 in Australia and 600 in New York City, New York State; Jersey City and New Jersey State); focus group research with museum and science centre audiences involving twelve groups in Sydney, Melbourne and Jersey City convened on the basis of six groups of older and younger families and six groups of adults (single income no children, 35–60 years; double income no children 25, 30 years). This research was complemented with one-on-one institutional interviews with CEOs from the five partnering institutions and a series of interviews with visitor services staff, curators, science communicators and educators (Cameron 2012, 318).

Deploying a gap analysis of the positioning statement data across all samples, a series of nine strategic positions, role extensions and changes emerge for the different forms of the museum and across the Australian and US samples. Table 1; Figs. 1, 2, 3 and 4 are detailed in the following discussion. The aim was to detail

Table 1 Top key role changes for science centres and museums—gaps between current and expected

	Australia	USA
Museums	Critical examination of climate change as a cultural, political, technological, economic and scientific issue (29%) Communicate the up-to-date science of climate change (25%) Provide a forum for discussion and debate for individuals, communities, organisations locally and globally to express views (25%) Present a range of views on climate change issues (e.g.) scientists, government, economists, industry leaders and diverse communities (23%) Promote collective action by bringing the sectors—government, business, scientific organisations, media and non-government organisations together with citizens to make decisions (22% gap) Acting as part of networks for individuals, communities and organisations with an interest in climate change (22%) Provide access to a range of resources on climate change (22%) Take a critical stance on climate change policy and decisions (20%) Lobbying on climate change (19%)	Communicate the up-to-date science of climate change (24%) Acting as part of networks for individuals, communities and organisations with an interest in climate change (22%) Provide a forum for discussion and debate for individuals, communities, organisations Locally and globally to express views (21%) promote collective action by bringing the sectors—government, business, scientific organisations, media and non-government organisations together with citizens to make decisions (21% gap) Critical examination of climate change as a cultural, political, technological, economic and scientific issue (20%) Provide access to a range of resources on Climate change (19%)

(continued)

Table 1 (continued)

	Australia	USA
Science centres	promote collective action by bringing the sectors—government, business, scientific organisations, media and non-government organisations together with citizens to make decisions (25% gap) Lobbying on climate change (24%) Acting as part of networks for individuals, communities and organisations with an interest in climate change (23%) Take a critical stance on climate change policy and decisions (23%) Present a range of views on climate change issues (e.g.) scientists, government, economists, industry leaders and diverse communities (22%) Critical examination of climate change as a cultural, political, technological, economic and scientific issue (21%) Communicate the up-to-date science of climate change (20%)	Acting as part of networks for individuals, communities and organisations with an interest in climate change (28%) Promote collective action by bringing the sectors—government, business, scientific organisations, media and non-government organisations together with citizens to make decisions (24% gap) Provide a forum for discussion and debate for individuals, communities, organisations locally and globally to express views (18%)

strategic priorities; how points raised in the manifesto might be instituted and how institutions might adapt and think in new ways about their roles in climate change (Cameron 2012, 318).

3   Roles and Agencies in Polycentric Governance

3.1   Places that Communicate the Up-to-Date Science of Climate Change

Without doubt the research confirmed the pivotal role museums and science centres have across all the samples as "Places that communicate the up-to-date science of climate change." Respondents however felt institutions should be doing more (Cameron 2012, 325).

The underlying rationale informing science communication expressed by staff, CEOs and audiences was that science can make discoverable objective facts that are socially and politically neutral; can adjudicate between competing claims to truth; and can determine the causes of climate change and climate sensitivity (Cameron 2012, 325–326). This belief according to climate scientist Hulme (2009) perpetrates

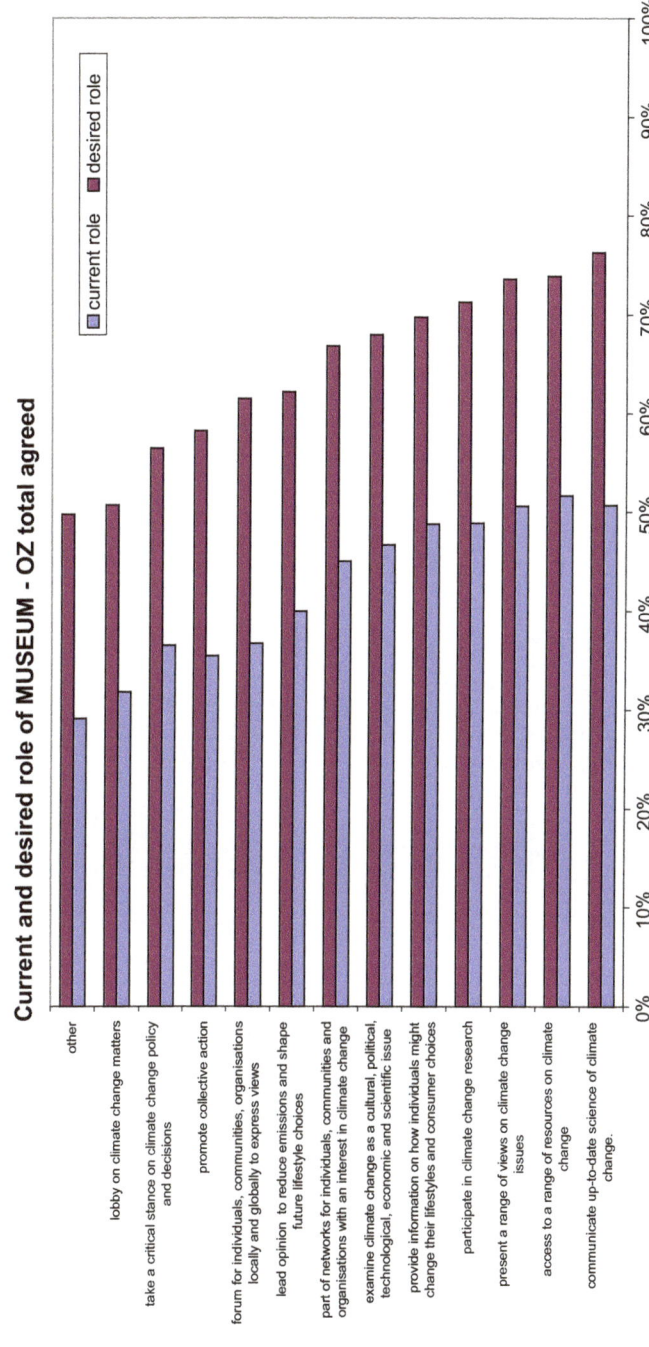

Fig. 1 Perceptions of current and potential/desired role of museums—Australian total agreed sample

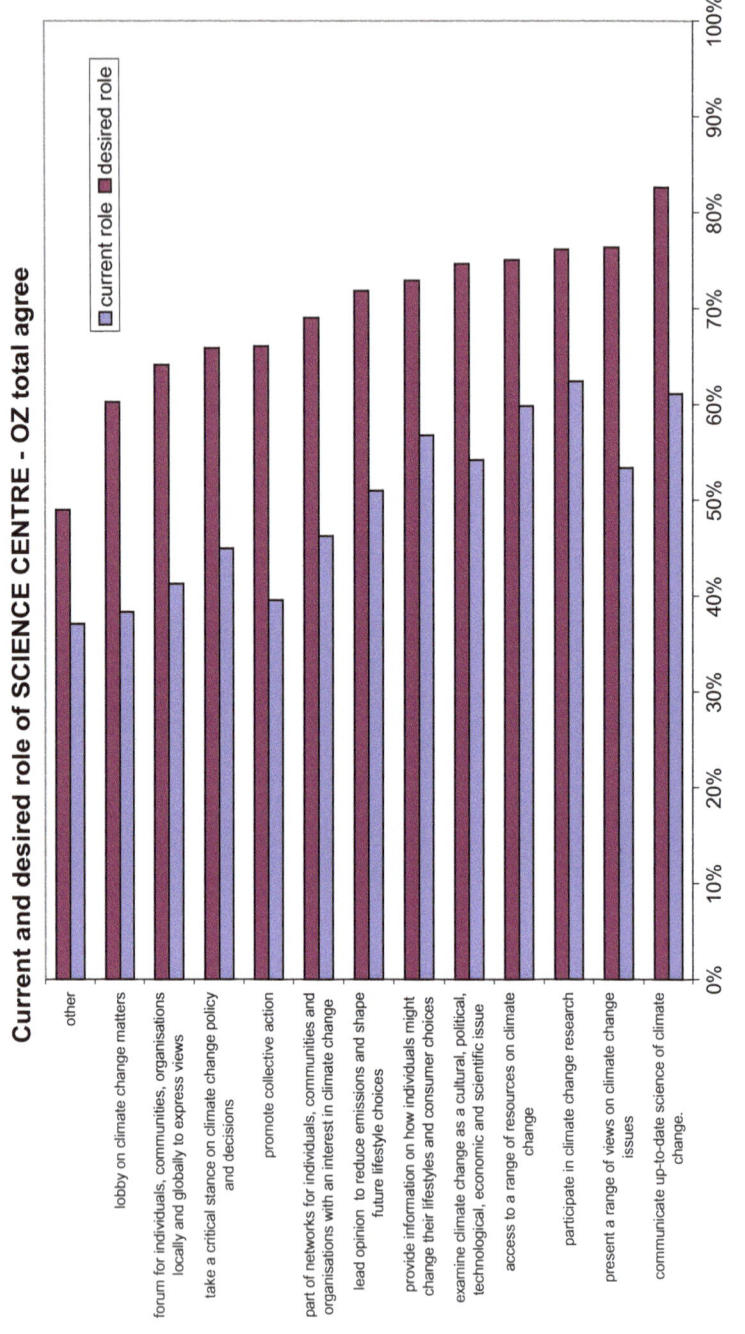

Fig. 2 Perceptions of current and potential/desired role of science centres—Australian total sample agreed

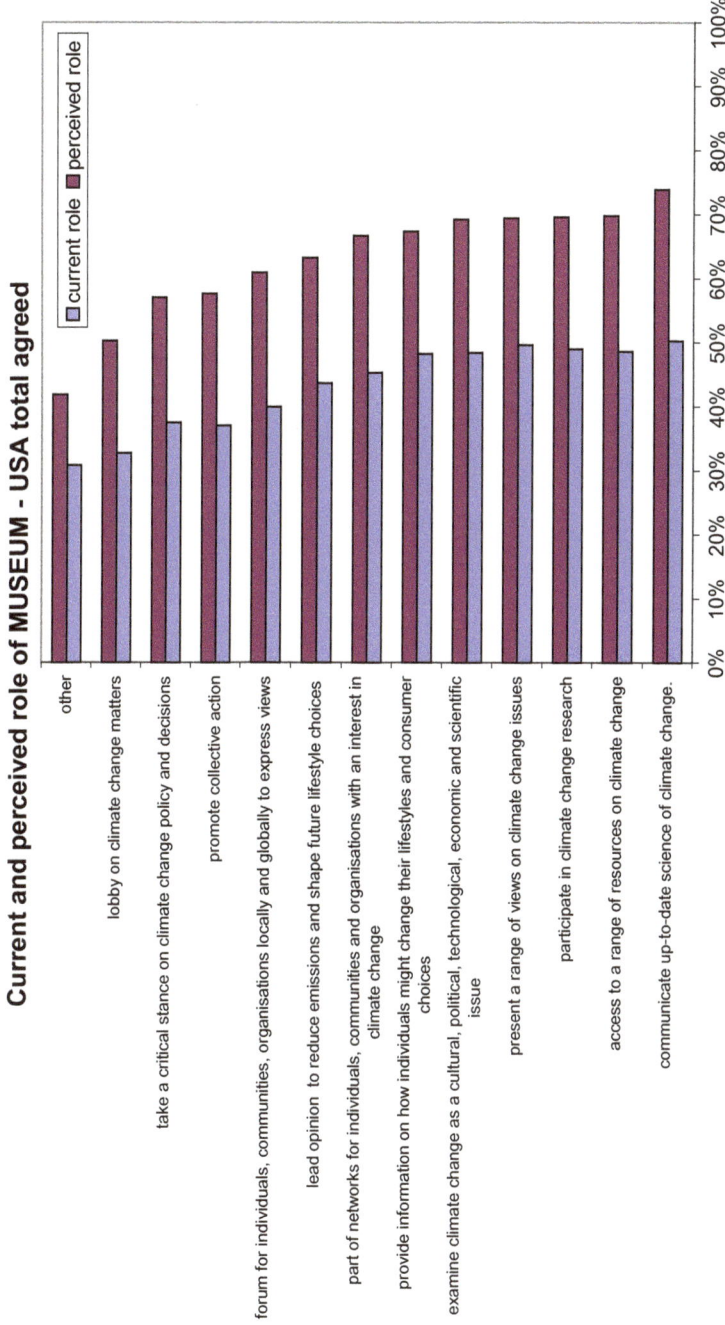

Fig. 3 Perceptions of current and potential/desired role of museums—US total sample agreed

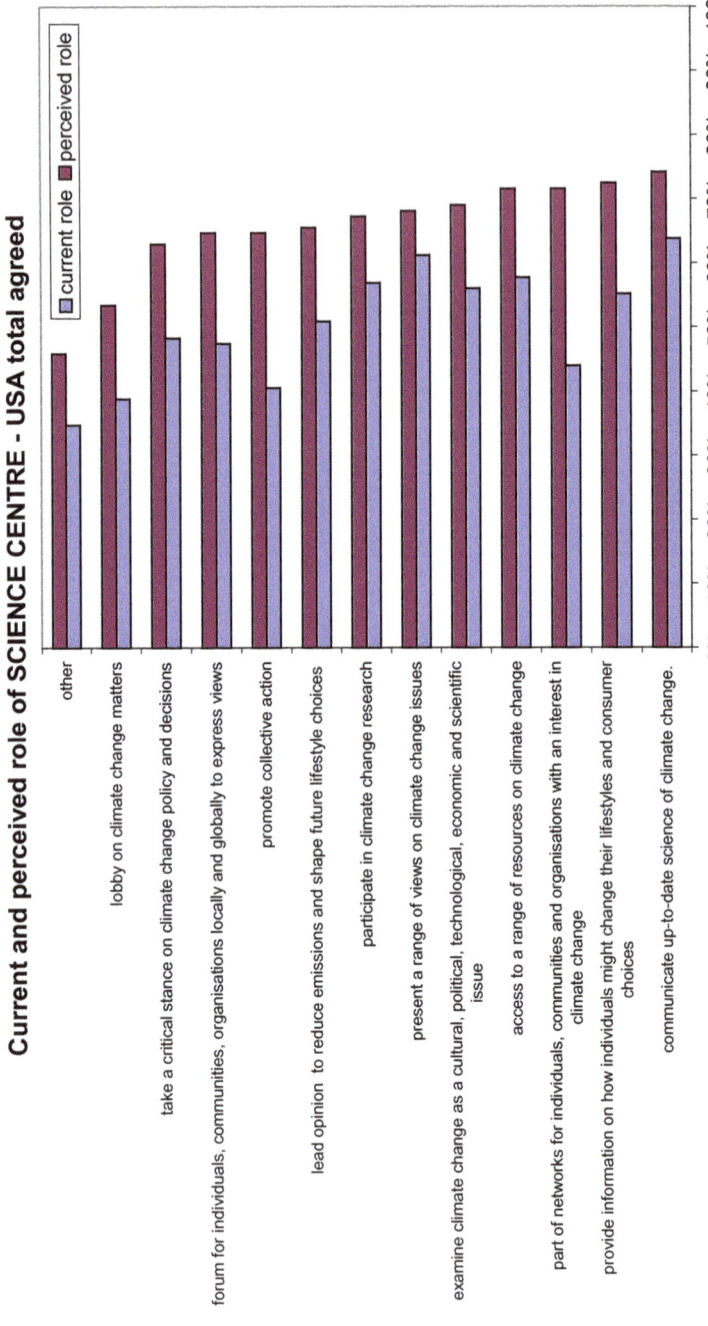

Fig. 4 Perceptions of current and potential/desired role of science centres—US total sample agreed

the view that science can produce definitive statements about what is and what not is dangerous for societies, and ways to solve the ecological crisis.

For these reasons, the findings demonstrate a strong demand for knowledge about new science to inform personal positions and actions with many stressing the political importance of science sector in solving the climate crisis (Cameron 2012, 326). Many saw museums as places for disseminating this information

> a place to learn all about the science behind climate change...what science is actually doing towards solving the problem. (HSGC Focus Group Transcript, AM#1)

Respondents strongly believed that institutions should be presenting all views on the science and where uncertainty or conflicts arose between different expert opinion and future predictions, the contextualisation of those positions was important (Cameron 2012, 326).

Because debates move quickly and are often conflicting, audiences cited judgements about the relative credibility of sources as new institutional forms of quality assurance as systems of peer review.

Systems of peer review in the museum context involves the presentation of climate change as a complex, social, cultural scientific and economic issue; the contextualisation of the research informing these debates; how knowledge under-pinning the debate is produced; weighting the various debates and sources according to their levels of acceptance and making judgements about the relative credibility of each, and what is at stake for each of the stakeholders (Cameron 2012, 326).

> you've got to take both sides into consideration...you can't unequivocally say one's right and one's wrong, you've got to lend different weights to different theories. (HSGC Focus Group Transcript, AM#3)

> I have a certain amount of confidence in the information ... there's a certain way it is researched, they've got to confirm where they've got their information from and to get it to a point it had to go through so many different educated professors and specialists in their field that there's a certain amount of credibility that goes along with it ... (HSGC Focus Group Transcript, AM#1)

Many respondents also wanted more information on how scientific knowledge is produced: how these are linked to the emergence of different world views and differing governing strategies; practices of expert deliberation around climate change and deep uncertainty, and how scientific knowledge gets used in society. Science accordingly becomes just one of the many kinds of knowledge informing climate change action and public policy debate rather than a discipline able to solve climate change (Cameron 2012, 326–327).

Public deliberation around climate science as assemblies of peer review can operate in conjunction with assemblies of public review as new forms of science education that can then be fed into the review, weighting and quality assurance processes.

New formulations of climate science through peer and open review has the potential to promote climate change as entanglements between nature and culture. An ontological and new materialism framing can also be used as an interpretative strategy to acknowledge the relation between the human and climate change as vital forces where the non-human such as the atmosphere, pollution and extreme weather operates as actants enfolded with human projects (Cameron 2012, 326, 2014a, b).

3.2 Places to Present a Range of Views on Climate Change Issues Including Those of Scientists, Government, Economists, Industry Leaders and Diverse Communities

The positioning statement, "Places to present a range of views on climate change issues including those of scientists, government, economists, industry leaders and diverse communities" articulated one of the most important gap roles particularly for Australian museums and science centres (Cameron 2012, 327). Participants in focus groups understood climate change as a complex and a highly controversial battleground of different ideologies and world views each having a profound influence on attitudes towards climate change and courses of action.

> there's so many variants...everyone's got different views...some views are more idealistic than realistic... (HSGC Focus Group Transcript, AM#2)

> don't just show me your point of view, show me all the points of view... show me the ones that matter. (HSGC Focus Group Transcript, AM#2)

For these reasons participants expressed a desire to hear about differing views, practices and courses of action. They wanted to know about the competing interests and agendas that cross cultural divides, sectors, scales, and disciplines, and for institutions to weight these views and values as part of peer review processes. Many felt that these views must be presented in a way that leaves space for visitors to come "to their own decisions" so they can formulate their own values, moral position and emotional responses to the topic (Cameron 2012, 327).

Here institutional impartiality and balance is reworked into a deliberative frame. Deliberation refers to processes of public consultation whereby participants are presented with the range of opinion and information that may take the form of a deliberative poll or citizen's assembly. It was also seen as important that the museum expressed their own position on any contesting representations (Cameron 2012, 327).

3.3 Places that Take a Critical Stance on Climate Change Policy and Decisions

The statement, "Places that take a critical stance on climate change policy and decisions" was a significant gap role for Australian museum and science centre respondents with females polling higher than males in role support and caused some concern among respondents because it challenged their conception of the museum as a place for unbiased information (Cameron 2012, 327–328). While some saw this role as too political, many including politicians said it was important for institutions to critically review policy but at a distance in the context of generic scenarios rather than specific policy: In this sense the institution would be a legitimate contributor to policy discussions as a 'secondary association' in 'institutionalising policy deliberation' (Cameron 2012, 327–328). This gap role was significant for Australian museum and science centre respondents with females polling higher than males in role support.

> I think rather than actually saying the government could decide A but we did B,…they could say best science thinks we should do A and people realise the government's done B. (HSGC Focus Group Transcript, AM#2)

> A discussion about carbon pollution reduction is fine and talking about the various ways of reducing carbon emissions but if you start talking about the scheme before Parliament, then you're getting into areas of political debate and disputation, you are bound to make some of your audience unhappy and court controversy. Institutions can use the science to say we need a cut of carbon emissions by 60 per cent by 2050 or 80 per cent by 2050 or whatever it is, I'm not saying in all cases controversy is to be avoided but court controversy of a kind where the downside is larger than the upside. (HSGC Focus Group Transcript, FP#1)

3.4 Places that Lobby on Climate Change Matters

Unsurprisingly "Places that lobby on climate change matters" mobilised a range of views in regards the relation of the museum to grass-roots movements and government (Cameron 2012, 328). While respondents were divided as to whether a lobbying role was desirable, strong support for this was evident from the Australian museum and science centre sample. Participants were not generally averse to museums playing a more active role in the public debate, but stressed the importance of institutions remaining impartial. Institutions were seen as having the potential to mobilise a range of policy inputs and responses that go beyond the dominant economic or scientific expertise (Cameron 2012, 328).

> I think it would be good for us to make decisions…what we feel is important to us and then they gather that information … and we say 'Listen we think this is a problem. (HSGC Focus Group Transcript, LSC#4)

> Providing information is a grass roots agency so that they can go in lobbying the government, that's exactly what the museum should be doing. (HSGC Focus Group Transcript, MV#4)

> When you go to a museum people are coming from all over, so you're getting all their opinions. (HSGC Focus Group Transcript, AM#1)

Here institutions were seen as a collection point for assembling and gathering diverse and contending opinions on policy akin to processes of collective intelligence and for museums to advocate to government on behalf of audiences by canvassing their views, and in an extended role as an information source. Collective intelligence describes a shared, group intelligence that emerges through the communication, collaboration and competition of many individuals and is seen to lead to better strategizing and decision making (Cameron 2012, 328).

3.5 Places that Provide a Forum for Debate and Discussion for Individuals, Communities, Organisations Locally and Globally to Express Their Views on Climate Change

Critically the statement, "Places that provide a forum for debate and discussion for individuals, communities, organisations locally and globally to express their views on climate change" represented one of the key gap roles across all Australian and US samples. The majority of respondents saw museums and science centres as having agency in networking and connecting communities across large distances (Cameron 2012, 329). They wanted to hear differing views, practices and different courses of action to tackle climate change based on different ways of framing the issue. The networked potential of the museum sector was acknowledged in two forms. The first mode was geographic dispersal as the sharing of local knowledge and experience across globally scattered institutions (Cameron 2012, 329).

The second mode was as a cosmopolitan congregation—the assembling of peoples from many different national and cultural backgrounds through the use of social media to feed into a broader debate on the scientific and social dimensions of climate change.

> when you go to a museum people are coming from all over, you're getting all their opinions. Everybody has their own situation, it's a good meeting place. (HSGC Focus Group Transcript, LSC#2)
>
> They could just explain how it is affecting them…Museums are placed all around the world…they can see how collectively it's affecting everyone. (HSGC Focus Group Transcript, MV#4)

3.6 Places that Act as Part of Networks for Individuals, Communities and Organisations with an Interest in Climate Change

"Places that act as part of networks for individuals, communities and organisations with an interest in climate change" was seen as one of the key gap roles across all samples. Focus group participants frequently cited climate change as an issue that demanded collective action. In their role as nodes in networks, museums and science centres were seen as places to disseminate information on which decisions might be made, by clarifying, synthesizing and presenting options using peer and open review processes; building grassroots movements by strengthening collaborations with other organizations such as schools and communities; and as a platform for conversations and decision-making (Cameron 2012, 329–330). Networked relations were framed as having a twofold benefit: (i) 'impacting on individual behaviour' and (ii) 'empowering communities'

> spread the word and get people actively involved in making a change. (HSGC Interview Transcript, LSC#1m)

> it's being with the community, trying to understand what particular concerns there are, and providing them with options. (HSGC Interview Transcript, MV#4m)

3.7 Places that Provide Access to a Range of Resources on Climate Change

"Places that provide access to a range of resources on climate change" was a significant gap role for Australian and US museums. Many expressed feelings of frustration and confusion by the conflicting and highly mobile nature of the debates and their inability to get clarity and access to sources that first detail the range of opinions and weight them, and second present clear information on what to do.

They saw museums as playing a stronger role as a reference base on climate change matters: for the coordination of resources across sectors and scales as systems of peer review: as places of documentation, collecting and archiving, and as contemporary civic documentation sites (Cameron 2012, 330).

> museums are places where hard evidence for example biodiversity loss is kept. (HSGC Focus Group Transcript, AM#4)

> documenting everything that's happening in recent times... (HSGC Focus Group Transcript, AM#1)

Through the coordination of resources, they were seen as key sites for raising awareness of climate change; empowering the individual to weigh up the options, informed decision-making and in facilitating broad-based mobilization and global initiatives (Cameron 2012, 330).

> You start to educate yourself and you start to educate your family and everybody moves together in that way ... it's our nature, it kind of comes automatically. We want to work together in this way so then we can have a sustainable life for everyone. (HSGC Focus Group Transcript, AM#1).

Some participants however, were more circumspect drawing a distinction between knowing about something and doing something about it.

Institutions were also seen as places that can provide different perspectives from the media, opening up debates to include other points of view beyond mainstream positions. New forms of quality assurance, trust and legitimacy can be framed around an institution's agency as resource bases in systems of peer review, and as expert reviewers along with others in complex debates (Cameron 2012, 330).

3.8   Places to Critically Examine Climate Change as a Cultural, Political, Technological, Economic and Scientific Issue

"Places to critically examine climate change as a cultural, political, technological, economic and scientific issue" represents one of the largest gap roles particularly for Australian, US museums and science centres. Many acknowledged that this new subject matter was inherently complex and political and noted that the cultural, political, technological, economic and scientific dimensions of climate change were contingent on each other (Cameron 2012, 330–331).

> They are all parts of climate change, they are intertwined, you can't really look at one without examining the others. (HSGC Focus Group Transcript, MV#3)
>
> by 'show[ing] the process, the background and how conclusions are made, it teaches you step by step how they got to that process ... how they came to that conclusion and how to get it to work. (HSGC Focus Group Transcript, AM#1)

Participants expressed a desire for institutions to offer a more critical and deeper contextualization of climate change debates by presenting the historical background and a contextualized view of climate change. This agency also maps onto climate geographer, Mike Hulme's (2009) view that climate change, as a metaphor, has done its work and now offers an opportunity to connect with the deeper ideological issues, values and power relations about what is at stake for the various actors, as a means to make progress. These critical, reflective roles orientate an institution's role in social change as facilitating new ways of thinking about climate change (Cameron 2012, 331).

3.9   Places to Promote Collective Action in Climate Change Debates and Decisions

"Places to promote collective action in climate change debates and decisions" is one of the key gap roles for institutions across all samples thereby expressing new forms of agency beyond individual behavioural reform.

Participants welcomed the idea of museums facilitating exchanges of opinion from different social locations, from science, government, business. A range of opinions and cross sectoral views was seen as a means of accessing significant ideas and a range of options from which negotiations can be made, and also to allow a range of creative options to emerge. This again connects with the idea of collective intelligence where museums are 'employed as a platform' for parties to perform these conversations and decision processes, while being careful not "to come out too heavily on either side" (Cameron 2012, 331–332).

In this role institutions have the task of negotiating and mediating among various, diverse and interconnected networks and multi-party stakeholders including communities, government and businesses.

> it would be great if that would happen. It would be amazing. (HSGC Focus Group Transcript, AM#3)

> to see quite a few different views will make people think ... why do these people have these views, why do the scientists have these views ... a whole range of views from different backgrounds. (HSGC Focus Group Transcript, 1 AM#1)

The refashioning of expertise along plural lines opens up a space to consider climate change as a contemporary social, cultural condition from diverse governmental positions (Cameron 2012, 332)

Inputs into cross-sectoral conversations of this kind according to political theorist Chris Ford (2011) can be transformational contributing to the cognitive fitness of policy over time as well as bringing about changes in lifestyles, relations, visions for the future and political forms.

> ... all these parties have good ideas you can collect to give them an option. (HSGC Focus Group Transcript, AM#4)

> They can collect ideas from the communities or individuals and then they can respond [and] maybe develop a plan or policy. (HSGC Focus Group Transcript, 1 AM#1)

Social media and alternative reality gaming network technologies can be used to assemble the ideological positions and interests of stakeholders and audiences, activate systems of peer review around the debates and positions in conjunction with systems of public review through interactive discussions that can be fed into the review, weighting and quality assurance processes (Cameron 2011, 2012, 332).

Museums need to reject the conceptual distinction between nature and culture. Reading cross-sectoral collective action according to Bruno Latour's actor networks, new materialisms and other ontological proposals that rework human subject positions and promote new types of social collectives that includes other than human actors has the potential to shift the museum towards new types of political frameworks (Cameron 2012, 332). The deliberative, performative platforms that people spoke about can be reworked according to Latours' 'Parliament of Things', as congregations of humans and non-humans. Things such as the atmosphere, oceans and ice therefore must be brought into museum programming and forums as stakeholders and actants entangled with human designs and governing strategies (Cameron 2012, 332).

3.10 Places that Promote Climate Change Action by Providing Information on How Individuals Might Change Their Lifestyles and Consumer Choices

"Places that promote climate change action by providing information on how individuals might change their lifestyles and consumer choices" had its strongest support for Australian closely followed by US museums and came from the older age group (55–65 year olds) and reflects the dominant approach to climate change action.

This positioning statement invoked questions about new forms of self-regulation and the museum's role in promoting an ecological citizenship where people are asked to curtail their own personal freedom and to adjust their personal lifestyles and choices. For these reasons respondents were more comfortable in museums providing advice on lifestyle choices rather than being told what to do.

> …people don't like being told what to do. It's just basic communication…it's the words that you use when you're given that action. (HSGC Focus Group Transcript, AM#2)

> [Museums ought to provide] the know-how, just the facts … you can [then] make your own choice on how to change your lifestyle. (HSGC Focus Group Transcript, LSC#2)

> Institutions motivate people…they come here they see it, they say what can I do, this is what you can do this is… (HSGC Focus Group Transcript, AM#1)

To this end, institutions were seen as having a motivational role in behaviour change that could then lead to broad-based mobilization by linking actions to causes and potential effects, and highlighting the cost benefits of change.

> …being from a museum it's a good perspective hopefully…you can go out there and find out tonnes of information on anything you want … [on] how to change your lifestyle … how to recycle better and all these things. (HSGC Focus Group Transcript, AM#2)

Others saw individual action as a way of countering government and corporate inaction and as a means to pressure businesses to produce more sustainable goods and services. This position suggests a tentative acceptance of the politics of grassroots movements, NGOs and community groups in museum activities and where collective action takes the form of the swarm—individual behavioural change leading to collective systemic change. Many cited recycling as an example.

> all of a sudden that little thing becomes routine and then when its habit…and all of a sudden one thing leads to another. Bad habits start that way, good habits start that way too and that will rub onto your family. (HSGC Focus Group Transcript AM#4)

Furthermore, many argued that ecological citizenship displaces social responsibility away from governments and corporations and onto individuals. We might ask what is the museum's complicity in the displacement of responsibility to the individual as a means of solving the climate crisis.

> It is a case of showing cause and effect, the immediate, the long term, the medium term. It's up to you to use the information… (HSGC Focus Group Transcript AM#2)

People want to know what they're going to get, I think individuals what they're going to get out of if they make these changes and therefore what their family are going to get out of it, what their community, the future generations. (HSGC Focus Group Transcript AM#2)

3.11 Places that Promote Climate Change Action by Providing Information on How Individuals Might Change Their Lifestyles and Consumer Choices

"Places that participate in climate change research (e.g. science, new technologies)" was most strongly supported for Australian museums and by female respondents although the research function of the museum appears to have a low public profile. When people did talk of research they focused on bio-diversity research. Overall audiences especially the older age groups could be drawn further on new possibilities for museums and science centres in relation to both scientific, technological research and public policy-making.

Documenting everything that's happening in recent times that's out of the normal and keeping records of that so people can see the result…something that can be as a result of global warming and make this document a record of it, for reference. Then people could go there to sort of gain more knowledge about global warming or climate change and also compare it with the past. (HSGC Focus Group Transcript AM#2)

Are they in the business of doing R&D? Are they the right people to [do this work?] … Is this the right place, do they have the right skills, technology to do research? (HSGC Focus Group Transcript, LSC#3; HSGC Focus Group Transcript, LSC#2)

3.12 Places that Lead Opinion on Ways to Reduce Emissions and Shape Future Lifestyle Choices

Places that lead opinion on ways to reduce emissions and shape future lifestyle choices was strongly supported for Australian museums and science centres and to a slightly lesser degree for US museums. Female respondents in the Australian museum sample were particularly supportive of this role change. Governing climate change demands future-forward thinking and challenges institutions in regards to the way they situate themselves in time and formulate themselves as places to offer a certainty.

In focus group discussions, museums and science centres were positioned in the future as places for detailing future lifestyles: as places to link the past to the far future through projections of what might happen; offer practical governing options; present long term temporal trajectories to counter short term thinking and the failure of the government to act; and as places to facilitate the expression of many governing options offering a mediated view of the future as a series of creative pathways.

Depends on the topic…if it's a scientific topic then the opinion is usually based on a lot of fact and scientific evidence. (HSGC Focus Group Transcript, AM#1)

…we know a lot of research goes into what museums develop. (HSGC Focus Group Transcript, AM#1)

I don't see how you can avoid that, how do you educate without having an opinion on something? (HSGC Focus Group Transcript, LSC#3)

While some respondents were uncomfortable about the term lead, leading opinion was routinely read in terms of: the best information/research on the various options; facilitating reflective processes through the critical interrogation of past relations between nature and society; detailing the link between lifestyles and environmental issues and how these relations lead to the unintended consequences of climate change taking account of many human and non-human actants and their entangled relations; and through the facilitation of creative, imaginative processes and experiences that interrogate the larger questions about how we might live in the world differently, and what future lifestyle options might look like as a series of governing proposals.

Further to this creativity is emerging as a popular mode of governing because it engages a capacity for cultural improvisation under conditions of complexity and uncertainty. Here imagination alongside reflective analysis becomes the basis for action in the future-present.

Because policy is a creative process for detailing future lifestyles, museums can help to forge connections in debates on generic policy options (as opposed to specific policy proposals that might be seen as too political) by critically reviewing the debates and options against the research and by examining their implications for various social futures scenarios through systems of peer, open review and quality assurance processes (Cameron 2012, 330). They can act as congregational spaces bringing cross-sectoral stakeholder groups and audiences together with the research and broker deliberations around the various options, testing these against various disciplinary, lay expertise and local knowledge taking account of the non-human as actants (Cameron 2012, 332).

3.13 Instituting Ecologizing Experimentations: A New Role for Museums

Climate change and the Anthropocene are the unintended consequences of an erroneous and dangerous detachment from our environment and others, and of various forms of exploitation, injustice and inequality founded on the illusion that the human is the centre of all things, a dominant view in western thought (Braidotti 2013: 14). Furthermore, the newly visible powers and agencies of a more-than-human world are becoming increasingly apparent as demonstrated for example through the escalating bush fire risk in Australia as we confront the material consequences of advanced capitalism and two centuries of industrialisation. These events and situations tell us that the world that we are bound to live in is no longer solely ours, and that our human capacities to drive change and control outcomes have been substantially curtailed (Schmidt 2013: 174; Protevi 2013). If we are to avoid compounding existing problems and creating new ones, we require new cognitive frames and practices of life that promote respectful, ethical thinking and forms of conjoint action that acknowledge our entanglements, shared vulnerabilities and futures not only with other humans, but also with the more-than-human world (Braidotti 2013).

Through their collections and exhibitions, museums are powerful pedagogical institutions, instrumental in shaping ideas about culture, identity and cultural difference, and about human relations with nature, technology and science (Cameron 2015, 2018). They are therefore ideally placed to frame and promote new theories and practices of life (Cameron 2015). In an effort to secure more viable futures and a habitable planet, Museum scholars, professionals and artists can progress real-world and scholarly change by first critiquing modern ways of thinking and acting by re-working familiar humanisms and modern dualisms such as "nature" and "culture," "human" and "nonhuman," "social" and "natural" "subject" and "object", "self and other" that prevent us from thinking ecologically (Cameron 2015). And then by undertaking what Cameron calls a series of "ecologizing experimentations" as another type of agency, museums have the potential to re-work the possible relations between things and people via new types of museum practices, interpretive storytelling, communicative strategies and embodied experiences. The ecologizing approach detailed in Cameron's 2015 manifesto is to promote relational and more-than-human ways of thinking and acting in the world through museum collection, exhibition and institutional practices. This approach is uniquely concerned with concretely reconstituting human-centred concepts and practices based on the human subject, and developing practices that have the potential to signal an entangled and more-than-human approach to the make-up and composition of the world (Cameron 2015, 16–33, 2018).

Ecological thinking is no longer the sole domain of the biological sciences through evolutionary thinking, rather we need to embed ecological thinking across museum practice (Cameron 2015, 2018). In doing so we have the capacity to promote new ways to represent, talk about humans, non-humans, culture and cultural diversity, heritage objects, the environment and climate change as alternative narratives, sets of practices and concepts (Cameron 2015, 16–33).

Over the last almost 10 years Cameron has been conducting ecologizing experimentations across domains of museum practice from climate change narratives to institutional forms; to collections and documentation, and to digital heritage. These ecological experiments have been strategic, directed towards promoting a new ethics of care that might better encourage respect for various forms of animate and inanimate things; nourish new forms of multispecies s connections, intercultural relations; social inclusiveness and interaction, and that can regard humanity as part of a larger dynamic living system (Domanska 2010) that ultimately has the potential to support the continuation of life itself (Cameron 2015, 16–33).

The theoretical coordinates Cameron draws upon for framing these "ecologizing experimentations" (Cameron 2015, 16–33) are derived from anthropologist Latour's (1993, 1998, 2013) notion of "ecologize." For Latour (1998, p. 22), "ecologizing", as opposed to our preoccupation with "modernizing", is a political project that seeks to inform new notions of the social by specifying that natural and social entities are bound together in complex interrelations, and that relational and ecological principles bear on every type of connection. Latour's (2010) idea of composition is referenced here as a way to think about how we might compose or arrange different museum interpretive worlds and narratives that illustrate the many alternative ways we can entangle ourselves with places, non-humans, technologies, and the material world (Cameron 2015, 16–33).

New knowledge and art practices are emerging in the humanities and social sciences that can do this work and that is aimed at comprehending and formulating culturally-intelligent ways to re-work modern humanism, dualistic ways of thinking and anthropocentric social collectives, in ways that are better able to deal with real-world complexities, and the climate crisis (Cameron 2015, 16–33). This shift represents a move from modern epistemology that sought to discover and represent the empirical world through rational investigation according to a predetermined set of rules as seen in natural history museums to ontology, a practice that seeks to address more directly the composition of the world (Woolgar and Lezaun 2013). These new theories or ontologies focus on histories and events as processes and seek to re-institute new types of relational and agentic connections between humans, animate life and non-animate things as part of one dynamic system (Deleuze and Guattari 1987; Haraway 2007; Wolfe 2010; Nayar 2014; deLanda 2006; Bennett 2010; Morton 2010; Latour 1993; Harvey 2007).

The application of ecologizing principles to museum practices and narratives has the potential to dissolve dualistic, hierarchical and hubris humanisms. It also breaks the idea of the existence of one-world ontologies and knowledge practices; the privileging of human intentionality as well as time and change as knowable, linear and progressive. This approach acts as a lever to consider how we might connect entities in collectives; formulate inter-relations between human and non-humans; found new formulations of history and change; propose new concepts for objects and collections including those defined as virtual and how multiple world views might exist and interact (Cameron 2015, 16–33). The objective here is to set up new forms of more-than human civic life that have the potential to be made manifest in museum practices and narratives that invite non-human others into social collectives, acknowledges, and is more respectful of, the diversity of forms and modes of thought and worlding categories that are existent in the world (Cameron 2015).

Furthermore, multi-naturalism (Viveiros de Castro 2005) can be used to re-work cultural diversity as a diversity of natures and a diversity of entities that include non-human others, thus breaking the culture/nature and self/other division. Ecologizing principles can also be directed towards the revision of one-world ontologies and replace them with concepts that allow institutional staff to compose multiple, divergent worlds (Stengers 2015, 994), to negotiate different realities and co-produce shared worlds through documentation practices and exhibition narratives between different groups within the museum (Cameron 2015, 16–33).

Ecologizing experimentations on natural history collections and exhibition narratives have the potential most broadly to re-work human-focused and hubristic perceptions of the world; build new social collectives that can acknowledge and work with the inter-and complementary relations between humans and non-humans; and promote concepts of social inclusiveness and an ethics of care for *beyond the human world* (Descola 2013, p. 11; Harvey 2007) as a new position from which interspecies transactions can be made (Cameron 2015, 16–33).

The author's first experiment was to refashion institutions as liquid (Cameron 2010). This experiment was motivated by responses to the positioning statements outlined above and how we can think about institutions as agential in global networks and polycentric, multi-stakeholder alliances including that of the non-human. Inspired by Deleuze and Guattari (1987) and deLanda's (2006) assemblages, here the museum's agency as an institutional form is refashioned as a series of heterogeneous assemblages made up of material and expressive forms enmeshed in diverse collectives comprising many human and nonhuman elements (Cameron 2010, 2011, 2015, 345–359). As an assemblage, the 'liquid' museum is distributed, multi- and interscalar, crossing sectors, scales, and transnational boundaries. The institution is comprised of a series of components that are mobile, cohere with others, and also disperse or disassemble according to certain conditions and events. Change, therefore, becomes a rhizomic process, both intended and unintended, in which agency is rethought in terms of relational resources and institutional capacities (Cameron 2010, 2011, 2015). By thinking about institutions in these terms, the museum emerges as more nimble, flexible and adaptive, better able to meaningfully engage topics as multi-stakeholder alliances for collective action on topics such as global environmental change, and consider institutional roles and capacities in broader political and media ecologies.

Ecologizing climate change narratives can be put to work to direct us to ways we might break modern human-centred views of climate change; gesture towards ways we might collapse and individuate modern nature into an array of coordinates some of whom were previously invisible; re-work relations between things as natural-cultural hybrids and fold the human and non-human into dynamic, non-linear and complex systems.

The second experiment was on the narratives in *Atmosphere: Climate Change* at the Science Museum, London (Cameron 2015, 51–77). Here Cameron composed climate change narratives and action differently, in ways that signal our embeddedness in a climate-changed world. Here Cameron suggested ways in which we might collapse and then individuate big Nature, the atmosphere, science and society into an array of coordinates some of whom were previously invisible and rework the relations among all these things as natural-cultural hybrids (Cameron 2015, 51–77). These ontologies also gesture towards ways in which the linear notions of cause and effect inherent in mitigation and stabilisation narratives can be reworked as non-linear complex systems involving the actions and agencies of many human, non-human and technological actors, and earthly processes (Cameron 2015, 51–77).

The third experiment settles on a melted green plastic bucket from Museum Victoria's Black Saturday bushfire collection. The Black Saturday bushfires of 7 February 2009 was Australia's most devastating natural disaster. Importantly, the green plastic bucket operates in the dramatic interface between the worlds of humans, more-than-humans and other-than humans (Cameron 2018).

The bucket is framed around Bill Putt's accounts of survival where it is used to recall and recite his last ditch attempts to save his house Rosewood at Strathewen from the impending furnace. Here Cameron refashioned museum-centred notions of objecthood from familiar human subject-object based positions to a consideration of them as 'socio-material compositions'. The aim here was to move beyond the social construction of the 'object', 'specimen' and 'collection' and a sole focus on social and cultural conventions of language—describing, naming and narrating histories and personal accounts as bounded objects—to a relational reconfiguration that takes account of the active and entangled life of all an object's components (Cameron 2018, 349–352). At the same time, the author sought to "put humans back in their ecological place"—knitted together and becoming with not just biological species, but all manner of animate and inanimate things including non-human beings, entities, materials and earthly processes as a new type of narration or storytelling. In doing so the bucket's narrative is reworked as one centred on Bill Putt's experiences and reconstituted in more-than-human and emergent terms that acknowledge and work with the different types of material, discursive, technological, biological, and non-human agencies and enactions that are embodied within it (Cameron 2018, 349–352).

Critically the bucket is used to illustrate the lively interdependences between humans, practices, decisions and the multifarious animate and inanimate things and materialites (Australian bush, fire, metal, plastic, heat, wind, technology, CO_2, human acts) that make up our world in ways that improve our understanding of bushfires and other events as entangled and dynamic socio-ecological processes (Cameron 2018, 349–352).

4   Conclusions

To conclude, the current, perceived roles for Australian museums focus on the provision of resources; communicating science; the presentation of a range of views; research and individual privatised action. The gap analysis undertaken between current and desired roles placed a greater emphasis on the critical examination of climate change as a complex phenomenon; science communication; a greater emphasis on network building and the provision of forums; the facilitation of collective action and critical positions on policy and lobbying.

The current perceived roles for US museums were somewhat similar and included: communicating science; the presentation of a range of views; research; individual behavioural reform and resource bases with a greater emphasis on the critical examination of climate change as a complex phenomenon than Australian museums. Role changes were as a greater emphasis on science production and communication; networks; collective action and more critical content.

For Australian science centres, current perceived institutional roles were as places for research; as information sources and resources; as places that promote individual attitudinal behavioural change; and communicating climate science Gap roles and adaptations, collective action was predominant and included decision-making forums, developing networks; building trans-national communities of risk; the facilitation of a range of views, lobbying; providing critical information for reflective analysis and to do more in regards climate change science.

The current perceived roles of US science centres were somewhat similar to Australia and were in this order: communicating science; presenting a range of views; research; resource bases; individual behavioural reform and again with the notable exception in comparison to the Australian science centres, the critical examination of climate change as a complex phenomenon. In terms of role changes a greater emphasis was placed on building networks and facilitating collective action. The desire for institutions to promote collective action, metropolitan New York City participants were much more likely to endorse this position than participants from New Jersey or New York State.

Generational differences were significant. Older respondents (55–65 year olds) generally having stronger views across all samples. Older respondents in the Australian and US museum sample strongly supported more climate science; forums and access to resources. The presentation of a range of views was strongly supported by this age group for Australian museums and lobbying for Australian science centres. The desire for critical information scored highly for both the 35–54 and 55–65 year olds in the Australian science centre sample with the 35–54 sample strongly supporting collective action across all samples.

The data showed a noticeable difference between males and females in all samples. Female respondents, somewhat more so than males, favoured an activist/ politically engaged role for museums and science centres. Overall the findings showed a greater emphasis on networks and collective action, and for political advocacy and building trans-national communities of risk in Australia.

Critically ecologizing experimentations can provide an empirical and theoretical case to support the argument that museum projects can, through the recognition of more-than human social collectives, and the blending of different ontologies, together make an important contribution to producing knowledge that can promote long-term sustainability discourses and strategies for action (Cameron 2015).

The findings illustrate the important roles that museums can play in polycentric governing arrangements in support of the Paris Agreement and the SDG Goals. Our specifically detailed empirical research in the partnering institutions also detailed the particular capacities and roles that museums can play in such arrangements. This research is not reported in this document. Some of the limitations of the study is the size of the data sets, restricted to the eastern sea board of the US and across Australia due to funding limitations. The research would benefit from a more global perspective, across sectors and other institutional forms including other civil society institutions such as libraries and NGOs.

Acknowledgements This chapter is an output from the Australian Research Council funded Linkage project, *Hot science, global citizens: the agency of the museum sector in climate change interventions*, led by the author. Parts of this chapter have been published in Cameron, FR. 2012, Climate change, agencies, and the museum for a complex world. October, vo Museum Management and Curatorship, 27 (4): 317–339. Special thanks to Dr. Ben Dibley and Dr. Ann Deslandes for their contributions to focus group data analysis and Carolyn Meehan for quantitative gap role analysis.

References

Bennett J (2010) Vibrant matter: a political ecology of things. Duke University Press, Durham
Braidotti R (2013) The posthuman. Polity Press, Cambridge
Cameron FR (2010) Liquid governmentalities, liquid museums and the climate crisis. In: Cameron FR, Kelly L (eds) Hot topics, public culture, museums. Cambridge Scholars, Newcastle upon Tyne, pp 112–128
Cameron FR (2011) From mitigation to complex reflexivity and creative imaginaries—museums and science centres in climate governance. Museum and society special issue: hot science, global citizens: the agency of the museum sector in climate change interventions, vol 9, no 2. http://www.le.ac.uk/ms/museumsociety.html. Last Accessed 3 June 2018
Cameron FR (2012) Climate change, agencies, and the museum for a complex world. Mus Manag Curatorship 27(4):317–339
Cameron FR (2014a) Ecologizing experimentations: A method and manifesto for museums and science centres. In: Cameron FR, Neilson B (eds) Climate change, and museum futures. Routledge museum research series, pp 16–33
Cameron FR (2014b) We are on nature's side? Experimental work in re-writing narratives of climate change for museum exhibitions. In: Cameron FR, Neilson B (eds) Climate change, and museum futures. Routledge Museum Research Series, New York, pp 51–77
Cameron FR (2015) The liquid museum: new ontologies for a climate changed world. In: Witcomb A, Message K (eds) Museum theory: an expanded field. Blackwell, UK, pp 345–362
Cameron FR (2018) Posthuman museum practices. In: Braidotti R, Hlavajova M (eds) Posthuman glossary. Bloomsbury Academic, London, pp 349–352
Cameron FR, Deslandes, A (2011). Museums and science centres as sites for deliberative democracy on climate change. Museum and society special issue: hot science, global citizens: The agency of the museum sector in climate change interventions, vol 9, no 2. http://www.le.ac.uk/ms/museumsociety.html. Last Accessed 3 June 2018
deLanda M (2006) A new philosophy of society: assemblage theory and social complexity. Continuum, London
Deleuze G, Guattari F (1987) A thousand plateaus. University of Minnesota Press, Minneapolis
Descola P (2013) Beyond nature and culture. University of Chicago Press, Chicago
Domanska E (2010) Beyond anthropocentrism. Historical studies. Historein 10:118–130
Haraway D (2007) When species meet. University of Minnesota Press, Minnesota
Harvey G (2007) Animism: respecting the living world. Columbia University Press, New York
Hodge B (2011) Climate change and the museum sector: 10 reflections from the 'Hot Science, Global Citizens' symposium. Available at http://ccr.uws.edu.au/2011/05/18/climate-change-the-museum-sector-10-reflections-from-the-hot-science-global-citizens-symposium/. Last Accessed 18 Sept 2011

Hulme M (2009) Why we disagree about climate change: understanding controversy, inaction and opportunity. Cambridge University Press, Cambridge

Latour B (1993) We have never been modern. Harvester Wheatsheaf, Hertfordshire

Latour B (1998) To modernise or ecologize? That's the question. In: Castree N, Willems-Braun B (eds) Remaking reality: nature at the millennium. Routledge, London, pp 221–242

Latour B (2010) An attempt at a compositionist manifesto. New Literary Hist 41:471–490

Morton T (2010) The ecological thought. Harvard University Press, Cambridge

Nayar PK (2014) Posthuman. Polity, London

Orstrom E (2010) Polycentric systems for coping with collective action and global environmental change. Glob Environ Change 20:550–557

Protevi J (2013) Life war earth. University of Minnesota Press, Minneapolis

Schmidt J (2013) The empirical falsity of the human subject: new materialism, climate change and the shared critique of artiface. Resilience Int Politics Pract Discourses 1(3):174–192

Stengers I (2015) In catastrophic times: resisting the coming barbarianism. Open Humanities Press, http://openhumanitiespress.org/books/download/Stengers_2015_In-Catastrophic-Times. pdf. Last Accessed 3 June 2018

United Nations Development Programme, Sustainable Development Goals (2015) http://www. undp.org/content/undp/en/home/sustainable-development-goals.html. Last Accessed 3 June 2018

United Nations Framework Convention on Climate Change (2016) Report on the Conference of the Parties held in Paris, 30th November to 13th December 2015. http://www.un.org/en/development/desa/population/migration/generalassembly/docs/globalcompact/FCCC_CP_2015_10_Add.1.pdf. Last Accessed 3 June 2018

Viveiros de Castro E (2005) Perspective and multi-naturalism in indigenous America. In: Surrallés A, García Hierro P (eds) The land within: indigenous territory and the perception of the environment IWGIA. Denmark, Copenhagen, pp 36–75

Wolfe C (2010) What is posthumanism?. University of Minnesota Press, Minneapolis

Woolgar S, Lezaun J (2013) The wrong bin bag: a turn to ontology. Soc Stud Sci 43:321–340

Professor Fiona Cameron is based as Linkoping University, Norrkoping, Sweden. Fiona is a Senior Research Fellow at the Institute for Culture and Society, Western Sydney University, Australia. Her research and writing focuses on museums and their roles in contemporary societies on topics of societal importance and digital heritage. Fiona is a leading figure in museums and climate change research and action and led a large Australian Research Council international grant, Hot Science, Global Citizens: the agencies of museums in climate change interventions (2008–2011). Cameron was also a museum practitioner and has worked in the sector as a museum director, a social history curator and as a curatorial consultant on major exhibition projects. In March 2011 Fiona led an Australian federal parliamentary briefing A Climate for Change on the findings of Hot Science, Global Citizens to parliamentarians, government department employees, academics and the museum sector. Fiona has published 80 articles and book chapters and 5 books with leading publishers on these topics.